Value-Added Biocomposites

Value-Added Biocomposites
Technology, Innovation, and Opportunity

Edited by
Malinee Sriariyanun, Sanjay Mavinkere Rangappa,
Suchart Siengchin, and Hom Nath Dhakal

CRC Press is an imprint of the
Taylor & Francis Group, an **informa** business

First edition published 2022
by CRC Press
6000 Broken Sound Parkway NW, Suite 300, Boca Raton, FL 33487-2742

and by CRC Press
2 Park Square, Milton Park, Abingdon, Oxon, OX14 4RN

© 2022 Taylor & Francis Group, LLC

CRC Press is an imprint of Taylor & Francis Group, LLC

Reasonable efforts have been made to publish reliable data and information, but the author and publisher cannot assume responsibility for the validity of all materials or the consequences of their use. The authors and publishers have attempted to trace the copyright holders of all material reproduced in this publication and apologize to copyright holders if permission to publish in this form has not been obtained. If any copyright material has not been acknowledged please write and let us know so we may rectify in any future reprint.

Except as permitted under U.S. Copyright Law, no part of this book may be reprinted, reproduced, transmitted, or utilized in any form by any electronic, mechanical, or other means, now known or hereafter invented, including photocopying, microfilming, and recording, or in any information storage or retrieval system, without written permission from the publishers.

For permission to photocopy or use material electronically from this work, access www.copyright.com or contact the Copyright Clearance Center, Inc. (CCC), 222 Rosewood Drive, Danvers, MA 01923, 978-750-8400. For works that are not available on CCC please contact mpkbookspermissions@tandf.co.uk

Trademark notice: Product or corporate names may be trademarks or registered trademarks and are used only for identification and explanation without intent to infringe.

ISBN: 978-0-367-67926-2 (hbk)
ISBN: 978-0-367-68439-6 (pbk)
ISBN: 978-1-003-13753-5 (ebk)

DOI: 10.1201/9781003137535

Typeset in Times
by codeMantra

Contents

Preface vii
Editors ix
Contributors xiii

Chapter 1 Introduction to Biomass and Biocomposites 1

Tibor Alpár, K. M. Faridul Hasan, and Péter György Horváth

Chapter 2 Influence of Natural Fibers and Biopolymers on the
Biocomposites Biodegradation 35

L. Joana Rodríguez and Carlos E. Orrego

Chapter 3 Property Analysis and Characterization of Biomass-Based
Composites 65

Tejas Pramod Naik, Ujendra Kumar Komal, and Inderdeep Singh

Chapter 4 Tensile Properties Analysis and Characterizations
of Single Fiber and Biocomposites: A Weibull Analysis and
Future Trend 87

*Mohamad Zaki Hassan, Mohamad Ikhwan Ibrahim,
and SM Sapuan*

Chapter 5 Crashworthiness Measurement on Axial Compression Loading
of Biocomposite Structures: Prospect Development 103

*Mohamad Zaki Hassan Zainudin A. Rasid, Rozzeta Dolah,
SM Sapuan, and Siti Hajar Sheikh Md. Fadzullah*

Chapter 6 Structure and Surface Modification Techniques for Production
of Value-Added Biocomposites 125

*Chaniga Chuensangjun, Sarote Sirisansaneeyakul, and
Takuya Kitaoka*

Chapter 7 Design and Fabrication Technology in Biocomposite
Manufacturing 157

*K. M. Faridul Hasan, Péter György Horváth, Kovács Zsolt,
and Tibor Alpár*

Chapter 8 Progress in Development of Biorefining Process: Toward
Platform Chemical-Derived Polymeric Materials 189

*Malinee Sriariyanun, Prapakorn Tantayotai, Yu-Shen Cheng,
Peerapong Pornwongthong, Santi Chuetor, and
Kraipat Cheenkachorn*

Chapter 9 Biocomposite Production from Ionic Liquids (IL)-Assisted
Processes Using Biodegradable Biomass 213

*Marttin Paulraj Gundupalli, Kittipong Rattanaporn,
Santi Chuetor, Wawat Rodiahwati, and Malinee Sriariyanun*

Chapter 10 Deep Eutectic Solvent-Mediated Process for Productions of
Sustainable Polymeric Biomaterials 251

*Elizabeth Jayex Panakkal, Yu-Shen Cheng,
Theerawut Phusantisampan, and Malinee Sriariyanun*

Chapter 11 Chitosan-Based Biocomposites for Biomedical Application:
Opportunity and Challenge 289

Chong-Su Cho, Soo-Kyung Hwang, and Hyun-Joong Kim

Chapter 12 Starch-Based Biocomposites: Opportunity and Challenge 319

Ankit Manral, Ranjana Mishra, and Rahul Joshi

Chapter 13 Current Status of Utilization of Agricultural Waste and
Prospects in Biocomposites 341

Pervinder Kaur and Harshdeep Kaur

Chapter 14 Applications of Biocomposites in Reduction of
Environmental Problems 401

*Chukwuma Chris Okonkwo, Francis Odikpo Edoziuno,
Adeolu Adesoji Adediran, and Kenneth Kanayo Alaneme*

Index 427

Preface

This book describes the current market situation, commercially competition and society impacts of biocomposites, including natural fibers and bioplastics. Different aspects in manufacturing and processing procedures to improve and develop the physical, mechanical, thermal, electrical, chemical and biological properties of biocomposites to achieve the required specification of downstream industries and customers. Chapters then gather the current situation of a wide range of various base materials and fillers of biocomposites and bioplastics, including cellulose, lignin chitin, rubber, polylactic acid, polyhydroxyalkanoate, polypropylene carbonate, starch, and protein in terms of the strength and weaknesses of materials, current research and potential in economical market. This book illustrates the valuable sources for readers, researchers and engineers in different related industries to understand the demand, situation and progress in biocomposite industries and demonstrations of biocomposite applications to produce high-value products in sustainable fashion.

The main focus of this book is the exploration and demonstration of current improvement on research, processing and manufacturing and novel applications of various biocomposites and bioplastics. This book also dedicates the contents to compile the market situation, environmental impact, and society position of biocomposites to substitute the use of conventional composites and to be a unique option of user's selection in a special application. The content of this book also illustrates and showcases the visions and insight of expert scientists and engineers who have first-hand experiences on working with biocomposites in various industries.

The content of this book provides the current progress of biocomposite processing and manufacturing in different industries leading to knowledge and updates to the readers to seek out for future opportunity for research and development activities. The reviews of current market situation, strength and weakness in the market, economical values and society impacts of biocomposites in this book provide the insight in different corners to readers, scientists, engineers and policy-makers to get the comprehensive views of biocomposites industries for further activities and policy development.

Currently, the research and report on production and property analysis of biocomposites are published in private and public literatures. As a rising requirement of industrial use, the processing and manufacturing procedures have been also improved and transferred to commercialization. However, the special and unique properties of biocomposites in particular sectors and future are still in demand to provide the opportunity of novel application and market sectors. The risk of new investment of research, product development and industrial processing practically relies on different factors, not only the technical issue but also economical, environment and society issue. Thus, the content of this book covers the different point of views of different aspects, not only the technical aspects, to allow readers from different sectors have comprehensive information required for future biocomposites.

Editors

Assoc. Prof. Dr. Malinee Sriariyanun is a Vice-Director of Science and Technology Research Institute (STRI), the main research institute of King Mongkut's University of Technology North Bangkok (KMUTNB), Thailand. She received her Ph.D. in Plant Pathology from the University of California, Davis, USA, in 2011. Currently, she works as a lecturer for Chemical and Process Engineering Program at The Sirindhorn International Thai-German Graduate School of Engineering (TGGS), KMUTNB. She has been promoted to Associate Professor at KMUTNB and became the consultant to the president office of KMUTNB in 2018. Her research interests in biorefinery and biocomposite materials. She has been working as the principle investigator (PI) and Co-PI of more than 55 research projects supported by governmental and industrial sectors in Thailand and other countries with values of more than 2.2 million USD. She has been supervised and co-supervise more than 40 graduate students. She is an editorial member of KMUTNB International Journal of Applied Science and Technology, Applied Science and Engineering Progress and Oriental Journal of Chemistry and the author of more than 45 peer-reviewed Journal Articles. She has participated with presentations, plenary speaker and invited speaker in more than 65 International Conferences and has contributed as reviewer for more than 25 peer-reviewed international journals.

Dr. Sanjay Mavinkere Rangappa is currently working as Research Scientist and also Advisor within the office of the President for University Promotion and Development toward International goals at King Mongkut's University of Technology North Bangkok, Bangkok, Thailand. He has received B.E. (Mechanical Engineering) in 2010, M.Tech. (Computational Analysis in Mechanical Sciences) in 2013, Ph.D. (Faculty of Mechanical Engineering Science) from Visvesvaraya Technological University, Belagavi, India, in 2017 and Post-Doctorate from King Mongkut's University of Technology North Bangkok, Thailand, in 2019. He is a Life Member of Indian Society for Technical Education (ISTE) and an Associate Member of Institute of Engineers (India). He acts as a Board Member of various international journals in the fields of materials science and composites. He is a reviewer for more than 85 international Journals (for Nature, Elsevier, Springer, Sage, Taylor & Francis, Wiley, American Society for Testing and Materials, American Society of Agricultural and Biological Engineers, IOP, Hindawi, NC State University USA, ASM International, Emerald Group, Bentham Science Publishers, Universiti Putra, Malaysia), also a reviewer for book proposals and international conferences. He has published more than 125 articles in high-quality international peer-reviewed journals, 5 editorial corners, 35 book chapters, 1 book, 15 books as an Editor, and also presented research papers at national/international conferences. In addition, he has filed one Thailand patent and three Indian patents. His current research areas include natural fiber composites, polymer composites, and advanced material technology. He is a recipient of the DAAD Academic exchange–PPP Program (Project-related Personnel Exchange) between Thailand and Germany to Institute of Composite Materials, University of Kaiserslautern, Germany. He has received a Top Peer-Reviewer 2019 award, Global Peer-Review Awards, Powered by Publons, Web of Science Group. The KMUTNB selected him for the "Outstanding Young Researcher Award 2020." He has been recognized by Stanford University's list of the world's Top 2% of the Most-Cited Scientists in Single Year Citation Impact 2019.

Prof. Dr.-Ing. habil. Suchart Siengchin is President of King Mongkut's University of Technology North Bangkok. He has received his Dipl.-Ing. in Mechanical Engineering from University of Applied Sciences Giessen/Friedberg, Hessen, Germany in 1999, M.Sc., in Polymer Technology from University of Applied Sciences Aalen, Baden-Wuerttemberg, Germany in 2002, M.Sc., in Material Science at the Erlangen-Nürnberg University, Bayern, Germany, in 2004, Doctor of Philosophy in Engineering (Dr.-Ing.) from Institute for Composite Materials, University of Kaiserslautern, Rheinland-Pfalz, Germany, in 2008 and Postdoctoral Research from Kaiserslautern University and School of Materials Engineering, Purdue University, USA. In 2016 he received the habilitation at the Chemnitz University in Sachen, Germany. He worked as a Lecturer for Production and Material Engineering Department at The Sirindhorn International Thai-German Graduate School of Engineering (TGGS), KMUTNB. He has been a full Professor at KMUTNB and became the President of KMUTNB. He won the Outstanding Researcher Award in 2010, 2012 and 2013 at KMUTNB. His research interests in Polymer Processing and Composite Material. He is Editor-in-Chief: KMUTNB International Journal of Applied Science and Technology and the author of more than 250 peer-reviewed Journal Articles. He has participated with presentations in more than 39 International and National Conferences with respect to Materials Science and Engineering topics. He has recognized and ranked among the world's top 2% scientists listed by prestigious Stanford University.

Prof. Dr. Hom Nath Dhakal is a Professor of Mechanical Engineering at the School of Mechanical and Design Engineering, University of Portsmouth, UK. In addition, he is also a Docent (visiting) Professor of bio-based materials at the Faculty of Textiles Engineering and Business, University of Borås, Sweden. He leads the Advanced Materials and Manufacturing (AMM) Research Group within the School of Mechanical and Design Engineering. His principal research interest lies in the design, development, testing and characterization of sustainable lightweight composites, nanocomposites, natural fiber composites and biocomposites, including their mechanical (tensile, flexural, low-velocity impact and fracture toughness), thermal and environmental properties (dimensional stability under various environmental conditions). He is a Fellow of the Higher Education Academy (FHEA), Chartered Engineer (CEng), a Fellow of the Institution of Engineering and Technology (FIET), Fellow of the Institute of Materials, Minerals and Mining (IOM3) (FIMMM) and a member of the American Society for Composites (MASC).

Professor Dhakal is the author/co-author of over 150 publications in the area of light-weight sustainable composite and biocomposites that have attracted well over 3,700 citations with an h-index of 30; i10-index of 58 (Google Scholar); ResearchGate score of 36; and higher than 92.5% of ResearchGate members. Professor Dhakal has led multiple research projects within the Advanced Materials and Manufacturing Research Group, with particular interest in the development of sustainable lightweight composites and biocomposites (Interreg SeaBioComp, Interreg FLOWER, Flax composites, low weight, End-of Life and recycling), alongside multiple KTPs into packaging design, composition and sustainability/circularity. He has successfully supervised many PhDs as a Director of Studies; been an external examiner for numerous PhDs nationally and internationally.

Contributors

Adeolu Adesoji Adediran
Department of Mechanical Engineering
Landmark University
Omu-Aran, Nigeria

Kenneth Kanayo Alaneme
Department of Metallurgical and
 Materials Engineering
Federal University of Technology Akure
Akure, Nigeria
and
Centre for Nanoengineering and
 Tribocorrosion
School of Mining, Metallurgy, and
 Chemical Engineering
University of Johannesburg
Johannesburg, South Africa

Tibor Alpár
Simonyi Károly Faculty of
 Engineering
Institute of Wood-based Products and
 Technologies
University of Sopron
Sopron, Hungary

Kraipat Cheenkachorn
Faculty of Engineering
Department of Chemical Engineering
King Mongkut's University of
 Technology North Bangkok
 (KMUTNB)
Bangkok, Thailand

Yu-Shen Cheng
Department of Chemical and Materials
 Engineering
National Yunlin University of Science
 and Technology
Douliu, Taiwan

Chong-Su Cho
Research Institute of Agriculture and
 Life Sciences
Seoul National University
Seoul, Republic of Korea

Chaniga Chuensangjun
Science and Technology Research
 Institute
King Mongkut's University of
 Technology North Bangkok
Bangkok, Thailand

Santi Chuetor
Faculty of Engineering
Department of Chemical Engineering
King Mongkut's University of
 Technology North Bangkok
Bangkok, Thailand

Rozzeta Dolah
Razak Faculty of Technology and
 Informatics
Universiti Teknologi Malaysia
Kuala Lumpur, Malaysia

Francis Odikpo Edoziuno
Department of Metallurgical &
 Materials Engineering
Nnamdi Azikiwe University
Awka, Nigeria
and
Department of Metallurgical
 Engineering
Delta State Polytechnic
Ogwashi-Uku, Nigeria

Siti Hajar Sheikh Md. Fadzullah
Universiti Teknikal Malaysia Melaka
Durian Tunggal, Malaysia

Marttin Paulraj Gundupalli
Chemical and Process Engineering
 Program
The Sirindhorn International
 Thai-German Graduate School of
 Engineering
King Mongkut's University of
 Technology North Bangkok
Bangkok, Thailand

K.M. Faridul Hasan
Simonyi Károly Faculty of
 Engineering
Institute of Wood-based Products and
 Technologies
University of Sopron
Sopron, Hungary

Mohamad Zaki Hassan
Razak Faculty of Technology and
 Informatics
Universiti Teknologi Malaysia
Kuala Lumpur, Malaysia

Péter György Horváth
Simonyi Károly Faculty of
 Engineering
Institute of Wood-based Products and
 Technologies
University of Sopron
Sopron, Hungary

Soo-Kyung Hwang
Research Institute of Agriculture and
 Life Sciences
Seoul National University
Seoul, Republic of Korea
and
Laboratory of Adhesion &
 Bio-Composites
Department of Agriculture, Forestry
 and Bioresources
Seoul National University
Seoul, Republic of Korea

Mohamad Ikhwan Ibrahim
Razak Faculty of Technology and
 Informatics
Universiti Teknologi Malaysia
Kuala Lumpur, Malaysia

Rahul Joshi
ME Department
Netaji Subhas University of Technology
New Delhi, India

Harshdeep Kaur
Department of Chemistry
Punjab Agricultural University
Ludhiana, India

Pervinder Kaur
Department of Agronomy
Punjab Agricultural University
Ludhiana, India

Hyun-Joong Kim
Research Institute of Agriculture and
 Life Sciences
Seoul National University
Seoul, Republic of Korea
and
Laboratory of Adhesion &
 Bio-Composites, Department of
 Agriculture, Forestry and Bioresources
Seoul National University
Seoul, Republic of Korea

Takuya Kitaoka
Faculty of Agriculture
Department of Agro-Environmental
 Sciences
Kyushu University
Fukuoka, Japan

Ujendra Kumar Komal
Department of Mechanical and
 Industrial Engineering
Indian Institute of Technology Roorkee
Roorkee, India

Contributors

Ankit Manral
MPAE Division
Netaji Subhas Institute of Technology
New Delhi, India

Ranjana Mishra
ME Department
Netaji Subhas University of Technology
New Delhi, India

Tejas Pramod Naik
Department of Mechanical and Industrial Engineering
Indian Institute of Technology Roorkee
Roorkee, India

Chukwuma Chris Okonkwo
Department of Agricultural & Bioresources Engineering
Nnamdi Azikiwe University
Awka, Nigeria

Carlos E. Orrego
Department of Physics and Chemistry
Universidad Nacional de Colombia
Manizales, Colombia

Elizabeth Jayex Panakkal
Chemical and Process Engineering Program
The Sirindhorn International Thai-German Graduate School of Engineering (TGGS)
King Mongkut's University of Technology North Bangkok (KMUTNB)
Bangkok, Thailand

Theerawut Phusantisampan
Faculty of Applied Science
Department of Biotechnology
King Mongkut's University of Technology North Bangkok (KMUTNB)
Bangkok, Thailand

Peerapong Pornwongthong
Faculty of Applied Science
Department of Agro-Industry Environment and Environmental Technology (AFET)
King Mongkut's University of Technology North Bangkok (KMUTNB)
Bangkok, Thailand

Zainudin A. Rasid
Malaysia-Japan International Institute of Technology
University Teknologi Malaysia
Kuala Lumpur, Malaysia

Kittipong Rattanaporn
Faculty of Agro-Industry
Department of Biotechnology
Kasetsart University
Bangkok, Thailand

Wawat Rodiahwati
Department of Chemistry
University of New England
Armidale, Australia

L. Joana Rodríguez
Department of Industrial Engineering
Universidad Nacional de Colombia
Manizales, Colombia

S.M. Sapuan
Department of Mechanical and Manufacturing Engineering
Advanced Engineering Materials and Composites Research Centre
Universiti Putra Malaysia
Serdang, Malaysia

Inderdeep Singh
Department of Mechanical and
 Industrial Engineering
Indian Institute of Technology Roorkee
Roorkee, India

Sarote Sirisansaneeyakul
Faculty of Agro-Industry
Department of Biotechnology
Kasetsart University
Bangkok, Thailand

Prapakorn Tantayotai
Faculty of Science
Department of Microbiology
Srinakarinwirot University
Bangkok, Thailand

Kovács Zsolt
Simonyi Károly Faculty of
 Engineering
University of Sopron
Sopron, Hungary

1 Introduction to Biomass and Biocomposites

Tibor Alpár, K. M. Faridul Hasan, and Péter György Horváth
University of Sopron

CONTENTS

1.1 Introduction 1
1.2 Materials for Biocomposites 5
    1.2.1 Biopolymers as Matrix Materials 6
    1.2.2 Other Matrix Materials 8
        1.2.2.1 Fossil-Based Thermoplastics 8
        1.2.2.2 Resins 9
        1.2.2.3 Inorganic Binders 10
    1.2.3 Natural Reinforcers—Biomass 11
        1.2.3.1 Plant-Based Fibers 11
        1.2.3.2 Animal-Based Fibers 13
    1.2.4 Production Techniques 14
1.3 Projects at University of Sopron—Some Examples 14
    1.3.1 Effect of Shape of Wood Element on Properties of Wood-PLA Biocomposites 14
    1.3.2 Continuous Fiber Reinforced PLA Biocomposite 16
    1.3.3 Development of Wood Plastic Composite with Optimized Inertia 17
    1.3.4 Wood Wool Cement Boards Produced with Nano Minerals 18
    1.3.5 PLA Nano Composite 19
1.4 Application of Biocomposites 20
    1.4.1 Medical Use 20
    1.4.2 Automotive Industry 22
    1.4.3 Construction Industry 23
        1.4.3.1 Structural Biocomposite 23
        1.4.3.2 Nonstructural Biocomposite 25
1.5 Future Potential 26
Notes 27
References 28

1.1 INTRODUCTION

Biomass is a popular expression of 21st century, but in many cases, it is misleadingly defined. The US Energy Information Administration (EIA) explains biomass substances as "organic material that comes from plants and animals, and it is a renewable source of energy." The EIA simplifies biomass to plants which contain stored energy from the sun and gives samples like wood and agricultural wastes, domestic garbage-based wastes, and animal manure and human sewage. Also, many publications [1–5] considered biomass as fuel for energy production, but they are actually talking about wood or other ligno-cellulosic materials, so dendromass or phytomass expressions should be used instead of biomass. It would be more precise to apply in industrial point of view; this is incorrect hence these materials are much worthier than fuel. These could be used as raw materials to produce various value-added products like furniture, building and construction materials, clothes, parts for automotive industry in their original form, or after processing into more complex applications like biocomposites.

Cambridge Dictionary (https://dictionary.cambridge.org/dictionary/english/biomass) explains biomass as "dead plant and animal material suitable for using as fuel" (Engineering) or "the total mass of living things in a particular area" (Biology).

Encyclopedia Britannica (https://www.britannica.com/topic/Encyclopaedia-Britannica-English-language-reference-work) describes it in a more comprehensive way:

> **Biomass**, the weight or total quantity of living organisms of one animal or plant species (species biomass) or of all the species in a community (community biomass), commonly referred to an unit area or volume of habitat. The weight or quantity of organisms in an area at a given moment is the standing crop. The total amount of organic material produced by living organisms in a particular area within a set period of time, called the primary or secondary productivity (the former for plants, the latter for animals), is usually measured in units of energy, such as gram calories or kilojoules per square meter per year. Measures of weight—e.g., tons of carbon per square kilometer per year or gigatons of carbon per year—are also commonly recorded.

It should clarify some definitions to avoid common misinterpretations. Biomass covers all the living life forms and their dead materials from bacteria through plants to animals even us, humans. Phytomass covers all the plants and zoomass covers all the animal lifeforms. Among phytomass can be found the woody species called dendromass. The global biomass distribution on Earth by taxa is shown in Figure 1.1, where total amount of biomass is estimated to be 550 Gt carbon (C) [6].

Forests (dendromass) cover 30.8%—4.06 billion hectares—of the global land area on earth with uneven distribution (Figure 1.2), according to Global Forest Resources Assessment (FRA) 2020. Forests supply water, store Carmon, affect climate change and provide habitats for essential pollinators. It is estimated that 75% of the leading food-based crops in the world, representing 35% of global food production. All these benefit from animal pollination—like bees—for fruit, vegetable, or seed production [7].

In 2018, the total production of fiber crops in global agriculture was 11.5 Gt [7]. Among these there are many annual plants with appropriate natural fibers suitable for composite production.

Polymers are organic, large molecules of repeating units, monomers, in which the units are bonded by chemical bonds. Polymers can also be further classified as natural polymers (e.g., cellulose, protein, hemicellulose, and so on), natural-based artificial materials (like viscose, PLA (poly (lactic acid)), thermoplastic starch (TPS), etc.) or plastics synthetized from artificial chemicals (e.g., PE (polyethylene), PET

Introduction

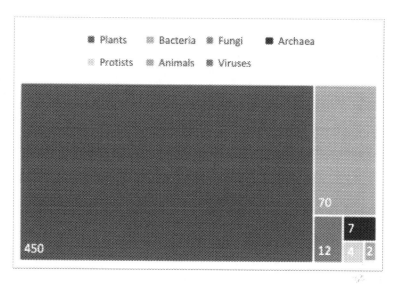

FIGURE 1.1 Biomass on Earth in Gt C.

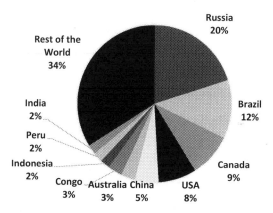

FIGURE 1.2 Distribution of forest in countries with the largest forest areas. (From Food and Agriculture Organization of the United Nations, Production (crops). 2018 [cited 2018 18th November]; Available from: http://www.fao.org/3/I8429EN/i8429en.pdf.) [7]

(polyethylene terephthalate), PP (polypropylene), and so on). First polymers were created by nature itself. The most abundant natural polymer in our world is cellulose and further important organic polymers are lignin, chitin, and protein or DNA. In 1811, Henri Braconnot created cellulose derivatives. In 1824, Friedrich Wöhler created urea, which later became the basic monomer of widely used urea—formaldehyde. Later in 19th century durability of rubber—an important natural polymer—was improved by vulcanization. In 1907, the first synthetic polymer created by Leo Baekeland, the Bakelite, through reacting formaldehyde and phenol under precisely controlled pressure and temperature [8].

Biocomposites are composite materials formed by a polymeric matrix (e.g., resin) and natural fiber reinforcements. The matrix phase can be formed by mineral binders like cement or gypsum, or by polymers derived from renewable or nonrenewable resources. The matrix defends the natural fibers from environmental impacts and mechanical deformations or damage; it keeps the reinforcing phase together, hence transferring the loads on it. Furthermore, biofibers are the main components of biocomposites, which are usually derived from plants, e.g., fibers from crops of hemp, flax, kenaf, etc., wood, used paper (cellulose), various byproducts [9]. A variety of biocomposites are shown in Figure 1.3. Nowadays the interest in biocomposites is growing fast regarding new industrial applications like automotive industry, aerospace applications, packaging, construction, and medical applications. The main advantages of using biocomposites are renewability, sustainability, less carbon footprint, relatively cheaper, recyclability, biodegradability, economical, and after all environment-friendly.

Wood–plastic composites (WPCs) have been rapidly developed in recent years due to their numerous advantages [10–12]. There are three key technologies to manufacture fiber-reinforced polymer composites: (a) short fiber, (b) long fiber, and (c) continuous fiber reinforcements. While long and short fibers are reinforced with the thermoplastics, continuous fibers (woven and nonwoven) are mostly embodied in thermosetting polymer matrix [13–15]. Both natural short and continuous fiber-reinforced fossil-based polymeric composites are used in automotive industry [16–18]. Polymer-reinforced composites are often modified by diverse functional additives, by using fillers, or by synthetic fibers such as aramid and glass [19,20], carbon, or even biofibers.

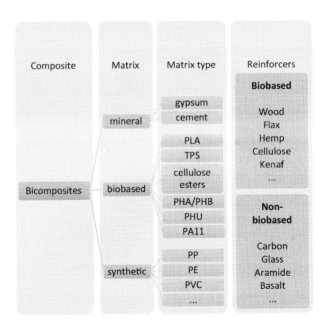

FIGURE 1.3 Different matrix, reinforcers, and type of matrix used for biocomposites formation. PVC, polyvinyl chloride; PHA, polyhydroxyalkanoates; PHB, poly(3-hydroxybutyrate); PHU, polyhydroxyurethanes; PA, polyamide.

Introduction

Thermoplastics are generally fossil-derived polymer materials (like natural gas, coal, and mineral oil); however, there are also existences of bio-based materials. The thermoplastic polymer materials structure may be partially crystalline, amorphous, or crystalline, and this structure affects the possible application and processing technology of the polymers. Thermoplastic polymers may be formed by heating at the beginning and then cooling through different polymer processing methods like injection molding or extrusion [21,22]. Natural fibers are also used as reinforcing materials in biocomposites to increase the toughness and strength of polymers [23,24]. The fibers are called as the "reinforcing phase" and the embodying polymeric materials as the "matrix phase." Frequently it is needed to apply a compatibilizer additive for increasing the interactions between phases of material [25,26].

The thermoplastic polymeric material is compounded short fiber through using extruders technology especially for the fibers with certain limited ranges up to ~2 mm length, which are generally below the pelletized compound length. The natural fibers are assimilated in a polymer randomly [27]. In case of long fiber technology fibers run parallelly from each other across the pellet length. These have among others better impact resistance [28–30], creep resistance [31,32], and better dimensional steadiness even at high or nearly sub-zero temperatures [29,33]. Publications often do not differentiate among types of elements as micro-particles, wood flour, and the real fibers. This paper will introduce the differences of these element types in their morphology like slenderness, specific surface, which are affecting the strength of the composites. To understand the specialties and role of morphology of wood elements the basic anatomical structure of wood should be recognized.

Natural plant—also called lignocellulosic materials—based reinforcers can be wood or annual plants [34,35]. Wood also function as a natural polymer in the composite system, as it is developed mainly from the polymers like partially crystallized or crystallized cellulose, amorphous lignin, and other extractives especially where the matrix is lignin and the reinforcing material is cellulose. The cross-linking degree of such systems is possible for lignin, which depends on the substitution degree, and the rigidity of the cellulose-lignin structure also varies as per the report from Thakur et al. [36]. These two materials are connecting each other by hemicellulose polymers, which ensures flexible but still a strong system [10]. Regarding the anatomical structure of plant cells have various tasks: support the plant's structure or transport of water and nutrients. There are various types of cells present in softwood and hardwood species [37].

1.2 MATERIALS FOR BIOCOMPOSITES

Biodegradable polymers are the polymers, usually produced from a renewable resource, which can be composted in the soil or by placing in a biotic environment for the enzymatic degradability of fungi, bacteria, or algae; they are decomposing into invisible parts during months or a few years and the decomposition products do not contaminate the environment or the compost.

Compostable polymers are such kind of polymers that are capable of biodegradation in compost. During its decomposition, it is converted into water, inorganic substances and biomass, in addition to the formation of carbon dioxide and—in an

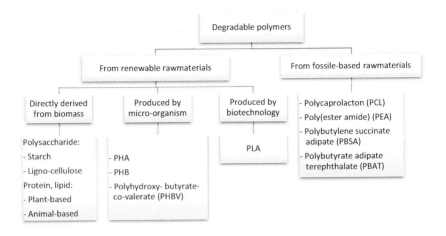

FIGURE 1.4 Grouping of biopolymers.

oxygen-free environment—methane. The decomposition process takes months, or up to a year.

Biodegradable polymer is a polymer that is able to degrade into water, inorganic compounds and biomass in a biotic environment, or compost due to the enzymatic degradation of micro-organisms, with the formation of carbon dioxide and—in an oxygen-free environment—methane, and the degradation process takes months to a maximum of 1 year.

Bio-erodible polymer is a polymer that is not capable of enzymatic degradation. These polymers are usually fragmented by heat and/or oxygen and/or ultraviolet (UV) aging, but the fragments are not capable of further degradation. They are not degradable in the traditional sense; they are just "disintegrating." Among others, these are a source of increasing amounts of micro-plastics that burden the environment.

A grouping of biodegradable polymers is shown in Figure 1.4.

1.2.1 Biopolymers as Matrix Materials

In Table 1.1, the main features of the most common thermoplastic biopolymers are listed.

Main biopolymers are the followings:

- *Cellulose-based plastics* are among others like cellulose esters and cellulose acetate (packaging blister) or nitrocellulose, and their associated derivatives, e.g., celluloid or cellophane.
- *Starch-based plastics* have a great market share. TPS (thermoplastic starch) represents ~50% of bioplastics market. It has various applications, like drug capsules or in different blends with biodegradable polyester starch, which is used for industrial purposes and these are also compostable naturally.
- *Protein-based plastics* are made from different raw materials such as wheat gluten, and casein. These are also biodegradable products.

TABLE 1.1
Biopolymers—Raw Materials, Features, Substitution [38–41]

Basis of Biopolymer	Feedstock	Raw Materials of Biopolymer	Properties	Substitute for
Starch based	Potato, corn, tapioca, and wheat	Starch	**TPS:** • Cheaper • Lower water vapor barrier • Functioning as a good oxygen (O_2) and carbon-di-oxide (CO_2) barrier • Higher shrinkage • Poor mechanical properties • Water solubility • Fastest decomposition • Bad processability • Brittleness	Polystyrene (PS)
Cellulose based	Pulp of wood	Cellulose	• Lower water vapor barrier • Poor mechanical performance • Bad processing capability • Brittleness	PP
Polyhydroxy-alkanoates (PHA) and (PHB)	Maize, corn, tapioca, potatoes, and vegetable oils	Starch	**PHAs** • Ranged within brittle and stiff to semi rubber-like as polymer **PHB** • Possess better oxygen barrier features compared to both PET and PP • Better water vapor resistance properties compared to PP • Exist odor and fat barrier characteristics that are enough to use for food packaging	• PP • PE
PLA	Potatoes, corn, sugar beet, maize wheat, and tapioca	Lactic acid	• Higher tensile strength and modulus • Low heat resistance • Water vapor and gas barrier worse than PET • Odor and fat resistance capabilities are excellent • UV resistant • Its low crystallinity and brittleness led to limited applications with lower thermal stability • Very stable at room temperature, but can be composted in a few months	• High-density and low-density polyethylene (HDPE and LDPE) • PP • PE • PET • PS

- *PLA* is an emerging transparent polymer made from sugar (dextrose). It is comparable with conventional plastics, and it can be processed with the same equipment (extrusion, injection mould, etc.). Main uses of PLA-based plastic composites are films, containers, cups, and bottles.
- PHA are natural linear polyesters and are generated by bacterial fermentation of lipids or sugars. PHA is less elastic and more ductile flexible than other plastics. These non-degradable plastics exist in extensive applications in medical sector.
- *Poly(3-hydroxybutyrate) (PHB)* is a well-known polyester generated polymeric material obtained through bacteria processing of corn starch, glucose, or even wastewater. PHB is also likewise other plastics as PP, and its processing methods are also the same. It is commonly used to produce biodegradable transparent foils.
- *PA 11 (Polyamide 11)* is a derivative of naturally originated oil and it is a non-biodegradable technical polymer. It is applied for high-performance usages like pneumatic airbrake tubing, electrical cable, automotive fuel lines, anti-termite sheathing, sports shoes, flexible oil and gas pipes, electronic device components, and catheters.
- *Bio-based PE* is derived from ethylene, which could also be obtained from the ethanol as well; can be manufactured from agricultural raw materials like sugar cane or corns fermentation processes. It is physically identifiable to conventional PE—as it does not degrade biologically.
- *PHUs* are bio-based and isocyanate-free polyurethanes. In contrary to traditional polyurethanes, cross-linked polyhydroxyurethanes are recyclable through chemical reactions like dynamic transcarbamoylation.
- Several bioplastics (e.g., polyurethanes, epoxy, polyesters, etc.) are synthesized from animal- and plant-derived oils and fats. These are similar to crude oil-based materials but have a huge potentiality for the growth in this area based on micro-algae-derived oils.
- *Genetically modified bioplastics* are under development to be used for genetically modified bacteria or crops for optimizing efficiency of polymer production [42–44].

1.2.2 Other Matrix Materials

Natural reinforcers can be embodied or connected by non-bio-based materials also, which can be both organic and inorganic.

1.2.2.1 Fossil-Based Thermoplastics

To combine wood or other bio-based reinforcing materials with thermoplastics has several challenges, such as surface compatibility and element size. The main compatibility factor is melting temperature of plastics, as lignocellulosic materials will degrade at high temperatures (from 200°C). However, polymers which can be processed below 200°C are generally appropriate. Change of market share of thermoplastics in WPC (Wood Polymer Composites) production is shown in Figure 1.5 [15,45,46].

PE is the most frequently used polymer for WPC's using both new and recycled material. All types of PEs are fairly easy to be stabilized against thermal degradations during production and in-use. All types of PEs are used for the manufacturing

Introduction

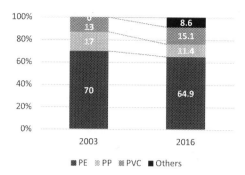

FIGURE 1.5 Change of thermoplastic's market share for WPC production.

of WPC like LDPE (low-density polyethylene: 920 kg/m³), MDPE (medium density polyethylene), and HDPE (high density polyethylene: 0.96 g/cm³). Still PE can be deformed at relatively lower temperature; permanent heat load capacity is between 60°C and 80°C and has melting temperature 110°C–130°C. PE has a low crystallinity (40%). The major advantages of PE are: (a) easy to process, (b) good impact resistance, and (c) good resistance against chemicals. Conversely, major disadvantages are: (a) UV degradation and (b) crackings by stress.

PP (density: 0.9 g/cm³) has various types: either in amorphous form or in semi-crystalline (70%) form. It is also widely used but requires more stabilizing additives to prevent the degradation compared to polyethylene. The significant properties of PP are: average impact resistance, stiffness, and melting point higher than HDPE, and limited cold resistance. Besides, melting temperature is within 162°C–165°C, and permanent heat load capacity is between 70°C and 90°C.

Polyvinyl chloride (PVC) is also another common plastic—cheaper and has very good mechanical properties—with various industrial and consumer uses. It has many forms like it could be flexible, such as automotive upholstery fabric and vinyl hose, or rigid, such as drainpipe and window profiles. Its melting and processing temperature is around 150°C. It was one of the first plastics in WPC production. Density is 1.2 g/cm³ for soft PVC and 1.4 g/cm³ for hard PVC [8,45].

1.2.2.2 Resins

Classic adhesives, or resins, are thermosetting polymers, such as phenol formaldehyde (PF), melamine formaldehyde (MF), and urea formaldehyde (UF), and any kind of their combinations, e.g., MUPF (melamine-urea-phenol-formaldehyde) and MUF (melamine-urea-formaldehyde) [47,48]. All these are creating adhesion (H-bridge and van der Waals) and mechanical bond between lignocellulosic elements (veneers, particles, strands, and fibers). Polymeric methylene diphenyl diisocyanate (pMDI) is a relatively newer type of adhesive which adds chemical bond, hence its –NCO groups can create urethane bridge bond with the free –OH groups of cellulose. These adhesives are widely used in wood-based panel and beam production:

- Panel boards:
    - Particleboard (PB)
    - Fiberboard (HB, LDF, MDF, HDF)[1]

- Oriented strand board (OSB), waferboard
- Plywood
- Solid wood panels (SWP)
- Cross laminated timber (CLT)
- Composite beams:
  - LVL (laminated veneer lumber)
  - Laminated strand lumber (LSL)
  - Parallel strand lumber (PSL)
    - I-joists
- *UF*: More than 90% of MDF and PBs are produced with UF, but it is also used in plywood and parquet production. It is made by condensation reaction of urea and formaldehyde in watery solution by means of acidic catalyst. Condensation is stopped by alkaline additions. Usual dry matter content is around 66%. It is a white color, cheaper adhesive; but it is rigid, not water- and heat-resistant—above 70°C, but above 90% relative humidity it is degrading—and has a significant formaldehyde emission. Recent molar ratio of urea and formaldehyde is 1 to 1.05, …, 1.15. To reduce free formaldehyde in the products additional urea or ABS (ammonium-bisulfite) is added during panel board productions. Usual amount in wood-based composites: 8–12 wt%.
- *MF*: Usually combined with UF to improve the latter's properties. It has good resistance against water and heat. Usual dry matter content is around 65%. In MDF production, usually MUF is used with 30%–50% MF content. It is a white color adhesive and also needs acidic catalyst (hardener) to control condensation reaction. The free formaldehyde content should be minimized similarly as in case of UF. Usual amount in wood-based composites: 10–14 wt%.
- *PF*: It is a dark reddish-brown adhesive with ~50% dry matter content. It is manufactured by using condensation reaction formaldehyde and phenol. It creates a water- and boiling-resistant bond. Often used for hardboard (wet process fiberboard), plywood, and OSB production, as well as for producing PSL. Usual amount in wood-based composites: 6–8 wt%.
- *MDI or pMDI*: Di-phenyl-4,4′-di-isocyanate monomer (or partially polymerized) solution, without any kind of solvent. It can add an exceptionally higher dimensional stability for wood-based composites, hence it creates chemical bond with wood: the isocyanate root (–NCO) reacts with free –OH (hydroxyl groups of cellulose) and results in strong urethane-bridge bond. Usual amount in wood-based composites: 2–6 wt%. Still very expensive and it has a limited availability [12,49–51].

1.2.2.3 Inorganic Binders

Portland cement is manufactured through heating clay and limestone in a rotating kiln to shrinkage. It results clinker, and it contains a few percent of gypsum added for controlling the curation of cement. These are grinded together and its properties are primarily determined by the clinker. It hardens by hydration, where clinker minerals and water are transformed into several components, e.g., calcium silicate hydrate. The main challenge in cement-bonded wood-based product (CBWP) production is to compatibilize cement and wood, hence wood contains various inhibitors, such as sugars, tannin,

Introduction

and polyoses, which are hindering the hydration of cement [52–54]. Not every wood species is suitable to produce CBWP, and there are numerous additives to enhance the wood-cement compatibility (e.g., magnesium chloride ($MgCl_2$), sodium silicate (Na_2SiO_3), etc.). For complete hardening of cement 28 days is necessary, but in CBWP production with use of accelerators and special climate (~60°C and ~100% relative humidity) the products need only ~10 h to reach the half of their full strength [54,55].

Gypsum ($CaSO_4 + 2H_2O$) can be obtained from various sources such as by mining, desulfurizing of flue gas of fossil-based power plants and as byproduct of organic acid manufacturing (e.g., phosphorus acid). Gypsum hardens in a very quick reaction, so retarders, plasticizers, and other additives are necessary to produce gypsum-bonded wood-based products.

1.2.3 Natural Reinforcers—Biomass

1.2.3.1 Plant-Based Fibers

Bledzki and Gassan [26] determine six types of natural plant-based (phytomass) fibers such as bast fibers (hemp, jute, flax, kenaf, and ramie), leaf fibers (sisal, abaca, and pineapple), seed fibers (kapok, cotton, and coir), core fibers (jute, hemp, and kenaf), reed and grass fibers (rice, wheat, and corn), and different types of other materials (like roots and wood). A similar classification of natural fibers was also reported by other researchers [9].

In composite production, some widely used plants are: hemp or flax shive and fiber (*Cannabis sativa, Linum usitatissimum*), bagasse (*Gramineae Saccharum officinarum* L.), China grass (*Mischantus* spp.), straw (any cereal's), cotton stalk (*Gossypium hirsutum*), kenaf (*Hibiscus cannabinus*), bamboo (*Bambusoideae* spp.), abaca (*Musa textilis*), and corn stalk (*Zea mays* spp.) [29,50,56–58].

Faruk et al. [59] lists various natural fibers, which are or can be used as reinforcers in composites:

- *Flax* (*Linum usitatissimum*) is a widely used bast fiber and it is harvested in temperate parts of the world. This fiber type is found from ancient times in the biosphere. Originally flax is the most commonly used bast fiber as a prominent textile raw material. Nowadays, flax is extensively used for manufacturing biocomposites, in both applications of woven and non-woven form [9,51,58–60].
- Another important bast fiber crop is *hemp* (*Cannabis* spp.). This plant is gown annually especially in temperate climates. "Hemp is currently the subject of European Union subsidy for non-food agriculture, and a considerable initiative in currently underway for their further development in Europe" [59].
- *Jute* is produced from *Corchorus* spp., which embraces nearly100 species. Jute is another cheapest biofiber entailing the highest production volumes among all the bast fibers. Largest providers are Bangladesh, China, and India.
- *Kenaf* belongs to the *Hibiscus* spp. Kenaf is relatively a newer crop found in the United States in the last decades and shows promising potentiality as a composite products raw material.

- *Sisal* (*Agave sisalana*) is an agave fiber and produced in East Africa and Brazil. The conventional market for fibers is decreasing because of increase on use of synthetic substitutes.
- *Pineapple* belongs to *Ananas comosus* which is a tropical plant grown in Brazil. The leaf fiber of Pineapple is enriched with cellulose, which is cheaper and widely available. Besides, it has good potentiality to reinforce with polymers.
- The *abaca* (banana) fiber is resistant to seawater with enhanced durability. Abaca is the strongest cellulosic fiber compared to other commercially available fibers in nature, ant it is produced mainly in Ecuador and Philippines. Formerly, it was the preferred cordage fiber used for naval purposes.
- *Coir* (*Cocos nucifera*) husk materials are located between the coconut husk and the outer shell [61]. However, coir is considered as a prominent by-product material. Its availability depends on the market needs.
- *Ramie* belongs to *Urticaceae (Boehmeria)* family. Ramie is popular as a prominent textile fiber [62] but based on its chemical composition its production is relatively difficult.
- *Rice* is an important cereal grain, which is suitable for producing hull fibers. Recently corn, wheat, rye, oat, and other cereal crops are also used for producing fibers. All are subjects of composite researches [63–69].
- *Bagasse* is a fibrous residue of sugarcane stalks after crushing them to extract their juice. It is increasingly used to produce composites materials [70,71].

Typically, wood-based reinforcers or filler materials (wood micro-particles, wood flour, and wood fibers) are used for wood-polymer composites production. However, all of the three material types are produced with completely diverse technologies and they vary in morphology and sizes as well (Figure 1.6) [72].

Based on an environmental push, the automotive industry has initiated research on substituting non-renewable fiber materials by natural fibers in the beginning of the 2000s. To replace glass or carbon fiber fabrics, various renewable, natural fibers came into focus. The main advantages of natural fibers over artificial glass and carbon fibers are their low density, low cost, good specific tensile properties, not abrasive to machining, and they are harmless to health, renewable, recyclable, and biodegradable [72].

There are several annual plants suitable for producing woven or non-woven fabrics, among others flax (*Linum usitatissimum*), hemp (*Cannabis sativa*), kenaf (*Hibiscus cannabinus*), and abaca (*Musa textilis*). Annual plants have usually long

FIGURE 1.6 Image of (a) wood flour, (b) micro-particles, and (c) fibers (actually fiber bundles). (Adapted with permissions from Ref. [72]. Copyright, Willey & Sons (2017).)

Introduction

fibers in their bast and the core of their stems. Kenaf bast fiber is 2–6 mm long, while hemp bast fiber is much longer and stronger, between 15 and 50 mm, averaging 25 mm [73,74]. Flax fibers are 5–77 μm wide and 4–77 mm long, or 19 μm and 33 mm on average, respectively [57]. Abaca has its valuable fiber content in the leaf sheaths, where fibers ranging in length from 180 to 370 cm [75,76].

In general, during weaving, two sets of threads are interlaced at right angles: the warp runs longitudinally and the weft in crosswise directions of the fabrics. This is the classic biaxial fabric made by plain weaving. There are derived weaving structures for reinforced composites: uni-directional (only a few wefts hold the warps together), three- or multi-axial fabrics to make strength which can be designed for specific purposes [37,77,78].

1.2.3.2 Animal-Based Fibers

Animal fibers (belongs to zoomass) (second) are most significant source of naturally originated fiber materials for reinforcing composites. But the most common animal fibers are silk, wool, and chicken feathers. These fibers are usually made up of different kinds of proteins.

Silk has a great potential due to their structure and properties as for it is made up from highly structured proteins, which give higher tensile strength, higher elongation, and good resistance against chemicals. However, silk is acquired from various sources with different attractive features [79,80]. A number of insects and mostly butterfly species larvae (around 140,000 *Lepidoptera*) produce cocoon silks during metamorphosis in labial gland. Silk is also gathered from spiders (around 40,000 species) and they could produce seven dissimilar types of silk with exclusive characteristics in their whole lifespan as Ramamoorthy et al. [81] has mentioned. Besides these, dragline silk (*Nephila*) from spider and Mulberry silk (*Bombyx mori*) from silkworms are widely reported. The mechanical strengths (tensile) of silk fiber (dragline) are reported higher compared to polyamide fiber and a number of synthetic fibers [81] (Table 1.2).

The animal fiber from Wool is a frequently used textile materials [77] which is obtained from sheep, camel, goat, rabbit, horse, and several other mammals. The most

TABLE 1.2
Main Characteristics of Common Natural Fibers [81–87]

Fiber	Density of Natural Fibers (g/cm^3)	Tensile strength (MPa)	Young's Modulus (GPa)
Flax (*Linum usitatissimum*)	1.45	500–900	50–70
Hemp (*Cannabis* **spp.**)	1.48	350–800	30–60
Coir (*Cocos nucifera*)	1.2	150–180	4–6
Sisal (*Agave sisalana*)	1.5	300–500	10–30
Jute (*Corchorus* **spp.**)	1.3	300–700	20–50
Wool (sheep)	1.3	120–200	2.5–3.5
Silk (*Bombyx mori*)	1.33	650–750	16
Dragline (spider silk)	1.3	1,100	10.5–13.7 (at 25% RH)
Chicken feather	0.89	221 (at 0% RH)	2.5

Relative humidity.

promising thing for wool is that these are hydrophilic in nature—after removing "wool grease"—like other plant-based fiber materials and absorbs water nearly one-third of its own weight. It is stronger than silk, but when moisture content increases it loses about up to 25% of its strength. Longer fiber results in higher strength of the yarn. Wool fiber diameter ranging from 16 to 40 µm depending on animal—e.g., the mean diameter of alpaca fiber—is around 0.4 µm, while that of wool fiber is ~1.0 µm [81,88].

Chicken feather generates the higher volume of feathers which could also be used by composite manufacturers since cannot be utilized by several other industries [84,89]. Kock [84] has reported in his thesis that the chicken feathers show outstanding characteristics such as lower density, good thermal, and acoustic insulation. The quill has an aspect ratio of approximately 15–30. Similarly, wool poultry feathers are consisting of 91% keratin (protein-based), 1% lipid, and 8% water [81].

1.2.4 Production Techniques

Main production techniques are the followings:

- *Inorganic-bonded (cement or gypsum) lignocellulose reinforced composites* are produced with wet process (similar to paper manufacturing) or by semi-dry process for mat forming. Both are then pressed in stack presses and after closing the stacks the compression time takes 8–12 h to harden the boards. The ratio of wood to cement is typically 1:2.5–3.0.
- *Resin-bonded biocomposites* are usually produced using thermosetting polymers (e.g., UF, MF, PF, and MDI) and are compressed in hot presses at high specific pressure (3–5 MPa) and temperature (180°C–200°C). Conventionally, multi-daylight presses with up to 30+ daylights. Nowadays, continuous hot presses are more popular for their higher productivity. Most common types are double belt flatbed presses up to 50+ m length (Dieffenbacher builds the longest continuous press with 80 m²) and calender presses.

WPCs are produced by various techniques such as injection molding, extrusion, and calendering or 3D molding.

See for more details in Chapter 10: *Design and Fabrication Technology in Biocomposite Manufacturing.*

1.3 PROJECTS AT UNIVERSITY OF SOPRON—SOME EXAMPLES

In the following some projects, publications are cited from the portfolio of Wood Sciences and Applied Arts, Institute of Wood Based Products and Technologies, Simonyi Károly Faculty of Engineering of University of Sopron, Hungary. The samples are focusing on biocomposites: mineral-bonded, WPC, and nanocomposite.

1.3.1 Effect of Shape of Wood Element on Properties of Wood-PLA Biocomposites

Many publications do not make distinction between particles used in wood polymer composites; they just call all of them as fibers. However, these materials are

FIGURE 1.7 Microscopic photographs of (a) wood flour, (b) micro particle, and (c) wood fibers. (Photos by Sándor Fehér.)

not always fibers in terms of anatomical structural aspects of wood, as the structure of fiber is destroyed by mechanical processes. Wood particles, flours, chips, flakes, and strands can be designed by pure mechanical size reduction processing. But these fibrous elements are derived from the cut pieces of woods comprising all anatomical and chemical components at miscellaneous dimensions. Definitely, all of these materials comprise fibers and bunches of fiber but in different fragmentations reliant on the elemental size caused by the size reduction operations. Fibers can be derived from wood by hydro-thermal plastifying of chemical components—especially lignin—and then a mechanical process called defibrating can separate the anatomical fibers (libriform cells). Figure 1.7 shows clearly the significant differences between the flours of wood (mechanically processed), micro-particles of wood, and the thermo-mechanically derived wood fibers.

Both wood flours and wood micro-particles are produced through mechanical size reduction methods which are appropriate as fillers in thermoplastic WPCs or for "short fiber" technologies. Wood fibers are mostly suitable to be used for long fiber-based technologies, through replacing other materials (like non-wood fibers such as glass or others).

$$\Psi_{LS} = \frac{D_F^{max}}{D_p (\Psi_{WS})^2} \quad (1.1)$$

where

Ψ_{LS}: longitudinal extended by width slenderness,
D_F^{max}: Feret diameter,
Ψ_{WS}: width slenderness, and
D_P: chord length diameter.

Bledzki [90] determined a definite limit to form factor, for deciding when a wood particle could be considered/mentioned as fiber. According to our previous research results [37] and Equation 1.1, the elements having form factor $\Psi_{LS} = 1$ are considered as wood flour elements, but if the form factor is within $1 < \Psi_{LS} \leq 10$, then the elements are rather micro-particles and for the case of $10 < \Psi_{LS}$ those are certainly fibers. However, these are termed as technical fibers with substantial slenderness, which indicate their longitudinal dimension, exceeding their other two transverse dimensions.

Many researchers stated that wood flours are suitable only as a filler material in WPC systems. Contrary to this wood fibers and even micro-particles exert an influential reinforcing effect. Besides, the length of fibers and corresponding fiber loads possess direct effects on the values of strengths of such composites. It is advantageous to apply suitable coupling agents or surface modification on wood-based materials to enhance the interaction between the wood and associated matrix polymers as many scientists have reported [72,91].

1.3.2 Continuous Fiber Reinforced PLA Biocomposite

A continuous fiber (natural) reinforced with the bio-polymer matrix was developed which is also compatible with the typical industrial processing methods, however with improved mechanical properties. The outcomes of this study were partly described by Markó et al. [58]. The biocomposite was based on renewable, biodegradable, recyclable, and after all natural raw materials deprived of adding any harmful components as well. One of the most important fiber plants was selected for research—flax (*Linum usitatissimum*) woven fabrics in biaxial and UD form (0° to 90°). As a polymeric matrix (PLA film) with 0.2–0.25 mm thickness and PP film (control materials) were selected for this study. The fiber to matrix (m/m) proportion was 40%/60%.

During conducting the experiment, a total of seven layers of flat panels (three layers entailing flax woven fabric and another four layers contained polymeric films) were laminated and hot-pressed in a Siempelkamp hot-pressing machine. The composite panels were then cooled quickly for avoiding crystallization of the PLA films.

Scanning electron microscopy (SEM) photographs of breaking portions of developed composite samples obtained after the tensile strength test exhibited better surface adhesion between the matrix in PLA and flax fiber compared to the PP (Figure 1.8). This is happened for the differences in surface energy. Furthermore, better interfacial interaction properties were noticed with PLA than in PP when a 5% additive (maleic anhydride) was added.

As a consequence of biaxial and UD (uni-directional) flax woven fabric reinforced with PLA and PP biocomposites, it was seen that failure of PLA matrix in

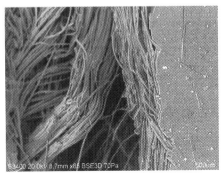 

FIGURE 1.8 Breakage parts of PLA-flax composites after tensile properties test. (Photos by Zoltán Börcsök.)

Introduction

terms of pulling out of the threads is comparatively shorter, which is indicating a better connection between the thermoplastic polymers and flax woven fabrics. The pulling out for PP is comparatively longer, thus a local delamination occurred at the failed section and no micro-impregnation happened during the manufacturing, opposing to the biocomposites with PLA films.

Better surface interaction was noticed through SEM images between the flax woven fabric and matrix for PLA films compared to PP. The surface interaction property was induced with neat PLA films (without any additives) in contrast with PP for a 5% maleic anhydrides additive solution. The bleaching treatment of flax fabrics with the hydrogen peroxide reagent degrades cellulose polymer from fiber, which in turn causes a decrease in strength and surface adhesion characteristics, consequences on weaker composite panels.

Moreover, we also have noticed further that yarns were arranged differently in two upright directions, which provides advantages into the design. Besides, with synthetic yarns, mechanical properties of the composites may be affected by the type and quantity of the yarns present in composite structures especially in fabrics. The weaving parameters of woven fabrics could be designed according to the flexural properties of composites using flax woven fabrics. Hereafter, the modulus of elasticity and strain could be adjusted within a wide range from 10 to 40 GPa, and the modulus of rupture (MOR) differs only within 10%–15% range. This provides extra benefits for 3D or in energy absorbing composite products [72].

1.3.3 Development of Wood Plastic Composite with Optimized Inertia

Markó et al. [92] reported that most of the companies (wood processing industries in Hungary) are small- and medium-sized enterprise. Typical wastes generated from those companies are solid wood or wood-based production saw dust, sanding dust, and shavings. These are considered as the byproducts or potential secondary raw material, which also could be utilized in other manufacturing processes. Plastic-based wastes from associated manufacturing industries could also be recognized as a secondary raw material for newer products. In some cases, novel wood-plastic composites were developed for the packaging and furniture industries, combining these two kinds of byproducts. However, the main aim of this study (supported by the European Union, European Social Funds, TÁMOP 4.2.1.B-09/1/KONV-2010-0006, and WPC_TECH BAROSS-ND07-ND_INRG5_07-2008-0087) was to invent an easy production methods for processing both types of wastes (byproducts) into a feasible lighter weight composites with higher flexural properties. In this regard, finite element simulations were performed initially for determining the optimal inertia of composite structures. Additionally, optimal mixing proportion and sizes of particles were also determined for producing a convenient wood plastic material. Lastly, aluminum press plates were manufactured for being capable for producing the inertia optimized tentative products.

The major findings of this research are: PP-wood mixtures were produced with 60% of the wood. The wood was used in two dissimilar forms like (a) saw dust (micro particles) and (b) sanding dust or wood flour. The major advantages for inertia optimized products are: (a) production is depending on by-products (previously

FIGURE 1.9 Experimental inertial-optimized components from compounds. (Photos by Tibor L. Alpár.)

considered as the wastes), (b) simple manufacturing methods (dry blending also existing besides the compounding), (c) low specific density than "clean" plastics, (d) higher strengths, and so on.

The mechanical features of experimental products (Figure 1.9) were also performed. Load-bearing capacity test resulted 70.0 Nmm/g in case of honeycomb structure and 117.1 Nmm/g in case of square structure. In both cases the WPC was compounded from 50 wt% micro-particles and 50 wt% PP.

The inertia-optimized boards advantages could be summarized as follows:

- Feasible and simple production methods (with cold blending)
- Higher compounding quality
- Using secondary raw materials
- Lower specific density
- Higher specific flexural strength
- Easier to handle the post processing operations
- Larger area for utilization

1.3.4 Wood Wool Cement Boards Produced with Nano Minerals

Alpár et al. [54] reported that conventional utilization of wood concerning cement typically is as formwork for concrete. Wood is preferred for concrete formwork of building foundations because of wood's natural advantages, such as availability, easy to work with, and simple removability and reusability after setting of the concrete. However, most species of woods are less or more non-compatible with the hydration of cement.

Among others the final strength of WCC is depending on the moisture content of wood. The water-soluble chemicals of wood are dissolved in water—some of these act as inhibitory ingredients for cements like hemicelluloses, tannins, and sugars. The inhibition process is the following: wood sugars absorb Alit (tricalcium silicate) on the surface, then create a gel around the Alit, after which the water cannot reach the Alit so hydration can no longer occur.

For this research, 20 poplar clones (as possible raw materials for wood wool boards) were examined regarding cement compatibility [93]. The project applied new

Introduction

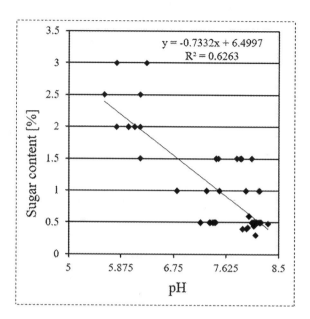

FIGURE 1.10 Sugar content as a function of pH.

additives to wood wool boards for increasing the adhesion strength between wood (*Populus* spp.) and ordinary Portland cement, thus increasing the strength of final composite products. The additives applied were montmorillonite nanoparticles and poly(diallyldimethylammonium chloride) (PDDA) for modifying the surface charges of wood materials. Results showed an increase in bending strength above 20%.

Also, the effect of fungi attack was researched. Hence fungi of brown rotting decomposes cellulose and hemicelluloses, so monosaccharides are produced, which turn partly into saccharide acids, and the pH value of wood will be slightly acidic. The pH value of wood can be determined by a fast hot water extraction, and if the pH of wood is above 7, the wood (*Populus euramericana* cv. I214) should be appropriate for board production (Figure 1.10).

Results of the use of montmorillonite nanoparticles and PDDA on sound I-214 poplar were very satisfactory. Higher strength values could be reached than in case of using a conventional additive such as $MgCl_2$. It was found that there are significant differences with the sugar content, tannin content, and pH between the sound and unsound poplar (I-214) wood. Wood attacked by brown rotten fungi has higher tannin and sugar content; pH drops below 7. A montmorillonite through combining with PDDA additive provides the best outcomes especially for bending tests of composite samples made from fresh poplar woods, enhancing the MOR by more than 20% in contrast to the control samples.

1.3.5 PLA NANO COMPOSITE

During a research by Halász et al. [94] the applied materials were montmorillonite nano-plates and cellulose nano-crystal. Production of nano-composite foils based

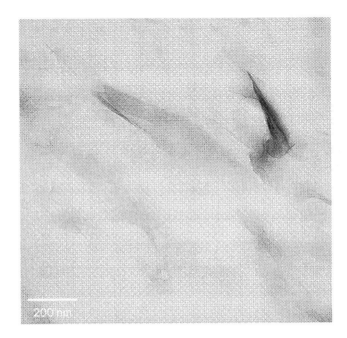

FIGURE 1.11 Montmorillonite in PLA matrix—TEM image. (Photo by Zoltán Börcsök.)

on layered silicate and nano- and micro-reinforced hybrid foils were produced. Real nano-composites were produced by determining optimal process parameters. Delaminated and embodied structures can be made as shown in Figure 1.11.

Based on their size reinforcing materials provide special features (increase of transparency, mechanical, thermal and barrier features, increase of decomposition, and anti-bacterial effect).

1.4 APPLICATION OF BIOCOMPOSITES

Different natural fibers display numerous advantages in terms of lower environmental impact, biodegradability, and so on than the artificial fibers likes glass, aramid, or carbon fibers, summarized by [36]. Nowadays some the naturally originated fibers are also used for diverse industrial applications. Numerous aspects of surface and structural properties of natural fiber/polymer biocomposites are already investigated but there are further still many issues worth noting to be considered for research. Figure 1.12 shows the market share of biocomposites, as Brunel University's study describes.[3]

1.4.1 MEDICAL USE

Medical applications are an emerging field of using biocomposites. Human bone and tissues are natural and anisotropic composite materials. The design criterion for implants made from composite biomaterial is a novel porous ceramic-polymeric biocomposite material, with mechanical resistance and morphology similar to those

Introduction

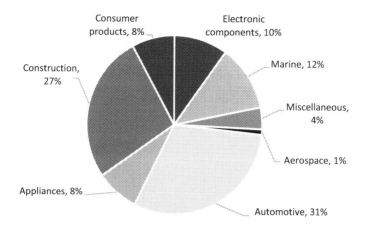

FIGURE 1.12 Market share of biocomposites (2009).

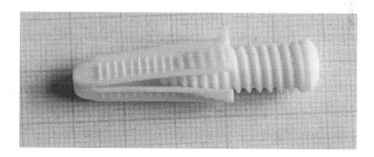

FIGURE 1.13 β-TCP/PLA-based decomposing interference screw. (From Kovesi, A., E. Bognar, and G. Erdelyi, Bio-intrafix interferenciacsavar vizsgalata ramanspektroszkopiaval. In *Fiatal Muszakiak Tudomanyos Ulesszaka XVIII*, Kolozsvar, Romania, 2013.) [95]

of the natural cancellous bones. Surgeons could cut the graft directly in surgery room for adapting its shape with the defect. For some uses like in dental implants, biopolymers give better aesthetic behavior. Biocomposites are usable for hard tissues, including dental post, prosthetic socket, bone plate, external fixator, total hip replacement, orthodontic archwire, composite screws, orthodontic bracket, and pins. Benefits are faster bone healing without any risk of pathogen transfer in contrast to allograft; cheaper and faster surgery providing less pain than auto grafting. Rapid prototyping is also a potential way to produce medical screws to fix insured tissue. These screws may be produced from PLA-based biocomposites and these are decomposing in human body. Most frequently applied fixation approaches for anterior crucial ligament reconstruction with four harmstring grafting is the so-called interference screw. These screws are usually made from the bioabsorbable materials [95] tested according to the BIO-INTRAFIX composite screw system, which has two bioabsorbable β-TCP/PLA (beta-tricalcium-phosphate/PLA) components (Figure 1.13). β-TCP acts as a prominent osteoconductive bioceramic material, which initiates bone growth into sheath screw. However, PLA is later simply absorbed by body. In their study Kövesi et al. [95] researched the long-term clinical outcome of use of β-TCP/PLA. They investigated the explanted samples compared to the sterile reference screws with Raman spectroscopy [95].

Medical biocomposites should be biocompatible. Bioactivity of biocomposites is a major characteristic for improving bonding at the biomaterial-bone interface. Within the period of time, the biodegradable biocomposites may be disappeared by biodegradation in human body [96–98].

1.4.2 Automotive Industry

To use bio-based composites in automotive industry has a long history. The famous East German car, Trabant's (originally P70 by VEB Automobilwerk Zwickau from 1955) complete bodywork was made from cotton fiber-reinforced phenolic resin. Today more and more car manufacturers are using natural fiber reinforced molds or biocomposites in car interiors like BMW, Ford, Mercedes, and so on (Figure 1.14). FlexForm Technologies produces lightweight natural fiber composites from non-woven hemp, sisal, kenaf, flax, and jute blended with thermoplastic polymers such as polyester and polypropylene.[4] The ratio of fiber and polymer is 50–50 wt%, available density, tensile strength, and flexural modulus ranges are 1,200–1,800 g/m^2, 20–42 MPa, 1,800–3,800 MPa, respectively.

The technique of rapid manufacture of LFT moulding is a direct process of in-line compounding using two twin extruders where the resin is mixed in with additives and the fibers are fed in. During a project of ICT Fraunhofer, this technique was used to produce natural fiber and resin panels using a mould previously used to produce an in-board foot-well panel suitable for a car. Several composites were manufactured using glass/PLA, flax/PLA, hemp/PLA, and cellulose fibres/PLA [37,99–111].

To evaluate new composites and production techniques, the automotive industry often tests generic structural elements, like a "structural beam" (Figure 1.15).

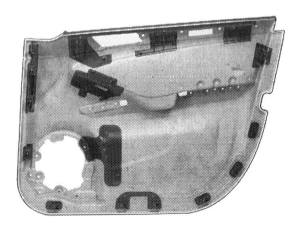

FIGURE 1.14 Non-woven mat from bast fiber and polymer fiber for car's door interior carpeting (FlexForm).

Introduction

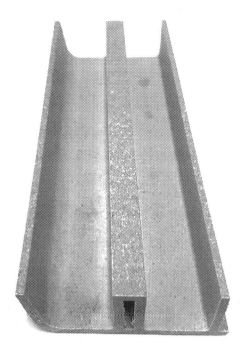

FIGURE 1.15 Structural beam for testing of new composites. (Photo by Tibor L. Alpár.)

Two materials were examined experimentally at Institute for Chemical Technology (ICT) Fraunhofer: a glass/filled polypropylene and flax/PLA. The test specimens were injection molded and supplied to MERL (Materials Engineering Research Laboratory Ltd.) for testing. The additional stiffness and failure load of the biocomposite materials show that these materials are promising candidates in structural applications. Also fewer raw materials are necessary in case of PLA/flax biocomposite material to achieve the same stiffness and maximum load [112].

Major benefits of biocomposites in automotive industry are higher tensile strength, higher impact strength, lower weight, greater durability, good dimensional stability, rapid parts production, acoustic benefits, superior 3D molding, larger versatility, lower volatile organic compound emission, recyclable, based on renewable materials, carbon pool, and reducing climate change.

1.4.3 Construction Industry

1.4.3.1 Structural Biocomposite

Biocomposites are becoming an interesting subject to many researches, also in construction and building industry for their favorable advantages like low weight, higher specific strength, and lower production costs. The so-called Green Building structures have gained significant global attentions since past few years. Biocomposites have an increasing role in green building constructions. Biocomposites could be classified into two key groups: one is structural and another one is nonstructural biocomposites [113].

Structural biocomposites should be able to carry a load during the usage, for example, load-bearing walls, roof systems, stairs, and sub-flooring structures.

One sort of use is roof structure, where structural beams have been designed, made, and tested. Conventional beams are LVL (Figure 1.16), PSL, LSL, and of course glulam (Figure 1.17).[5] Experimentally cellulose fibers and soy oil-based resin, in the paper sheet forms, made from the recycled cardboard boxes may be used to produce biocomposite structures [114].

Stay-in-place (SIP) bridge forms are used for spanning distance between the bridge girder beams. Biocomposite-based SIPs have many benefits than the steel

FIGURE 1.16 Kerto-Q LVL used in Metropol Parasol in Sevilla, Spain.

FIGURE 1.17 Glulam construction of a swimming hall in Sopron, Hungary. (Photo by Tibor L. Alpár.)

Introduction

structures, such as porosity. Consequently, this phenomenon lets water for evaporating which facilitates to avoid corrosion. It is also considered as a biodegradable product. Furthermore, the biocomposite form is lighter than the steel, facilitating fast, feasible, and cheap installations [98].

1.4.3.2 Nonstructural Biocomposite

Nonstructural biocomposite materials need not to carry a load throughout the service. Matrix materials can be inorganic (cement and gypsum), thermoplastics (bio-based fossil), thermosets, and reinforcers can be wood or other lignocellulosic elements (particles, fibers, and flour), and textiles (woven or non-woven). Non-structured composites have applications for products like roofing, ceiling tiles, siding, windows, furniture, doors, and outdoor products. Some studies on structural biocomposites are introduced below.

Wood fiber plastic composites (WPC) are made across the cross-section of standard lumber profile dimensions which could be used for exterior constructions. These products are used for dock surface boards, picnic tables, deck, industrial flooring, and landscape timbers (Figure 1.18).

Composite panels have a long history: particleboard, mineral-bonded panels, and fiberboard. Various fibers—see in Figure 1.6—are applied for fiberboards, particleboards, and composite panels manufacturing. Cereal straw (from rice, wheat, etc.) is the second highest agro-based fiber used usually in panels production. Also reed hat has another potentiality as Alpar et al. [115] reports:

> In terms of the high density panels, we did not reach the standard specifications. Further research is needed for improving the mechanical and physical characteristics. On the other hand, additional tests were performed by adding nano materials like PDDA and Montmorilonite, both panel characteristics still did not reach the values specified in the standards. The effect of other accelerator agents and different types of cements needs to be examined. The real challenge in terms of compliance with the relevant standard strength values is the presence of cement bonding inhibitors in reed.

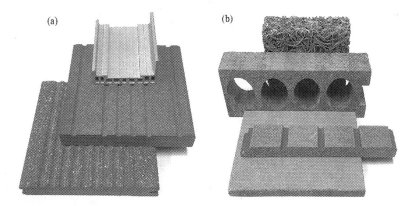

FIGURE 1.18 Inorganic materials bonded composite products are fiber-cement boards, cement-bonded particle boards, or wood wool boards are used in light- weight or panel-based constructions. (a) Extruded WPC products; (b) cement-bonded wood-based products—from top: wood-wool board, extruded wall panel, sidings. (Photosby Tibor L. Alpar.)

The insulation board showed good performance in terms of thermal conductivity. Although there is no standards for insulation board strength, this product also needs further development to meet the user expectations. If successful, this will create a new, environment friendly product that combines the positive characteristics of reed and cement bonded composites.

The higher percentages of silica in reed or in cereal straw give the advantage of a higher fire resistance. On other hand, rice husks presence in building panels helps to enhance thermal and acoustic properties [98].

There are several advantages of natural fibers and biocomposites to substitute other less sustainable building materials. These benefits can be categorized as follows [37,98]:

- *Environmental perspective:* Plant-based fiber materials are originated from renewable resources available in nature. They need low energy requirements during production. They and their production have a small carbon footprint. They are storing carbon. Biomass cascading is a sustainable way to handle these products. These sustainable technologies and products have a positive effect on influencing climate change.
- *Biological aspects*: Biocomposites are considered as natural organic product material (where cellulose is the common and most available natural polymers on the Earth). There are no skin irritation issues for their handling/usage in contrast with the glass fibers. They are not becoming a bio-hazard wastes when disposed. At the end of use the stored solar energy (based on photosynthesis) can be recovered at almost zero CO_2 balance by heat and electric power co-generation.
- *Weight/strength issues*: Natural fibers are lightweight and their specific strength is also high.
- *Production perspectives*: Natural fibers are also non-abrasive materials and provide great formability. They are easy to process in various ways.
- *Financial perspectives*: Natural fibers are cheaper compared to other synthetic materials like glass or carbon fibers.
- *General perspectives*: Natural fibers possess good acoustic and thermal insulation properties for their porous structures.

1.5 FUTURE POTENTIAL

Natural fiber-reinforced polymeric composites contribute to enhance the biocomposites development in terms of sustainability and performance. Biocomposites have reached a significant market position for value-added attractive product materials especially in the automotive sectors. For expanding the other markets, like novel construction techniques and consumer goods, biocomposites need to reach high-quality performance with enhanced durability, cascading, and reliability. Although biocomposites have an increased interest recently, the main challenge is still in replacing conventional composites. Biocomposites should show comparable or better functional and structural characteristics during storage, usage, and environmental degradations upon disposal. Based on extensive research, the biocomposite markets are anticipating expansions in the near future as social, environmental awareness, and sustainability issues are forcing such substitutions. Products and technologies are evolving rapidly. There are still challenges regarding the acceptance criteria, performance-cost issues, despite the sustainability of them (recyclability and renewability of the matrix and reinforcing phases) is attractive. New standards are required to characterize these new materials.

Biocomposites are sustainable material and can be entirely recycled, but also could be more costly as derived from fully/partially natural-based biodegradable materials and they are also sensitive to the moisture, temperature, and biological pests. Also, biocomposites designed for the structural purposes should meet various regulations such as emissions, strength, and the management of large amount of waste. New generations of the biocomposite materials should be used for diverse applications especially for mass-produced consumer product items in both indoor and outdoor products. Also, medical applications have a promising future as researches are unveiling new features of biocomposites. Biocomposites show good specific properties, but their properties are varying highly. Their weakness should and will be overcome with recent and future developments of advanced processing of natural fiber/matrix based composites [42,116].

In near future, besides others, the use in structural applications of biocomposites will also increase. If an appropriate matrix is applied, biocomposites could be nearly 100% biodegradable, but controlling their biodegradation is a challenging task. Biocomposites showed poor long-term performance, non-linear mechanical behavior, and low impact strength. It will be important to assess the life-cycle of biocomposites for retaining the main advantages during high-performance biocomposites development. New markets would emerge if these products become more dimensionally stable, durable, strong, fire resistant, and moisture proof [59].

Future potentials are shown in nanotechnology. It gives numerous promising opportunities for improving biocomposite products. Not only nanocomposites should be considered, but also improvements by coating nanomaterials for decreasing water uptake, reducing biodegradation and VOCs, and even flame retardancy. The application of nanocrystalline cellulose is being investigated for a variety of usage as it is stronger compared to steel and even stiffer compared to aluminum. Moreover, nanocrystalline cellulosic material reinforced composites will deliver advanced performance, value, durability, and service-life at a fully sustainable technology [94,117].

NOTES

1. Hardboard, low density fiberboard, medium density fiberboard, high density fiberboards, respectively
2. https://www.woodworkingnetwork.com/news/woodworking-industry-news/-dieffenbacher-manufacture-worlds-longest-continuous-press (11.15.2020)
3. http://sites.brunel.ac.uk/grow2build/knowledge-database/fibre-composites
4. http://flexformtech.com/Downloads/FlexForm-Brochure.pdf (11.05.2020)
5. Laminated Veneer Lumber, Parallel Strand Lumber, Laminated Strand Lumber, glued laminated wood respectively

REFERENCES

1. Baird, D.G. and D.I. Collias, *Polymer Processing: Principles and Design*. 2014, John Wiley & Sons: New York.
2. McKendry, P., Energy production from biomass (part 1): Overview of biomass. *Bioresource Technology*, 2002, **83**(1): pp. 37–46.
3. Demirbas, A., Combustion characteristics of different biomass fuels. *Progress in Energy and Combustion Science*, 2004, **30**(2): pp. 219–230.
4. Cheng, J., *Biomass to Renewable Energy Processes*. 2017, CRC Press: Boca Raton, FL.
5. Wang, S., et al., Lignocellulosic biomass pyrolysis mechanism: A state-of-the-art review. *Progress in Energy and Combustion Science*, 2017, **62**: pp. 33–86.
6. Bar-On, Y.M., R. Phillips, and R. Milo, The biomass distribution on Earth. *Proceedings of the National Academy of Sciences*, 2018, **115**(25): pp. 6506–6511.
7. Food and Agriculture Organization of the United Nations, Production (crops). 2018 [cited 2018 18th November]; Available from: http://www.fao.org/3/I8429EN/i8429en.pdf.
8. Czvikovszky, T., P. Nagy, and J. Gaal. *A Polimertechnika Alapjai (Fundaments of Polymertechnics)*. 2006, Technical University Publishing House: Budapest, Hungary.
9. Hasan, K.M.F., P.G. Horváth, and T. Alpár, Potential natural fiber polymeric nanobiocomposites: A review. *Polymers*, 2020, **12**(5): pp. 1–25.
10. Rowell, R.M., *Handbook of Wood Chemistry and Wood Composites*. 2012, CRC Press: Boca Raton, FL.
11. Klyosov, A.A., *Wood-Plastic Composites*. 2007, John Wiley & Sons: New York.
12. Winkler, A., *Farostlemezgyártás (Fiberboard Production)*. 1999, Farostlemezek, Mezőgazdasági Szaktudás Könyvkiadó: Budapest, Hungary, p. 11.
13. Czvikovszky, T., P. Nagy, and J. Gaal, A polimertechnika alapjai. *Műegyetemi Kiadó, Budapest*, 2000, 132: p. 453.
14. Bánhegyi, G., Hosszú szállal erősített műanyagok az autóiparban (continuous fiber reinforced plastics in automotive industry). *Muanyagipari Szemle*, 2007, **02**. https://-quattroplast.hu/muanyagipariszemle/2007/02/hosszu-szallal-erositett-muanyagok-az--autoiparban-02.pdf (accessed 05.17.2021).
15. Vogt, D.-G.D., D.-I.C. Schmidt, and D.-G.C. Gahle, Wood-Plastic-Composites (WPC). 2006, Nova Institut fur okology und Innovation.
16. Sobczak, L., R.W. Lang, and A. Haider, Polypropylene composites with natural fibers and wood–general mechanical property profiles. *Composites Science and Technology*, 2012, **72**(5): pp. 550–557.
17. Anandakumar, P., M.V. Timmaraju, and R. Velmurugan, Development of efficient short/continuous fiber thermoplastic composite automobile suspension upper control arm. *Materials Today: Proceedings*, 2020, 215: 108767.
18. Kroll, L., et al., Highly rigid assembled composite structures with continuous fiber-reinforced thermoplastics for automotive applications. *Procedia Manufacturing*, 2019, **33**: pp. 224–231.
19. Hasan, K.F., et al., Coloration of aramid fabric via in-situ biosynthesis of silver nanoparticles with enhanced antibacterial effect. *Inorganic Chemistry Communication*, 2020, **119**: pp. 1–8.
20. Zhou, D., et al., Design and synthesis of an amide-containing crosslinked network based on diels-alder chemistry for fully recyclable aramid fabric reinforced composites. *Composites Science and Technology*, 2020, 197: p. 108280.
21. Otto Schwarz, F.W.E., H. Huberth, H. Schirber, N. Schlör, *Kunststoffkunde (Plastics)*. 2007, Vogel Business Media: Würzburg, Germany.
22. Béla Pukánszky, J.M., *Műanyagok (Plastics)*. 2011, Budapesti Műszaki Egyetem, Budapest, Hungary: Typotex Kiadó.

23. Alpár, T., G. Markó, and L. Koroknai, Natural fiber reinforced PLA composites: Effect of shape of fiber elements on properties of composites. In M.R. Kessler, et al. (Ed.) *Handbook of Composites from Renewable Materials*. 2017, John Wiley & Sons: Hoboken, NJ, pp. 287–312.
24. Dominici, F., et al., Improved toughness in lignin/natural fiber composites plasticized with epoxidized and maleinized linseed oils. *Materials*, 2020, **13**(3): p. 600.
25. Pilla, S., et al., Polylactide-pine wood flour composites. *Polymer Engineering & Science*, 2008, **48**(3): pp. 578–587.
26. Bledzki, A. and J. Gassan, Composites reinforced with cellulose based fibres. *Progress in Polymer Science*, 1999, **24**(2): pp. 221–274.
27. Calumby, R., Long fiber reinforced thermoplastics a lightweight solution for engineering applications. *Sampe Brazil Conference 2014* [cited 2021 15th February]; Available from: http://www.feiplar.com.br/materiais/palestras/SAMPE/apresentacao/Celanese.pdf.
28. Thomason, J. and M. Vlug, Influence of fibre length and concentration on the properties of glass fibre-reinforced polypropylene: 4. Impact properties. *Composites Part A: Applied Science and Manufacturing*, 1997, **28**(3): pp. 277–288.
29. Scherübl, B. An innovative composite solution in the new Mercedes A class-a successful story about the natural fiber "Abaca". *In 6th Global Wood and Natural Fibre Composites Symposium*, Kuala Lumpur, Malaysia, 2006.
30. Holbery, J. and D. Houston, Natural-fiber-reinforced polymer composites in automotive applications. *JOM*, 2006, **58**(11): pp. 80–86.
31. Silverman, E.M., Effect of glass fiber length on the creep and impact resistance of reinforced thermoplastics. *Polymer Composites*, 1987, **8**(1): pp. 8–15.
32. Zhou, T.H., et al., A novel route for improving creep resistance of polymers using nanoparticles. *Composites Science and Technology*, 2007, **67**(11–12): pp. 2297–2302.
33. Bogoeva-Gaceva, G., et al., Natural fiber eco-composites. *Polymer Composites*, 2007, **28**(1): pp. 98–107.
34. Hasan, K.M.F., P.G. Horváth, and T. Alpár, Lignocellulosic fiber cement compatibility: A state of the art review. *Journal of Natural Fibers*, 2021, 4: pp. 1–26.
35. Hasan, K.M.F., P.G. Horváth, and T. Alpár, Development of lignocellulosic fiber reinforced cement composite panels using semi-dry technology. *Cellulose*, 2021, **38**: pp. 3631–3645.
36. Thakur, V.K., M.K. Thakur, and R.K. Gupta, Raw natural fiber–based polymer composites. *International Journal of Polymer Analysis and Characterization*, 2014, **19**(3): pp. 256–271.
37. Alpár, T., G. Markó, and L. Koroknai, Natural fiber reinforced PLA composites: Effect of shape of fiber elements on properties of composites, In M.R. Kessler, et al. (Ed.) *Handbook of Composites from Renewable Materials, Design and Manufacturing*. 2017, John Wiley & Sons: Hoboken, NJ, pp. 287–309.
38. Tábi, T. *Lebomló Polimerek. (Deagradable Polymers)*. 2012, Technical University of Budapest: Budapest, Hungary.
39. Rose, M. and R. Palkovits, Cellulose-based sustainable polymers: State of the art and future trends. *Macromolecular Rapid Communications*, 2011, **32**(17): pp. 1299–1311.
40. Bioplastics guide (the bioplastic strategy expert). 2020 [cited 2020 17th November]; Available from: http://www.bioplastics.guide/ref/bioplastics/biodegradable-bioplastics/.
41. Tamas, T., *Keményítőből és politejsavból előállított fröccsöntött lebomló polimerek feldolgozásának és feldolgozhatóságának elemzése (Analysis of the Processing and Processability of Injection-Molded Degradable Polymers Made from Starch and Polylactic Acid)*. 2010, Technical University of Budapest: Budapest, Hungary.
42. Gurunathan, T., S. Mohanty, and S.K. Nayak, A review of the recent developments in biocomposites based on natural fibres and their application perspectives. *Composites Part A: Applied Science and Manufacturing*, 2015, **77**: pp. 1–25.

43. Song, J., et al., Biodegradable and compostable alternatives to conventional plastics. *Philosophical Transactions of the Royal Society B: Biological Sciences*, 2009, **364**(1526): pp. 2127–2139.
44. Faruk, O., et al., Biocomposites reinforced with natural fibers: 2000–2010. *Progress in Polymer Science*, 2012, **37**(11): pp. 1552–1596.
45. Optimat, L. and L. Merl, *Wood Plastic Composites Study: Technologies and UK Market Opportunities*. The Waste and Resources Action Programme: Banbury, Oxon, 2003.
46. Businesswire, A.B.H.C. Growth in housing and construction industries will drive the wood plastic composites market, says Technavio. 2020 [cited 2020 18th November]; Available from: https://www.businesswire.com/news/home/20170310005591/en/Growth-in-Housing-and-Construction-Industries-Will-Drive-the-Wood-Plastic-Composites-Market-Says-Technavio (11.11.2020).
47. Mahmud, S., et al., Comprehensive review on plant fiber-reinforced polymeric biocomposites. *Journal of Materials Science*, 2021, **12**: pp. 1–34.
48. Hasan, K.F., et al., Thermo-mechanical properties of pretreated coir fiber and fibrous chips reinforced multilayered composites. *Scientific Reports*, 2021, **11**: p. 3618.
49. Deppe, H.-J. and K. Ernst, *Mitteldichte Faserplatten: MDF*. 1996, DRW-Verlag: Germany.
50. Dunky, M. and P. Niemz, *Teil II Bindemittel Und Verleimung (Part 2: Adhesives and Adhesive Bonding). Holzwerkstoffe Und Leime—Technologie Und Einflussfaktoren (Wood Composites and Glues—Technology and Influencing Facors)*. 2002, Springer: Berlin, Germany.
51. Hasan, K.M.F., H. Péter György, and A. Tibor, Thermo-mechanical behavior of MDI bonded flax/glass woven fabric reinforced laminated composites. *ACS Omega*, 2021, **6**(9): pp. 6124–6133
52. Hasan, K.M.F., P.G. Horváth, and T. Alpár, Lignocellulosic fiber cement compatibility: A state of the art review. *Journal of Natural Fibers*, 2021, **4**: pp. 1–26.
53. Alpara, T.L., M. Schlosser, I. Hajdu, and L. Bejó, Developing building materials from cement-bonded reed composite based on waste materials, *In International Scientific Conference on Sustainable Development & Ecological Footprint*, Sopron, Hungary, 2012, pp. 1–7.
54. Alpár, T., et al. Wood wool cement boards produced with nano minerals. *In Proceedings 3rd International Scientific Conference on Hardwood Processing (ISCHP32011) I: Peer Reviewed Contributions*, Blacksburg, VA, 2011.
55. Balázs, G., *A Cement Szilárdulása (Hardening of Cement)*. 1984, Budapesti Műszaki Egyetem, Budapest, Hungary.
56. Mohanty, A., M.A. Misra, and G. Hinrichsen, Biofibres, biodegradable polymers and biocomposites: An overview. *Macromolecular Materials and Engineering*, 2000, **276**(1): pp. 1–24.
57. Baley, C., Analysis of the flax fibres tensile behaviour and analysis of the tensile stiffness increase. *Composites Part A: Applied Science and Manufacturing*, 2002, **33**(7): pp. 939–948.
58. Markó, G., K. Halász, and T. Alpár. Natural fibre reinforced PLA composite. in Kasza Gy. *International Conference on Bio-Friendly Polymers and Polymer Additives: From Scientific Aspects to Processing and Applications: Program and Book of Abstracts*, Budapest, 2014.
59. Faruk, O., et al., Biocomposites reinforced with natural fibers: 2000–2010. *Progress in Polymer Science*, 2012, **37**(11): pp. 1552–1596.
60. Hasan, K.M.F., et al., Thermo-mechanical characteristics of flax woven fabric reinforced PLA and PP biocomposites. *Green Materials*, 2021: pp. 1–9. https://doi.org/10.1680/jgrma.20.00052.
61. Hasan, K.M.F., P.G. Horváth, and T. Alpár. Effects of alkaline treatments on coconut fiber reinforced biocomposites. *In 9th Interdisciplinary Doctoral Conference*, Pecs, Hungary, Doctoral Student Association of the University of Pécs, 2020.

62. Mahmud, S., et al., In situ synthesis of green AgNPs on ramie fabric with functional and catalytic properties. *Emerging Materials Research*, 2019, **8**(4): pp. 623–633.
63. Masłowski, M., J. Miedzianowska, and K. Strzelec, Natural rubber biocomposites containing corn, barley and wheat straw. *Polymer Testing*, 2017, **63**: pp. 84–91.
64. Ibrahim, M., et al., Potential of using multiscale corn husk fiber as reinforcing filler in cornstarch-based biocomposites. *International Journal of Biological Macromolecules*, 2019, **139**: pp. 596–604.
65. Kremensas, A., et al., Mechanical performance of biodegradable thermoplastic polymer-based biocomposite boards from hemp shivs and corn starch for the building industry. *Materials*, 2019, **12**(6): pp. 845.
66. Bruni, G.P., et al., Biocomposite films based on phosphorylated wheat starch and cellulose nanocrystals from rice, oat, and eucalyptus husks. *Starch-Stärke*, 2020, **72**(3–4): p. 1900051.
67. Masłowski, M., J. Miedzianowska, and K. Strzelec, Influence of wheat, rye, and triticale straw on the properties of natural rubber composites. *Advances in Polymer Technology*, 2018, **37**(8): pp. 2866–2878.
68. Beigmohammadi, F., Z.M. Barzoki, and M. Shabanian, Rye flour and cellulose reinforced starch biocomposite: A green approach to improve water vapor permeability and mechanical properties. *Starch-Stärke*, 2020, **72**(5–6): p. 1900169.
69. Guo, D., Effect of electron beam radiation processing on mechanical and thermal properties of fully biodegradable crops straw/poly (vinyl alcohol) biocomposites. *Radiation Physics and Chemistry*, 2017, **130**: pp. 202–207.
70. Correa-Aguirre, J.P., et al., The effects of reprocessing and fiber treatments on the properties of polypropylene-sugarcane bagasse biocomposites. *Polymers*, 2020, **12**(7): p. 1440.
71. Lila, M.K., et al., Accelerated thermal ageing behaviour of bagasse fibers reinforced poly (lactic acid) based biocomposites. *Composites Part B: Engineering*, 2019, **156**: pp. 121–127.
72. Alpár Tibor, M.G. and L. Koroknai, Natural fiber reinforced PLA composites: Effect of shape of fiber elements on properties of composites, In V.J. Thakur, M.K. Thakur, and M.R. Kessler (Eds), *Handbook of Composites from Renewable Materials, Design and Manufacturing*. 2017, John Wiley & Sons: Hoboken, NJ, pp. 287–309.
73. Nanko, H., D. Hillman, and A. Button, *The World of Market Pulp [Resource Électronique]*. 2005, WOMP LLC: Atlanta, GA.
74. Rymsza, T.A. Kenaf and hemp-identifying the differences. Vision paper 2010 [cited 2021 15th February]; Available from: http://www.visionpaper.com/PDF_speeches_papers/Rymkenafhemp.pdf.
75. Bailey, L.H., *The Standard Cyclopedia of Horticulture*, vol. 2. 1947, Macmillan: New York.
76. Munawar, S.S., K. Umemura, and S. Kawai, Characterization of the morphological, physical, and mechanical properties of seven nonwood plant fiber bundles. *Journal of Wood Science*, 2007, **53**(2): pp. 108–113.
77. Hasan, K.F., et al., Wool functionalization through AgNPs: Coloration, antibacterial, and wastewater treatment. *Surface Innovations*, 2020, **9**(1): pp. 25–36.
78. Hasan, K.F., et al., Colorful and antibacterial nylon fabric via in-situ biosynthesis of chitosan mediated nanosilver. *Journal of Materials Research and Technology*, 2020, **9**(6): 16135–16145.
79. Jin, H.-J., et al., Electrospinning Bombyx mori silk with poly (ethylene oxide). *Biomacromolecules*, 2002, **3**(6): pp. 1233–1239.
80. Kaplan, D.L., M.C., Arcidiacono, S., Fossey, S., Senecal, K., Muller, W., *Silk*, In K.D.K. Mcgreth (Ed.) *Protein-Based Materials*. 1997, Birkhäuser: Boston, MA, pp. 103–131.
81. Ramamoorthy, S.K., M. Skrifvars, and A. Persson, A review of natural fibers used in biocomposites: Plant, animal and regenerated cellulose fibers. *Polymer Reviews*, 2015, **55**(1): pp. 107–162.

82. Żak, K., et al. The mechanical properties of fibres and yarns in different group of animals. *In 9th Youth Symposium on Experimental Solid Mechanics, Trieste*, Italy, 2010. Citeseer.
83. Bongarde, U. and V. Shinde, Review on natural fiber reinforcement polymer composites. *International Journal of Innovative Science Engineering and Technology (IJISET)*, 2014, **3**(2): pp. 431–436.
84. Kock, J.W., *Physical and Mechanical Properties of Chicken Feather Materials*. 2006, Georgia Institute of Technology: Atlanta, GA, p. 114.
85. Shubhra, Q.T., et al., Characterization of plant and animal based natural fibers reinforced polypropylene composites and their comparative study. *Fibers and Polymers*, 2010, **11**(5): pp. 725–731.
86. Ko, F.K. and L.Y. Wan, Engineering properties of spider silk, In A.R. Bunsell (Ed.), *Handbook of Properties of Textile and Technical Fibres*. 2018, Woodhead Publishing: Duxford, United Kingdom, pp. 185–220.
87. Bongarde, U. and V. Shinde, Review on natural fiber reinforcement polymer composites. *International Journal of Engineering Science and Innovative Technology*, 2014, **3**(2): pp. 431–436.
88. Liu, X. and X. Wang, A comparative study on the felting propensity of animal fibers. *Textile Research Journal*, 2007, **77**(12): pp. 957–963.
89. Chen, S., et al., Compatibilities and properties of poly lactide/poly (methyl acrylate) grafted chicken feather composite: Effects of graft chain length. *Journal of Applied Polymer Science*, 2020, **137**(34): p. 48981.
90. Bledzki, A.K.S.V.E. Mechanische und thermomechanische Verfahren zur Herstellung von Holzpartikeln für WPC (Mechanical and thermo-mechanical process to produce wood particles for WPC). *In 6th Global Wood and Natural Fibre Composites Symposium*, Kassel, 2006, University of Kassel.
91. Horváth, R.Z., Fa részecskék alakiságának hatása fa-PLA kompozitok tulajdonságaira. 2015, NYME.
92. Markó, G., K.G., Á. Ott, L. Koroknai, T.L. Alpár. Development of wood plastic composite with optimized inertia. *In International Scientific Conference on Sustainable Development and Ecological Footprint*, Sopron, Hungary, University of West Hungary, 2012.
93. Cs., K., *Effect of Different Wood Species on Hydration of Cement*. 2008, University of West Hungary: Sopron, Hungary, p. 19–45.
94. Halász, K., Y. Hosakun, and L. Csóka, Reducing water vapor permeability of poly (lactic acid) film and bottle through layer-by-layer deposition of green-processed cellulose nanocrystals and chitosan. *International Journal of Polymer Science*, 2015: pp. 1–6.
95. Kövesi, A., E. Bognár, and G. Erdélyi, Bio-intrafix interferenciacsavar vizsgálata ramanspektroszkópiával. In *Fiatal Muszakiak Tudomanyos Ulesszaka XVIII*, Kolozsvar, Romania, 2013.
96. Furtos, G., et al. Biocomposites for orthopedic and dental application. In *Key Engineering Materials*, Switzerland, Trans Tech Publications, 2015.
97. Oroszlány, Á.I., Gyors prototípusgyártási technológiával előállított orvosi csavarok jellemzése, 2013.
98. Sharma, R.S., Raghupathy V.P., Rao, A.A., and Shubhanga, P. Review of recent trends and developments is biocomposites. *In International Conference on Recent Developments in Structural Engineering (RDSE-2007)*, Manipal, India, 2007.
99. Troster, S.H.F., Geiger, O., and Eyerer, P. Added value for long-fiber reinforced thermoplastic components by in-line compounding in the LFT-D-ILC process. *In Proceedings of 62nd Annual Technical Conference*, Chicago's Navy Pier, Chicago, Illinois, Brookfield, Society of Plastics Engineers – SPE, 2004.
100. Xiao, W., et al., Tough and strong porous bioactive glass-PLA composites for structural bone repair. *Journal of Materials Science*, 2017, **52**(15): pp. 9039–9054.

101. Kandola, B.K., W. Pornwannachai, and J.R. Ebdon, Flax/PP and flax/PLA thermoplastic composites: Influence of fire retardants on the individual components. *Polymers*, 2020, **12**(11): p. 2452.
102. Manral, A., F. Ahmad, and V. Chaudhary, Static and dynamic mechanical properties of PLA bio-composite with hybrid reinforcement of flax and jute. *Materials Today: Proceedings*, 2020, **25**: pp. 577–580.
103. Senthilkumar, K., et al., Performance of sisal/hemp bio-based epoxy composites under accelerated weathering. *Journal of Polymers and the Environment*, 2020, **29**(10): pp. 1–13.
104. Sankhla, D., M. Sancheti, and M. Ramachandran. Review on mechanical, thermal and morphological characterization of hemp fiber composite. *In IOP Conference Series: Materials Science and Engineering*, Bristol, IOP Publishing, 2020.
105. Wang, G., et al., Strong and thermal-resistance glass fiber-reinforced polylactic acid (PLA) composites enabled by heat treatment. *International Journal of Biological Macromolecules*, 2019, **129**: pp. 448–459.
106. Mahmud, S., et al., The consequence of epoxidized soybean oil in the toughening of polylactide and micro-fibrillated cellulose blend. *Polymer Science, Series A*, 2019, **61**(6): pp. 832–846.
107. Martin, R., et al., Biocomposites in challenging automotive applications. *In 17th International Conference Proceedings ICCM*, Edinburgh, 2009.
108. Eyerer, P. et al. (Ed.), Opportunities and risks involved in designing structural components made of polymers, In *Polymers-Opportunities and Risks I*, 2010, Springer: Berlin, Heidelberg, pp. 293–361.
109. Courgneau, C., et al., Characterisation of low-odour emissive polylactide/cellulose fibre biocomposites for car interior. *Express Polymer Letters*, 2013, **7**(9): 787.
110. Jain, V., M. Mittal, and R. Chaudhary. Design optimization and analysis of car bumper with the implementation of hybrid biocomposite material. *In IOP Conference Series: Materials Science and Engineering*, Chennai, IOP Publishing, 2020.
111. Shahinur, S. and M. Hasan, Hybrid sustainable polymer composites, In R. Rahman (Ed.), *Advances in Sustainable Polymer Composites*. 2020, Elsevier: Amsterdam, pp. 209–229.
112. Martin, R.H., Giannis, S., Mirza, S., Hansen, K. Biocomposites in challenging automotive applications. In *Proceedings of 17th International Conference on Composite Materials*, Edinburgh, UK, 2009.
113. Rowell, R.M., A new generation of composite materials from agro-based fiber, In J.E. Mark, P.N. Prasad, and T.J. Fai (Eds), *Polymers and other Advanced Materials*. 1995, Springer: Berlin/Heidelberg, pp. 659–665.
114. Dweib, M., et al., Bio-based composite roof structure: manufacturing and processing issues. *Composite Structures*, 2006, **74**(4): pp. 379–388.
115. Alpar, T.L., et al. Developing building materials from cement-bonded reed composite based on waste materials. *In International Scientific Conference on Sustainable Development & Ecological Footprint*, Sopron, Hungary, 2012.
116. Mohanty, A.K., M. Misra, and L. Drzal, Sustainable biocomposites from renewable resources: Opportunities and challenges in the green materials world. *Journal of Polymers and the Environment*, 2002, **10**(1–2): pp. 19–26.
117. Sierra-Fernandez, A., et al., Application of magnesium hydroxide nanocoatings on cellulose fibers with different refining degrees. *RSC Advances*, 2016, **6**(57): pp. 51583–51590.

2 Influence of Natural Fibers and Biopolymers on the Biocomposites Biodegradation

L. Joana Rodríguez and Carlos E. Orrego
Universidad Nacional de Colombia

CONTENTS

Abbreviations 35
2.1 Introduction 36
2.2 Basic Concepts 37
2.3 Degradation Methods 38
     2.3.1 Soil Burial Degradation Test: Buried in Soil Compost 39
     2.3.2 Biodegradability in Soil by ASTM 5988-12 40
     2.3.3 Accelerated Weathering 41
2.4 Influence of Natural Fibers on the Biocomposites Biodegradation 42
2.5 Influence of Biopolymer on the Biocomposites Biodegradation 47
2.6 Conclusions 58
References 58

ABBREVIATIONS

ASTM: American Society for Testing and Materials
BF: Banana Fiber
HDPE: High-Density Polyethylene
HCL: Hydrochloric Acid
ISO: International Organization for Standardization
KOH: Potassium Hydroxide
LDPE: Low-Density Polyethylene
PLA: Polylactic Acid
PHA: Polyhydroxyalkanoate
PHB: Polyhydroxybutyrate
PBS: Polybutylene Succinate
PVA: Poly(vinyl alcohol)
PCL: Polycaprolactone
PET: Polyethylene Terephthalate

DOI: 10.1201/9781003137535-2

PP: Polypropylene
TPS: Thermoplastic Starch

2.1 INTRODUCTION

Plastics are lightweight, low-cost, processable materials with unique and versatile properties such as transparency, protection and preservation, safety and hygiene. In 2018, a global production of 360 million tons was recorded, accumulating a total volume of 8.3 billion tons since 1950 (Greenpeace, 2020). However, due to the plastics come from non-renewable raw materials and have resistance to microbial degradation—their biodegradability is minimal and polluted water sources and oceans (see Figure 2.1) (Shah et al., 2008). Their conventional production faces increasing regulatory restrictions that favor the production of alternative materials more environmentally friendly (Heimowska and Krasowska, 2019; Iwańczuk et al., 2015). Recycling plastics is an option to consider, but since some products are difficult to collect and recycle, it involves additional processes and costs (see Figure 2.1). For this reason, the use of biomass from renewable resources instead of petro-fossil carbon is preferable to achieve both the reduction of the carbon footprint at the beginning of the life cycle, as well as to guarantee safety and low resistance to the environment at the end of the life cycle (Shah et al., 2008). Consequently, the most viable options include the manufacture of plastics from renewable or conventional sources modified with additives intended to improve their biodegradability, such as bio-based plastics—bioplastics—and natural fiber composite materials—biocomposites (Narayan, 2006; Song et al., 2009).

The bioplastics and the biocomposites are wholly or partially produced from renewable sources such as polylactic acid, chitosan, thermoplastic starch, and natural fibers such as sisal, jute, and lignocellulosic residues (see Figure 2.1). The biocomposites market increased 12.3% in 2019 with US $5.83 billion and the bioplastics 15% in the same year with 2.1 million tons. They are used in a variety of products in the packaging, textile, agriculture, automotive, construction, and electronics sectors (Ali et al., 2016; Mundo_plast, 2020).

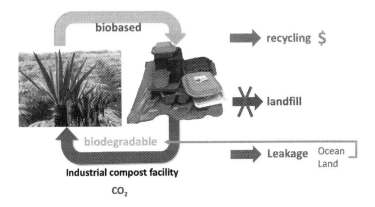

FIGURE 2.1   Alternatives of disposal and production of plastic materials.

Biodegradation is a characteristic of some substances that can be substrates of microorganisms, to produce energy by two metabolic routes: aerobic and anaerobic. A polymer degrades when it undergoes chain cleavage, resulting in a decrease in molar mass, and a release of molecules accessible by microorganisms (Bátori et al., 2018). All organic carbon plastics are biodegradable, however, not during an acceptable period (Folino et al., 2020). The biodegradation can occur both in natural and industrial environments, under aerobic conditions (compost, soil and some aquatic environments) and/or anaerobic (anaerobic digestion plants). The biodegradation of plastic, bioplastic or biocomposite depends on the characteristics of the polymer such as mobility, crystallinity, molecular weight, functional groups and substituents present in its structure, added additives and pretreatments (Stelescu et al., 2017). The natural fibers, which are common filler of biocomposites, degrade rapidly in the presence of fungi, similar to the biopolymers, which are also susceptible to microorganisms (Stelescu et al., 2017).

The term biodegradability has been misused in different studies that do not provide direct experimental evidence based on established and certified standards (Zumstein et al., 2019). The biodegradation requires a complete microbial assimilation in a short period of time, without consequences for the environment or human health (ASTM_D6400, 2019). According to ISO_17088 (2013), a plastic can be considered biodegradable if a significant change occurs in the chemical structure. The exposed biodegradable materials result in carbon dioxide, water, inorganic compounds and biomass without visible residues or toxic in composting conditions. For biodegradable polymeric materials are also required to convert 90% to carbon dioxide within 6 months as a condition for composting in the presence of oxygen (ASTM_D6400, 2019). For anaerobic degradation, the conversion requirement is a minimum of 50% of the substance into biogas (based on the theoretical) within 2 months (ASTM_D5526, 2018). According to the standards and regulations, for a qualified biodegradability claim, it is also necessary to define the disposal environment, time/rate and degree of biodegradation (ASTM_D6400, 2019).

This chapter presents a list of some international standards to evaluate the biodegradation of plastic materials. Based on published research, the main effects of including natural fibers and biopolymers on the biodegradability of the synthetic plastic are reported. Finally, some studies on biodegradation of different blends of synthetic polymer/biopolymer reinforced with natural fibers according to the soil burial test and the ASTM_D5988 (2012) standard are presented.

2.2 BASIC CONCEPTS

Aerobic biodegradation is usually assimilated to composting under industrial conditions. It is carried out in a high-oxygen environment (not <6%), the microorganisms use the polymer as a source of carbon and energy and produce carbon dioxide, water and compost (Bátori et al., 2018).

Anaerobic biodegradation is the process in which microorganisms break down biodegradable material in the absence of oxygen. As a result of metabolic interactions by different microorganisms, organic matter is converted to methane gas, carbon dioxide, water, hydrogen sulfide, ammonia and hydrogen (Bátori et al., 2018).

Degradable plastic is a plastic designed to undergo a significant change in its chemical structure under specific environmental conditions, resulting in a loss of some properties that may be measured by standard test methods appropriate to the plastic and the application in a period of time that determines its classification (ASTM_D883, 2012).

Biodegradable plastic is a degradable plastic in which the degradation results from the action of naturally occurring microorganisms such as bacteria, fungi and algae (ASTM_D883, 2012).

Compostable plastic is a plastic that undergoes degradation by biological processes during composting to yield CO_2, water, inorganic compounds and biomass at a rate consistent with other known compostable materials and leave no visible, distinguishable or toxic residue (ASTM_D883, 2012).

2.3 DEGRADATION METHODS

Plastic is composed of different chains and some links susceptible to breaking (see Figure 2.2a). When plastic is disposed in the environment (soil, compost, fill or water) by the action of hydrolytic, oxidative and enzymatic mechanisms it breaks into fragments (see Figure 2.2b). Afterward, the fragmented waste must be consumed by microorganisms as a source of food and energy. Under aerobic conditions, carbon is biologically oxidized to CO_2, releasing energy that is used by microorganisms for their life process (see Figure 2.2c). There must be 90% assimilation in an estimated time and, at the end, 100% undergoes microbiological assimilation so that the material/product is biodegradable (ASTM_D6400, 2019).

To determine the biodegradability of plastics, organizations such as the American Society for Testing and Materials (ASTM) and the International Organization for Standardization (ISO), have published a series of biodegradability and compostability standards in different exposure environments. Generally, these standards describe definitions, test guidelines, time frames, procedures, conditions, limits, and interpretation of results.

The biodegradability of synthetic plastic materials with natural fibers and with biopolymers has been evaluated by different methods, such as soil burial in normal garden soils or composts, natural weathering, accelerated weathering, degradation

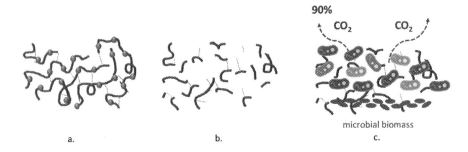

FIGURE 2.2 Polymeric biodegradation sequence: (a) weakening of links; (b) breaking of links; (c) microbial action. (Adapted from Zumstein et al. (2019).)

Influence of Natural Fibers and Biopolymers

by moisture/chemicals/microorganisms and other controlled degradation system by specific standards. Some of the standards corresponding to these methods are the following:

- *Compostability*: ISO 16929, ISO 20200 and ASTM 6400.
- *Industrial composting*: EN 13432, EN 14995, ISO 18606 and ISO 17088.
- *Home composting*: Australian Norm AS 5810 and the French Norm NF T 51-800.
- *Biodegradability in soil*: ASTM 5338, ISO 846:1997, ASTM G160-12, EN 17033, ASTM D5988-03, DIN 53739
- *Biodegradability in anaerobic environment*: ISO 11734, ISO 14853, ISO 15985, ASTM D5210–92, ASTM D5511-02 ASTM D5526-94D.
- *Biodegradability in marine environments*: ASTM D708, ASTM D6691, ASTM D6692, ASTM D7473, OECD 306 and ISO 16221.
- *Outdoor weathering test*: ASTM 1435-13, ASTM G7/G7M-13, ASTM G154-06, ASTM G151-10
- *Accelerated weathering*: ISO 877.2.
- *Resistance of synthetic polymers to fungal attack*: ASTM 6866-2010

Three of the abovementioned biodegradation methods are described below with some of the conditions reported in different articles.

2.3.1 Soil Burial Degradation Test: Buried in Soil Compost

The process of soil burial degradation is one of the most reported in the literature. It consists of burying samples in the ground and recording their weight loss over time. The process can be carried out in natural or controlled soil conditions, outdoors or in the laboratory (see Figure 2.3a and c). Some of the conditions that usually must be reported are the depth at which samples are buried, the size of the samples and the humidity of the soil (see Figure 2.3c and Table 2.1).

Typical steps in this procedure are:

1. In laboratory test cases, the soil must be prepared with organic soil, organic fertilizer and field soil.
2. Samples are weighed, and initial data recorded.

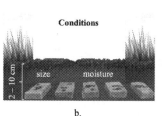

FIGURE 2.3 Soil burial degradation: (a) outside test; (b) conditions; (c) inside test.

TABLE 2.1
Selected Soil Burial Degradation Conditions from Literature

Days	Size	Soil Moisture (%)	Reference
60	10×5 cm²	36	Behera et al. (2020)
30	35×25×0.3 mm³	35–50	Wu et al. (2020)
80	1×2 cm²	40–45	Kusumastuti et al. (2020)
90	Film	60	Kochkina and Lukin (2020)
180	3×10 cm²		Mittal et al. (2020)
90	3×3 cm²	50–60	Lal et al. (2020)
360	3×3 cm²	40–50	Abdullah and Dong (2019)
45	20×20×3 mm³	10	Raj et al. (2018)

3. Containers with soil are prepared and samples are buried.
4. To weight the samples, the soil is removed with little water and gently blotted with paper.
5. The samples are placed in an oven at a temperature of ~70°C and left there for 24 h.
6. After 24 h, the weight of the samples is recorded.
7. *Typical additional studies*: morphology (scanning electron microscope SEM), thermal analyzes (Differential Scanning Calorimetry-DSC and Thermogravimetric analysis-TGA) and measurement of structural changes based on X-Ray Diffractometry-XRD and Fourier-transform infrared spectroscopy-FTIR.

2.3.2 Biodegradability in Soil by ASTM 5988-12

Among the standards mentioned above, one of the most recognized standards is the standard ASTM_D5988 (2012). Its objective is to obtain the percentage of biodegradability based on the amount of CO_2 produced by the microorganisms when using the carbon from the samples. The experimentation time depends on the degree of biodegradation suffered by the samples. Some of the more important steps are mentioned below.

An arrangement of the elements to do this test is shown in Figure 2.4a which includes containers (usually laboratory Vessels), KOH (0.5 N) and HCl (0.25 N) solutions, and soil preparation.

1. To ensure good soil activity a pH below 6.0 is required and it is advisable to mix field soil, compost and organic soil. In some cases, microorganisms have been added.
2. Vessels are prepared for monitoring, namely:
   a. *Control*: no sample, no soil.
   b. *Blanks*: containing only soil
   c. *Positives*: soil + a sample that does biodegrade like cellulose:
   d. Samples

Influence of Natural Fibers and Biopolymers

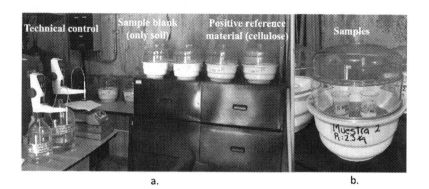

FIGURE 2.4 Aerobic biodegradability in soil: (a) arrangement of the method; (b) samples.

3. The theoretical carbon value of the samples must be calculated to ensure that it is sufficient for the production of CO_2 by the microorganisms.
4. The soil and the sample were added to each vessel.
5. A beaker with 50 mL of distilled water and another with 20 mL of KOH are placed; the vessels are closed ensuring that they are well sealed as shown in Figure 2.4b.

The amount of CO_2 produced was measured by means of titration of the KOH solution, conducted using HCl to a phenolphthalein endpoint. During the first few days more CO_2 will be released; therefore, the first 3 weeks should be titrated every 2, 3 or 4 days. After the 3rd week, it should be titrated every 1, 2 or 3 weeks approximately. For the titration, phenolphthalein is added to the KOH solution and the amount of HCl used is recorded. The desiccators are weighed, and the missing water is added, the beaker is recharged with KOH for the next measurement and it is sealed again. Finally, the calculation is made to convert them to CO_2 and a percentage of biodegradation, in accordance with the standard.

2.3.3 Accelerated Weathering

The objective of this test is to evaluate the decomposition of plastic products under the effect of oxidation through heat and/or UV light. The UV radiation can disrupt polymer chains, since the radiation can be absorbed by oxygen-containing components to initiate a primary degradation; these polymers are known as photodegradable polymers (Yousif and Haddad, 2013). In other words, sunlight combined with oxygen can lead to photo oxidative degradation, whereas sunlight combined with heat causes oxidative-degradation (Folino et al., 2020). The rate of weight loss has been used to predict how long the plastic will last (Qin et al., 2021).

The weathering test can be performed under natural and controlled environmental conditions. Under natural conditions it has been carried out for periods of up to 2 years (Siakeng et al., 2019). Weight loss is evaluated in time intervals (e.g., 240, 480, 720 and 960 h). According to the ISO_877-2 (2009) standard, the study must contain conditions such as dates, exact location (latitude and longitude) and the

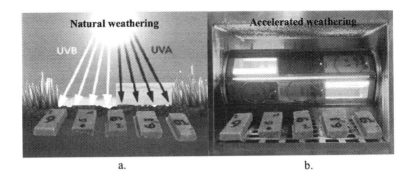

FIGURE 2.5 Accelerated weathering: (a) outside test; (b) inside test.

TABLE 2.2
Some Reported Accelerated Weathering Test Conditions

Sunlight Radiation	Temperature	Radiation	Time (h)	Standard	References
280–700 nm		478 W/m^2	1,970	ISO 877.2.	Moriana et al. (2014)
UVA-340 radiation for 90 cycles	60°C	0.89 mW/cm^2/day	2,160	ASTM G154-06 and ASTM G151-10	Costa et al. (2018)
0.89 W/m^2 at 340 nm	60°C		960	ASTM G154	Peng et al. (2020)
313 nm	60°C/50°C	0.63 W/m^2	1,200	ASTMG154	Torres-Huerta et al. (2014)
365 to 295 nm	60°C	0.89 W/m^2	200	ASTM G154	Lizárraga-Laborín et al. (2018)

climatic conditions of the place (mean temperature, precipitation and relative humidity) must be reported (see Figure 2.5a). Under controlled conditions, it is carried out in accelerated aging chambers that simulate solar radiation that includes the visible light spectrum and a part of the ultraviolet (UV) spectrum (see Figure 2.5b). The natural environmental conditions are replied and the destructive effects of prolonged outdoor exposure by exposing composite samples to UV radiation, humidity, and temperature in a controlled manner. The method is faster and reproducible compared to natural aging. Table 2.2 shows some of the reported conditions for accelerated weathering tests.

2.4 INFLUENCE OF NATURAL FIBERS ON THE BIOCOMPOSITES BIODEGRADATION

The biodegradation of biocomposites depends on their chemical structure such as reactivity, hydrophilicity and functional groups (Triwulandari et al., 2019). It is

Influence of Natural Fibers and Biopolymers

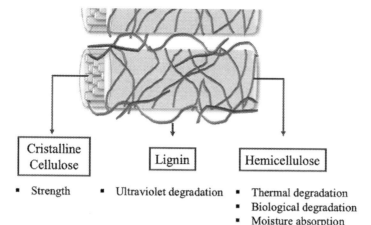

FIGURE 2.6 Polymers of the cell wall of natural fibers and their influence on the properties and degradability.

influenced by environmental conditions such as humidity, temperature, UV radiation and microbial and enzymatic activity (Hidayat and Tachibana, 2012).

In Figure 2.6, the different polymers of the cell wall of natural fibers with their intervention in biocomposite properties and degradability are outlined (Azwa et al., 2013). For example, the presence of the cellulose improves mechanical resistance, the hemicellulose is associated with thermal, biological and moisture degradation, and the lignin with UV degradation (Azwa et al., 2013). The lignocellulose present in natural fibers can be partially degraded by microorganisms and by photochemical action when exposed to UV light. However, as lignocellulose contains complex aromatic heteropolymers of high molecular weight, it is difficult to hydrolyze, and to solubilize in water. So, it can only be completely degraded using specific enzyme systems (Triwulandari et al., 2019). The fibers with a high hemicellulose content absorb more moisture and thermally degrade at lower temperature (Azwa et al., 2013). Also due to the high content of carbohydrates and amorphous regions present in the fibers, they can be attacked by microorganisms in their cell walls (Ali et al., 2016; Yussuf et al., 2010).

The fiber content in the biocomposite also makes it susceptible to water absorption due to its hydrophilicity (see Figure 2.7a) (Takagi, 2019). By absorbing water, the fibers promote dimensional changes in the matrix, which leads to cracking and fissuring of the polymer surface (see Figure 2.7b) (Tazi et al., 2018). This phenomenon with the consequent greater penetration of light accelerates the decomposition of the biodegradable plastic matrix on the surface of the sample (Takagi, 2019). The surface detriment favors leaching by the force of water during exposure to the open air and deterioration by the action of microorganisms and fungi embedded in surface discontinuities (see Figure 2.7c) (Zaaba and Ismail, 2017). The distribution and dispersion of the fibers lead to greater heterogeneity and roughness of the surface, facilitating the access of the biodegradable component to microbial attack and the consequent enzymatic degradation of the cellulosic chains (See Figure 2.7d) (Gunning et al., 2013; Iwańczuk et al., 2015).

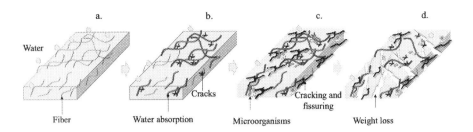

FIGURE 2.7 Biodegradation process of biocomposites with natural fibers: (a) water absorption; (b) superficial cracks; (c) cracking and fissures and microorganism attack; and (d) weight loss.

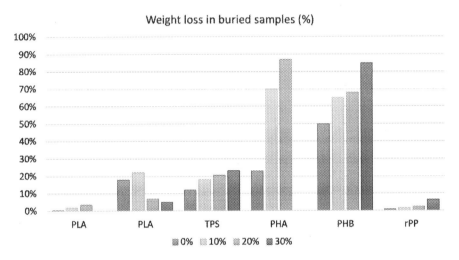

FIGURE 2.8 Weight loss in buried samples PLA (Gunti et al., 2018): PLA (Dong et al., 2014), TPS (Behera et al., 2020), PHS (Wu et al., 2020), PHB (Gunning et al., 2013; Zaaba and Ismail, 2017).

Figure 2.8 and Tables 2.3 and 2.4 show published examples of biodegradation tests of plastic matrix and fiber biocomposites. Figure 2.8 shows the percentage of degradation in a period of 30 days of biopolymers and petrochemical-based polymers with different percentages of fiber. It is observed how the weight loss rises with the increase of the amount of fiber in the period observed. As seen in Table 2.3, the biocomposites degrade much faster than pure PLA, registering a maximum weight loss of 34.9% in biocomposites reinforced with treated coconut fiber, compared to 18% in PLA (Siakeng et al., 2019). The inclusion of the natural fibers in a hydrophobic matrix causes a poor adhesion or weakness in the chemical interaction that generates voids around the fiber and porosity (Luthra et al., 2020; Tazi et al., 2018). These spaces also allow the absorption of water in the biocomposite that facilitates microbial growth (Behera et al., 2020).

Treatments to improve the interaction of the fiber with the plastic matrix affected the biodegradability of the material. Chemical or physical surface treatments carried out on the fibers in many cases decrease hydrophilicity and improve

TABLE 2.3
Percentage of Biodegradation Results from Literature for Biocomposites of Biopolymers and Natural Fiber

Biopolymer	TPS		PHS	PLA		PHB			PLA		
Days/% Fiber	30	60	30	30	90	30			30	180	25	18
0	12.2	16.6	23	0.5	1.5	50	52	58			60	18.1
10	18.1	22.8	70	1.8	4.2	65	58	61				22.6
20	20.5	34.2	87	3.6	12	68	64	67	1.2			7
30	23.2	43.9				85	80	90				5
40	25.5	45.6									99	
70										12	48	
Fiber	Jute	Jute	Pineapple			Jute	Lyocell	Hemp	Kenaf	kenaf	Banana	Coir
References	Behera et al. (2020)	Behera et al. (2020)	Wu et al. (2020)	Gunti et al. (2018)	Gunti et al. (2018)	Gunning et al. (2013)	Gunning et al. (2013)	Gunning et al. (2013)	Yussuf et al. (2010)	P. ostreatus (Hidayat and Tachibana, 2012)	B. cepacia strain (Jandas et al., 2011)	Dong et al. (2014)

TABLE 2.4
Biodegradation Results from Literature for Biocomposites of Synthetic Polymer and Natural Fiber

Days/% Fiber	PP	LDPE		rPP		HDPE		PP/TPS	rPP		rPP	
	45	50	120	180		45	45×1	90	180	90	180	
0			0.42	0.84					0.84	0.84	0.84	
5		2.23										
10	2.2	3.53	1.5	1.56					1.56	0.6	1.56	
15		3.03	1.4									
20	2.8			2.96					2.96	2.04	2.96	
25								19				
30				6.52		17	11.2		6.52	4.55	6.52	
40	3.7			9.24					9.24	8.36	9.24	
Fiber	Banana peel	Wood	Cellulose			Jute			PSP		PSP peanut shell powder	
References	Luthra et al. (2020)	Nuryawan et al. (2020)	Mubarak and Abdulsamad (2018)	Zaaba and Ismail (2017)		Tazi et al. (2018)	Pseudomonas putida (Chaudhuri et al., 2013)	Yang et al. (2020)	Triwulandari et al. (2019)		Zaaba et al., 2017	

adhesion with the matrix. This reduces porosity, therefore, less microbial action and decrease in the degradation rate (Jandas et al., 2011; Zaaba and Ismail, 2019). The radiation treatments allowed the creation of a crosslinking network, which generates resistance to the leaching of hydrophilic components (Zaaba and Ismail, 2017). In other cases, the use of chemicals in treatments such as silane, not friendly to the environment, partially resisted bacterial growth on the surface of the PLA matrix (Jandas et al., 2011). The compatibilizers improved the interaction between both phases, which influenced the reduction of the degradation of biocomposites (Tazi et al., 2018). For example, the use of maleic anhydride with 40% fiber reduced the percentage of weight loss of the biocomposite up to 45 times (Tazi et al., 2018). The addition of the compatibilizer improved the interaction between the pendant carboxylic acid groups of the acrylic acid of the polypropylene and the hydroxyl groups of the fibers, which resulted in an increase in interfacial adhesion and better resistance to microorganisms and photodegradation by 23% (Zaaba et al., 2017).

On the other hand, in some cases the fiber treatments provided the degradation of the compound. In the case of NaOH treatment could break the PLA polymer chains, which led to a faster degradation rate than untreated ones (Dong et al., 2014). In other cases, the removal of lignin after treatment hinders the attack of microorganisms, which slows down the rate of biodegradation (Luthra et al., 2020).

Accelerated weathering studies can determine environmental factors that affect biodegradation of biocomposites. Weathering photodegradation of the main components of natural fibers/polymeric compounds results from the combined effects of light, water, oxygen and heat. The absorption of UV radiation, the formation of quinoid structures, the Norrish reactions and the photoyellowing reactions of the lignin are responsible for aging or degradation (Azwa et al., 2013). In biocomposites of HDPE with rice husks had a greater number of surface holes/cracks after accelerated aging by radiation. The fibers expose to sunlight with a wavelength >290 nm in the presence of oxygen resulted in the rapid deterioration of these materials, as they become extremely brittle and susceptible to degradation after short periods of exposure (Costa et al., 2018). Moreover, the photodegradation of polymers due to photooxidation is promoted by UV irradiation. The UV radiation absorbed by the polymers modified the chemical structure, providing surface oxidation, cleavage of the molecular chain and breaking of the molecules (Catto et al., 2017). In biocomposites of HDPE with wood fiber could physically impede the ability of HDPE to crosslink, resulting in the possibility of cleavage of the HDPE chain, consequently, the degradation (Stark and Matuana, 2004).

The environmental conditions for the biodegradation tests influenced the level of biodegradation of the samples, which indicated that to guarantee the biodegradation of the materials, special conditions of the soil, climate, variety of microorganisms and enzymatic environment must be met (Fazita et al., 2015; Gunti et al., 2018).

2.5 INFLUENCE OF BIOPOLYMER ON THE BIOCOMPOSITES BIODEGRADATION

The study of synthetic polymers and biopolymers blends is carried out with the aim of developing materials with intermediate properties. The biopolymers provide a

TABLE 2.5
Biodegradability in Percentage of Weight Loss of Some Biopolymers

Test Conditions	PLA %	PLA days	PHA %	PHA days	TPS-Based %	TPS-Based days	Cellulose Based %	Cellulose Based days	PBS-Based %	PBS-Based days	PCL-Based %	PCL-Based days
Soil environment	10–60	98	8–98	365	3.4–96	110	100	154	1.2–24.4	28	5–90	270
Simulated or field composting environments	9–100	130	40–80	110	30–85	90	100	154	90	160	38–88	44
Anaerobic conditions	29–90	277	67–91	175	23–73	50			2	100	3–85	277
Aquatic environments	2–100	365	8.5–100	365	1.5–68.9	236					2–80	365

Influence of Natural Fibers and Biopolymers

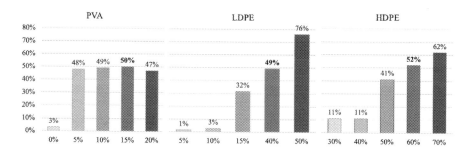

FIGURE 2.9 Results of the percentage of weight loss of mixtures of biopolymers with starch measured by the soil burial test: PVA (30 days) (Lal et al., 2020), LDPE (30 days) (Abioye and Obuekwe, 2020), HDPE (56 days) (Nurhajati et al., 2019).

higher degree of biodegradability, while synthetic polymers impart processability, low density and reduce costs (Zhong et al., 2020). Table 2.5 reports data on the biodegradability as a percentage of weight loss over time of some biopolymers (Folino et al., 2020). As can be seen biopolymers can lose 100% of their weight in <1 year, under suitable conditions.

Figure 2.9 shows the evolution of weight loss of blends of 3 synthetic polymers (PVA, LDPE and HDPE) with different percentages of thermoplastic starch (TPS). Similar to the effect of using natural fibers in a plastic matrix, the higher the percentage of biopolymer, the greater the weight losses. Apparently, PVA shows a higher percentage of biodegradation compared to the other two polymers, since with only 20% addition of starch it reaches around 50% loss percentage, while LDPE and HDPE reached the same percentage with 40% and 60% addition of starch, respectively. HDPE has a lower weight loss percentage even with the higher addition of starch of 70%. Other data on biodegradation of mixtures can be found in Table 2.6.

Torres-Huerta et al. (2014) made blends of polyethylene (PET) and recycled polyethylene (R-PET) with different percentages of PLA and chitosan. After the accelerated aging test, they predicted the biodegradation of the blends, measured in years. As seen in Figure 2.10, the blends with a higher amount of PLA biodegrade faster. The addition of 15% of PLA reduced the biodegradation of PET and R-PET by 45 and 32 years, respectively. The addition of 5% chitosan reduced the biodegradation of PET and R-PET in 31 and 47 years more than with the addition of PLA. The R-PET decomposes faster than PET; this is associated with the amorphous structure of R-PET with respect to semi-crystalline PET, which makes it more susceptible to degradation.

The structure of the polymer blends is made up of crystalline and amorphous zones as is depicted in Figure 2.11a. According to the literature, the degradation of polymer blends with TPS begins with the attack of microorganisms to the amorphous regions of the mixture and the water absorption due to their hygroscopicity (see Figure 2.11b) (Berruezo et al., 2013; Vieyra et al., 2013). In these regions, both microbes and fungi are present that, using water as a medium, penetrate the surface, generating porosity (see Figure 2.11c) (Raj et al., 2018). Finally, the surface degrades, the material changes its color and its weight decreases (see Figure 2.11d).

TABLE 2.6
Report of Percentage of Weight Loss by Soil Burial Test of Synthetic Polymer with Biopolymers

Synthetic Polymer	PVA		LDPE				PS	LDPE
%TPS Days	30	90	60	180	30	180	90	80
0		12.9	0.04	0.9	0.0	1	0.8	0.15
5		67.8			0.1	2.6		
10		68.1	0.07	3.2	0.1	4.1		1.3
15		67.7			1.0	5.6		
20		67.5	0.08	8	1.0	5.6		1.4
30			0.09	9.7	2.3	8.6		1.8
35							38.9	
45							44.7	
50	45							
Biopolymer References	Starch Singha and Kapoor (2014)	Abioye and Obuekwe (2020)	Pichaiyut et al. (2018)		Kormin et al. (2018)		Chaudhary and Vijayakumar (2020)	Chitosan Kusumastuti et al. (2020)

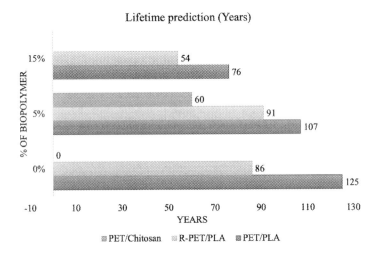

FIGURE 2.10 Lifetime prediction of biodegradation years of biopolymer/synthetic polymer blends (Torres-Huerta et al., 2014).

Influence of Natural Fibers and Biopolymers

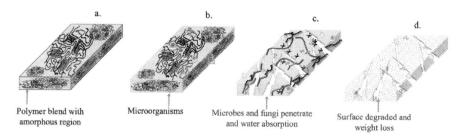

FIGURE 2.11 Biodegradation process of synthetic polymer/biopolymer blends: (a) structural composition of blends; (b) microorganism attack; (c) cracking and fissures; (d) degradation and weight loss.

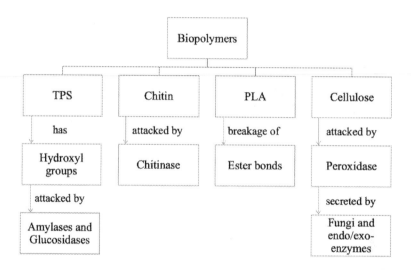

FIGURE 2.12 Modes of biodegradation of some biopolymers.

Figure 2.12 is a diagram of some of the degradation types caused by the enzymes released by microorganisms on TPS, chitin, cellulose and PLA. As these biopolymers are derived from natural raw materials, they can be metabolized by a great variety of microorganisms (Bercini Martins and Campomanes Santana, 2019). The attack of microorganisms on the surface of the polymer forms a biofilm. In accelerated weathering processes, the films began to lose mass under the action of microorganisms and fungi, which is evidenced by the appearance of characteristic color spots on the surface of the films after exposure to the soil between 20 and 40 days (Kochkina and Lukin, 2020). After consuming the biopolymers, it has been observed that polymeric chains of polyolefins decreased their molecular weight and were finally consumed by microorganisms due to the enzymatic oxidation of their hydroxyl and carbonyl groups, followed by hydrolysis (Abdullah and Dong, 2019). Then, the polymers become brittle, favoring degradation under natural environmental conditions (Mittal et al., 2020).

The degradation process in the soil also depends on the amount of water absorbed. In different studies, a relationship was observed between the water absorption—evidenced by swelling and contact angle tests—with greater degradation (de Oliveira et al., 2020; Pichaiyut et al., 2018). At higher values of the contact angle, there is less surface tension, which at the same time encourages the attack of fungi that causes the appearance of cracks in the samples, promoting an increase in the roughness of the material (de Oliveira et al., 2020). Roughness and cracks lead to an increase in the diffusion of water within the film and, consequently, an increase in the cleavage of polyolefin molecules by hydrolysis and the formation of free radicals (Lal et al., 2020). This hydrolysis process produces different fragments of chains of lower molecular weight, which serve as food for the bacteria of the compost; for example, PLA contains ester groups that have hydrolysable covalent bonds ($C=O$) (Torres-Huerta et al., 2018). Consequently, the molecular weight of the polymer is reduced, and oligomers are obtained (Queirós et al., 2013). For this reason, the greater addition of biopolymers increases hydrophilicity and promotes water absorption, as observed in the tests described in Figure 2.9 and Table 2.6 (Kusumastuti et al., 2020).

In the weathering accelerated test, plastics are subject to deterioration by solar radiation, UV rays and heat. In polymer blends, the degradation was attributed to the breaking of the polymer chains, causing an increase in the amount of carbonyl groups possibly responsible for the absorption of UV rays and the photoinitiation reactions that facilitate the degradation of the polymer (Torres-Huerta et al., 2018). However, this degradation process is very slow, so the action of microorganisms was necessary (Singh et al., 2012). The browning toward yellow to brown tones after 200h of samples of polymeric mixtures with PLA has been attributed to the appearance of carbonyl functional groups, which are one of the chromophore groups produced by radiation (Lizárraga-Laborín et al., 2018). In another study, the degradation of polymeric mixtures with PLA, the loss of weight and the embrittlement of the samples was evidenced only after 600h of the test (Torres-Huerta et al., 2014).

The degradation rate and microbial activity are also highly dependent on the type of synthetic polymer and fabrication process. Some polyolefins with long chains reduce the mobility of biopolymers and their dispersibility (Nurhajati et al., 2019). The agglomeration of biopolymers due to the lack of compatibility with polyolefins generates small areas for microorganisms to attack (Bercini Martins and Campomanes Santana, 2019). This is a reasonable explanation of the behavior during degradation of the mixtures with HDPE that presented a lower percentage of biodegradation than those made with LDPE, as observed in Figure 2.9. The type of polymer processing and the use of compatibilizers improved the distribution of the components and thus affect the biodegradation (Singh et al., 2011). This allows the diffusion of O_2 and H_2O, beginning their degradation within disorganized molecular regions that were initially in lower proportions (Torres-Huerta et al., 2018).

Finally, as in the case of natural fibers, the compatibilizers improve phases interaction, the formation of cracks, micro-holes between the interfaces and the diffusion of water molecules are reduced. This is attributed to the substitution of hydroxyl groups on the surface by acetyl moieties, increasing resistance to attack by microorganisms and therefore less biodegradability (Ahmadi et al., 2018).

Influence of Natural Fibers and Biopolymers

Case Study 6.1: Biodegradation of Synthetic Polymer/Biopolymer Blends with the Addition of Natural Fibers

This section presents the biodegradability evaluation of a biocomposite material manufactured with a HDPE/PLA matrix, reinforced with banana fibers (BF). The BF were extracted from pseudo-stems of banana plantations located in Manizales, Colombia, with a fiber decorticator (see Figure 2.13a). The raw and wet BF were dried in an oven to ~8% moisture content and then milled. HDPE (trade name DMDA-8920 NT 7) and polylactic acid (trade name PLA Ingeo™ Biopolymer 2003D) were obtained from Dow (Colombia) and NatureWorks (USA), respectively. Maleic anhydride grafted polyethylene (PE-g-MA) also used as a compatibilizing agent was supplied by Dupont-Colombia (Fusabond® E226). The blend of the three components was carried out in a Haake rotor (Model RHEODRIVE 7 HAAKE POLYLAB OS, Germany). Biocomposite samples were manufactured by an injection molding equipment (LIEN YU Machinery Co. Ltd., Taiwan) (see Figure 2.13b).

Two biodegradability tests were performed on the sample jars. One biodegradation assay was done by soil burial test method according to Gunti et al. (2016). The specimen was buried in the soil, away from UV light, at room temperature (18°C ± 2°C) and 70% relative humidity. Water was supplied constantly to keep constant the initial weight and moisture content. Rectangular biocomposite specimens of size 20 × 10 × 4 mm approximately were prepared (see Figure 2.14a). A total of 30 specimens were buried for each type of biocomposite, and five of those specimens were removed from the soil without replacement at 30, 60, 90, 120 and 150 days for characterization (see Figure 2.14b.). For monitoring and weight loss analysis, each sample was periodically extracted, washed with distilled water, and dried at 70°C in a hot air oven for 24 h. A photographic record was also made of the entire process.

The second biodegradation test was aimed at evaluating the anaerobic biodegradation rate by measuring CO_2 release (ASTM_D5988, 2012). The soil was collected from three different agricultural fields. Soil mixture was enriched with compost 100

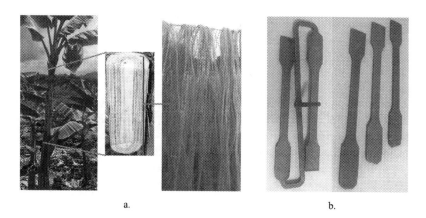

FIGURE 2.13 Banana fibers extracted from the pseudo-stem and samples of biocomposite materials.

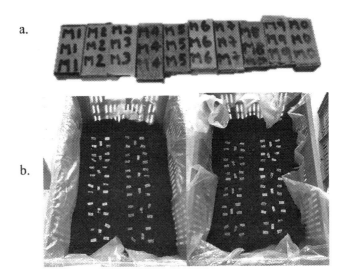

FIGURE 2.14 (a) Biocomposite specimen and (b) compost preparation for soil burial test.

TABLE 2.7
Soil and BF Characterization

Test	Parameter	Result
Soil		
Moisture retention curves	Retention capacity (RC) (%)	32.7
Mixture of soil and deionized water ratio of 1:1 (w/v)	pH	7.26
Walkley-Black spectrometry	Organic matter (g/kg)	117.99
Oxidizable carbon W:black	Carbon (g/kg)	51.58
BF		
Carbon combustion	Carbon (%)	33.2

g/kg (40 g/kg). Soil characterization and the carbon content of the fiber results are shown in Table 2.7.

The water content was measured as weight loss and the final soil moisture was adjusted at 14.6% (about 90% of the Water Holding Capacity). The assays were developed at the Instituto de biotecnologia y agroindustria Manizales (Colombia).

The test was set up with blank jars (without material) and with reference material jars (1 g of cellulose). Blank and reference jars were similarly treated as the samples. Two replicates were carried out for each polymer particle size, for blank and for reference, and incubated in the dark at 20°C ± 2°C. A 50 mL beaker filled with 30 mL of 0.5 M KOH, as a CO_2 trapping solution, was placed in each jar. The amount of CO_2 produced was measured by means of titration of the KOH solution with 0.3 N HCl with a Mettler Toledo (T50) potentiometric titrator. The measurement was conducted every 2–3 days during the first 2 weeks, when the degradation rate was expected to be maximal, and weekly or biweekly thereafter. Moisture content was maintained constant by adding deionized water throughout the biodegradation test whenever the KOH solution was titrated and replaced with a fresh one.

The net CO_2 production evolved from the test materials was calculated by subtracting the average amount of CO_2 produced in the blank soils to the amount of CO_2 produced in the polyester jars. The percentages of biodegradation were calculated from the ratio between the net CO_2 production and the theoretical CO_2 production calculated on the basis of polyester carbon content.

RESULTS

The weight loss of samples of HDPE biocomposites blends after 150 days is presented in the *y*-axis of Figure 2.15 as a function of their biodegradable material content (BF, PLA and the sum of both). Regarding the fiber content (BF), the higher weight loss was measured g for the samples that with 20% of BF. For those with a higher proportion of BF, the measured weight losses were slightly lower. The reduction in biodegradability for biocomposites with higher natural fiber percentage could be due to a heterogeneous distribution of it in the polymeric matrix and possibly a lower concentration of fiber on the surface. This leads to a reduction in water absorption and the consequent proliferation of microorganisms (Bercini Martins and Campomanes Santana, 2019). The weight loss on the biocomposites regarding the PLA content presented a similar behavior respect the BF, which could be due to the use of HDPE, which restricts the distribution of PLA and the lack of compatibility between these plastics (Nurhajati et al., 2019). In addition, as it was observed in the tests described in Figure 2.6, PLA was the biopolymer that presented the lowest weight loss compared to other biopolymers such as TPS, which could be another reason the low biodegradability. Pang et al. (2017) reported similar reductions for soil burial tests of LLDPE/PVOH blends with kenaf fibers after 6 months (11% for 40% fiber content samples). They concluded that the attack of microorganisms only took place on the surface of the compounds and oxygen penetration through the soil was limited. In contrast, when the sum of the concentrations of fiber (BF) and PLA is considered, the biodegradability is greater as the content of BF + PLA increases (Figure 2.15).

The low percentage of weight loss compared to other published works for the same test and analogous materials may be due to the use of the compatibilizer—PE-g-MA, which improved the interaction between HDPE, PLA and BF (Ahmadi et al., 2018;

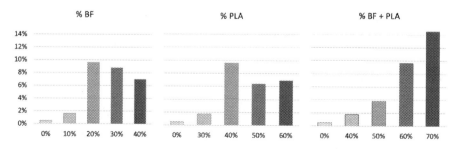

FIGURE 2.15 Results of the percentage of weight loss observed in the soil burial test.

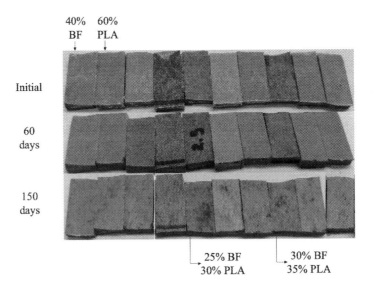

FIGURE 2.16 Photographs of biocomposites before and after the soil burial test.

Zaaba and Ismail, 2019). A similar case was reported by Gunti et al. (2018), who by using surface treatments on natural PLA reinforcing fibers reduced their hygroscopicity in such a way that at the end of the test they could only detect a 3% weight loss of biocomposite samples.

Photographs of the samples are shown in Figure 2.16. After 60 days, a slight color change is observed. On day 150, the areas with a dark brown color are clearly observed, especially in the mixtures with 25% and 30% BF and 30% and 35% PLA. It is possible that these types of blends, with similar proportion of three components, provide a better environment for microorganisms (Kochkina and Lukin, 2020). As already mentioned, the color change is associated with the formation of fungi due to the activation of microorganisms and the water absorption (Kochkina and Lukin, 2020).

The results of the biodegradation test using the ASTM D5338 standard are shown in Figure 2.17. The microcellulose powder was used as positive control reference material. The biocomposites evaluated contained 40% BF, 30% PLA and 30% HDPE with three fiber particle sizes: 600 µm, 1.0 and 1.7 mm. In Figure 2.17 the percentage of biodegradation is expressed as the emission of CO_2 released from the different components of the materials by the action of the microorganisms present in the compost. The percentages of biodegradation observed were 21.04%, 7.13%, 6.23% and 1.43%, which correspond in their order to the biocomposite samples that contained fibers from smaller to larger particle size. Similar results were found by Pinheiro et al. (2017) in which after 180 days, the cellulose showed a biodegradation of 52%. The low average biodegradability may be due to the prevailing external conditions in the test and the selection of the soil that defines the consortium of microorganisms present in the compost, which in turn may lead to the production of insufficient quantities of lipases and cellulases to degrade cellulose of the fibers (Madhu et al., 2016).

Influence of Natural Fibers and Biopolymers

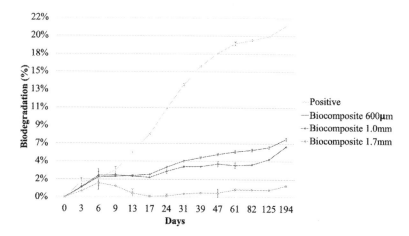

FIGURE 2.17 Biodegradation profiles of positive and biocomposites at three powder particle sizes.

In addition, as already mentioned, the biocomposites manufactured with the compatibilizers show better interactions between the components, generating more stable structures. This result suggests that a better dispersion and homogeneity of the fibers in the matrix provides some resistance to microbial attack because the fibers are covered and protected by the matrix polymer (de Campos et al., 2013). Chattopadhyay et al. (2011) reported a better interaction between the hydroxyl groups of the BF and the maleic anhydride groups of the compatibilizer MA-g-PP, generating covalent ester bonds between fiber and plastic matrix. However, in the smaller particle size biocomposites samples the microorganisms access to biological material is straightforward, and therefore, the biodegradability is higher. Other authors had reported low levels of biodegradation of 5%, 15% and 10% in biocomposites with PP and fibers from banana, bamboo and pineapple leaves, respectively (Chattopadhyay et al., 2011). In other studies, the test was extended up to 2 years, due to the low organic matter and microbial biomass of the soil used as incubation medium, so in real-field conditions the biodegradation kinetics would be even slower (Touchaleaume et al., 2016).

Finally, it is important to note that in these tests the cellulose did not reach 70% biodegradation in 180 days (minimum biodegradation required according to ASTM 5988). Therefore, the test cannot be validated according to such standard. However, despite the low biodegradation, a comparative profile of the biodegradability of biocomposites with PLA/HDPE and BF evaluated in different sizes can be made. Although the biodegradation process is slow, the biodegradation process is expected to continue. It should also be taken into account that as mentioned in Folino et al. (2020), the degradation process depends on a combination of abiotic (UV, temperature, moisture, pH) and biotic processes and parameters (microbial activity). Therefore, in waste disposal processes, optimal conditions for the biodegradation process must be guaranteed and this is a difficult situation to obtain in landfills.

2.6 CONCLUSIONS

This chapter discussed the different methods for measuring the biodegradability of pure plastic matrices and their blends with plastics and fibers that are commonly considered biodegradable. The biodegradability of a plastic material or a biocomposite is highly dependent on its polymeric matrix, and the rate of biodegradation depends on many environmental factors, such as humidity, radiation, temperature, and the microbial population in its environment. Although the simple exposure of plastic to the environment seems to be the best method to demonstrate its biodegradability, it is a method discarded as impractical. On the other hand, the laboratory tests recommended by accepted standards have limitations to emulate the natural environment for several reasons, among them the types and concentrations of microorganisms can significantly affect the rates of deterioration. Thus, current testing methodologies run the risk of promoting development of "biodegradable plastics" on a laboratory scale but with low or uncertain biodegradability in the natural environment. For this reason, it may be worth considering in regulations materials that are slowly biodegradable but that eventually become fully integrated into the natural environment.

In general, the biocomposites show more degradation than pure polymers due to the hydrophilicity of the natural fiber filler. Exposure of the biocomposites to light leads to the absorption of UV radiation, which in turn leads to browning, a change in surface roughness, an increase in brittleness and a deterioration in mechanical properties that increases in the presence of moisture that spreads through the fibers. The bioomposites that use compatibilizers or whose fibers have been treated with physical or chemical means to reduce their hydrophilicity tend to show less biodegradation than counterparts that do not use them.

REFERENCES

Abdullah, Z.W., Dong, Y., 2019. Biodegradable and water resistant poly(vinyl) alcohol (PVA)/Starch (ST)/Glycerol (GL)/Halloysite Nanotube (HNT) nanocomposite films for sustainable food packaging. *Front. Mater.* doi: 10.3389/fmats.2019.00058.

Abioye, A.A., Obuekwe, C.C., 2020. Investigation of the biodegradation of low-density polyethylene-starch Bi-polymer blends. *Results Eng.* 5, 100090. doi: 10.1016/j.rineng.2019.100090.

Ahmadi, M., Behzad, T., Bagheri, R., Heidarian, P., 2018. Effect of cellulose nanofibers and acetylated cellulose nanofibers on the properties of low-density polyethylene/thermoplastic starch blends. *Polym. Int.* 67, 993–1002. doi: 10.1002/pi.5592.

Ali, A., Shaker, K., Nawab, Y., Jabbar, M., Hussain, T., Militky, J., Baheti, V., 2016. Hydrophobic treatment of natural fibers and their composites: A review. *J. Ind. Text.* 47, 2153–2183. doi: 10.1177/1528083716654468.

ASTM_D5526, 2018. ASTM D5526-18 standard test method for determining anaerobic biodegradation of plastic materials under accelerated landfill conditions.

ASTM_D5988, 2012. ASTM D5988-12 standard test method for determining aerobic biodegradation of plastic materials in soil.

ASTM_D6400, 2019. ASTM D6400-19 standard specification for labeling of plastics designed to be aerobically composted in municipal or industrial facilities.

ASTM_D883, 2012. ASTM D883 standard terminology relating to plastics.

Azwa, Z.N., Yousif, B.F., Manalo, A.C., Karunasena, W., 2013. A review on the degradability of polymeric composites based on natural fibres. *Mater. Des.* 47, 424–442. doi: 10.1016/j.matdes.2012.11.025.

Bátori, V., Åkesson, D., Zamani, A., Taherzadeh, M.J., Sárvári Horváth, I., 2018. Anaerobic degradation of bioplastics: A review. *Waste Manag.* 80, 406–413. doi: 10.1016/j.wasman.2018.09.040.

Behera, A.K., Mohanty, C., Pradhan, S.K., Das, N., 2020. Assessment of soil and fungal degradability of thermoplastic starch reinforced natural fiber composite. *J. Polym. Environ.* doi: 10.1007/s10924-020-01944-z.

Bercini Martins, A., Campomanes Santana, R.M., 2019. Structure-properties correlation in PP/thermoplastic starch blends containing sustainable compatibilizer agent. *Mater. Res. Express* 6, 95336. doi: 10.1088/2053-1591/ab0f73.

Berruezo, M., Ludueña, L.N., Rodriguez, E., Alvarez, V.A., 2013. Preparation and characterization of polystyrene/starch blends for packaging applications. *J. Plast. Film Sheeting* 30, 141–161. doi: 10.1177/8756087913504581.

Catto, A.L., Montagna, L.S., Santana, R.M.C., 2017. Abiotic and biotic degradation of postconsumer polypropylene/ethylene vinyl acetate: Wood flour composites exposed to natural weathering. *Polym. Compos.* 38, 571–582. doi: 10.1002/pc.23615.

Chattopadhyay, S.K., Singh, S., Pramanik, N., Niyogi, U.K., Khandal, R.K., Uppaluri, R., Ghoshal, A.K., 2011. Biodegradability studies on natural fibers reinforced polypropylene composites. *J. Appl. Polym. Sci.* 121, 2226–2232. doi: 10.1002/app.33828.

Chaudhary, A.K., Vijayakumar, R.P., 2020. Synthesis of polystyrene/starch/CNT composite and study on its biodegradability. *J. Polym. Res.* 27, 187. doi: 10.1007/s10965-020-02164-8.

Chaudhuri, S., Chakraborty, R., Bhattacharya, P., 2013. Optimization of biodegradation of natural fiber (Chorchorus capsularis): HDPE composite using response surface methodology. *Iran. Polym. J.* 22, 865–875. doi: 10.1007/s13726-013-0185-8.

Costa, C.C., Andrade, G.R.S., Almeida, L.E., 2018. Biodegradation in simulated soil of HDPE/pro-oxidant/rice husk composites: application in agricultural tubes. Matéria (Rio Janeiro).

de Campos, A., Tonoli, G.H.D., Marconcini, J.M., Mattoso, L.H.C., Klamczynski, A., Gregorski, K.S., Wood, D., Williams, T., Chiou, B.-S., Imam, S.H., 2013. TPS/PCL composite reinforced with treated sisal fibers: Property, biodegradation and water-absorption. *J. Polym. Environ.* 21, 1–7. doi: 10.1007/s10924-012-0512-8.

de Oliveira, T.A., Barbosa, R., Mesquita, A.B.S., Ferreira, J.H.L., de Carvalho, L.H., Alves, T.S., 2020. Fungal degradation of reprocessed PP/PBAT/thermoplastic starch blends. *J. Mater. Res. Technol.* 9, 2338–2349. doi: 10.1016/j.jmrt.2019.12.065.

Dong, Y., Ghataura, A., Takagi, H., Haroosh, H.J., Nakagaito, A.N., Lau, K.-T., 2014. Polylactic acid (PLA) biocomposites reinforced with coir fibres: Evaluation of mechanical performance and multifunctional properties. *Compos. Part A Appl. Sci. Manuf.* 63, 76–84. doi: 10.1016/j.compositesa.2014.04.003.

Fazita, M.R.N., Jayaraman, K., Bhattacharyya, D., Hossain, M.S., Haafiz, M.K.M., Abdul Khalil H.P.S., 2015. Disposal options of bamboo fabric-reinforced poly(lactic) acid composites for sustainable packaging: Biodegradability and recyclability. *Polymers.* doi: 10.3390/polym7081465.

Folino, A., Karageorgiou, A., Calabrò, P.S., Komilis, D., 2020. Biodegradation of wasted bioplastics in natural and industrial environments: A review. *Sustainability.* doi: 10.3390/su12156030.

Greenpeace, 2020. Datos sobre la producción de plásticos [WWW Document]. URL https://es.greenpeace.org/es/trabajamos-en/consumismo/plasticos/datos-sobre-la-produccion-de-plasticos/.

Gunning, M.A., Geever, L.M., Killion, J.A., Lyons, J.G., Higginbotham, C.L., 2013. Mechanical and biodegradation performance of short natural fibre polyhydroxybutyrate composites. *Polym. Test.* 32, 1603–1611. doi: 10.1016/j.polymertesting.2013.10.011.

Gunti, R., Ratna Prasad, A.V, Gupta, A.V.S.S.K.S., 2016. Preparation and properties of successive alkali treated completely biodegradable short jute fiber reinforced PLA composites. *Polym. Compos.* 37, 2160–2170. doi: 10.1002/pc.23395.

Gunti, R., Ratna Prasad, A.V., Gupta, A.V.S.S.K.S., 2018. Mechanical and degradation properties of natural fiber-reinforced PLA composites: Jute, sisal, and elephant grass. *Polym. Compos.* 39, 1125–1136. doi: 10.1002/pc.24041.

Heimowska, A., Krasowska, K., 2019. Influence of different environments on degradation of composites with natural fibre. *IOP Conf. Ser. Earth Environ. Sci.* 214, 12060. doi:10.1088/1755-1315/214/1/012060.

Hidayat, A., Tachibana, S., 2012. Characterization of polylactic acid (PLA)/kenaf composite degradation by immobilized mycelia of Pleurotus ostreatus. *Int. Biodeterior. Biodegrad.* 71, 50–54. doi: 10.1016/j.ibiod.2012.02.007.

ISO_17088, 2013. ISO 17088:2012(en) specifications for compostable plastics.

ISO_877-2, 2009. ISO 877-2:2009: Plastics: Methods of exposure to solar radiation: Part 2: Direct weathering and exposure behind window glass.

Iwańczuk, A., Kozłowski, M., Łukaszewicz, M., Jabłoński, S., 2015. Anaerobic biodegradation of polymer composites filled with natural fibers. *J. Polym. Environ.* 23, 277–282. doi: 10.1007/s10924-014-0690-7.

Jandas, P.J., Mohanty, S., Nayak, S.K., Srivastava, H., 2011. Effect of surface treatments of banana fiber on mechanical, thermal, and biodegradability properties of PLA/banana fiber biocomposites. *Polym. Compos.* 32, 1689–1700. doi: 10.1002/pc.21165.

Kochkina, N.E., Lukin, N.D., 2020. Structure and properties of biodegradable maize starch/chitosan composite films as affected by PVA additions. *Int. J. Biol. Macromol.* 157, 377–384. doi: 10.1016/j.ijbiomac.2020.04.154.

Kormin, S., Kormin, F., Dalour Hossen Beg, M., 2018. Study of natural weathering exposure on properties of ldpe incorporated with sago starch. *Mater. Today Proc.* 5, 21636–21643. doi: 10.1016/j.matpr.2018.07.013.

Kusumastuti, Y., Putri, N.R.E., Timotius, D., Syabani, M.W., Rochmadi, 2020. Effect of chitosan addition on the properties of low-density polyethylene blend as potential bioplastic. *Heliyon* 6, e05280. doi: 10.1016/j.heliyon.2020.e05280.

Lal, S., Kumar, V., Arora, S., 2020. Eco-friendly synthesis of biodegradable and high strength ternary blend films of PVA/starch/pectin: Mechanical, thermal and biodegradation studies. *Polym. Polym. Compos.* doi: 10.1177/0967391120972881.

Lizárraga-Laborín, L.L., Quiroz-Castillo, J.M., Encinas-Encinas, J.C., Castillo-Ortega, M.M., Burruel-Ibarra, S.E., Romero-García, J., Torres-Ochoa, J.A., Cabrera-Germán, D., Rodríguez-Félix, D.E., 2018. Accelerated weathering study of extruded polyethylene/poly (lactic acid)/chitosan films. *Polym. Degrad. Stab.* 155, 43–51. doi: 10.1016/j.polymdegradstab.2018.06.007.

Luthra, P., Vimal, K.K., Goel, V., Singh, R., Kapur, G.S., 2020. Biodegradation studies of polypropylene/natural fiber composites. *SN Appl. Sci.* 2, 512. doi: 10.1007/s42452-020-2287-1.

Madhu, G., Bhunia, H., Bajpai, P.K., Nando, G.B., 2016. Physico-mechanical properties and biodegradation of oxo-degradable HDPE/PLA blends. *Polym. Sci. Ser. A* 58, 57–75. doi: 10.1134/S0965545X16010077.

Mittal, A., Garg, S., Bajpai, S., 2020. Thermal decomposition kinetics and properties of grafted barley husk reinforced PVA/starch composite films for packaging applications. *Carbohydr. Polym.* 240, 116225. doi: 10.1016/j.carbpol.2020.116225.

Moriana, R., Strömberg, E., Ribes, A., Karlsson, S., 2014. Degradation behaviour of natural fibre reinforced starch-based composites under different environmental conditions. *J. Renew. Mater.* 2, 145–156. doi: 10.7569/JRM.2014.634103.

Mubarak, Y.A., Abdulsamad, R.T., 2018. Thermal properties and degradability of low density polyethylene microcrystalline cellulose composites. *J. Thermoplast. Compos. Mater.* 32, 487–500. doi: 10.1177/0892705718766387.

Mundo_plast, 2020. El mercado mundial de bioplásticos crecerá más de un 15% en los próximos años [WWW Document]. https://mundoplast.com/crecimiento-mercado-bioplastic os/#:~:text=Comoasegura Hasso von Pogrell, en los próximos cinco años.

Narayan, R., 2006. Biobased and biodegradable polymer materials: Rationale, drivers, and technology exemplars, in: *Degradable Polymers and Materials*, ACS Symposium Series. American Chemical Society: Washington, DC, pp. 18–282. doi: 10.1021/bk-2006-0939.ch018.

Nurhajati, D.W., Setyadewi, N.M., Indrajati, I.N., 2019. Biodegradability of cassava starch/high density polyethylene reactive blend during compost burial. *IOP Conf. Ser. Mater. Sci. Eng.* 553, 12051. doi: 10.1088/1757-899x/553/1/012051.

Nuryawan, A., Hutauruk, N.O., Purba, E.Y.S., Masruchin, N., Batubara, R., Risnasari, I., Satrio, F.K., Rahmawaty, Basyuni, M., McKay, D., 2020. Properties of wood composite plastics made from predominant Low Density Polyethylene (LDPE) plastics and their degradability in nature. *PLoS One* 15, e0236406.

Pang, A.L., Ismail, H., Abu Bakar, A., 2017. Degradation of linear low-density polyethylene/poly(vinyl alcohol)/kenaf composites. *Iran. Polym. J.* 26, 703–709. doi: 10.1007/s13726-017-0555-8.

Peng, Y., Li, X., Wang, W., Cao, J., 2020. Photodegradation of wood flour/polypropylene composites incorporated with carbon materials with different morphologies. *Wood Mater. Sci. Eng.* 15, 104–113. doi: 10.1080/17480272.2018.1496359.

Pichaiyut, S., Nakason, C., Wisunthorn, S., 2018. Biodegradability and thermal properties of novel natural rubber/linear low density polyethylene/thermoplastic starch ternary blends. *J. Polym. Environ.* 26, 2855–2866. doi: 10.1007/s10924-017-1174-3.

Pinheiro, I.F., Ferreira, F.V., Souza, D.H.S., Gouveia, R.F., Lona, L.M.F., Morales, A.R., Mei, L.H.I., 2017. Mechanical, rheological and degradation properties of PBAT nanocomposites reinforced by functionalized cellulose nanocrystals. *Eur. Polym. J.* 97, 356–365. doi: 10.1016/j.eurpolymj.2017.10.026.

Qin, J., Jiang, J., Tao, Y., Zhao, S., Zeng, W., Shi, Y., Lu, T., Guo, L., Wang, S., Zhang, X., Jie, G., Wang, J., Xiao, M., 2021. Sunlight tracking and concentrating accelerated weathering test applied in weatherability evaluation and service life prediction of polymeric materials: A review. *Polym. Test.* 93, 106940. doi: 10.1016/j.polymertesting.2020.106940.

Queirós, Y.G.C.., Machado, K.J.A.., Costa, J.M.., Lucas, E.F., 2013. Synthesis, characterization, and in vitro degradation of poly(lactic acid) under petroleum production conditions. *Braz. J. Pet. Gas* 7, 57–59.

Raj, M., Savaliya, R., Joshi, S., Raj, L., 2018. Studies on blends of modified starch–LDPE. *Polym. Sci. Ser. A* 60, 805–815. doi: 10.1134/S0965545X18060081.

Shah, A.A., Hasan, F., Hameed, A., Ahmed, S., 2008. Biological degradation of plastics: A comprehensive review. *Biotechnol. Adv.* 26, 246–265. doi: 10.1016/j.biotechadv.2007.12.005.

Siakeng, R., Jawaid, M., Ariffin, H., Sapuan, S.M., Asim, M., Saba, N., 2019. Natural fiber reinforced polylactic acid composites: A review. *Polym. Compos.* 40, 446–463. doi: 10.1002/pc.24747.

Singh, G., Bhunia, H., Rajor, A., Choudhary, V., 2011. Thermal properties and degradation characteristics of polylactide, linear low density polyethylene, and their blends. *Polym. Bull.* 66, 939–953. doi: 10.1007/s00289-010-0367-x.

Singh, G., Bhunia, H., Bajpai, P.K., Choudhary, V., 2012. Thermal degradation and physical aging of linear low density polyethylene and poly(l-lactic acid) blends. *J. Polym. Eng.* 32, 59–66. doi: 10.1515/polyeng-2011-0106.

Singha, A.S., Kapoor, H., 2014. Effects of plasticizer/cross-linker on the mechanical and thermal properties of starch/PVA blends. *Iran. Polym. J.* 23, 655–662. doi: 10.1007/s13726-014-0260-9.

Song, J.H., Murphy, R.J., Narayan, R., Davies, G.B.H., 2009. Biodegradable and compostable alternatives to conventional plastics. *Philos. Trans. R. Soc. Lond. B. Biol. Sci.* 364, 2127–2139. doi: 10.1098/rstb.2008.0289.

Stark, N.M., Matuana, L.M., 2004. Surface chemistry changes of weathered HDPE/woodflour composites studied by XPS and FTIR spectroscopy. *Polym. Degrad. Stab.* 86, 1–9. doi: 10.1016/j.polymdegradstab.2003.11.002.

Stelescu, M.-D., Manaila, E., Craciun, G., Chirila, C., 2017. Development and characterization of polymer eco-composites based on natural rubber reinforced with natural fibers. *Materials*. doi: 10.3390/ma10070787.

Takagi, H., 2019. Review of functional properties of natural fiber-reinforced polymer composites: Thermal insulation, biodegradation and vibration damping properties. *Adv. Compos. Mater.* 28, 525–543. doi: 10.1080/09243046.2019.1617093.

Tazi, M., Erchiqui, F., Kaddami, H., 2018. Influence of SOFTWOOD-fillers content on the biodegradability and morphological properties of WOOD–polyethylene composites. *Polym. Compos.* 39, 29–37. doi: 10.1002/pc.23898.

Torres-Huerta, A.M., Palma-Ramírez, D., Domínguez-Crespo, M.A., Del Angel-López, D., de la Fuente, D., 2014. Comparative assessment of miscibility and degradability on PET/PLA and PET/chitosan blends. *Eur. Polym. J.* 61, 285–299. doi: 10.1016/j.eurpolymj.2014.10.016.

Torres-Huerta, A.M., Domínguez-Crespo, M.A., Palma-Ramírez, D., Flores-Vela, A.I., Castellanos-Alvarez, E., Del Angel-López, D., 2018. Preparation and degradation study of HDPE/PLA polymer blends for packaging applications. *Rev. Mex. Ing. Química* 18. doi: 10.24275/uam/izt/dcbi/revmexingquim/2019v18n1/Torres

Touchaleaume, F., Martin-Closas, L., Angellier-Coussy, H., Chevillard, A., Cesar, G., Gontard, N., Gastaldi, E., 2016. Performance and environmental impact of biodegradable polymers as agricultural mulching films. *Chemosphere* 144, 433–439. doi: 10.1016/j.chemosphere.2015.09.006.

Triwulandari, E., Ghozali, M., Sondari, D., Septiyanti, M., Sampora, Y., Meliana, Y., Fahmiati, S., Restu, W.K., Haryono, A., 2019. Effect of lignin on mechanical, biodegradability, morphology, and thermal properties of polypropylene/polylactic acid/lignin biocomposite. *Plast. Rubber Compos.* 48, 82–92. doi: 10.1080/14658011.2018.1562746.

Vieyra, H., Aguilar-Méndez, M.A., San Martín-Martínez, E., 2013. Study of biodegradation evolution during composting of polyethylene–starch blends using scanning electron microscopy. *J. Appl. Polym. Sci.* 127, 845–853. doi: 10.1002/app.37818.

Wu, C.-S., Wu, D.-Y., Wang, S.-S., 2020. Preparation, characterization, and functionality of bio-based polyhydroxyalkanoate and renewable natural fiber with waste oyster shell composites. *Polym. Bull.* doi: 10.1007/s00289-020-03341-x.

Yang, F., Long, H., Xie, B., Zhou, W., Luo, Y., Zhang, C., Dong, X., 2020. Mechanical and biodegradation properties of bamboo fiber-reinforced starch/polypropylene biodegradable composites. *J. Appl. Polym. Sci.* 137. doi: 10.1002/app.48694.

Yousif, E., Haddad, R., 2013. Photodegradation and photostabilization of polymers, especially polystyrene: Review. *Springerplus* 2, 398. doi: 10.1186/2193-1801-2-398.

Yussuf, A.A., Massoumi, I., Hassan, A., 2010. Comparison of polylactic acid/kenaf and polylactic acid/rise husk composites: The influence of the natural fibers on the mechanical, thermal and biodegradability properties. *J. Polym. Environ.* 18, 422–429. doi: 10.1007/s10924-010-0185-0.

Zaaba, N.F., Ismail, H., 2017. Comparative study of irradiated and non-irradiated recycled polypropylene/peanut shell powder composites under the effects of natural weathering degradation. *BioResources*, 13(1), 487–505.

Zaaba, N.F., Ismail, H., 2019. Effects of natural weathering on the degradation of alkaline-treated peanut shell filled recycled polypropylene composites. *J. Vinyl Addit. Technol.* 25, 26–34. doi: 10.1002/vnl.21655.

Zaaba, N.F., Ismail, H., Jaafar, M., 2017. A study of the degradation of compatibilized and uncompatibilized peanut shell powder/recycled polypropylene composites due to natural weathering. *J. Vinyl Addit. Technol.* 23, 290–297. doi: 10.1002/vnl.21504.

Zhong, Y., Godwin, P., Jin, Y., Xiao, H., 2020. Biodegradable polymers and green-based antimicrobial packaging materials: A mini-review. *Adv. Ind. Eng. Polym. Res.* 3, 27–35. doi: 10.1016/j.aiepr.2019.11.002.

Zumstein, M.T., Narayan, R., Kohler, H.-P.E., McNeill, K., Sander, M., 2019. Dos and do nots when assessing the biodegradation of plastics. *Environ. Sci. Technol.* 53, 9967–9969. doi: 10.1021/acs.est.9b04513.

3 Property Analysis and Characterization of Biomass-Based Composites

Tejas Pramod Naik, Ujendra Kumar Komal, and Inderdeep Singh
Indian Institute of Technology Roorkee

CONTENTS

3.1	Introduction	66
3.2	Chemical Characterization	67
	3.2.1 FTIR	68
	3.2.2 Nuclear Magnetic Resonance	68
	3.2.3 X-Ray Diffraction	69
3.3	Physical Characterization	70
	3.3.1 Optical Microscope	70
	3.3.2 Scanning Electron Microscope	70
	3.3.3 Density	71
	3.3.4 Void Fraction	71
	3.3.5 Water Absorption Test	72
	3.3.6 Hardness Test	72
3.4	Mechanical Characterization	73
	3.4.1 *Tensile Strength*	73
	3.4.2 Compression Test	73
	3.4.3 Flexural Test	74
	3.4.4 Impact Test	75
	3.4.5 Single-Fiber Tensile Test	76
	3.4.6 Tensile Shear Strength Test for Single Lap Joints	77
3.5	Thermal Characterization	77
	3.5.1 Differential Scanning Calorimetry	77
	3.5.2 Differential Thermal Analysis	78
	3.5.3 Thermogravimetric Analysis	78
	3.5.4 Dynamic Mechanical Analysis	78
3.6	Durability Characterization	80
	3.6.1 *Creep Testing*	80
	3.6.2 Fatigue Testing	81

DOI: 10.1201/9781003137535-3

3.6.3　Wear Testing 81
　　3.6.4　Environmental Testing 82
3.7　Test Standards 83
3.8　Conclusions 83
References 84

3.1　INTRODUCTION

The origin of fiber-reinforced plastics (FRPs) was found in the early 1960s with a target to deliver lightweight materials with improved properties. The scope of FRP composites is broad from structural to non-load-bearing applications. These days' polymer composites are used to construct ships, pressure vessels, automobile parts, plane boards, etc., because of the advantages like lightweight, high strength, corrosion and UV resistance, and long service life. Nowadays, polymer composites are in their development stage, and research and industrial fraternity are continuously working on solving the challenges associated with their fabrication, characterization, and applications. FRPs consist of two elements: matrix and reinforcement. The polymer utilized as a matrix in manufacturing FRPs can be thermoplastic or thermoset. The traditional FRPs use synthetic fibers as reinforcement, which offers exceptional properties making them suitable for various engineering applications. However, the non-biodegradability and non-renewability of synthetic fibers have encouraged the researchers to look for alternate sustainable materials. The biomass-based natural fibers have been identified as an alternative to synthetic fibers (Pickering, Efendy, and Le 2016). Several investigators established that the natural fibers could be used as reinforcement and offer excellent properties making their composites suitable for various engineering applications (Pickering, Efendy, and Le 2016).

　　The quality of composites relies upon different variables, including the properties of the individual constituents, the interfacial bonding between the constituents, the processing route, and its parameters. The various fabrication techniques explored for these composites are hand lay-up, injection molding, compression molding, autoclave, extrusion-compression molding, and extrusion-injection molding, vacuum bagging, microwave curing, etc. (Biswas and Anurag 2019; Biswas 2017). Over the past decade, various materials (polymers and natural fibers) are being used to fabricate the composites. However, there have been very few changes in the testing standards used for characterizing these composites. The methodologies and practices used to evaluate and interpret should increasingly be the best in class as materials and designs become theoretically troubling. To evaluate the properties of composites, compare two different composites, and understand the failure mechanism in material fracture and reverse engineering, characterization plays an important role. Characterization of the bio-based composites helps us resolve various issues and challenges during the design and development, in-service, recycling, and degradation of these composites. The importance of the characterization can be seen in Figure 3.1. Characterization helps the designers/scientists predict the material's performance in various environmental conditions, thus preventing the component failure during operation (Cross et al. 2015). This chapter highlights the essential characterization techniques employed by researchers worldwide to investigate the behavior of the biomass-based composites.

Property Analysis of Biomass-Based Composites

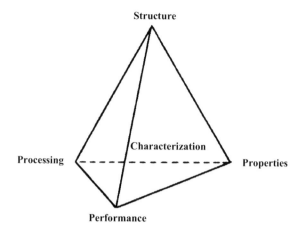

FIGURE 3.1 Material science tetrahedron.

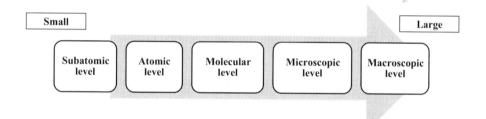

FIGURE 3.2 Different levels of characterization.

Different levels of characterization starting from subatomic to macroscopic are shown in Figure 3.2. These characterization techniques are used to investigate the behavior of biomass-based composites during the development phase (Lila, Komal, and Singh 2019).

The characterization techniques employed for the biomass-based composites are broadly classified into the following groups:

a. Chemical characterization
b. Physical characterization
c. Mechanical characterization
d. Thermal characterization
e. Durability characterization

3.2 CHEMICAL CHARACTERIZATION

Some standard techniques used for the chemical characterization of biomass-based composites are mentioned below.

3.2.1 FTIR

FTIR stands for Fourier Transform Infrared Spectroscopy. It is the most preferred method of spectroscopy utilizing Infrared (IR) radiation. A rapid and precise, qualitative and quantitative study of polymer constituents can be given by FTIR spectroscopy. Including design, failure analysis, and even waste classification, FTIR spectroscopy can be extended to all phases of the product lifecycle. The primary application of FTIR is to classify products such as fillers, resins, compounded plastics, adhesives, and paints, easily and conclusively. When IR rays passed over the sample, some of the radiations are absorbed by the material, and some are transmitted by it, which is detected by the detector. The energy is absorbed by the sample when there is stretching or bending vibration in the bonds. The graph plotted in FTIR spectroscopy is transmittance (%) v/s wavenumber (cm^{-1}), and it consists of two regions known as functional group region (4,000 to 1,450 cm^{-1}) and fingerprint region (1,450 to 500 cm^{-1}). The functional group region is used to detect the functional group present in the sample. The fingerprint region is unique to all materials. We can match this region with the known substance in the database, as shown in Table 3.1. One can use this data to identify the functional group present in the sample, which can be used to identify the sample type. Komal, Lila, and Singh (2020) used FTIR to detect the presence and type of bond in the banana fiber/polylactic acid (PLA) biocomposites. Chaitanya, Singh, and Song (2019) used it to investigate the degradation of PLA after recycling of the sisal/PLA-based composites.

3.2.2 Nuclear Magnetic Resonance

Nuclear Magnetic Resonance (NMR) spectroscopy has become the most frequently used analytical tool to find chemical structures. It is mainly used for carbon-based chemical compounds, also known as organic compounds. It is a nuclear-level characterization technique. It detects the nucleus and gives information about the overall structure of the molecule. There are two types of NMR techniques: (a) Hydrogen NMR (H NMR) and (b) C-13 NMR. In the H NMR technique, a number of hydrogen atoms (protons) are identified, while in C-13 NMR, carbon atoms are identified in a structure. NMR spectroscopy can be performed only on

TABLE 3.1
Infrared Spectroscopy Correlation for Some Common Functional Group

Functional Group	Wavenumber (cm^{-1})
C–H	3,000–3,100
C=O	1,685–1,740
C–O	1,000–1,300
O–H (acids)	2,500–3,300

Source: Kennepohl, Farmer, and Reusch (2020).

Property Analysis of Biomass-Based Composites

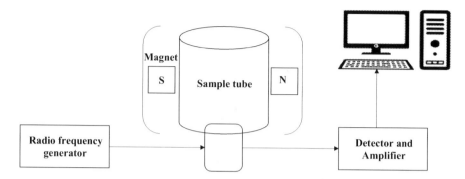

FIGURE 3.3 Schematic diagram of NMR.

NMR active material (means materials having an odd number of protons). NMR spectroscopy's basic components are the strong magnet, radiofrequency generator, detector, amplifier, and computer, as shown in Figure 3.3. The sample to be tested for its chemical structure is kept in a sample tube. In the case of H NMR, initially, all the protons rotate about their axis. As a magnetic field is applied, all the protons reorient themselves due to the high magnetic field. After this, radio waves are supplied to the protons. The protons' energy level increases, and then the proton tries to get stable by releasing this energy, which the detector can detect. The signal is amplified, which is then seen in the form of spectra. Analysis of these spectra would help to identify the chemical structure of the unknown sample. It can also be used to see the variation in the chemical structure of cellulose-based composites before and after degradation (Sain et al. 2014). Perremans et al. (2018) studied the effect of chemical treatment on the crystal structure and degree of crystallinity of cellulose in the fibers.

3.2.3 X-Ray Diffraction

The degree of crystallinity for semi-crystalline, amorphous polymers and their composites is determined by X-ray diffraction (XRD) analysis. Polymers may be highly crystalline, semi-crystalline, and amorphous. The presence and relative quantities of these types depend on how the polymer has been processed. The crystallinity of the biomass-based composites affects the mechanical properties significantly. Thus, crystallinity is an important property to be accurately determined through analysis of the XRD peaks. The higher the crystallinity for the polymer or composite better will be the mechanical properties and vice-versa. The degree of crystallinity can be determined using the expression (Equation 3.1) proposed by Vonk (1973). This expression uses amorphous and crystalline regions to determine the crystallinity of the polymers and their composites. A typical XRD spectrum of polypropylene (PP) is shown in Figure 3.4.

$$\text{Crystallinity}(\%) = \frac{\text{Crystalline region area}}{\text{Total area under the curve}} \times 100 \quad (3.1)$$

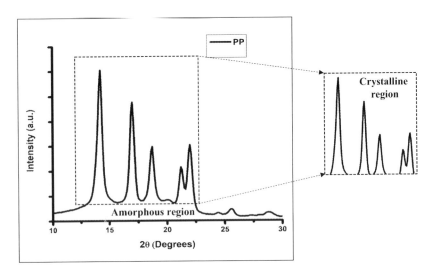

FIGURE 3.4 Typical XRD curve for polypropylene.

Several authors have determined the crystallinity of natural fiber-based composites using XRD analysis (Komal, Lila, and Singh 2020; Lila et al. 2019).

3.3 PHYSICAL CHARACTERIZATION

The standard techniques used for the physical characterization of biomass-based composites are discussed in the following section.

3.3.1 Optical Microscope

Optical microscopy is the conventional method of microscopy. New developments to this technique are used in confocal microscopy, depending on specific layers within the specimen to achieve depth resolution. Several studies have suggested the use of an optical microscope to study natural fibers' orientation within the polymeric matrices (Chaitanya and Singh 2016b; Komal, Lila, and Singh 2020).

3.3.2 Scanning Electron Microscope

A scanning electron microscopy (SEM) is a type of electron microscope that is used to examine the sample at a microscopic level. SEM provides an in-depth and high-resolution image. The various components of SEM are the source (tungsten filament), an anode (+ve charged electrode), scan coil, objective lens, sample, detector, and sensor. The working principle of SEM is that the tungsten filament is heated to produce electrons that are initially scattered. In order to make these electrons alien in a single beam, generally, a very high voltage of 20 kV is used. This single electron beam is then made to fall on the sample area, which is to be examined. The scattered electrons from the sample surface are then detected by the detector, which is then shown in the form of an image as an output. More electrons escape from large/high

Property Analysis of Biomass-Based Composites

surfaces on the sample, and from the lower surface, fewer electrons escape. As a result of high- and low-electron escape, a differential image can be captured by the complementary metal oxide semiconductor (CMOS) sensor and projected as an image. Sample preparation is required to examine the surface of non-conductive materials. The polymer-based specimens are usually coated with gold or carbon using the sputtering machine to make the sample conductive. Scanning electron microscopy (SEM) is used to analyze the fractured surface, degradation of fibers, matrices, matrix-fiber interfaces, etc.

3.3.3 Density

It is possible to calculate the density of polymer composites using Archimedes theory. Archimedes' theory states that "The weight of an object which is immersed in a liquid is equal to the weight of the displaced liquid." Therefore, the density of these composites can be calculated using Equation 3.2.

$$\rho_c = \frac{m_a}{m_a - m_l} \rho_l \qquad (3.2)$$

where m_a = mass of material in air, m_l = mass of material in liquid, ρ_l = density of liquid.

For calculating the bulk density of the polymer composites, Equation 3.2 is used. However, to find actual density, little modification is required in Equation 3.2 as the polymer composites consist of two phases having different densities. The actual density for polymer composites can be measured in terms of volume fraction by using Equation 3.3 and in terms of weight fraction using Equation 3.4.

$$\rho_c = \rho_f v_f + \rho_m v_m \qquad (3.3)$$

where ρ_c = density of the composite, ρ_f = density of fiber, v_f = fiber volume fraction, ρ_m = density of the matrix, v_m = matrix volume fraction.

$$\rho_c = \frac{1}{\frac{m_f}{\rho_f} + \frac{m_m}{\rho_m}} \qquad (3.4)$$

where ρ_c = density of the composite, m_f = mass of fiber, ρ_f = density of fiber, m_m = mass of the matrix, ρ_m = density of the matrix.

3.3.4 Void Fraction

Polymer-based composites are porous solids, which contain cavities and voids due to the fiber, foreign particles, and dust particles. Air gets trapped into these voids/pores; these voids/pores are categorized as macropores (>500 Å), mesopores (200–500 Å), and micropores (<20 Å) according to diameter. The presence of the voids inside the biomass-based composites, as shown in Figure 3.5, results in a decrease in the mechanical properties, greater susceptibility to water absorption, and poor

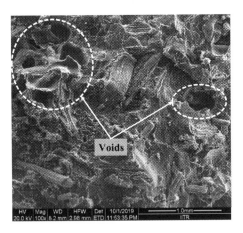

FIGURE 3.5 Typical SEM image showing voids content.

mechanical interlocking between the fiber and the matrix. Therefore, the void fraction in the manufactured composites needs to be studied. Analysis of the void fraction helps the researchers to take corrective measures in the fabrication process, reducing the void fraction in the composite, and improving the performance and properties. It is possible to determine the void content in the composites using Equation 3.5.

$$\%v_{void} = \frac{\rho_{bulk} - \rho_{actual}}{\rho_{bulk}} \times 100 \qquad (3.5)$$

where ρ_{bulk} = bulk density of composites, ρ_{actual} = actual density of composites.

3.3.5 Water Absorption Test

ASTM D570 is the standard used to measure the water absorption of polymers. The specimens are usually dried in an oven for a specified time and temperature for the water absorption test and then put in a desiccator to cool. The specimens are weighed once they have cooled in a desiccator. The sample is then kept for a defined period underwater at ambient temperature (25°C). Then the specimen is dried with a cloth and weighed again. Water absorption is expressed (Equation 3.6) as an increase in weight %.

$$\text{Percentage of water absorption} = \frac{\text{Wet sample weight} - \text{Dry sample weight}}{\text{Dry sample weight}} \times 100 \quad (3.6)$$

3.3.6 Hardness Test

The hardness of the soft materials such as rubber, polymer, and elastomers is measured using an instrument called a durometer. Durometer is a small, compact, easy to use, a hand-operated instrument used to measure the hardness of soft materials. Shore A hardness tester is used to measure the hardness of soft polymers and elastomers with an

Property Analysis of Biomass-Based Composites

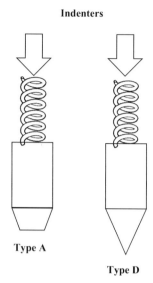

FIGURE 3.6 Indenter Type A and Type D used in durometer for hardness testing.

indenter tip diameter of 0.79 mm. Shore D type hardness tester (an indenter tip diameter of 0.1 mm) can be used to measure the hardness of rigid polymers and biomass-based composites. Durometer measures the depth of an indentation in the material created by an applied force. A higher number on the durometer scale indicates excellent resistance to indentation (rigid material) and vice-versa. Figure 3.6 shows the two different types of indenters used in the durometer for hardness testing. The use of Shore D tester has been suggested to measure the hardness of polymers, bagasse/PP, and banana fiber/PLA composites (Lila et al. 2018a; Komal, Lila, and Singh 2020).

3.4 MECHANICAL CHARACTERIZATION

3.4.1 Tensile Strength

One of the most widely used standards for measuring the tensile properties of the polymers and polymer composites is ASTM D3039M-17. According to this standard, the cross-section of the specimen is rectangular. The tensile test is carried out on a universal testing machine (UTM), as shown in Figure 3.7. The tensile properties of biomass-based composites depend on various factors such as processing technique, fiber weight fraction, fiber length, individual properties of both the constituents. Generally, three to five specimens are tested, and the average value is reported. The fractured specimens are examined under the SEM to understand the failure mechanisms.

3.4.2 Compression Test

The compression test is performed to observe the composites' behavior under the gradually applied compressive load. The same UTM machine used to perform the

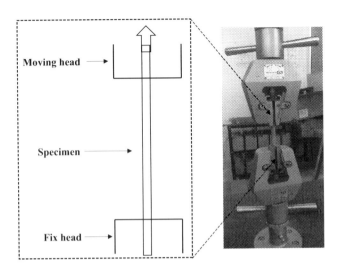

FIGURE 3.7 Tensile testing of composite specimen.

tensile test can be used to perform a compression test on composites with little modification in the fixture. Test specimen usually of a cuboid or cylindrical geometry is placed in between the head, as shown in Figure 3.8 to perform the test. The use of specially designed University of California, Santa Barbara (UCSB) fixture has also been suggested to conduct a compression test of rectangular specimens (Chaitanya and Singh 2016a). Under compression, the composite may experience various failure modes, including elastic buckling, fiber crushing, and shear splitting, which can be analyzed using SEM.

3.4.3 Flexural Test

ASTM D7264M-15 is one of the most commonly used test standards for measuring the flexural properties of biomass-based composites. The common purpose of performing the flexural test is to measure the flexural strength and modulus. It measures the force required to bend the composite specimens and determines the stiffness of the material before permanent deformation. The flexural test is usually conducted using UTM in a three-point bending mode (Figure 3.9). It is generally seen that the thermoset-based composite usually breaks into two parts because of its brittle nature. In contrast, thermoplastics-based composites show a U-bend due to their highly ductile nature. The load and the subsequent deflection are determined, the stress-strain curve is drawn, and the intensity is estimated under the three-point bending mode using Equation 3.7.

$$\sigma_f = \frac{3FL}{2bd^2} \qquad (3.7)$$

where σ_f = flexural stress, F = maximum load, L = support span, b, and d = width and thickness of the test specimen, respectively.

Property Analysis of Biomass-Based Composites

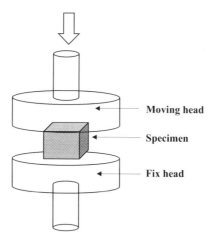

FIGURE 3.8 Compression testing of composite specimen.

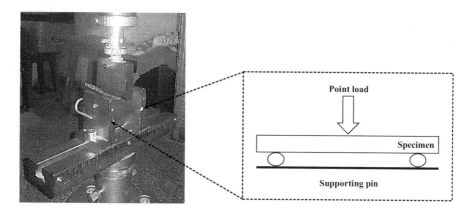

FIGURE 3.9 Flexural testing of composite specimen.

3.4.4 Impact Test

The toughness of the biomass-based composites is evaluated using the impact test methodology. In the short impact, the test is performed to understand the effect of suddenly applied load on the behavior of the composites. The amount of energy absorbed in breaking the sample in a single hammer blow is the measure of impact energy. There are two types of impact tests generally performed for composites: (a) Izod impact test; (b) Charpy impact test. A specimen with either a notch (V/U) or an unnotched specimen is used in both the testing methods. Both the methods are similar to each other, and the only variation is in the dimension and the clamping of the test specimen. During the Izod test, the specimen is clamped as a cantilever beam, and in the Charpy test, the specimen is clamped as a supported beam, as shown in Figure 3.10. The specimen is fixed in a fixture, and then it is struck from

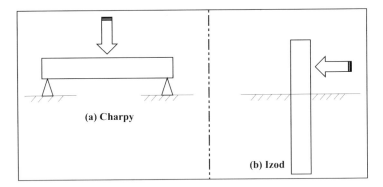

FIGURE 3.10 (a) Charpy test. (b) Izod test.

a particular height with a hammer with a known weight. The hammer possesses potential energy at height, and this energy gets converted into kinetic energy on release. Several articles are available focusing on the mechanical characterization of the biomass-based composites (Komal et al. 2018a,b; Chaitanya and Singh 2016a; Lila et al. 2020).

3.4.5 Single-Fiber Tensile Test

ASTM D3822 is used to determine the tensile properties of the fibers. Depending on the fiber modulus, there are two types of single-fiber tests. Fibers with low modulus are directly gripped into the jaws of the UTM machine, and fibers with a high modulus (e.g., carbon, natural fibers) are first pasted on a paper and then tested on UTM as shown in Figure 3.11. The tensile test is carried out until the failure at a lower loading rate. Around 50 single natural fibers are usually tested to obtain the average strength of the natural fiber.

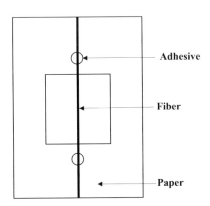

FIGURE 3.11 Single-fiber tensile test.

Property Analysis of Biomass-Based Composites

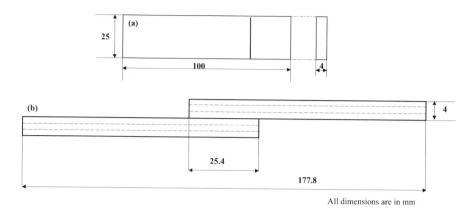

FIGURE 3.12 (a) Joining specimen dimension and (b) joint specimen.

3.4.6 Tensile Shear Strength Test for Single Lap Joints

ASTM D5868 is used to determine the lap shear adhesion strength for biomass-based composites. In the jaws of an UTM, the test specimens are placed and pulled until failure. Two specimens of 25 × 100 mm each are bonded together with an adhesive (Melese, Naik, and Singh 2020) as seen in Figure 3.12, and then the joint is clamped in the fixtures of the UTM machine, and the load is applied gradually until the specimen fails at the adhesive.

3.5 THERMAL CHARACTERIZATION

3.5.1 Differential Scanning Calorimetry

Differential Scanning Calorimetry (DSC) is a thermo-analytical technique used to study polymers and biomass-based composites' behavior when they are heated. It is a very popular technique due to its easy and fast operation and requires a small sample size. It gives valuable information about polymers' thermal properties that help to develop the best methods for processing the materials into a useful product and predicting product performance. DSC measures the energy absorbed or released by a sample as it is heated or cooled, or held isothermally. When a sample is tested, it will undergo one or more phase changes on heating or cooling. The phase transition occurs as a result of energy exchange with the environment. DSC measures the energy needed for phase change. DSC consists of a reference pan and sample pan along with the heaters. The sample and reference pan are heated from the bottom, and the sensor measures the heat flow. Whether more or less heat will flow to the sample depends on whether the process is exothermic or endothermic. In the exothermic transition, heat flows out of the sample. During the cooling cycle, crystallization, curing, and oxidation take place. In these transitions, the DSC plot shows an increase in heat flow, and the sensor measures a higher temperature for the sample pan compared to the reference pan.

In endothermic transition, heat flows into the sample. During the heating cycle, glass transition, melting, and evaporation take place. In these transitions, the DSC plot shows a decrease in heat flow, and the sensor measures a lower temperature for the sample pan compared to the reference pan. The critical information that can be drawn from the DSC analysis are thermal stability, crystallization time and temperature, oxidative stability, melting point, degree of cure, glass transition temperature, degree of crystallinity, purity, cure rate, etc. (Chaitanya, Singh, and Song 2019; Komal, Lila, and Singh 2020; Lila et al. 2018b).

3.5.2 Differential Thermal Analysis

Differential Thermal Analysis (DTA) is similar to DSC. The composite specimen under analysis and the reference material is configured to undergo similar thermal cycles, capturing any variation in temperature between the specimen and the reference pan (Stamm 2008). This differential temperature is then plotted against time or temperature, and any phase transitions in the sample, either exothermic or endothermic, can then be observed in contrast to the reference. Thus, the thermogram provides data on the various transformations, such as glass transition temperature (T_g), crystallinity, melting, and sublimation.

3.5.3 Thermogravimetric Analysis

Thermogravimetric Analysis (TGA) is a thermal analysis process in which changes in physical and chemical properties of the composite samples are calculated as a function of increasing temperature or time (Selvey et al. 2001). TGA contains physical changes such as desorption, absorption, and phase transformation and chemical changes such as solid-gas reactions, chemisorption, and thermal decomposition (Maus 2008). The TGA analysis can be used to

a. analyze the material by analyzing the decomposition pattern
b. study the degradation mechanism
c. determine the organic and inorganic contents in the specimens

The typical TGA, DTA, and DTG curves obtained for the PP are shown in Figure 3.13.

3.5.4 Dynamic Mechanical Analysis

Dynamic Mechanical Analysis (DMA) is a technique in which composites' elastic and viscous response under oscillating load is monitored against temperature, frequency, and time. In this analysis, the sample can be in the form of powder, fiber, films, and sheets kept in the furnace, which is kept at fixed or elevated temperature. As soon as the oscillating strain energy is applied to the sample, stress is developed into the sample, which is detected by the sensor. DMA measures variations in the composites' viscoelastic behavior resulting from a change in time, temperature, and amplitude. DMA is a robust tool to measure the glass transition temperature of the polymers and biomass-based polymeric composites. It can measure all the minor

Property Analysis of Biomass-Based Composites

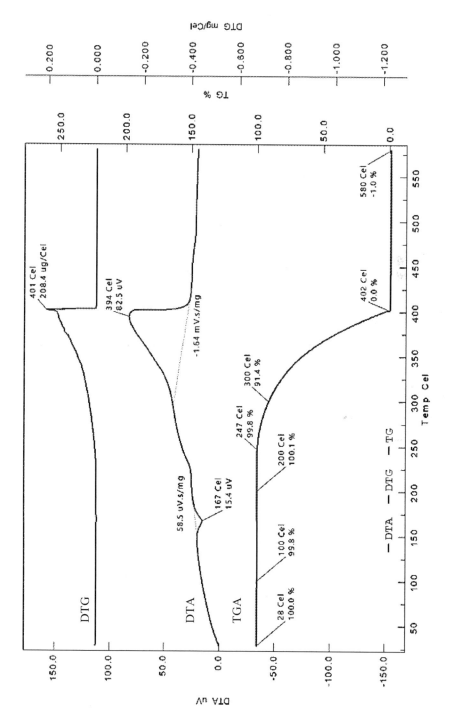

FIGURE 3.13 Typical DTA, DTG, and TGA curves for polypropylene.

transitions, such as sub-glass transition, which are very difficult to measure using DSC. DMA can be used

- to examine the viscoelastic behavior of the material as a function of stress, strain, time, frequency, and temperature;
- to measure the toughness of the composites;
- to measure the storage and loss modulus of the composites.

3.6 DURABILITY CHARACTERIZATION

3.6.1 *Creep Testing*

Polymer composites have a higher strength to weight ratio compared to other conventional materials. This advantage of PMCs finds its use in structural applications, particularly in the automotive industries. Therefore, it is necessary to check the variation in properties of the PMCs with time to predict the in-service life/performance of a component. It was found that the properties of the PMCs degrade when subjected to static and dynamic loads. The loading condition, creep time, and fatigue loading could cause permanent deformation in the PMCs (Sullivan 2017). Creep is a polymer's tendency to deform under external loads as the temperature increases. Initially, the polymer chains unwind, and under the application of the constant stress, as shown in Figure 3.14, these chains try to slip over one another. This phenomenon is temperature-dependent; with the increase in temperature, the secondary bonding between the chains decreases, and chain mobility increases. Crystalline polymers have excellent creep resistance than amorphous polymers as the secondary bond between chains is more significant in the crystalline polymer. Cross-linking chains also contribute to reducing creep. Also, the incorporation of fibers in the polymers

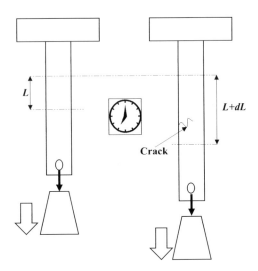

FIGURE 3.14 Creep testing of polymer composites.

TABLE 3.2
Different Stages in Creep Testing

Stages of Creep	Starting Rate	Terminating Rate
Primary	Rapid	Slow
Secondary	Relatively uniform	Relatively uniform
Tertiary	Accelerated	Material breaks

reduces creep because of chain entanglement. Creep takes place at an elevated temperature. Therefore, testing is usually performed in a chamber where temperature can be set. Creep generally occurs in three stages: (I) primary, (II) secondary, and (III) tertiary (Table 3.2).

3.6.2 Fatigue Testing

Fatigue testing is used to analyze the sturdiness of the composites. It is the procedure of progressive, confined permanent structural variation that occurs in a composite subjected to conditions that cause fluctuating stresses and strains. After an adequate number of fluctuations, the cracks are developed, which results in a complete fracture of the composites. The primary difference between the creep and fatigue is the loading condition, as shown in Figure 3.15. Similar to creep, fatigue failure also occurs in three stages: (I) initial fatigue damage, (II) crack propagation, (III) final fracture.

The information obtained from fatigue analysis is plotted as an S-N curve. Generally, ASTM D671 and ASTM D3479 are used for fatigue testing of polymer composites in flexural and tension-tension mode, respectively.

3.6.3 Wear Testing

The wear testing is usually conducted to investigate the friction and tribological behavior of the materials. Whenever there is an interaction between the two composite components, it becomes imperative to study the tribology of these composites. The process parameters, such as applied load, sliding distance, and sliding speed, significantly affect the friction and wear characteristics. The wear mechanisms of the composites can be broadly divided into adhesion, abrasion, and surface fatigue. Out of these, abrasive wear is the most common type of wear in composites due to the penetration of the rigid material into the soft material in contact, resulting in material removal. Fretting wear, corrosive wear, and delamination wear are also some other forms of wear that can be noticed in the polymeric composites (Bajpai, Singh, and Madaan 2013). The different tribological techniques are available for carrying out the wear test of composites. Pin on the disc is a widely adopted technique for composites. A sample in the form of a cylinder or flat is made to slide over a circular disc, which is continuously rotated with the motor's help, as shown in Figure 3.16. The sliding distance, sliding speed, and applied load can be fixed, and the experiments are carried out. The test's output can be the material removal rate, coefficient of friction, and specific wear rate in terms of mass or volume.

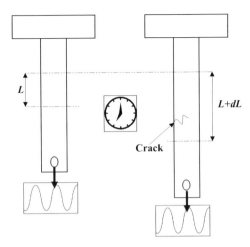

FIGURE 3.15 Fatigue testing of polymer composites.

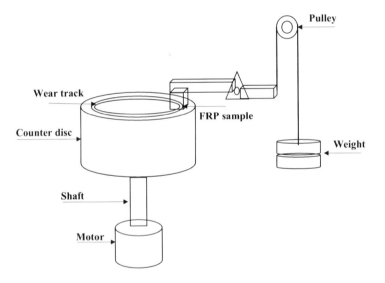

FIGURE 3.16 Typical pin-on-disc tribometer.

3.6.4 Environmental Testing

Nowadays, biomass-based composites have been used in many applications right from domestic to industrial. These composites are exposed to various environmental conditions like moisture, water, humidity, heat, radiations, soil, acidic, cold environment during their service life. Various researchers worldwide have studied the effect of the environments based on their intended application. The in-service life of these composites can be predicted by studying and analyzing their biodegradability and aging behavior. During aging and biodegradation studies, the composite specimens

Property Analysis of Biomass-Based Composites

TABLE 3.3
Various Test Standards

Test	ASTM Standards	Specimen Dimension (mm)	References
Tensile test	D3039-17	150×15×4	Chaitanya et al. (2019)
Flexural test	D790-10	100×15×4	Komal et al. (2018a)
Compression test	D3410	100×15×4	Chaitanya and Singh (2016a)
Impact test	D256-10	55×10×4	Komal et al. (2018b)
Lap shear tensile strength	D5868-01	177.8×25×8	Melese et al. (2020)
Water absorption test	D570-98	30×28×3	ASTM D570 (2014)
Hardness test	D2240-15	–	Komal, Lila, and Singh (2020)
Tensile testing of single fiber	D3822-01	–	Chaitanya and Singh (2018)
Dynamic mechanical analysis	D4065-12	60×15×4	Lila et al. (2020)

are subjected to various environmental conditions for several months or years. The investigators also perform experiments for a shorter period by exposing these composite specimens in accelerating environmental conditions (Lila et al. 2019). The composites incorporating natural fibers absorb moisture from the environment, which causes degradation and ultimately decreases its mechanical properties (Lila et al. 2018b; Komal et al. 2018a). Thus, the environmental aging characterization is essential to predict the in-service performance of the biomass-based composites.

3.7 TEST STANDARDS

The various standards followed by researchers worldwide to characterize the polymers and biomass-based composites are shown in Table 3.3.

3.8 CONCLUSIONS

The biomass-based composites possess a tremendous potential to be used in various structural and non-structural applications. The application spectrum of these composites is increasing exponentially. Therefore, the researchers are continuously working and suggesting efficient and accurate methods for the characterization of these composites. This chapter highlights the crucial characterization techniques employed by researchers worldwide to investigate the behavior of biomass-based composites. Characterizations help the researchers/engineers to resolve the various issues and challenges during the design, development, in-service, recycling, and degradation of these composites. It also helps the scientists/engineers predict the composite performance in various environmental conditions, thus preventing component failure during operation. The characterization techniques used for conventional materials may not be suitable for biomass-based composites. Therefore, there is still enormous scope for the advancement of characterization techniques to get an insight into the biomass-based composites.

REFERENCES

ASTM D570. 2014. "Standard test method for water absorption of plastics." ASTM Standards, vol. 98. doi:10.1520/D0570-98R18.2.

Bajpai, Pramendra Kumar, Inderdeep Singh, and Jitendra Madaan. 2013. "Tribological behavior of natural fiber reinforced PLA composites." *Wear* 297 (1–2). Elsevier: 829–40. doi:10.1016/j.wear.2012.10.019.

Biswas, Sandhyarani. 2017. "Primary manufacturing of thermosetting polymer matrix composites." In *Primary and Secondary Manufacturing of Polymer Matrix Composites*, edited by Kishore Debnath and Inderdeep Singh, pp. 1–16. Boca Raton, FL: CRC Press. doi:10.1201/9781351228466.

Biswas, Sandhyarani, and Jasti Anurag. 2019. "Fabrication of composite laminates." In *Reinforced Polymer Composites*, edited by Pramendra K. Bajpai and Inderdeep Singh, pp. 39–53. Weinheim, Germany: Wiley-VCH Verlag GmbH & Co. KGaA. doi:10.1002/9783527820979.ch3.

Chaitanya, Saurabh, and Inderdeep Singh. 2016a. "Novel aloe vera fiber reinforced biodegradable composites: Development and characterization." *Journal of Reinforced Plastics and Composites* 35 (19): 1411–23. doi:10.1177/0731684416652739.

Chaitanya, Saurabh, and Inderdeep Singh. 2016b. "Processing of PLA/sisal fiber biocomposites using direct and extrusion-injection molding." *Materials and Manufacturing Processes* 32 (05). Taylor & Francis: 468–74. doi:10.1080/10426914.2016.1198034.

Chaitanya, Saurabh, and Inderdeep Singh. 2018. "Ecofriendly treatment of aloe vera fibers for PLA based green composites." *International Journal of Precision Engineering and Manufacturing: Green Technology* 5 (1): 143–50. doi:10.1007/s40684-018-0015-8.

Chaitanya, Saurabh, Inderdeep Singh, and Jung Il Song. 2019. "Recyclability analysis of PLA/sisal fiber biocomposites." *Composites Part B: Engineering* 173 (May). Elsevier Ltd: 106895. doi:10.1016/j.compositesb.2019.05.106.

Cross, Julie O., Robert L. Opila, Ian W. Boyd, and Elton N. Kaufmann. 2015. "Materials characterization and the evolution of materials." *MRS Bulletin* 40 (12): 1019–33. doi:10.1557/mrs.2015.271.

Kennepohl, Dietmar, Steven Farmer, and William Reusch. 2020. "Infrared spectra of some common functional groups functional groups containing the C-O bond." https://chem.libretexts.org/Bookshelves/Organic_Chemistry/Map%3A_Organic_Chemistry_(McMurry)/12%3A_Structure_Determination_-_Mass_Spectrometry_and_Infrared_Spectroscopy/12.10%3A_Infrared_Spectra_of_Some_Common_Functional_Groups.

Komal, Ujendra Kumar, Vivek Verma, Tarachand Ashwani, Nitin Verma, and Inderdeep Singh. 2018a. "Effect of chemical treatment on thermal, mechanical and degradation behavior of banana fiber reinforced polymer composites." *Journal of Natural Fibers* 17 (7). Taylor & Francis: 1026–38. doi:10.1080/15440478.2018.1550461.

Komal, Ujendra Kumar, Vivek Verma, Tarachand Aswani, Nitin Verma, and Inderdeep Singh. 2018b. "Effect of chemical treatment on mechanical behavior of banana fiber reinforced polymer composites." *Materials Today: Proceedings* 5 (9–1). Elsevier Ltd: 16983–89. doi:10.1016/j.matpr.2018.04.102.

Komal, Ujendra Kumar, Manish Kumar Lila, and Inderdeep Singh. 2020. "PLA/banana fiber based sustainable biocomposites: A manufacturing perspective." *Composites Part B: Engineering* 180. Elsevier Ltd: 107535. doi:10.1016/j.compositesb.2019.107535.

Lila, Manish Kumar, Arpit Singhal, Sukhwant Singh Banwait, and Inderdeep Singh. 2018a. "A recyclability study of bagasse fiber reinforced polypropylene composites." *Polymer Degradation and Stability* 152. Elsevier Ltd: 272–79. doi:10.1016/j.polymdegradstab.2018.05.001.

Lila, Manish Kumar, Brijendra Singh, Bahadur Singh Pabla, and Inderdeep Singh. 2018b. "Effect of environmental conditioning on natural fiber reinforced epoxy composites."

Materials Today: Proceedings, 5. Elsevier Ltd: 17006–11. doi:10.1016/j.matpr.2018. 04.105.

Lila, Manish Kumar, Kartikeya Shukla, Ujendra Kumar Komal, and Inderdeep Singh. 2019a. "Accelerated thermal ageing behaviour of bagasse fibers reinforced poly (lactic acid) based biocomposites." *Composites Part B: Engineering* 156. Elsevier: 121–27. doi:10.1016/j.compositesb.2018.08.068.

Lila, Manish Kumar, Ujendra Kumar Komal, and Inderdeep Singh. 2019b. "Characterization techniques of reinforced polymer composites." In *Reinforced Polymer Composites: Processing, Characterization and Post Life Cycle Assessment*, edited by Pramendra K. Bajpai and Inderdeep Singh, pp. 119–45. Weinheim, Germany: Wiley-VCH Verlag GmbH & Co. KGaA. doi:10.1002/9783527820979.ch7.

Lila, Manish Kumar, Ujendra Kumar Komal, Yashvir Singh, and Inderdeep Singh. 2020. "Extraction and characterization of munja fibers and its potential in the biocomposites." *Journal of Natural Fibers*: 1–19. doi:10.1080/15440478.2020.1821287.

Maus, Andreas. 2008. "Characterization and analysis of polymers." *Macromolecular Chemistry and Physics* 209 (14). John Wiley & Sons: 1515. doi:10.1002/macp.200800286.

Melese, Kassahun Gashu, Tejas Pramod Naik, and Inderdeep Singh. 2020. "Adhesive joining of sisal/jute/hybrid composites with drilled holes in lap area." *Proceedings of the Institution of Mechanical Engineers, Part L: Journal of Materials: Design and Applications*. doi:10.1177/1464420720959808.

Perremans, Dieter, Koen Hendrickx, Ignaas Verpoest, and Aart Willem Van Vuure. 2018. "Effect of chemical treatments on the mechanical properties of technical flax fibres with emphasis on stiffness improvement." *Composites Science and Technology* 160: 216–23. doi:10.1016/j.compscitech.2018.03.030.

Pickering, Kim Louise, Aruan Efendy Mohd Ghazali, and Tan Minh Le. 2016. "A review of recent developments in natural fibre composites and their mechanical performance." *Composites Part A: Applied Science and Manufacturing* 83: 98–112. doi:10.1016/j.compositesa.2015.08.038.

Sain, Sunanda, Shubhalakshmi Sengupta, Abhirupa Kar, Aniruddha Mukhopadhyay, Suparna Sengupta, Tanusree Kar, and Dipa Ray. 2014. "Effect of modified cellulose fibres on the biodegradation behaviour of in-situ formed PMMA/cellulose composites in soil environment: Isolation and identification of the composite degrading fungus." *Polymer Degradation and Stability* 99 (1). Elsevier Ltd: 156–65. doi:10.1016/j.polymdegradstab.2013.11.012.

Selvey, Saxon, Erik W. Thompson, Klaus I. Matthaei, Rodney A. Lea, Michael G. Irving, and Lyn R. Griffiths. 2001. "β-actin: An unsuitable internal control for RT-PCR." *Molecular and Cellular Probes* 15 (5). Noyes Publications: 307–11. doi:10.1006/mcpr.2001.0376.

Stamm, Manfred (Ed.). 2008. "Polymer surface and interfaces characterization techniques." In *Polymer Surfaces and Interfaces: Characterization, Modification and Applications*, pp. 1–324. Berlin, Heidelberg: Springer Berlin Heidelberg. doi:10.1007/978-3-540-73865-7.

Sullivan, John Lorenzo. 1991. "Measurement of Composite Creep." *Experimental Techniques* 15 (5): 32–37. doi:10.1111/j.1747-1567.1991.tb01210.x.

Vonk, Chris G. 1973. "Computerization of Ruland's X-ray method for determination of the crystallinity in polymers." *Journal of Applied Crystallography* 6: 148–52. doi:10.1107/S0021889873008332.

4 Tensile Properties Analysis and Characterizations of Single Fiber and Biocomposites
A Weibull Analysis and Future Trend

Mohamad Zaki Hassan and
Mohamad Ikhwan Ibrahim
Universiti Teknologi Malaysia

SM Sapuan
Universiti Putra Malaysia

CONTENTS

4.1	Introduction	88
4.2	Vegetable Fiber Classifications	88
4.3	Tensile Properties of Single Fibers and Biocomposites	90
	4.3.1 Weibull Distribution Characteristics of Tensile Strength	92
4.4	Monotonic Tensile Properties of Single Fiber	92
	4.4.1 Effect of Gauge Length	92
	4.4.2 Effect of Chemical Treatment	94
4.5	Weibull Distribution of Biocomposites	95
4.6	Future Development on Weibull Statistical Distribution of Single Fiber and Biocomposite Properties	96
4.7	Conclusions	98
	Acknowledgments	99
	References	99

DOI: 10.1201/9781003137535-4

4.1 INTRODUCTION

Today, most engineering constructions and technologies are made up of composites. Composite is a mixture of two or more of individual materials with notably different physical or chemical characteristics. When combined, it forms a new material with properties that are far superior from their constituent behavior. Composite is generally composed of reinforcing fibers (dispersed phase) that are held within the matrix (continuous phase). Reinforcing fibers have stiffness in their longitudinal direction, and they can either be synthetic fibers, such as carbon [1], glass [2], and metal or natural fibers [3]. On the other hand, the matrix is robust and chemically inert as compared to fibers. Therefore, it works as a binder that holds the fibers in one place and eventually acts as a medium for transferring external forces to reinforcing fibers.

The excellent mechanical properties exhibited by synthetic fibers like high strength and stiffness make them suitable to be applied in many high-tech applications [4]. However, due to eco-friendly and green nature concepts, natural fiber-based composites are more preferred [5]. The utilization of natural fibers is a great approach for environmental protection due to their environmental-friendly properties such as biodegradable, renewable, sustainable, and abundantly available. In fact, the cost and energy needed to process natural fibers are much lower as compared to synthetic fibers. Late current industry research reported that in 2019, the worldwide market for cellulose fiber had achieved nearly USD10.5B, with 23% of compound annual growth rate within the last 5 years. Moreover, the market is expected to reach $36.3B by 2025 [6].

Natural fibers can be obtained from animal, plant or mineral resources. Among these resources, the combination of plant-based fibers with synthetic polymer matrices to form natural fiber composites (NFC) is extensively used. This consolidation has offered superior mechanical properties, high stiffness and is lightweight. The application of NFC in the automotive, building materials, furniture and wood plastic composite is a good demonstration to this affirmation. Table 4.1 lists the most common plant-based natural fibers and their world productions [7].

4.2 VEGETABLE FIBER CLASSIFICATIONS

Vegetable fibers are continuous filaments that are directly attainable from natural resources and can be converted into felt, yarns and woven fabric. Figure 4.1 illustrates the sources of natural fiber and classification of plant fibers. Vegetable fibers can be divided into five subgroups based on plant origin. These fibers mainly consist of polysaccharides cellulose, with various substances, including hemicellulose, aromatic polymer lignin, fat, pectins, and waxes that need to be removed or reduced while they are being processed as a composite reinforcement due to the adverse effect on fiber and matrix adhesion. Besides, vegetable fibers are obtained from different parts of the plant, such as fruit, stalk, bast, leaves and grass. As seen from Figure 4.1, cotton, milkweed, kapok and coir are classified as a seed-hair group, whereby each fiber contains long and narrow cell [8]. Many of them (kapok, milkweed, linter and dandelion) are not spun as yarn. However, they are used as an essential substance in mattresses and pillows. Bast fibers, i.e., ramie, hemp, jute [9], kenaf [10], banana [11,12] and ramie, are obtained from stems or dicotyledonous plants. In general,

TABLE 4.1
Natural Fibers and their World Production

Fiber Source	World Production (10³ ton)
Sugar cane bagasse	75,000
Bamboo	30,000
Jute	2300
Kenaf	970
Flax	830
Grass	700
Sisal	375
Hemp	214
Coir	100
Ramie	100
Abaca	70

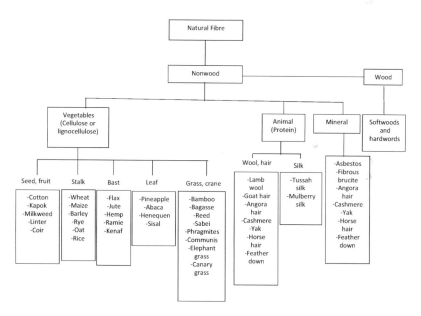

FIGURE 4.1 Classification of natural fibers. (From Petroudy S.D. Physical and mechanical properties of natural fibers. In M. Fan, F. Fu (eds.) *Advanced High Strength Natural Fibre Composites in Construction*. Amsterdam: Elsevier; 2017, pp. 59–83.) [8]

the fiber bundle is made up between 10 and 25 of elementary fibers, a diameter of 10–50 µm with an average length of 2–5 mm.

Leaf fibers are also referred to as hard fiber and have limited application because they are coarser and stiffer than bast fibers. The fibers are generally acquired from the fibrovascular system of leaves by mechanical equipment that removes the non-fibrous element. Sisal, abaca, pineapple and henequen are most commonly fibers of this group. Some natural fibers are obtained from the stalk, for example, corn, wheat, sorghum, rye, barley and oats. They are utilized as agro-based bio fibers and have potential use as textile, paper industry and biofuel applications. Grass and crane fibers are abundant and represent great potential. Predominantly, the elementary fiber cells of these fibers are bounded by pectin middle lamellae and the fibers bundle is separated with parenchyma cells. The most representative fibers in this group are bamboo [13], sugarcane bagasse, reed, sabei, elephant grass and canary grass.

4.3 TENSILE PROPERTIES OF SINGLE FIBERS AND BIOCOMPOSITES

Recently, the use of natural cellulose fibers is steadily increasing in application due to their affirmative in mechanical properties, low cost, high toughness, and high thermal resistance. The mechanical properties of the structure are closely related to the single fiber strength. The damage properties and strength behavior of single fiber in fibrous composites are also used in governing micromechanical damage modelling and prediction of product performance. Many attempts were made to evaluate the strength of a single fiber by using a universal testing machine (UTM). Table 4.2 shows the tensile properties of single fibers of selected fiber plants. Here, most experimental setups were conducted according to ASTM D3379-75 [14]. Figure 4.2 shows the experimental setup of a single fiber tensile test.

These investigations have mainly concentrated on three factors, namely, chemical treatment [15–17], cross-section morphology [18,19] and gauge length [20]. It can be seen from Table 4.2 that flax fiber offers the highest value of tensile strength, while coir shows the lowest value. The tensile value of each natural fibers is slightly different due to distinctness in microstructures that results from various species, and thus their growth cannot be fully controlled. Many parameters need to be considered while evaluating the tensile properties of single fiber.

Fidelis et al. [23] mentioned that tensile strength characters could be affected by the cross-sectional area, fiber flaws and morphology of structure. The weakest

TABLE 4.2
Average Value and Standard Deviation of Single Fiber Tensile Properties for Selected Natural Fibees

	Tensile Strength (MPa)	Tensile Modulus (GPa)	Strain (%)	References
Bamboo	262.21 ± 75.00	9.80 ± 1.60	2.70 ± 0.70	[15]
Kenaf	129.12 ± 40.00	9.02 ± 1.50	1.35 + 0.21	[18]
Hemp	277.45 ± 191.00	9.50 ± 5.70	2.3 + 0.80	[19]
Jute	450.21 ± 40.20	21.20 ± 1.90	1.7 + 1.00	[16]
Coir	195.21 ± 30.30	3.79 ± 0.30	40.95 + 20	[21]
Ramie	439.51 ± 114.91	37.18 ± 14.68	1.34 + 0.45	[20]
Sisal	447.20 ± 23.90	20.16 ± 3.52	3.00 + 1.00	[17]
Flax	945.00 ± 200.00	52.5 ± 8.60	2.07 + 0.45	[22]

Tensile Properties Analysis of Biocomposites

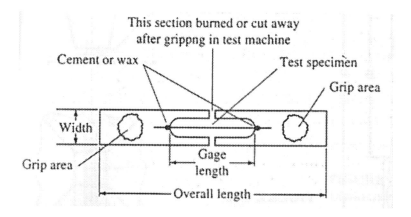

FIGURE 4.2 Schematic of the experimental set-up for a single fiber tensile test.

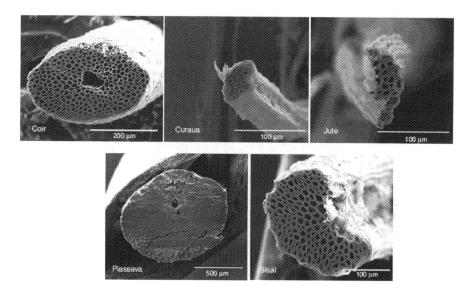

FIGURE 4.3 Microstructure of fibre cell with lumen of coir, curaua, jute, piassava and sisal [23], and temperatures.

point of the fiber is closely related to the distribution of flaws and at the smallest area or combination of both in the fiber volume. It was reported that the tensile strength is mainly influenced by the internal morphology of the fiber including real cross-sectional area, number of lumens, size of fiber cell and internal area of the lumen. Figure 4.3 shows the morphology of fiber cells corresponding to coir, curaua, jute, pissava and sisal that majorly affected the tensile properties of the fibers.

In general, the tensile strength of biocomposites was performed following ASTM D 3039. The stress-strain curves were obtained using the UTM. Extensive performance of the tensile parametric studies will be discussed in Section 4.5.

4.3.1 WEIBULL DISTRIBUTION CHARACTERISTICS OF TENSILE STRENGTH

In Table 4.1, a large inconsistency value of tensile strength and Young's modulus occurred in literature, in which their massive errors of measurement could arise in mechanical properties calculation. Due to that reason, a Weibull probability approach was promoted by researchers [18,23]. According to Weibull analysis, the probability distribution function of tensile strength, σ, is given by (4.1),

$$P(\sigma) = 1 - \exp\left[-V\left(\frac{\sigma}{\sigma_o}\right)^m\right] \qquad (4.1)$$

where $P(\sigma)$ is the probability of survival, V is the fiber volume, m is the Weibull modulus, σ and σ_0 are defined as the applied stress and characteristic strength. In general, the ranking of the fiber strengths is evaluated by using a probability index that is expressed by (4.2),

$$P(\sigma) = \frac{i - 0.3}{N + 0.4} \rightarrow \qquad (4.2)$$

where N is the total number of fibers tested. The prediction of tensile properties value can be determined by substituting Equation (4.2) into Equation (4.1) result,

$$\ln\ln\left[\frac{N+1}{N+1-i}\right] = m\ln\left[\frac{\sigma}{\sigma_o}\right] \qquad (4.3)$$

A corresponding curve of lnln $((N+1)/(N+1-i))$ versus ln (σ/σ_{0o}) yields the slope of the straight line equivalent to the m value.

4.4 MONOTONIC TENSILE PROPERTIES OF SINGLE FIBER

Table 4.3 shows the Weibull modulus and parametric study that is evaluated by using the tensile properties of the single fiber. As can be seen from Table 4.3, many researchers have mainly focused on the effect of the gauge length, chemical treatment and types of fibers on the tensile performance output characteristics.

4.4.1 EFFECT OF GAUGE LENGTH

Weibull probability statistic was used to evaluate the tensile strength of flax fiber by Zafeiropoulos et al. [24]. The test was examined for three different gauge lengths which were 5, 8, and 10 mm. Four types of probability index were also used to predict the probability of survival (P), Weibull modulus (m) and characteristic strength (σ_0) values. The test found that estimator, $P(\sigma) = (i - 0.5)/n$, yielded the highest value for Weibull modulus (m), while estimator $P(\sigma) = i/(n+1)$ offered the lowest value. In this study, linear regression method and the maximum likelihood method were also compared. It was suggested that both methods offered similar Weibull modulus (m) value.

Further, the probability of tensile strength of Vietnamese bamboo, coir fibers and Bangladesh's jute fibers was assessed by Defoirt et al. [25]. The variation in

TABLE 4.3
Reported Research on Weibull Modulus for the Range of Natural Fibers

Reference	Year	Fiber	Weibull Modulus (m)	Variable Parameter	Comments
Zafeiropoulos et al. [24]	2007	Flax	1.96–4.23	Gauge length	A similar value of Weibull modulus (m) obtained from maximum likelihood and the linear regression methods.
Defoirt et al. [25]	2010	Coir Bamboo Jute	5.8–6.0 3.5–9.3 2.7–4.6	Gauge length	Weibull modulus (m) deviated with various test length.
Fidelis et al. [23]	2013	Sisal Jute Curau Piassava Coir	3.70 2.74 2.22 3.68 2.74	Gauge length	Sisal fibers offered the massive Weibull modulus (m).
Guo et al. [26]	2014	Palm	5.42–8.56	Gauge length	Three-parameter Weibull obtained the lowest value of Weibull modulus (m)
Wang et al. [27]	2014	Bamboo	3.22–4.77	Gauge length	Modified Weibull distribution offered a close prediction value of tensile properties.
Naik et al. [28]	2016	Kenaf	2.42–6.91	Gauge length	Median rank, $P(\sigma) = i/(n+1)$ resulted in the good estimation of Weibull modulus, m.
Balaji et al. [29]	2017	Aloe vera	3.22–3.33	Chemical treatment	Weibull modulus (m) between 2 and 4 is reliable.
Senthmaraikannan et al. [30]	2018	*Coccinia grandis*	5.26–95.66	Chemical treatment	Huge range of Weibull modulus and unacceptable.
Tarres et al. [31]	2019	Henequen	2.60–2.88	Gauge length	Lower Weibull modulus (m) values indicate higher variability of the tensile strength.
Boumediri et al. [32]	2019	Date palm	3.3–6.6	Part of the palm tree	Fiber strands showed a lower Weibull modulus (m) in tensile strength and Young's modulus.
Kathiselvam et al. [33]	2019	Thespesia populnea	4.30–16.12	Chemical treatment	The highest value of Weibull modulus (m) for tensile strength and inconsistency was obtained.

(Continued)

TABLE 4.3 (*Continued*)
Reported Research on Weibull Modulus for the Range of Natural Fibers

Reference	Year	Fiber	Weibull Modulus (*m*)	Variable Parameter	Comments
Belaadi et al. [34]	2020	Flax	1.10–4.50	Chemical treatment	Two-parameter Weibull analysis to determine the values of the mechanical properties was proposed.
Hashim et al. [35]	2020	Kenaf	2.52–3.02	Chemical treatment	A low value of Weibull modulus (*m*) is due to non-homogenous fiber.

gauge length between 5 and 35 mm of those fibers was evaluated by using normal and Weibull distribution methods. It was suggested that the Weibull modulus (*m*) value for natural fiber was obtained between 1 and 6. These values were altered significantly for the various natural fiber lengths, but between 5 and 15 for the synthetic fibers. Also, the monotonic tensile behavior of palm leaf sheath fiber was conducted at four different gauge lengths by Guo et al. [26]. The study mentioned that the R^2 value of the three-parameter Weibull model was higher than those of the two-parameter model, suggesting that the three-parameter Weibull model was more reliable for evaluating the distribution of tensile properties. Wang et al. [27] evaluated the variability in tensile strength and its dependence on the length of bamboo fibers. The probability of fiber strengths was compared between the Weibull distribution model and modified Weibull model for fiber lengths ranging that ranged from 20 to 60 mm. It was shown that the modified Weibull distribution model offered a close prediction value of tensile properties and the value of the Weibull modulus extent was found from 3.22 to 4.77. This Weibull modulus showed the variability in fiber strength. A large Weibull modulus value refers to low variability strength. This can be believed as the flaw sizes and distributions may vary along the fiber and from fiber to fiber [23]. To evaluate Weibull modulus (*m*) and scale parameters (σ_0), the effect of the gauge length on the tensile modulus of the kenaf fiber by using Weibull distribution model was evaluated by Naik et al. [28]. The value of Weibull modulus (*m*) fluctuated between 2.42 and 6.91, which explained the variation in Young's modulus values. Experiment results showed that the estimator, $P(\sigma) = i/(n+1)$, resulted in the good estimation of Weibull modulus (*m*) for kenaf fibers. Another study on the effect gauge length on tensile strength property of henequen fibers was performed by Tarres et al. [31]. The study obtained a small value of Weibull modulus, m, resulting significantly in a higher distribution variability of tensile properties.

4.4.2 Effect of Chemical Treatment

Balaji et al. [29] obtained the Weibull parameter (*m*) and characteristic strength (σ_0) values that fit with the distribution of the experimental data by using two-parameter Weibull statistical models. Initially, the aloe-vera fibers were treated with 5% of sodium hydroxide (NaOH) solution. A scatter value of Weibull modulus (*m*) between 3.22 and 3.33 was recorded. Weibull modulus (*m*) between 2 and 4 was acceptable. Again, higher Weibull modulus (*m*) value is closely related to a lower variation in shape parameter character in a normal distribution. The tensile properties of raw and alkali-treated *Coccinia grandis* fibers were characterized by using the Weibull distribution method by Senthmaraikannan et al. [30]. A large value of Weibull modulus (*m*) was observed in these series of tensile tests. The Weibull modulus (*m*) value for the effect of diameter in the range between 80.60 and 95.66 was unsatisfactory. It was massively larger than Weibull modulus (*m*) range value for natural fiber than that mentioned by other researchers.

Recently, the effect of alkali treatment under various concentrations and soaking period was chosen as a *Thespesia populnea* bark fiber treatment method and a probability of survival of tensile strength was used by using a two-parameter Weibull distribution statistical approach [33]. It was calculated that Weibull modulus (m) for properties like diameter, strain and Young's modulus for tensile strength ranged between 4.3–4.6, 4.5–6.9, and 5.9–6.4, respectively. In addition, the varied data plot of fiber characteristics nearly fitted inside the Weibull distribution curve with 95% confidence level.

Recently, Belaadi et al. [34] examined the quasi-static tensile mechanical properties of a single flax fiber that was immersed in sodium bicarbonate solution of different concentrations (5%, 10% and 20%) by using Weibull statistical analysis. The authors measured that the tensile property values of the flax fibers obtain a large dispersion of Weibull modulus was influenced by different concentrations of sodium bicarbonate. They suggested that a two-parameter Weibull analysis to be a better choice to predict the values of the mechanical properties of a single fiber. Hashim et al. [35] characterized the monotonic tensile behavior of kenaf fiber that was treated with different concentrations of sodium hydroxide using Weibull probability approach. They found that the low value of Weibull modulus (*m*) was due to non-homogenous fiber. Fiber with large a cross-sectional area potentially has a large flaw as compared to a smaller one. They observed that the Weibull modulus range between 2.40 and 3.12.

4.5 WEIBULL DISTRIBUTION OF BIOCOMPOSITES

Towo et al. [36] evaluated the effect of chemical treatment on the sisal-fiber-reinforced polyester interface to ascertain the variability in tensile properties and interfacial strength using Weibull probability of failure. Here, the alkaline treatment decreases variability and improves shear strength; however, inconsistent values of shear strength were recorded. It has been shown that Weibull probability plots reduce the variations in the value of tensile strength due to the surface flaws. Weibull probability analysis offers an outstanding correlation between the shear strength and surface modification in droplet pull-out tests. However, fiber strength is slightly decreased due to excessive chemical treatment at the fiber surface. Suzuki [37] also used a two-parameter Weibull plot to evaluate the tensile strength of kenaf-reinforced polylactic acid (PLA). He used scanning electron microscopy to examine the correlation of the mechanical properties of the surface fracture initiation and propagation in the present composites. A huge bimodal distribution of Weibull modulus (m) values at a different level of volume fraction suggested that complicated and multiple effects in the surface fractography of the composites were obtained (Table 4.4).

Further, the tensile properties under extreme conditions including temperatures and impact and ballistic loadings of the basalt-fiber-reinforced polymer were evaluated by Zhang et al. [38]. In this study, the effects of strain rate sensitivity and elevated temperature on the tensile properties of a biocomposite laminate were characterized by a two-parameter Weibull distribution. The Weibull modulus (m) measured under quasi-static loading was higher than those for high strain rates. They revealed that the larger variabilities in dynamic tests resultant from the imperfections in yarn structure and the presence of flaws during processing. On the other hand, less scattered values of the Weibull modulus (m) were recorded under different temperatures except at 100°C. Closely similar work was also conducted by Ou et al. [39]. Basal yarn and basalt-fiber-reinforced epoxy (BFRP) coupons were examined under four different strain rates and four distinct temperatures using a floor-standing drop-weight impact tower. This found that Young's modulus increases with increasing strain rate but decreases with increasing temperature. Again, the Weibull modulus (m) of the biocomposite obtained from the lower strain rate was higher than those from dynamic loading. Figure 4.4 shows the cumulative failure probability against the tensile strength of (a and b) basalt yarn and (c and d) basalt-reinforced composites under different strain rates and temperatures. These plots were used to determine the characteristic strength (σ) and Weibull modulus (m).

Graupner et al. [40] examined the effects of gauge length on ductile regenerated cellulose fiber epoxy composites with different diameters. It was mentioned that the two-parameter, least-squares, Weibull statistics applicable for the prediction of the strength value accuracy is only sufficient at small sample volumes.

The tensile properties of a roselle-fiber-reinforced vinyl ester composite fabricated using the hand layup technique was evaluated by Navaneethakrishnan et al. [41] This mechanical property of the composite was also analyzed using two-parameter Weibull distribution. They concluded that using the two-parameter Weibull distribution to express the tensile strength distribution was reasonably well predicted. However, the Weibull modulus (m) value measured from this study is 22 shows a higher probability of the composite to fail for every unit increase in applied load. Further, the discreteness of the failure strength at various strain-rates is evaluated by Xu et al. [42] using Weibull distribution. The experimental investigation proved that the basalt-fiber-reinforced epoxy tendons are highly influenced by the strain rate. Two-parameter Weibull distribution is suitable to characterize the mechanical properties of NFC tendons in structural engineering applications.

4.6 FUTURE DEVELOPMENT ON WEIBULL STATISTICAL DISTRIBUTION OF SINGLE FIBER AND BIOCOMPOSITE PROPERTIES

The two-parameter Weibull distribution model is widely employed in statistical reliability analysis. However, many attempts were made to improve the prediction of

TABLE 4.4
Reported Research on the Weibull Modulus for a Range of Biocomposites

References	Year	Fiber	Matrix	Weibull Modulus (m)	Variable Parameter	Comments
Towo et al. [36]	2005	Sisal	Polyester	2.7–5.7	Chemical treatment	The Weibull modulus (m) value for the treated fibers is higher, as compared to untreated fibers. This reflects the reduction in flaws due to surface treatment of the fiber.
Suzuki [37]	2013	Kenaf	PLA	5.0–76.9	Volume fraction	The bimodal distribution of the Weibull modulus (m) probably reflects complicated and multiple sources for the fracture initiation and propagation present in the composites.
Zhang et al. [38]	2016	Basalt	Epoxy	8.4–51.4	Strain rate and temperature	The Weibull modulus obtained from quasi-static loading is much higher than those for high strain rates.
Ou et al. [39]	2016	Basalt	Epoxy	13.3–23.6	Strain rate and temperature	The shape parameter (m) of basalt composites obtained from a lower strain rate is higher than those from higher strain rates.
Graupner et al. [40]	2018	Regenerated cellulose fibre	Epoxy	6.1–17.3	Gauge length	A large number of samples leads to a clear overestimation of experimental strength values.
Navaneethakrishnan et al. [41]	2019	Roselle	Vinylester	22.3	Fiber content	Two-parameter Weibull distribution can be used to express the tensile strength and predict its values accurately.
Xu et al. [42]	2020	Basalt	Epoxy	8.2–22.9	Strain rate	Two-parameter Weibull distribution is suitable to characterize the mechanical properties of NFC tendon in structural engineering applications.

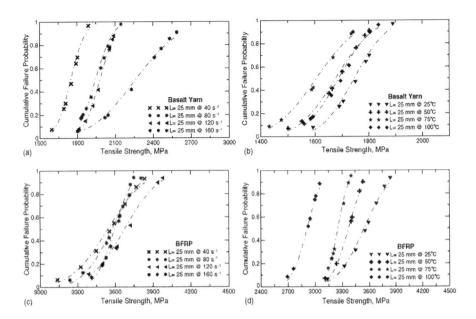

FIGURE 4.4 Cumulative failure probability against tensile strength of (a and b) basalt yarn and (c and d) basalt reinforced composites under different strain rates.

strength parameters. For example, least-squares estimation [40], method of moments (MoM) [43], and maximum likelihood estimation (MLE) [44] are often proposed to evaluate the distribution of survival rate parameters.

Besides the two-parameter Weibull distribution used, the three-parameter Weibull distribution is also applied to provide a robust and versatile in the characterization of the tensile strength data. Several studies have also attempted to compare the survival probability distribution between the two-parameter and three-parameter Weibull models based on the goodness-of-fit of the distribution by using Chi-square [45], Kolmogorov–Smirnov [46] and Anderson–Darling [47] tests. These goodness-of-fit tests calculated the square of the error between the actual data and hypothesized prediction at the tails portion of the distribution curve. It is a statistical hypothesis test to see how well the sample data fit a distribution from a population with a normal distribution.

A variety of probability index is employed to estimate the probability density function, Weibull modulus (m) and characteristic strength, (σ_0). These probability indices or also known as estimators are used to keep the statistics unbiased in comparison with the variability in the experimental strength of the material [48].

4.7 CONCLUSIONS

Natural fibers are fibers that are produced by plant or vegetables, animal-based or protein-based and mineral. In this review, the monotonic tensile behavior of natural fibers was explored. The variability of fibers strength that was significantly due to fiber cells, size of cell walls and morphology fiber was evaluated by using Weibull statistics. In general, the Weibull parameter (m) value for natural fiber was varied between 1 and 4. However, a huge standard deviation of m values for the biocomposite sample was obtained. Currently, several models were developed for formulating and improving the prediction of probability density function parameters.

ACKNOWLEDGMENTS

The authors wish to thank Universiti Teknologi Malaysia for supported extensively in order to complete the course of this research project. This work was funded by Universiti Teknologi Malaysia throughout the Contract Research (DTD), Q.K.130000.7656.4C342, Contract Research (DTD), Q.K.130000.7656.4C343, and "Geran Universiti Penyelidik" UTMFR Q.K.130000.2656.21H13 scheme.

REFERENCES

1. Hassan M.Z., Sapuan S., Rasid Z.A., Nor A.F.M., Dolah R., Md Daud M.Y. Impact damage resistance and post-impact tolerance of optimum banana-pseudo-stem-fiber-reinforced epoxy sandwich structures. *Applied Sciences.* 2020;10(2):684.
2. Zhou J., Hassan M.Z., Guan Z., Cantwell W.J. The low velocity impact response of foam-based sandwich panels. *Composites Science and Technology.* 2012;72(14):1781–90.
3. Roslan S., Hassan M., Rasid Z., Zaki S., Daud Y., Aziz S., et al. Mechanical properties of bamboo reinforced epoxy sandwich structure composites. *International Journal of Automotive & Mechanical Engineering.* 2015;12:2882–92.
4. Muhalim N., Hassan M., Daud Y. Elastic-plastic behaviour of ultrasonic assisted compression of polyvinyl chloride (PVC) foam. *IOP Conference Series: Materials Science and Engineering.* 2018;344(1):012009.
5. Hassan M.Z., Roslan S.A., Sapuan S., Rasid Z.A., Mohd Nor A.F., Md Daud M.Y., et al. Mercerization optimization of bamboo (Bambusa vulgaris) fiber-reinforced epoxy composite structures using a box–behnken design. *Polymers.* 2020;12(6):1367.
6. https://www.reportsinsights.com/industry-forecast/cellulose-fiber-market-analysis-1987. Cellulose Fiber Market to 2025- Global Analysis, Industry Growth, Regional Share, Trends, Competitor Analysis: The Rhode Island Publications Society; Jan 2020.
7. Fan M., Fu F. *Advanced High Strength Natural Fibre Composites in Construction.* Sawston: Woodhead Publishing; 2016.
8. Petroudy S.D. Physical and mechanical properties of natural fibers. In M. Fan, F. Fu (eds.) *Advanced High Strength Natural Fibre Composites in Construction.* Amsterdam: Elsevier; 2017, pp. 59–83.
9. Bakhori S.N.M., Hassan M.Z., Fadzulah S.H.S.M., Ahmad F. Tensile properties of jute-polypropylene composites. *Chemical Engineering Transactions.* 2018;63:727–32.
10. Bakhori S., Hassan M., Daud Y., Sarip S., Rahman N., Ismail Z., et al. Mechanical behavior of Kenaf/Epoxy corrugated sandwich structures. *IOP Conference Series Materials Science and Engineering.* 2015;100(1):012011.
11. Hassan M.Z., Roslan S.A., Rasid Z.A., Yusoff M.Z.M., Sapuan S., Muhammad-Sukki F. Optimizing the mercerisation effect on the mode I fracture toughness of bambusa vulgaris bamboo using surface response method. In S. M. Sapuan, M. T. Mastura (eds.) *Implementation and Evaluation of Green Materials in Technology Development: Emerging Research and Opportunities.* Hershey, PA: IGI Global; 2020, pp. 112–29.

12. Hassan M., Rasid Z. Thermal degradation and mechanical behavior of banana pseudo-stem reinforced composite. *International Journal of Recent Technology Engineering.* 2019;8(4):5889–5902.
13. Roslan S.A., Hassan M.Z., Rasid Z.A., Bani N.A., Sarip S., Daud M.Y.M., et al. Mode I fracture toughness of optimized alkali-treated bambusa vulgaris bamboo by box-behnken design. In M. Awang, S. Emamian, F. Yusof (eds.) *Advances in Material Sciences and Engineering.* Singapore: Springer; 2020. p. 565–75.
14. Standard A. D3379-75, "Standard Test Method for Tensile Strength and Young's Modulus for High-Modulus Single-Filament Materials"; ASTM International: West Conshohocken, PA, 1975.
15. Zhang K., Wang F., Liang W., Wang Z., Duan Z., Yang B. Thermal and mechanical properties of bamboo fiber reinforced epoxy composites. *Polymers.* 2018;10(6):608.
16. Chen H., Cui Y., Liu X., Zhang M., Hao S., Yin Y. Study on depositing SiO2 nanoparticles on the surface of jute fiber via hydrothermal method and its reinforced polypropylene composites. *Journal of Vinyl and Additive Technology.* 2020;26(1):43–54.
17. Ferreira S.R, de Andrade Silva F., Lima P.R.L., Toledo Filho R.D. Effect of hornification on the structure, tensile behavior and fiber matrix bond of sisal, jute and curauá fiber cement based composite systems. *Construction and Building Materials.* 2017;139:551–61.
18. Ibrahim M.I., Dolah R., Zuhri M., Sharif S., Hassan M. Chemical treatment evaluation of tensile properties for single Kenaf fiber. *Journal of Advanced Research in Applied Mechanics.* 2016;32:9–14.
19. Shahzad A. A study in physical and mechanical properties of hemp fibres. *Advances in Materials Science and Engineering.* 2013;2013: 325085.
20. Sarasini F., Tirillò J., Puglia D., Dominici F., Santulli C., Boimau K., et al. Biodegradable polycaprolactone-based composites reinforced with ramie and borassus fibres. *Composite Structures.* 2017;167:20–9.
21. Biswas S., Ahsan Q., Cenna A., Hasan M., Hassan A. Physical and mechanical properties of jute, bamboo and coir natural fiber. *Fibers and Polymers.* 2013;14(10):1762–7.
22. Baley C., Bourmaud A. Average tensile properties of French elementary flax fibers. *Materials Letters.* 2014;122:159–61.
23. Fidelis M.E.A., Pereira T.V.C., da Fonseca MartinsGomes O., de Andrade Silva F., Toledo Filho R.D. The effect of fiber morphology on the tensile strength of natural fibers. *Journal of Materials Research and Technology.* 2013;2(2):149–57.
24. Zafeiropoulos N., Baillie C. A study of the effect of surface treatments on the tensile strength of flax fibres: Part II. Application of Weibull statistics. *Composites Part A: Applied Science and Manufacturing.* 2007;38(2):629–38.
25. Defoirdt N., Biswas S., De Vriese L., Van Acker J., Ahsan Q., Gorbatikh L., et al. Assessment of the tensile properties of coir, bamboo and jute fibre. *Composites Part A: Applied Science and Manufacturing.* 2010;41(5):588–95.
26. Guo M., Zhang T., Chen B., Cheng L. Tensile strength analysis of palm leaf sheath fiber with Weibull distribution. *Composites Part A: Applied Science and Manufacturing.* 2014;62:45–51.
27. Wang F., Shao J. Modified Weibull distribution for analyzing the tensile strength of bamboo fibers. *Polymers.* 2014;6(12):3005–18.
28. Naik D.L., Fronk T.H. Weibull distribution analysis of the tensile strength of the kenaf bast fiber. *Fibers and Polymers.* 2016;17(10):1696–701.
29. Balaji A., Nagarajan K. Characterization of alkali treated and untreated new cellulosic fiber from Saharan aloe vera cactus leaves. *Carbohydrate Polymers.* 2017;174:200–8.
30. Senthamaraikannan P., Kathiresan M. Characterization of raw and alkali treated new natural cellulosic fiber from Coccinia grandis. L. *Carbohydrate Polymers.* 2018;186:332–43.
31. Tarrés Q., Vilaseca F., Herrera-Franco P.J., Espinach F.X., Delgado-Aguilar M., Mutjé P. Interface and micromechanical characterization of tensile strength of bio-based

composites from polypropylene and henequen strands. *Industrial Crops and Products.* 2019;132:319–26.
32. Boumediri H., Bezazi A., Del Pino G.G., Haddad A., Scarpa F., Dufresne A. Extraction and characterization of vascular bundle and fiber strand from date palm rachis as potential bio-reinforcement in composite. *Carbohydrate Polymers.* 2019;222:114997.
33. Kathirselvam M., Kumaravel A., Arthanarieswaran V., Saravanakumar S. Characterization of cellulose fibers in Thespesia populnea barks: Influence of alkali treatment. *Carbohydrate Polymers.* 2019;217:178–89.
34. Belaadi A., Amroune S., Bourchak M. Effect of eco-friendly chemical sodium bicarbonate treatment on the mechanical properties of flax fibres: Weibull statistics. *The International Journal of Advanced Manufacturing Technology.* 2020;106(5–6):1753–74.
35. Hashim M.Y., Amin A.M., Marwah O.M.F., Othman M.H., Hanizan N.H., Norman M.K.E. Two parameters Weibull analysis on mechanical properties of kenaf fiber under various conditions of alkali treatment. *International Journal of Integrated Engineering.* 2020;12(3):245–52.
36. Towo A.N., Ansell M.P., Pastor M.-L., Packham D.E. Weibull analysis of microbond shear strength at sisal fibre–polyester resin interfaces. *Composite Interfaces.* 2005;12(1–2): 77–93.
37. Suzuki K. A study on mechanical properties of short kenaf fiber reinforced polylactide (PLA) composites. *Journal of Solid Mechanics and Materials Engineering.* 2013;7(3): 439–54.
38. Zhang H., Yao Y., Zhu D., Mobasher B., Huang L. Tensile mechanical properties of basalt fiber reinforced polymer composite under varying strain rates and temperatures. *Polymer Testing.* 2016;51:29–39.
39. Ou Y., Zhu D., Li H. Strain rate and temperature effects on the dynamic tensile behaviors of basalt fiber bundles and reinforced polymer composite. *Journal of Materials in Civil Engineering.* 2016;28(10):04016101.
40. Graupner N., Basel S., Müssig J. Size effects of viscose fibres and their unidirectional epoxy composites: Application of least squares Weibull statistics. *Cellulose.* 2018;25(6):3407–21.
41. Navaneethakrishnan S., Sivabharathi V., Ashokraj S. Weibull distribution analysis of roselle and coconut-shell reinforced vinylester composites. *Australian Journal of Mechanical Engineering.* 2019:1–10. doi:10.1080/14484846.2019.1637089.
42. Xu X., Rawat P., Shi Y., Zhu D. Tensile mechanical properties of basalt fiber reinforced polymer tendons at low to intermediate strain rates. *Composites Part B: Engineering.* 2019;177:107442.
43. Zakaria M., Crosky A., Beehag A. Weibull probability model for tensile properties of kenaf technical fibers. *AIP Conference Proceedings.* 2018;2030(1):020015.
44. Li X., Wang F. Effect of the statistical nature of fiber strength on the predictability of tensile properties of polymer composites reinforced with bamboo fibers: Comparison of Linear-and Power-law Weibull models. *Polymers.* 2016;8(1):24.
45. Bourahli M. Uni-and bimodal Weibull distribution for analyzing the tensile strength of Diss fibers. *Journal of Natural Fibers.* 2018;15(6):843–52.
46. de Oliveira D.N., Claro P.I., de Freitas R.R., Martins M.A., Souza T.M., da S. e Silva B.M., et al. Enhancement of the Amazonian Açaí Waste Fibers through Variations of Alkali Pretreatment Parameters. *Chemistry and Biodiversity.* 2019;16(9):e1900275.
47. dos Santos J.C., de Oliveira L.Á., Vieira L.M.G., Mano V., Freire R.T., Panzera T.H. Eco-friendly sodium bicarbonate treatment and its effect on epoxy and polyester coir fibre composites. *Construction and Building Materials.* 2019;211:427–36.
48. Amroune S., Belaadi A., Bourchak M., Makhlouf A., Satha H. Statistical and experimental analysis of the mechanical properties of flax fibers. *Journal of Natural Fibers.* 2020: 1–15. doi: 10.1080/15440478.2020.1775751.

5 Crashworthiness Measurement on Axial Compression Loading of Biocomposite Structures
Prospect Development

Mohamad Zaki Hassan Zainudin A. Rasid and Rozzeta Dolah
Universiti Teknologi Malaysia

SM Sapuan
Universiti Putra Malaysia

Siti Hajar Sheikh Md. Fadzullah
Universiti Teknikal Malaysia Melaka

CONTENTS

5.1 Introduction	103
5.2 Crashworthiness Characteristics	104
5.3 Compression Behavior of Natural Fiber Composite Tubes	107
5.3.1 Effect of Tube Dimensions	107
5.3.2 Hybridization	109
5.3.3 Triggering	109
5.4 Failure Modes of Composite Tubes	110
5.5 Finite Element Analysis and Future Trends	111
5.6 Conclusions	120
Acknowledgments	120
References	120

5.1 INTRODUCTION

Biocomposites are increasingly attracting the attention of researchers due to benefits of low density per unit volume, recyclability, abundant availability, high specific strength to weight ratios, and renewable and degradable characteristics. Also, these

DOI: 10.1201/9781003137535-5

composites are a possible candidate for manufacturing highly durable consumer products including automotive compartments [1], boards for musical instruments [2], decking products [3], and composite flooring [4] which are easily recyclable. Compared to synthetic fibers, these fibers have a low emission level of toxic fumes when subjected to heat. However, they have a lower thermostability than glass fiber and putrefy at above 240°C [5], which is slightly higher than the processing temperature. Here, the use of natural fibers in a composite laminate reduces the flame retardant characteristic of the structures.

Despite their outstanding recyclability, natural fibers generally exhibit poor interfacial bonding due to the hydrophilic/hydrophobic effects of the fiber-matrix interaction, weak moisture resistance, and low durability, which tends to reduce their mechanical properties so hindering their long-term technological uses. In recent years, techniques to increase the interfacial bonding performance between vegetable fibers and polymer matrices have been developed. Among these, alkaline treatment is the most applied and low-cost technique as it not only improves the surface bonding performance of fiber-matrix interactions but also increases the final mechanical behaviors [6–8]. To reduce the potential damage associated with moisture sorption, hydrophobization treatments of vegetable fiber composites likely to experience environmental exposure have been promoted. In general, for natural fiber composites in service with a long-term exposure to moisture and an outdoor environment, there often appears detrimental effects such as strength reduction, structure color fading, distortion, mold growth, fungal fiber-decay, and warpage. The use of silane coupling agents including c-methacryloxypropyltrimethoxy silane, vinyltrimethoxy silane, and c-glycidoxypropyltrimethoxy silane to treat the fibers is the most suggested method [9]. Those chemical treatment techniques also improved the durability of the composites including their impact strength, flexural, and mechanical properties [10].

During a crash event, total energy absorbed by a thin wall tube structure is crucial. Some structural components must be able to sustain abnormal loading to meet stringent integrity requirements. Simultaneously, other components need to dissipate impact energy in a controlled manner that allows reducing a vehicle's speed to a required safety limit. Therefore, materials and designs have been developed that can combat wave, blast, and impact events to ensure the safety of people, machines, or equipment. Figure 5.1 shows the experimental work and numerical simulation of the crush tube on a military aircraft seat for infantry [11]. It was found that the crashworthiness behavior of the crush tube is useful for human protection and survivability.

Currently, the thin wall structures that are consolidated from natural fiber composites have been highlighted for their crashworthiness qualities. In this review, the characteristics of crashworthiness, the energy absorbed, failure modes, and a finite element analysis of composite tubes are highlighted.

5.2 CRASHWORTHINESS CHARACTERISTICS

Currently, the use of high-performance and lightweight materials for dynamic energy absorption and crashworthiness has intentionally increased, particularly in the construction and transportation industries. The study of crashworthiness focuses on the evaluation of structural integrity to combat the enormous or catastrophic damage

Crashworthiness Measurement

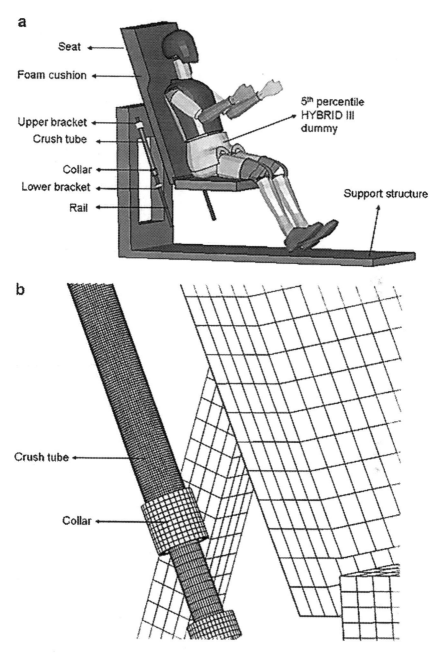

FIGURE 5.1 (a) Design of the military aircraft seat; (b) close up of the steel collar and crush tube. (From Tabiei A., Nilakantan G. Axial crushing of tubes as an energy dissipating mechanism for the reduction of acceleration induced injuries from mine blasts underneath infantry vehicles. *International Journal of Impact Engineering.* 2009;36(5):729–36.) [11]

of a collision and protecting human beings from its effects. Many researchers have studied the crashworthiness capability of fiber-reinforced composites and their beneficial properties including weight reduction, stiffness-to-weight ratio, and durability. It has been shown that these materials are able to absorb a massive amount of energy through structural deformation and collapsing progressively in a controlled manner [12]. Currently, the majority of experimental works were conducted for quasi-static [13–15], dynamic [14], and ultrasonic [15–17] conditions.

In general, the crashworthiness parameters of a material are measured from the load-displacement curve. Figure 5.2 shows the typical load-displacement of composites following quasi-static loading [18]. It can be divided into three distinct zones: the pre-crushing region which indicates the elastic crushing behavior, followed by the post-crushing zone where the failure spreads throughout the whole structure. The majority of the energy absorption capacity of the system occurs in this average crushing region. The final zone is then categorized as the compaction region, the final deformation phase that involves the densification of the structures, and a sharp increase in the compression loading.

In addition, the crashworthiness characteristics of the structures can be analyzed using eight important parameters as follows:

i. Peak load, F_{max} (kN), obtained from maximum loading in an elastic region.
ii. Post-crushing deformation, δ (mm), referring to the total displacement in the average crushing region.
iii. Maximum compression strength, σ_{max} (MPa), calculated from peak load, F_{max} (kN), over nominal cross-sectional area A (m^2),
iv. Absorbed energy, E (J), measured by the area under the load-displacement curve.

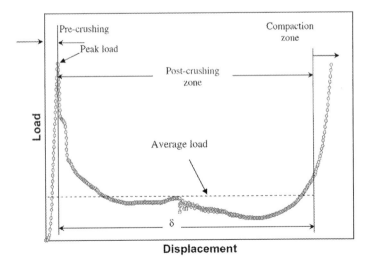

FIGURE 5.2 Typical load-displacement curve for composite structures. (From Ataollahi S., Taher S.T., Eshkoor R.A., Ariffin A.K., Azhari C.H. Energy absorption and failure response of silk/epoxy composite square tubes: Experimental. *Composites Part B: Engineering.* 2012;43(2):542–8.) [18]

Crashworthiness Measurement

v. Specific energy absorption (SEA) (J/kg) calculated as the ratio of the absorbed energy to the mass of the composite mass.

vi. Stroke efficiency (SE) (mm/mm), evaluated from the ratio of the total post-crushing deformation, δ (mm), to the specimen's length, L (mm).

vii. Average crushing load, $\bar{F}$ (kN), obtained from

$$\bar{F} = \frac{\int_0^\delta F d\delta}{\delta} \tag{5.1}$$

where F (kN) is the load and δ (mm) is the post-crushing displacement.

viii. Crash force efficiency (CFE) (kN/kN) referring to the ratio of the average crushing load, $\bar{F}$, to the peak load, F_{max}.

5.3 COMPRESSION BEHAVIOR OF NATURAL FIBER COMPOSITE TUBES

Based on previous studies, many attempts have been made to evaluate the crashworthiness of natural fiber composites. Table 5.1 illustrates the energy absorption performance characteristics of natural fiber tubes that were established by previous researchers. Some design strategies and configurations of the tubes have been implemented to enhance the energy absorption applications. As can be seen from Table 5.1, the dimensions of the composite tube including height and thickness have been of interest to researchers, while a few studies have considered the shape of the tube, any hybridization, and the type of cutting trigger as the main variable parameters.

5.3.1 Effect of Tube Dimensions

Extensive experimental work has been evaluated by researchers concerning the various structural dimensions of the composite tubes. The effect of tube length on the energy-absorbing capability of the ramie epoxy composite was investigated by Ghoushji et al. [19]. It was found that the values of peak load dropped significantly with the increase in the tube length, thus demonstrating that the shorter coupon exhibited high resistance to peak load when subjected to compression loading. Ding et al. [20] mentioned that a slenderness ratio of the composite structure that was lower than six contributed to the non-buckling effect and corresponded to the vertical micro-cracks on the outer surface leading to failure at maximum compression strength. In their study, the failure of a basalt fiber composite tube was initiated at the bottom surface when the fiber reached the maximum buckling load of the structure [20]. The larger load drop was also reported by Eshkoor et al. [21] when the span of the tube increased from 50 to 120 mm following the compression test of the silk/epoxy rectangular composites. Also, Yan et al. [22] stated that there was no clear relationship between the peak load and the diameter of the tube. However, they clarified that the maximum load was greatly affected by a combination of the thickness and the slenderness ratio of the composite flax/epoxy tube.

TABLE 5.1
Reported Research into the Crashworthiness Behavior of a Range of Natural Fibers

Reference	Year	Natural Fiber	Hybrid	Matrix	Shape	Height (mm)	Thickness/ Internal Diameter (mm)	Trigger	Peak load (kN)	Max. SEA (kJ/kg)
Eshkoor et al. [21]	2014	Silk	–	Epoxy	□	50–120	3.3	Flat surface 45° chamfer metallic trigger	~78.00	~10.00
Yan et al. [22]	2014	Flax	–	Epoxy	○	96 & 129	Ø64–86	Flat surface 45° chamfer	159.20	28.80
Ghoushji et al. [19]	2017	Ramie	–	Epoxy	□	50–120	1.8–4.5	Flat surface	140.70	15.70
Ude et al. [23]	2017	*B. Mori* Silk	–	Epoxy	○	–	33.3–80.6	Flat surface	135.80	~10.00
Albahash et al. [24]	2017	Jute	Glass Kevlar	Epoxy	○□	200	Ø50	Flat surface	24.86	26.51
Attia et al. [25]	2018	Jute	Glass Carbon	Epoxy	○	60	3.24–5.33	Flat surface	20.30	25.12
Sivagurunathan et al. [26]	2018	Jute	–	Epoxy	○	100	Ø44.3	Flat surface 45° chamfer Tulip 90°	~33.00	~32.00
Ding et al. [20]	2018	Basalt	–	Vinyl Ester	○	60–700	2	Flat surface	39.80	–
Mahdi et al. [27]	2019	Date Palm	–	Epoxy	□	100	3	Flat surface	8.90	10.30
Supian et al. [28]	2019	Kenaf	Glass	Epoxy	○	100	7.08–7.68	Flat surface	58.05	38.08
Mache et al. [29]	2020	Jute	–	Polyester	□	75	2.00–4.00	Flat surface	29.00	25.20
Zhu et al. [30]	2020	Basalt	–	Epoxy	○	30	Ø30	Flat surface	10.58	–
Kumar et al. [31]	2020	Kenaf	Glass	Epoxy	○	50	–	Flat surface	12.2	5.1
Tomlinson et al. [32]	2020	Flax	–	Epoxy pine resin	○	300	2.13–7.02	Flat surface	171	494
Ye et al. [33]	2020	Kenaf	Carbon	Epoxy	○	150	~1	Flat surface	23.01	15.50

This study showed that more energy was absorbed by a tube with a larger diameter. Further, adding thicker layers increases the peak load, where the larger tubes carried slightly more stresses than the thinner structures [32]. This result was similar to that previously measured for the peak load and the SEA agrees well with a circular composite tube under the layer/thickness effect [23]. The larger diameter tubes also displayed unstable collapsed behavior compared with the smaller sizes [24].

5.3.2 Hybridization

To improve the crashworthiness capacity of a composite structure, an approach of using natural fiber hybridized with a synthetic fiber was explored by Albahash et al. [24]. Significant improvements in the peak load and SEA of the tube structures were obtained following the hybridization of woven jute fabric with glass or Kevlar fibers. Besides, hybrid structures exhibited high energy capability and specific energy absorption due to splaying failure compared to the brittle breakage mode, especially tubes made from jute composites. Following axial quasi-static compression tests, Attia et al. [25] reported that a tube made from a combination of carbon/glass/jute offered the highest load resistance compared to virgin jute, glass, and carbon composite structures. This was probably due to the effects of the stacking sequence and positioning of the fabric in the outer and inner faces of the tube. They added that the integration between the glass and jute fibers has resulted in the tube collapsing under the shear and transverse conditions, while the combination of carbon fiber with jute laminate corresponds to buckling failure. A comparative study into the effect of hybridization on composite tubes was also discussed by Supian et al. [28]. Here, the peak load of kenaf/glass tubes was approximately 1.7–2.4 times greater than for non-hybrid tubes. Besides the matrix volume fraction, the crashworthiness characteristics of the hybrid tube were also influenced by the winding orientation parameter [28]. Similar conclusions were also reported by Kumar et al. [31], where the value of SEA for a kenaf/glass tube increased almost twofold compared to the cylindrical kenaf composite structure. Ye et al. [33] also found that a braided tube made from kenaf/carbon hybrid composites appeared to give the highest energy absorption characteristics as compared to kenaf and carbon tubes.

5.3.3 Triggering

Apart from the hybridized structures, the triggering effect of the tube has been recognized as a dominant factor in determining energy-absorbing performance. The crashworthiness behavior of four different types of triggering mechanisms, non-trigger, 45° single chamfered trigger, 45° double-chamfered trigger, and 90° edge-sharp tulip trigger, was evaluated under axial quasi-static compression by Sivagurunathan et al. [26]. They found that the tulip trigger mechanism offered the highest crashworthiness characteristics values compared to other trigger systems due to progressive axial compression failure. In contrast, the non-trigger specimen yielded the lowest energy absorbed value resulting from a catastrophic mode of fracture. This observation agrees with the findings by Yan et al. [22]. Here, a non-trigger specimen experienced initial unstable local cracks in the middle of the longitudinal axis of the tube height followed by the stable compaction mode. Also, the cracks began at the top of the specimen with the formation of fronds either inward or outward of the tube which indicated the existence of a progressive crushing mechanism in the 45° chamfer test sample.

5.4 FAILURE MODES OF COMPOSITE TUBES

In general, there are two types of collapse mechanism for a composite tube subjected to axial compression loading: catastrophic failure and progressive crushing mode. Catastrophic failure can occur as a sudden failure that leads to a sharp load drop and can be obtained due to the specimen being crushed at the mid-plane, thus lowering the energy absorbed. In progressive crushing mode, most of the energy absorption capability of the composite tube initially fails *due to* micro-fractures which become macro-fractures in the post-crushing region and can be measured by a greater area under the load/displacement curve. Mache et al. [29] reported that matrix cracking, fiber breakage, and matrix-fiber debonding were common failures for brittle jute/epoxy tubes undergoing progressive crushing mode. Figure 5.3 shows the typical buckling failure of tubes corresponding to catastrophic and progressive crushing failure [34].

Figure 5.4, shows four types of failure-mode of a circular tube obtained by Meidell [35] that occurred due to the axial compression load. This study suggested that the types of failure were predominantly dominated by global buckling, local buckling (wrinkling), catastrophic fracture, or progressive crushing mode. Zhu et al. [30] elucidated the stress concentration of a basalt/epoxy composite was mainly distributed in the middle region due to the untriggered specimen.

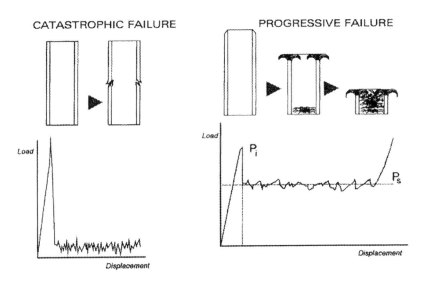

FIGURE 5.3 Load-displacement traces of catastrophic failure and progressive failure following axial compression loading. (From Jimenez M., Miravete A., Larrode E., Revuelta D. Effect of trigger geometry on energy absorption in composite profiles. *Composite Structures.* 2000;48(1–3):107–11.) [34]

Crashworthiness Measurement

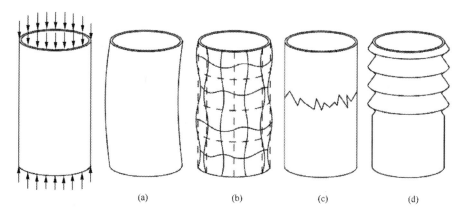

FIGURE 5.4 Failure mode of a cylindrical tube: (a) global buckling, (b) local buckling (wrinkling), (c) catastrophic fracture, (d) progressive crushing [35].

According to Mamalis et al. [36], three multi-failure mode combinations of tubular collapse—categorized as modes I, II, and III—were initiated during a series of static and dynamic axial compression tests. In Modes I and II, interlaminar fracture involving both splaying and sliding modes dominated at the edge of the respective composite tubes. This causes the tube to form the continuous outward and inward fronds and gave higher energy absorption due to the friction between laminates fiber fracture, shear cracking, and bending. Figure 5.5 shows the typical Mode I and Mode II interlaminar fracture that occurs during the crushing of the composite tube [37], whereas mode III is classified by the circumferential fracturing or mid-length collapse mode. The energy absorption is lower as compared to Modes I and II regarding the unstable collapse of the tube. Detailed discussion of these failure modes is in Mahdi et al. [27], Supian et al. [28], and Zhu et al. [30].

5.5 FINITE ELEMENT ANALYSIS AND FUTURE TRENDS

The simulation of the compressive behavior of composite structures using the finite element method (FEM) is an important and powerful approach that can reduce the high cost of the experimental method [38,39]. However, due to the complex nature of the non-linear behavior of composite structures that is considered in the finite element analysis (FEA), both FEA and an experimental method are needed for correlation and validation purposes [40–44]. The difficulty of the FEA study comes in the development of formulations that need to include the geometric nonlinearity of the structures when considering axial compression and buckling failures [45]. Further, the material non-linearity of the fibers and matrix that need to be considered as the material goes beyond its elastic limits requires the application of failure theories. These 2 non-linear situations, along with the non-linear contact problem at the common surfaces between the top and bottom plates and the composite, allow a complete FEM formulation that gives the load-displacement response of the compressive behavior of the composites corresponding to either a catastrophic or a progressive failure. However, in a progressive failure approach, besides using

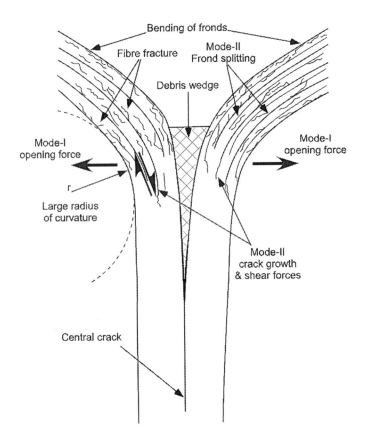

FIGURE 5.5 Schematic diagram of Mode I and Mode II failures that occur during crushing of composite tube. (From Savona S.C., Hogg P. Effect of fracture toughness properties on the crushing of flat composite plates. *Composites Science and Technology.* 2006;66(13):2317–28.) [37]

failure theory, such as the Tsai-Wu, Hashin, and Tsai-Hill theories to predict the onset of failure, the progress toward the failure of the material is examined through approaches such as stiffness degradation for elements that have already reached the failure limit—see Figure 5.6 for a material property upgrade in the ABACUS VUMAT user-subroutine [46]. With the availability of the FEM software that features all structural, material, and contact non-linearity while considering failure theories, the difficulties in developing a formulation are left to the software developers. Using ABACUS software, for example, has enabled the user to plot the force-displacement curve and determine the crashworthiness parameters while considering factors such as imperfections and several other geometric factors. The FEA studies on the crashworthiness of synthetic fiber-reinforced composites have been vast [47–49]. In the case of composite reinforced with natural fibers, since there are various suitable vegetables fibers for reinforcement that show different compression behaviors, the need for further investigation of the simulation of compressive behaviors of composites reinforced with such vegetable fibers is still high [50].

Crashworthiness Measurement

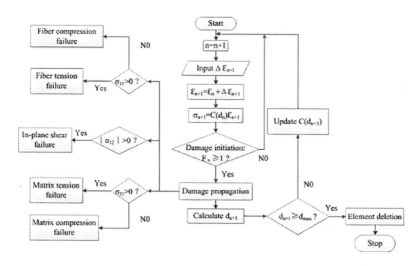

FIGURE 5.6 Flow chart of VUMAT's material property down-grade. (From Jiang H., Ren Y., Gao B. Research on the progressive damage model and trigger geometry of composite waved beam to improve crashworthiness. *Thin-Walled Structures*. 2017;119:531–43.) [46]

FEM codes were developed by Jiang et al. [46] to study the failure response of a carbon-epoxy composite waved beam loaded with quasi-static axial crushing, by applying the proposed non-linear progressive damage model. While considering the intra-laminar failure and delamination, the initiation of in-plane damage was detected using the maximum-stress failure criterion based on continuum damage mechanics. The progressive delamination was predicted by a cohesive stiffness degradation method and an exponential softening law based on a mix-mode fracture criterion. For crashworthiness improvement, the wedge-trigger, two types of W-triggers, and bevel-triggers with hybrid angles were used. The FEM model of the waved beam consisting of a fixed rigid wall, a moving crushing plate, and the composite wave beam with a 45° chamfer trigger is shown in Figure 5.7. The experiment was conducted and the simulation results were found to agree well with experimental results. The proposed bevel-triggers had a significant effect in decreasing the peak load while the angles of the chamfer-trigger make little difference to the crashworthiness.

Hussain et al. [48] used LSDYNA software to investigate the effect of 6 novel geometrically intrinsic triggers on the crashworthiness of glass fibre reinforced plastic (GFRP) composite crash box structures subjected to a low-velocity impact loading of 16km/h, following the standard RCAR test. Four cross-sections of the crash-box (square, cylindrical, hexagonal, and decagonal) were equipped with the six triggers. Material type 58 was used while applying Belystcho-Tsay shell elements with ELFORM 2 formulation. Hypermesh, LYDYNA, and Hyperview were the sequences applied in the software for pre-processor, analysis, and post-processor. Good correlations were found between the FEM results and the experimental ones [49] in terms of the force-displacement curve, the peak force, and the mean force. The 6 triggers for a square cross-section crush-box are shown in Figure 5.8 while the

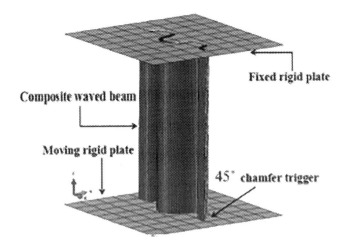

FIGURE 5.7 The FEM model of a waved beam. (From Jiang H., Ren Y., Gao B. Research on the progressive damage model and trigger geometry of composite waved beam to improve crashworthiness. *Thin-Walled Structures*. 2017;119:531–43.) [46]

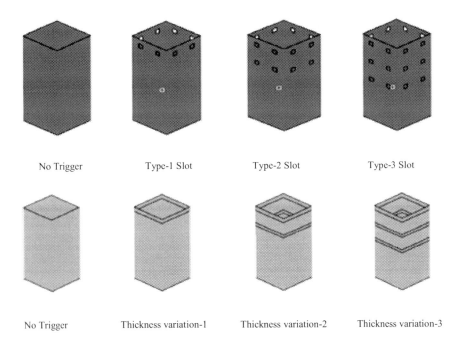

FIGURE 5.8 The square type crush-box with triggers before impact. (From Hussain N.N., Regalla S.P., Rao Yendluri V.D. Numerical investigation into the effect of various trigger configurations on crashworthiness of GFRP crash boxes made of different types of cross sections. *International Journal of Crashworthiness*. 2017;22(5):565–81.) [48]

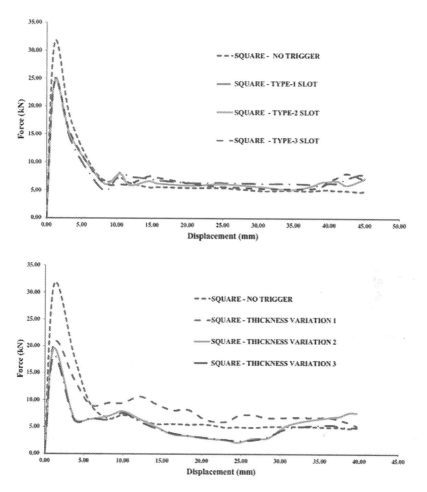

FIGURE 5.9 The force-displacement curve for a square cross-section crush box. (From Hussain N.N., Regalla S.P., Rao Yendluri V.D. Numerical investigation into the effect of various trigger configurations on crashworthiness of GFRP crash boxes made of different types of cross sections. *International Journal of Crashworthiness.* 2017;22(5):565–81.) [48]

corresponding force-displacement curves are shown in Figure 5.9. Table 5.2 shows the crashworthiness parameters for the square crush-box with the six types of trigger where the highest SEA was given by the Type-1 slot trigger. It was shown that the overall highest value of SEA was 8135.39J/kg, which corresponds to a decagonal geometry with thickness variation-1 trigger while the lowest value was 2813.64J/kg, belonging to square geometry with thickness variation-3.

In a recent study by Nia et al. [47], an optimization process was initially performed using the ABAQUS Topology Optimization Module on a composite frusto-conical structure with a 20° apex angle with the objective of volume minimization that will

TABLE 5.2
Crashworthiness Parameters for a Square Crush Box with Different Triggers [48]

Triggers Square	Energy Absorbed (J)	Primary Peak Force (kN)	Secondary Peak Force (kN)	Mass (gm)	SE (J/kg)
No trigger	313.292	31.35	7.2	79.68	3935.82
Type-1 slot	341.716	24.7	8.05	78.99	4375.364
Type-2 slot	296.822	24.3	7.8	78.31	3790.346
Type-3 slot	299.415	24.2	7.4	77.63	3856.949
Thickness variation 1	327.053	20.4	10.64	77.34	4230.957
Thickness variation 2	226.154	19.54	7.53	75.62	3519.622
Thickness variation 3	207.928	18.11	7.41	73.9	2813.64

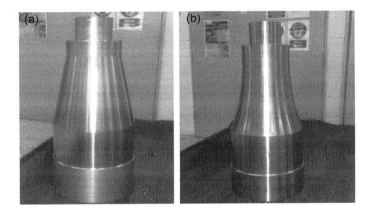

FIGURE 5.10 The mandrels for (a) CC specimen and (b) CV specimen. (From Nia A.B., Nejad A.F., Xin L., Ayob A., Yahya M.Y. Energy absorption assessment of conical composite structures subjected to quasi-static loading through optimization based method. *Mechanics & Industry.* 2020;21(1):113.) [47]

give the highest specific energy absorption capability to the conical structure. The optimization process conducted on the originally conical shell geometry (CC) has resulted in a concave (CV) shell geometry that has 15% less volume. Figure 5.10 shows the mandrel for the CC and CV specimens. Both structures were given a quasi-static compressive loading to investigate their energy absorption capability through FEA and experiments. Two methods of fabrication of the composites were hand lay-up (H) and filament winding (F). Using ABACUS software, Hashin's failure theory was applied to detect the onset of four failure types before material degradation was conducted. The results of the compression test experiments and FEA show good correlations, as shown in Figure 5.12.

Note from Figure 5.11 that the force-displacement curve for the conical structure shows a more stable curve without an instantaneous drop in load as seen in the

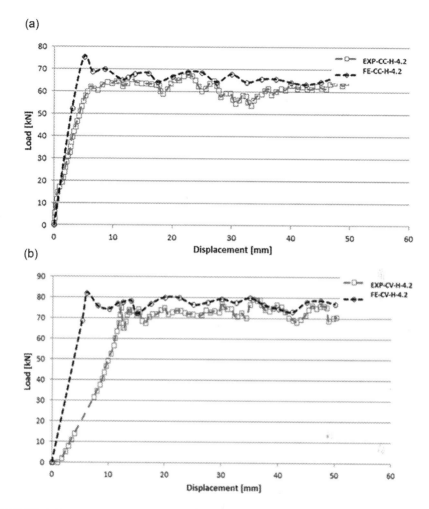

FIGURE 5.11 The experimental and FEM force-displacement curves for axial compression of a hand lay-up specimen of (a) CC and (b) CV. (From Nia A.B., Nejad A.F., Xin L., Ayob A., Yahya M.Y. Energy absorption assessment of conical composite structures subjected to quasi-static loading through optimization based method. *Mechanics & Industry*. 2020;21(1):113.) [47]

other cross-sections previously discussed. The curve can be divided into three stages: pre-buckling, buckling, and post-buckling. In the pre-buckling stage, a rather linear load-displacement curve can be seen until it reaches the first peak load where buckling occurs. A small drop in load follows and the curve experiences ups and downs in which a subsequent maximum value can be higher than the first.

Mahdi et al. [50] conducted a novel study on the highly non-linear crushing behavior of corrugated tubes with eco-friendly cotton fibers embedded in recyclable polypropylene using ABACUS/Explicit FEM software with an explicit dynamic finite element formulation. The initial crushing velocity was 6 m/s while the crushing time was 0.03 s. The effects of several parameters including the corrugation angle,

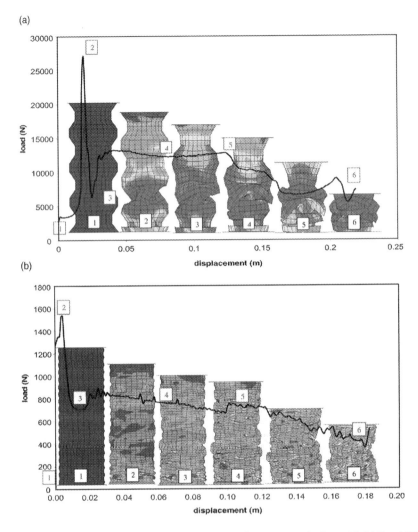

FIGURE 5.12 Load-displacement curve for composite corrugated tubes with (a) 5 and (b) 25 corrugations. (From Abramowicz W., Jones N. Dynamic progressive buckling of circular and square tubes. *International Journal of Impact Engineering.* 1986;4(4):243–70.) [51]

the number of corrugations, and the tube thickness on the energy absorption characteristics of the tubes were explored. In validating the simulation work, the crushing of a straight rectangular steel tube was conducted where the mean crush loads and the number of lobes produced were found to agree well with the experimental values [51]. The study found that, as shown in Figure 5.12, tubes with a higher number of corrugations crushed more predictably and absorbed more energy.

Oshkovr et al. [52] conducted a comprehensive non-linear explicit simulation study on the crashworthiness characteristics of natural silk/epoxy square tubes using MSC Patran and Dytran software. A series of quasi-static compression test simulations were conducted on short, medium, and long square tubes with 12, 24, and 30 layers of silk/epoxy. Structural response and crashworthiness characteristics including average crushing load, $\bar{P}$, peak load, P_{max}, crash force efficiency (CFE), specific absorbed energy (SAE), and stroke efficiency (SE) have been compared with experimental values. The silk woven fabric weft (90°) and wrap (0°) direction were used where the weft fiber direction was parallel to the tubes' axis. The shell element was used for both rigid plates while contact surface modeling was applied at the elements contacting the rigid plates and tube. The Tsai-Hill theory was applied to predict failure. In general, the simulation responses were found to closely agree with experimental values such as in the load-displacement curves in Figure 5.13. The different patterns shown by tubes with different lengths detected in experiments and through simulation can be seen in Figure 5.2 for short and long tubes.

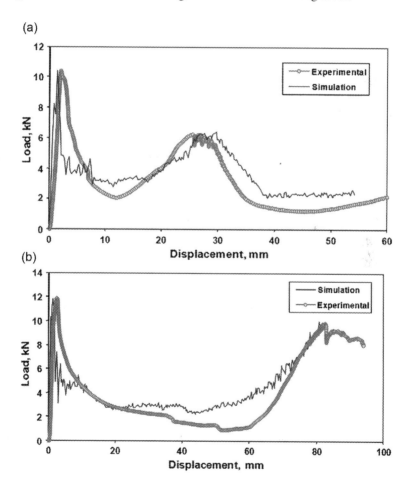

FIGURE 5.13 The comparison between the experimental and FEM load-displacement curves for (a) medium tube and (b) long tube. (From Mahdi E., Mokhtar A., Asari N., Elfaki F., Abdullah E. Nonlinear finite element analysis of axially crushed cotton fibre composite corrugated tubes. *Composite Structures*. 2006;75(1–4):39–48.) [50]

5.6 CONCLUSIONS

The developments of the experimental and numerical work on the crashworthiness of vegetable-fiber-reinforced composites have been elaborated. Several crashworthiness parameters were explained along with the factors that affect them. The use of in-house developed codes and FEM commercial software has managed to determine the crashworthiness characteristics of vegetable-fiber-reinforced composite tubes. In many cases, the force-displacement graphs determined numerically were in close agreement with the experimental ones. As such, the experimental-numerical combination is necessary for investigating the crashworthiness of vegetable-reinforced composites subjected to compression loading to reduce the high experimental cost. Thus, the accuracy of the numerical results arising from the complicated development of the FEM formulations can be confirmed.

ACKNOWLEDGMENTS

The authors wish to thank Universiti Teknologi Malaysia for the extensive support which allowed this research project to be completed. Throughout, this work was funded by Universiti Teknologi Malaysia with the Contract Research (DTD), Q.K.130000.7656.4C342, Contract Research (DTD), Q.K.130000.7656.4C343, the "Geran Universiti Penyelidik' UTMFR Q.K.130000.2656.21H13 scheme and "Geran Universiti Penyelidik' Tier 2 UTMFR Q.K130000.2643. 14J90.

REFERENCES

1. Holbery J., Houston D. Natural-fiber-reinforced polymer composites in automotive applications. *JOM*. 2006;58(11):80–6.
2. Sun Z. Progress in the research and applications of natural fiber-reinforced polymer matrix composites. *Science and Engineering of Composite Materials*. 2018;25(5):835–46.
3. Stark N., Gardner D. Outdoor durability of wood–polymer composites. In K.O. Niska, M. Sain (eds.) *Wood–Polymer Composites*. Amsterdam: Elsevier; 2008. pp. 142–65.
4. Dossche P.V., Erramuzpe P. Engineered waterproof plastic composite flooring and wall covering planks. Google Patents; 2018.
5. Hassan M., Salit S., Rasid Z.A. Thermal degradation and mechanical behavior of banana pseudo-stem reinforced composite. *International Journal of Recent Technology and Engineering*. 2019;8(4):5899–902.
6. Hassan M.Z., Roslan S.A., Sapuan S., Rasid Z.A., Mohd Nor A.F., Md Daud M.Y., et al. Mercerization optimization of bamboo (Bambusa vulgaris) fiber-reinforced epoxy composite structures using a box–behnken design. *Polymers*. 2020;12(6):1367.
7. Hassan M.Z., Roslan S.A., Rasid Z.A., Yusoff M.Z.M., Sapuan S., Muhammad-Sukki F. Optimizing the Mercerisation Effect on the Mode I Fracture Toughness of Bambusa Vulgaris Bamboo Using Surface Response Method. In: S.M. Sapuan, M.T. Mastura (eds.) *Implementation and Evaluation of Green Materials in Technology Development: Emerging Research and Opportunities*. Hershey, PA: IGI Global; 2020. pp. 112–29.
8. Hassan M.Z., Sapuan S., Roslan S.A., Sarip S. Optimization of tensile behavior of banana pseudo-stem (Musa acuminate) fiber reinforced epoxy composites using response surface methodology. *Journal of Materials Research and Technology*. 2019;8(4):3517–28.

9. Xie Y., Hill C.A., Xiao Z., Militz H., Mai C. Silane coupling agents used for natural fiber/polymer composites: A review. *Composites Part A: Applied Science and Manufacturing.* 2010;41(7):806–19.
10. Peças P., Carvalho H., Salman H., Leite M. Natural fibre composites and their applications: A review. *Journal of Composites Science.* 2018;2(4):66.
11. Tabiei A., Nilakantan G. Axial crushing of tubes as an energy dissipating mechanism for the reduction of acceleration induced injuries from mine blasts underneath infantry vehicles. *International Journal of Impact Engineering.* 2009;36(5):729–36.
12. Zhou G., Sun Q., Fenner J., Li D., Zeng D., Su X., et al. Crushing behaviors of unidirectional carbon fiber reinforced plastic composites under dynamic bending and axial crushing loading. *International Journal of Impact Engineering.* 2020;140:103539.
13. Zhou J., Guan Z., Cantwell W. The energy-absorbing behaviour of composite tube-reinforced foams. *Composites Part B: Engineering.* 2018;139:227–37.
14. Liu X., Belkassem B., Jonet A., Lecompte D., Van Hemelrijck D., Pintelon R., et al. Experimental investigation of energy absorption behaviour of circular carbon/epoxy composite tubes under quasi-static and dynamic crush loading. *Composite Structures.* 2019;227:111266.
15. Muhalim N.A.D., Hassan M.Z., Daud M.Y.M. Studies on the quasi-static and ultrasonic compression of copper tube filled with polyvinyl chloride (PVC) foam. *Journal of Advanced Research in Applied Mechanics.* 2017;32(2):1–8.
16. Muhalim N., Hassan M., Daud Y. Elastic-plastic behaviour of ultrasonic assisted compression of polyvinyl chloride (PVC) foam. *IOP Conference Series: Materials Science and Engineering.* 2018;344(1):012009.
17. Muhalim N., Daud M.Y., Hassan M.Z., Raman N., Sarip S. Compressive behaviour and energy absorption of copper tube under quasi-static and ultrasonic compression test. *Journal of Mechanical Engineering.* 2017;3(1):169–84.
18. Ataollahi S., Taher S.T., Eshkoor R.A., Ariffin A.K., Azhari C.H. Energy absorption and failure response of silk/epoxy composite square tubes: Experimental. *Composites Part B: Engineering.* 2012;43(2):542–8.
19. Ghoushji M.J., Eshkoor R.A., Zulkifli R., Sulong A.B., Abdullah S., Azhari C.H. Energy absorption capability of axially compressed woven natural ramie/green epoxy square composite tubes. *Journal of Reinforced Plastics and Composites.* 2017;36(14):1028–37.
20. Ding L., Liu X., Wang X., Huang H., Wu Z. Mechanical properties of pultruded basalt fiber-reinforced polymer tube under axial tension and compression. *Construction and Building Materials.* 2018;176:629–37.
21. Eshkoor R., Ude A., Oshkovr S., Sulong A., Zulkifli R., Ariffin A., et al. Failure mechanism of woven natural silk/epoxy rectangular composite tubes under axial quasi-static crushing test using trigger mechanism. *International Journal of Impact Engineering.* 2014;64:53–61.
22. Yan L., Chouw N., Jayaraman K. Effect of triggering and polyurethane foam-filler on axial crushing of natural flax/epoxy composite tubes. *Materials & Design (1980–2015).* 2014;56:528–41.
23. Ude A., Eshkoor R., Azhari C. Crashworthy characteristics of axial quasi-statically compressed Bombyx mori composite cylindrical tubes: Experimental. *Fibers and Polymers.* 2017;18(8):1594–601.
24. Albahash Z.F., Ansari M. Investigation on energy absorption of natural and hybrid fiber under axial static crushing. *Composites Science and Technology.* 2017;151:52–61.
25. Attia M.A., Abd El–Baky M.A., Hassan M.A., Sebaey T.A., Mahdi E. Crashworthiness characteristics of carbon–jute–glass reinforced epoxy composite circular tubes. *Polymer Composites.* 2018;39(S4):E2245–E61.

26. Sivagurunathan R., Way S.L.T., Sivagurunathan L., Yaakob M.Y. The effects of triggering mechanisms on the energy absorption capability of circular jute/epoxy composite tubes under quasi-static axial loading. *Applied Composite Materials.* 2018;25(6):1401–17.
27. Mahdi E., Ochoa D., Vaziri A., Eltai E. Energy absorption capability of date palm leaf fiber reinforced epoxy composites rectangular tubes. *Composite Structures.* 2019;224:111004.
28. Supian A., Sapuan S., Zuhri M., Zainudin E., Ya H. Crashworthiness performance of hybrid kenaf/glass fiber reinforced epoxy tube on winding orientation effect under quasi-static compression load. Defence Technology. 2019.
29. Mache A., Deb A., Gupta N. An experimental study on performance of jute-polyester composite tubes under axial and transverse impact loading. *Polymer Composites.* 2020;41(5):1796–812.
30. Zhu L., Zhang H., Guo J., Wang Y., Lyu L. Axial compression experiments and finite element analysis of basalt fiber/epoxy resin three-dimensional tubular woven composites. *Materials.* 2020;13(11):2584.
31. Kumar A.P., Shunmugasundaram M., Sivasankar S., Sankar L.P. Static axial crushing response on the energy absorption capability of hybrid Kenaf/Glass fabric cylindrical tubes. *Materials Today: Proceedings.* 2020;27:783–7.
32. Tomlinson D., Fam A. Axial response of flax fibre reinforced polymer-skinned tubes with lightweight foam cores and bioresin blend. *Thin-Walled Structures.* 2020;155:106923.
33. Ye H., Zhou X., Ma J., Wang H., You Z. Axial crushing behaviors of composite pre-folded tubes made of KFRP/CFRP hybrid laminates. *Thin-Walled Structures.* 2020;149:106649.
34. Jimenez M., Miravete A., Larrode E., Revuelta D. Effect of trigger geometry on energy absorption in composite profiles. *Composite Structures.* 2000;48(1–3):107–11.
35. Meidell A. Computer aided material selection for circular tubes designed to resist axial crushing. *Thin-Walled Structures.* 2009;47(8–9):962–9.
36. Mamalis A., Manolakos D., Demosthenous G., Ioannidis M. Analysis of failure mechanisms observed in axial collapse of thin-walled circular fibreglass composite tubes. *Thin-Walled Structures.* 1996;24(4):335–52.
37. Savona S.C., Hogg P. Effect of fracture toughness properties on the crushing of flat composite plates. *Composites Science and Technology.* 2006;66(13):2317–28.
38. Han H., Taheri F., Pegg N., Lu Y. A numerical study on the axial crushing response of hybrid pultruded and ±45 braided tubes. *Composite Structures.* 2007;80(2):253–64.
39. Huang J., Wang X. Numerical and experimental investigations on the axial crushing response of composite tubes. *Composite Structures.* 2009;91(2):222–8.
40. Rasid Z.A., Zahari R., Ayob A. The instability improvement of the symmetric angle-ply and cross-ply composite plates with shape memory alloy using finite element method. *Advances in Mechanical Engineering.* 2014;6:632825.
41. Mamalis A.G., Manolakos D.E., Ioannidis M., Papapostolou D. The static and dynamic axial collapse of CFRP square tubes: Finite element modelling. *Composite Structures.* 2006;74(2):213–25.
42. Roslan S., Yusof Z., Rasid Z.A., Yahaya M., Hassan M., Mahmud J. Dynamic instability response of smart composite material. *Materialwissenschaft und Werkstofftechnik.* 2019;50(3):302–10.
43. Yusof Z., Rasid Z.A. Numerical modelling of parametric instability problem for composite plate using finite element method. *AIP Conference Proceedings: AIP Publishing LLC*; 2016. p. 030044.
44. Rasid Z.A. The natural frequency of the shape memory alloy anti-symmetric angle-ply composite plates using finite element method. *Applied Mechanics and Materials.* 2015;695:52–55. Trans Tech Publ.

45. Naveen J., Jawaid M., Vasanthanathan A., Chandrasekar M. Finite element analysis of natural fiber-reinforced polymer composites. In M. Jawaid, N. Saba, M. Thariq (eds.) *Modelling of Damage Processes in Biocomposites, Fibre-Reinforced Composites and Hybrid Composites*. Amsterdam: Elsevier; 2019. pp. 153–70.
46. Jiang H., Ren Y., Gao B. Research on the progressive damage model and trigger geometry of composite waved beam to improve crashworthiness. *Thin-Walled Structures*. 2017;119:531–43.
47. Nia A.B., Nejad A.F., Xin L., Ayob A., Yahya M.Y. Energy absorption assessment of conical composite structures subjected to quasi-static loading through optimization based method. *Mechanics & Industry*. 2020;21(1):113.
48. Hussain N.N., Regalla S.P., Rao Yendluri V.D. Numerical investigation into the effect of various trigger configurations on crashworthiness of GFRP crash boxes made of different types of cross sections. *International Journal of Crashworthiness*. 2017;22(5):565–81.
49. Tabiei A., Yi W., Goldberg R. Non-linear strain rate dependent micro-mechanical composite material model for finite element impact and crashworthiness simulation. *International Journal of Non-Linear Mechanics*. 2005;40(7):957–70.
50. Mahdi E., Mokhtar A., Asari N., Elfaki F., Abdullah E. Nonlinear finite element analysis of axially crushed cotton fibre composite corrugated tubes. *Composite Structures*. 2006;75(1–4):39–48.
51. Abramowicz W., Jones N. Dynamic progressive buckling of circular and square tubes. *International Journal of Impact Engineering*. 1986;4(4):243–70.
52. Oshkovr S.A., Taher S.T., Oshkour A.A., Ariffin A.K., Azhari C.H. Finite element modelling of axially crushed silk/epoxy composite square tubes. *Composite Structures*. 2013;95:411–8.

6 Structure and Surface Modification Techniques for Production of Value-Added Biocomposites

Chaniga Chuensangjun
King Mongkut's University of Technology North Bangkok

Sarote Sirisansaneeyakul
Kasetsart University

Takuya Kitaoka
Kyushu University

CONTENTS

6.1 Introduction 126
    6.1.1 Physical Fiber Surface Modification Techniques 126
    6.1.2 Chemical Fiber Surface Modification Techniques 131
        6.1.2.1 Acid Hydrolysis 132
        6.1.2.2 Alkaline Treatment 132
        6.1.2.3 Acetylation 132
        6.1.2.4 Oxidizing Agents 133
        6.1.2.5 Other Chemical Treatments 133
    6.1.3 Biological Fiber Surface Modification Techniques 133
6.2 TEMPO-Mediated Oxidation of Natural Fiber 135
6.3 Surface Modified Cellulose Trends for Production of Value-Added Biocomposites 143
    6.3.1 Functional Biomaterials 144
    6.3.2 Catalytic Systems and Biochemical Process 145
    6.3.3 Biosensor Application 146
    6.3.4 Food Applications 146
    6.3.5 Medical Biomaterials and Drug Delivery 147
6.4 Concluding Remarks 148
Acknowledgments 149
References 149

DOI: 10.1201/9781003137535-6

6.1 INTRODUCTION

Biocomposite or natural fiber composite is defined as a combination of one or more phases of natural fibers and organic matrix or biopolymers, to replace the composites produced from a petroleum-based nonrenewable resource (Reddy et al., 2016). The natural fiber is classified to wood fibers such as soft and hardwoods, and nonwood fibers such as straw, bast leaf, seed, fruit and grass fibers (Bharath and Basavarajappa, 2016). Whereas, the most common matrix or biopolymers include natural biopolymers such as gelatin, corn zein and soy protein; synthetic biopolymers such as poly(lactic acid) (PLA); and microbial fermentation such as polyhydroxy-alkanoates (Avérous and Pollet, 2012). As well known that a majority of cell walls in plants consists of cellulose (35%–50%), hemicelluloses (20%–35%), lignin (10%–25%), and other extracts such as pectin and waxes (Sharma et al., 2018), natural fiber, or lignocellulosic fiber thus shows a compact structure of hydrogen bonding like the tightly packed networks in cellulose fibers. This results in some crystalline region which gives the toughness, strength, water or solvent impermeability, nontoxicity and biodegradability properties. Moreover, there are abundant hydroxy groups present in glucose monomer units in the cellulose chain which gives hydrophilic structures (Yildizhan et al., 2018).

Although cellulose-fiber-reinforced biopolymer composite has been presented much potential in many researches, it has still been less applied for industrial production processes because of some major drawbacks of cellulose structures like poor interfacial adhesion with hydrophobic matrices, high water absorption, and the presence of hydrophilic hydroxy groups, pectin and waxy substances exposed on cellulose surfaces (Sharma et al., 2018). Processes for treatment or modification of the natural fibers are thus necessary for the fabrication of cellulose/nanocellulose reinforced biocomposites to be applied for production of high value-added biocomposites. The various treatment methods for natural fibers surface modification are briefly summarized in Table 6.1.

6.1.1 Physical Fiber Surface Modification Techniques

Physical fiber surface modification techniques can be classified into two groups, i.e., thermal treatment and nonthermal treatment. Thermal treatments of cellulose surfaces, for example, heat treatment in a drying oven or an oil bath (Bhuiyan et al., 2000) and steam explosion (Auxenfans et al., 2017, Cui et al., 2012), were applied for improving the structure and mechanical properties of cellulose fibers. Bhuiyan et al. (2000) reported the changes of crystallinity in wood cellulose after heat treatment (180°C–220°C) using the drying oven (oven-dried condition) and oil bath (highly moist condition). The crystallinity of treated wood cellulose increased after heat treatment, corresponding to the thermal decomposition of the amorphous region in addition to crystallization. However, a high capability of apparatus has been developed for increasing the efficiency of thermal treatment technique such as steam explosion technology.

Steam explosion technology is distinctly developed for biorefinery application due to which it presents many advantages, i.e., low capital investment, moderate energy

TABLE 6.1
Treatment Methods for Natural Fibers Surface Modification

Treatment Technique	Fiber Matrix	Method/Condition	Results on Fiber Modification	References
Physical treatment	Wood powder (spruce, buna)	*Heat treatment:* • Oven-dried in a dying oven at 180°C–220°C • Highly moist condition carried out in an oil bath at 180°C–220°C.	Increase of crystallinity of wood cellulose	Bhuiyan et al. (2000)
	Tossa jute yarn/fiber	*UV treatment/Corona discharge:* • UV emission spectrum at a focal point 53 nm, energy output of 254 nm • Corona treatment at 40–100 W, 10–88 cm/min	• Higher polarity up to 200% • Increase of composite flexural strength, • Increase of polar component of free surface energy	Gassan and Gutowski (2000)
	Softwood bleached pulp	*Dielectric-barrier discharge/Cold plasma:* Using a ceramic-coated aluminum electrode at 1.0 and 5.0 kW m^{-2}/min, 23.5°C, atmospheric pressure	Increase of wettability	Wielen et al. (2006)
	Cellophane foil	*Oxygen plasma treatment:* Using 2.46 GHz, 600 W of microwave source, aluminum vacuum chamber, pressure of 0.16 mbar, oxygen flow of 10 cm^3/min, 80–160 s	• Increase of surface texture and mean roughness, • Forming aldehyde and carboxylate groups on the surface	Calvimontes et al. (2011)
	Spruce wood Beech sawdust	*Steam explosion:* Steam pressure of 11–31 bar, 180°C–235°C	• Better break up and decrease the particle size of fiber, • Increase of digestibility	Pielhop et al. (2016) He et al. (2020)
	Poplar Wheat straw	*Acid-catalyzed steam explosion*	Removal of hemicellulose, reduction of cross-linking phenolic acid and redistribution of lignin	Auxenfans et al. (2017)
	Eucalyptus pulp Pinus pulp Miscanthus fiber	*Corona discharge:* • 10 kV, 60 μA, 60 Hz, distance of 3 cm, for 30 s • 15 kV, 50 Hz	• Improved water vapor permeability, water absorption, tensile strength • Improved compatibility between fiber and matrix	Lopes et al. (2018) Ragoubi et al. (2012)

(Continued)

TABLE 6.1 (Continued)
Treatment Methods for Natural Fibers Surface Modification

Treatment Technique	Fiber Matrix	Method/Condition	Results on Fiber Modification	References
	Microcrystalline cellulose	*Plasma treatment/ Ultrasonication:* • Plasma jet was ignited in open air at 5000 sccm argon, 150 W, immersion time for 30 min • Using ultrasonic processor with titanium alloy probe of 19 mm, 500 W, 20 kHz for 30 min	Increase of defibrillation, surface functionalization, reinforcing ability	Vizireanu et al. (2018)
	Kapok fiber	*Cold plasma treatment:* Using coaxial electrode, electric current of 0.09–0.12 A, voltage of 400–500 V, 10 cm³/min gas flow rate, 1.5 mbar, 30–60 min, 72°C	Improved surface roughness, oil/water uptake	Macedo et al. (2020)
Chemical treatment	Green flax fiber	*Acetylation:* Incubating fiber in acetylating solution (toluene, acetic anhydride, catalyst perchloric acid) at 60°C, 1–3 h	• Improved surface morphology and moisture resistance of fiber • Increase of tensile and flexural strength of composites	Bledzki et al. (2008)
	Rice straw	*Acetylation:* Incubating fiber in a mixture of acetic anhydride, glacial acetic acid and sulfuric acid, 50°C, 2 h	Improved thermal properties of fiber	Huang et al. (2014)
	Makino bamboo	*Acetylation:* Incubating fiber in AA/dimethylformamide (DMF) at a solid/liquid ratio of 0.05 g/mL, 140°C, 2 h	Increase of thermal decomposition	Jhu et al. (2019)
	Coir fiber	*Alkali treatment:* • NaOH concentration of 5 wt.%, 72 h, RT[a]	High interfacial shear strength (55.6 %)	Nam et al. (2011)
	Borassus fruit fiber	*Alkali treatment:* • NaOH concentration of 5 wt.%, 8 h, 30°C	Improved tensile properties (strength, modulus, % elongation)	Reddy et al. (2013)

(Continued)

TABLE 6.1 (Continued)
Treatment Methods for Natural Fibers Surface Modification

Treatment Technique	Fiber Matrix	Method/Condition	Results on Fiber Modification	References
	Date palm fiber	*Alkali treatment:* • NaOH concentration of 6 wt.%, 24 h, RT • NaOH concentration of 5 wt.%, 1 h, RT	• High interfacial adhesion and strength with polymer matrix • Increase of tensile strength and interfacial property	Alsaeed et al. (2013) Oushabi et al. (2017)
	Ladies finger fiber	*Alkali treatment:* 2% NaOH, 70°C, 2.5 h *Single-stage treatment:* 4% $Cr_2(SO_4)_3 \cdot 12(H_2O)$ (pH 4), 3 h *Double stage treatment:* 4% $Cr_2(SO_4)_3 \cdot 12(H_2O)$ mixed with $NaHCO_3$, 2 h	• Improved rough surface of fiber • Increase of tensile strength and Young's modulus	Hossain et al. (2013)
	Sisal fiber	*Alkaline treatment:* 10 wt% $NaHCO_3$, RT, 120 h	Highest interfacial adhesion and mechanical properties with matrix	Fiore et al. (2016)
	Wheat straw Rye straw	*Oxidizing agents treatment:* Ozonolysis by using ozone concentration 2.7% w/w, 60 L/h, 2.5 h	Improved accessible structure to enzymatic hydrolysis	García-Cubero et al. (2010)
	Poplar wood	*Oxidizing agents treatment:* Incubating fiber in 4.4 mmol of H_2O_2 solution containing metal salts or 0.1 mmol $Mn(OAc)_3$, RT, 7 days	• Distribution of cellulose in the wood structure • Removal of lignin	Lucas et al. (2012)
	Softwood bleached kraft pulp Bagasse bleached pulp	*Oxidizing agents treatment:* Incubating wet pulp in the suspension of TEMPO, NaBr, NaClO at the maintained pH 10–11, RT, >8 h, followed by postoxidation using $NaClO_2$ in acetate buffer at pH 4.8, RT, 2 days	Improved interfacial adhesion between cellulose nanofibers and lactide and oligomer of PLA Improved dispersibility of biocomposites in a polar aprotic solvent	Chuensangjun et al. (2019a, b)
	Curaua fibers	*Oxidizing agents treatment:* Bleaching of fibers with NaClO and NaOH solution at 50°C for 2 h, followed by oxidizing with TEMPO in the suspension of NaBr and NaClO at the maintained pH 10–10.5.	Increased crystallinity and thermal stability Decreased degree of polymerization	Neves et al. (2020)

(Continued)

TABLE 6.1 (Continued)
Treatment Methods for Natural Fibers Surface Modification

Treatment Technique	Fiber Matrix	Method/Condition	Results on Fiber Modification	References
Biological treatment	Sisal fiber	*Microorganism:* Bacteria. *Brevibacillus parabrevis* was used as inoculum. Incubating condition: 0.5 g fiber in 90 mL culture medium, 10 mL inoculums, 150 rpm, 30°C, 1 week	• Smooth and shiny surface of sisal fiber • Enhanced thermal stability and crystallinity	Kalia and Vashistha (2012)
	Triploid poplar	*Microorganism:* White rod fungus, *Trametes velutina D10149* was used as inoculum. Statically incubating condition: 5 g poplar in 12.5 mL water, 5 mL inoculums, 28°C, 4–16 weeks	Modification of lignin structure Improved cellulose conversion	Wang et al. (2013)
	Hemp	*Enzyme:* Cellulase, hemicellulose/cellulase, cellulase/β-glucosidase were used. Reaction condition: 10 g fiber in 0.05 M acetate buffer (pH 4.8), 37°C, 24 h	• Change in accessibility • Increase of pores on surface structure	Buschle-diller et al. (1999)
	Wheat straw	*Enzyme:* Lipase from *Candida rugosa*, Cellulase from *Trichoderma Reseii* and *Aspergillus niger* were used. Reaction condition: 1 g of enzyme in 20 mL of water or citric acid (pH 4.8), 35°C, 60 min	• Reduction of surface wax • Increase of interfacial adhesion property	Zhang et al. (2003)
	Flax fiber	*Enzyme:* 1,4-β-xylanase, Pectinase, Xyloglucanase were used. Reaction condition: 0.5–10 wt.% enzyme, pH 5–5.5, 50°C, 2–6 h	• Removal of hemicellulose and pectin • Individualization of yarns • Improved interfacial adhesion • Increase of thermal stability	Seghini et al. (2020)

[a] RT means room temperature.

requirements, no acid-base or solvent chemicals requirements, cost-effective process, low environmental impacts, and widely used for various lignocellulosic fibers (e.g., wood, wheat straw, bagasse, grass crops, corn stalks) (Auxenfans et al., 2017; Pielhop et al., 2016). The study of cellulose surface modification by steam explosion treatment revealed the differences in the micro- and nanostructure of the thermally treated fibers, measured by scanning electron microscopy (SEM) analysis. There were no condensed droplet-like structures which corresponded to lignin residues on the surfaces, and also many holes were visible on the exploded fiber surfaces (Pielhop et al., 2016). Because the steam gun establishes the high-pressure saturated steam at high temperature (180°C–235°C) and then reducing pressure swiftly makes the fibers in the high-pressure reactor are under explosive decompression, resulting in the breakdown of the lignocellulosic fibers, removal of lignin and/or hemicellulose, and increasing the digestibility of cellulose (He et al., 2020).

Nevertheless, the thermal treatment produced high temperature which might affect the crucial properties of the cellulose fibers. Nonthermal treatments of cellulose surfaces, such as dielectric-barrier discharge (Wielen et al., 2006), corona discharge (Lopes et al., 2018), cold plasma treatment (Macedo et al., 2020), therefore, were extensively studied. The electric discharges, microwave or radiofrequency are also usually used as a plasma generation for nonthermal plasma treatment. This is related to changes in the polar and dispersive components of surface energy, including the formation of hydroxy, carbonyl, and other functional groups. Plasma treatment technology is recently adopted to open pores and remove oils and wax from the fiber surfaces, resulting in improving the wettability, water uptake and absorption of the cellulosic fibers (Macedo et al., 2020). Several protocols of plasma treatment have been widely reported according to the operating conditions (e.g., gas pressure, applied power and exposure time) (Camargo et al., 2017). For instance, low-pressure oxygen plasma generated by microwave source was applied to modify the cellulose surfaces which, the results disclosed to the increase of surface texture and mean roughness, forming aldehyde and carboxylate groups on the surface due to the decomposition of polymer chains and oxidation reaction (Calvimontes et al., 2011).

Additionally, UV treatment, mechanical comminution and ultrasonication could be combined with the corona discharge and plasma treatment for surface modification of biocomposites (Gassan and Gutowski, 2000, Vizireanu et al., 2018). Like for example, submerged liquid plasma combined with ultrasonication treatments used for modification of microcrystalline cellulose demonstrated the increase of defibrillation and surface functionalization, including its reinforcing ability for biocomposites applications (Vizireanu et al., 2018).

6.1.2 Chemical Fiber Surface Modification Techniques

The number of researches involved improving the process of reinforced biocomposites production using natural fibers, e.g., lignocellulosic fiber, has increased in recent years. Especially the study of surface modification of cellulose fibers because the presence of hydrophilic hydroxy groups, pectin and waxy substances exposed on cellulose surfaces is a barrier to reinforce with the hydrophobic matrices. The incompatibility between these two natural materials results in weakening bonding at the interface, poor resistance to moisture absorption which affects the poor mechanical properties of biocomposites (Amin et al., 2017, Kabir et al., 2012, Li et al., 2007). Chemical treatment of natural fibers has been widely employed in improving the interfacial properties between fiber and polymer matrix because it is more effective for surface modification (Fiore et al., 2016, García-Cubero et al., 2010, Hossain et al., 2013, Liu et al., 2019). Various chemical methods such as acid hydrolysis, alkaline treatment, acetylation and oxidizing agents have been discussed shortly.

6.1.2.1 Acid Hydrolysis

Pretreatment of lignocellulosic materials with acid hydrolysis, e.g., HCl, H_2SO_4, CH_3COOH, H_3PO_4, demonstrated the disruption of the van der Waals forces, hydrogen bonds and covalent bonds that function together for network structure assembly. The fiber components consisting of hemicellulose, xylan and glucomannan are hydrolyzed, and xylose, galactose, mannose and glucose are sugar products from the reaction. Furfural and hydroxy methyl furfural might be obtained from the dehydration of these sugars (Amin et al., 2017).

6.1.2.2 Alkaline Treatment

Alkaline solution especially NaOH has been often employed to alter the cellulose surfaces *via* alkalization process. The strongly compact crystalline cellulose is separated and changed to an amorphous region due to the forming of reactive molecules between fiber surfaces and alkaline reagents. This results in the removal of certain portions of hemicellulose, lignin, pectin, wax and oil exposed on cellulose surfaces (Kabir et al., 2012). Many researches focused on the effect of alkali concentration on the surface modification; for example, Alsaeed et al. (2013) and Oushabi et al. (2017) reported using 5–6 wt.% NaOH is the optimal concentration for improving the surface of date palm fibers with less damage to its strength, resulting in the enhancement of the interfacial adhesion between the fiber and polymer matrices. Like the study of alkaline treatment of coir fiber which revealed the significant improvement of fiber surface roughness, interfacial bonding strength and the wettability, when the coir fibers were soaked in 5% NaOH for 72 h (Nam et al., 2011). Not only NaOH used for alkaline treatment of fibers, but also $NaHCO_3$ could be used as an eco-friendly chemical for surface treatment of sisal fibers (Fiore et al., 2016).

6.1.2.3 Acetylation

Acetylation is one type of natural fiber treatment methods by esterification of an acetyl group (C_2H_3O) and the hydrophilic hydroxy groups (OH) on the fiber surfaces (Kabir et al., 2012). After the reaction, the existed moisture on the fiber is taken out, resulting in the decrease of hydrophilic nature of the fiber and increase of rough surfaces. The improved fiber presented better interfacial adhesion with the matrix, dimensional stability and mechanical property of the biocomposites (Kabir et al., 2012, Li et al., 2007). Flax fibers were acetylated with acetylating solution consist of toluene, acetic anhydride and catalyst perchloric acid, at 60°C for 1–3 h, resulting in the improvement of fiber surface morphology, moisture resistance properties and tensile strengths of biocomposites (The modified fiber reinforced polypropylene) (Bledzki et al., 2008). Bamboo fibers prepared from Makino bamboo (*Phyllostachys makinoi* Hayata) were acetylated with AA/dimethylformamide (DMF) using a solid/liquid at 140°C for 2 h. The increased thermal decomposition temperature of fiber was disclosed from this methodology (Jhu et al., 2019).

6.1.2.4 Oxidizing Agents

The oxidants such as ozone and hydrogen peroxide have received attention for application in the development of advanced surface modification research. The fiber surface treatment by ozone is known as ozonolysis method (Bensah and Mensah, 2013). The sparging of ozone which generated by ozone generator or plasma model into a mixture of natural fiber and water at room temperature leads to the leaching of lignin and hemicelluloses from the fiber surfaces, high dry matter concentration of modified cellulose fibers and low production of inhibitory products. Thus the use of ozone pre-treatment of grain straw in a fixed bed reactor under room conditions revealed the reduction of acid-insoluble lignin content in grain straw and provided a more accessible structure of cellulose to higher enzymatic hydrolysis yield (García-Cubero et al., 2010). Despite many benefits come from this process, the expense for the large requirements of ozone generation and operating system is distinctly high. Alternatively, other oxidizing agents have been studied for fiber surface treatment such as hydrogen peroxide (H_2O_2), peracetic acid ($C_2H_4O_3$), sulfur trioxide (SO_3) and chlorine dioxide (ClO_2) (Bensah and Mensah, 2013, Lucas et al., 2012).

Sodium chlorite ($NaClO_2$) can be acidified in acid solution and liberates choleric acid ($HClO_2$), which forms into chlorine dioxide (ClO_2) *via* an oxidation reaction. The ClO_2 can react with lignin constituents and remove them from the fiber surfaces. Moreover, it reacts with hydrophilic hydroxy groups of hemicelluloses, resulting in the removal of moisture from the fiber and enhancing the hydrophobic nature of the fiber. This treatment method is also known as bleaching method (Kabir et al., 2012). Because the lignin is removed, fibers present more flexible and have lower stiffness properties. This advantage is more attracted to further develop for the high potential surface modification of natural fibers, which will be discussed subsequently.

6.1.2.5 Other Chemical Treatments

Unless the chemical treatment methods mentioned above, silane treatment by using silane (SiH_4) as coupling agents, benzoylation by using benzoyl chloride in organic synthesis, peroxide treatment by using benzoyl or dicumyl peroxide, maleated coupling agents, stearic acid treatment, and permanganate treatment were also studied to reduce the hydrophilic hydroxy groups on the fiber surfaces and improve the adhesion of fibers and polymer matrix (Kabir et al., 2012, Li et al., 2007).

6.1.3 BIOLOGICAL FIBER SURFACE MODIFICATION TECHNIQUES

The thermochemical treatment with higher temperature and pressure process is generally used in the composite/plastic industry. However, some of the waste generation and high energy use might be a limitation procedure for the development of surface modification toward the environmentally friendly process. Biological treatments including microorganisms and enzymatic treatment have been proposed as the optional method for surface modification of cellulose fibers. Because it requires mild reaction conditions, low energy and formation of minimal toxic waste (Pramanik and Sahu, 2017). For few decades, enzymatic treatment has been studied for surface modification of cellulose fibers in order to application of the textile industry (Buschle-diller et al., 1999). Commercial cellulase, hemicellulase, β-glucosidase and lipase were usually used for the enzymatic treatment of cellulose fibers, such as hemp fabric, wheat straw and flax fabric, through hydrolysis reaction (Kan et al., 2007, Zhang et al., 2003). These researches revealed the capability of enzymes for eliminating the wax exposed on cellulose fiber surfaces, conduce toward an improvement of adhesion property of fiber matrix. The formation of pores structure on the surfaces upon enzymatic hydrolysis was also promoted obviously. Besides the cellulase, lipase and its derivative enzymes, recently, the commercially available and inexpensive enzyme preparations based on pectinase and xyloglucanase exhibited a synergistic effect on the removal of hydrophilic pectin and hemicellulose from flax yarns surfaces (Seghini et al., 2020). This enzymatic treatment method also increased thermal stability and reduced mechanical properties of fibers, depending on the enzyme concentration and treatment duration.

Microorganisms, fungi and bacteria can establish the enzyme systems to degrade lignocellulosic substrates which consist of a complex mix of cellulose, hemicellulose and lignin (Andlar et al., 2018, Thapa et al., 2020). The fungal enzymes involved in lignocellulose degradation are mostly extracellular enzymatic system which could be classified into two types of enzymes, i.e., a hydrolytic system for polysaccharide (i.e., cellulose and hemicellulose) degradation and oxidative system for lignin degradation together with opening phenyl rings. There are three groups of fungi including soft-rot (e.g., *Aspergillus* and *Neurospora*), brown-rot (e.g., *Gloephyllum trabeum, Coniophora puteana* and *Postia placenta*) and white-rot fungi (e.g., *Aspergillus niger, Streptomyces griseus* and *Trichoderma reesei*) have been studied and described in the different effects and lignocellulosic degradation mechanisms (Andlar et al., 2018, Saritha et al., 2012). The lignocellulosic enzymes produced by fungi such as cellulase (endoglucanases, exoglucanases, β-glucosidase) for cellulose degradation, endo-1,4-β-xylanase and exo-1,4-β-xylosidase for hemicellulose degradation, and laccases and peroxidases for lignin degradation. The study of cellulose surfaces modification by fungi treatment, for example, white-rot fungus *Trametes velutina* D10149 was incubated together with wood chips for biopretreatment of cellulose process (Wang et al., 2013). The study reported the significant effect on the digestibility of the lignocellulosic sample probably due to the removal of lignin and the crystallinities of the sample were rarely changed.

Additionally, bacteria could be cultured for potentially used in improving properties and surface modification of lignocellulosic materials, such as decomposition of lignin, enhanced thermal stability and crystallinity, better smooth and shiny surface (Kalia and Vashistha, 2012). These are the results from the mechanism of a group of cellulolytic and xylanolytic enzymes, like cellulases, β-glucosidase, endoglucanases, cellobiohydrolase, xylanases, β-xylosidases, etc. (Béguin et al., 1992, Thapa et al., 2020). Several strains of bacteria produce cellulolytic and xylanolytic enzymes like *Bacillus* sp. (e.g., *B. circulans, B. amyloliquefaciens, B. subtilis*), *Brevibacillus parabrevis, Clostridium* sp. (e.g., *C. thermocellum, C. longisporum, C. cellobioparum*), *Flavobacterium* sp., *Pseudoaltermonas* sp. (e.g., *P. haloplanktis*), *Pseudomonas* sp., *Shewanella* sp., *Streptomyces halstedii, Thermobacillus xylanolyticus* (Kalia and Vashistha, 2012, Maki et al., 2009; Thapa et al., 2020). In the recent year, both microorganisms and enzymatic treatments are extensively used as biopretreatments for enhancing the potential of bioethanol/biofuel production by saccharification and fermentation of lignocellulosic fibers (Prajapati et al., 2020, Schneider et al., 2020, Sekhon et al., 2018, Tang et al., 2018, Tramontina et al., 2020).

Among many advantages of physical, chemical and biological modification techniques, a limitation for developing an eco-friendly industry with the lowest-cost of the biocomposites production process still needs to be improved. This report focuses on the discussion of the recent potential technique for cellulose surface modification under the mild condition and low production cost, namely, 2,2,6,6-tetramethylpiperi dine-1-oxyl (TEMPO)-mediated oxidation.

6.2 TEMPO-MEDIATED OXIDATION OF NATURAL FIBER

TEMPO or 2,2,6,6-tetramethylpiperidine-1-oxyl ($C_9H_{18}NO$), water-soluble and stable nitroxyl radical which has been proposed as a catalytic agent to selectively convert primary hydroxy groups of polysaccharides to the corresponding aldehydes, ketones and carboxy groups under mild conditions (Isogai et al., 2011, Perez et al., 2003). TEMPO has been mostly used in catalytic oxidation, especially the surface modification of cellulose, wood fibers, and starch, etc., to be utilized in paper industries, chemistry or packaging applications (Azetsu et al., 2013, Kanomata et al., 2018, Bideau et al., 2017).

The mechanism of TEMPO-mediated oxidation was mostly proposed in the terms of TEMPO/NaBr/NaClO oxidation of cellulose in water (Figure 6.1). The TEMPO radicals are activated by NaClO into the nitrosonium ion (TEMPO+ ion), which subsequently oxidize the C-6 hydroxy groups exposed on cellulose surfaces to aldehyde and carboxy, respectively. The carboxy groups on cellulose surfaces are converted to the sodium salt form in an aqueous of sodium hydroxide (NaOH). Meanwhile, the reaction between sodium bromide (NaBr) and sodium hypochlorite (NaClO) generates hypobromide, which is used to regenerate the nitrosonium ion (Dang et al., 2007, Kato et al., 2003). The method for preparation of TEMPO-oxidized cellulose nanofibers (TOCNs) by TEMPO-mediated oxidation has been continually developed over the last two decades. Evidently, Nooy et al. (1995) first applied TEMPO-mediated oxidation to water-soluble polysaccharides, i.e., potato starch, amylodextrin and pullulan. The TEMPO/NaBr/NaClO oxidation system was carried out in the water for selective conversion of C6 primary hydroxy to carboxylate groups. The optimum pH for the reaction was reported between 10 and 11. Native celluloses have also been investigated for producing individual microfibrils by TEMPO-catalyzed oxidation; for example, Saito et al. (2006) developed the oxidation of bleached sulfite wood pulp, cotton, tunicin and bacterial cellulose by using TEMPO radical followed by a homogenizing mechanical treatment. They reported the successful disintegration of those native celluloses into long individual microfibrils with a regular width of 3–5 nm. Many related researches, thereafter, have been extensively studied in TEMPO-mediated oxidation, particularly, the study on preparation of TOCNs from various native celluloses and the optimal condition of oxidation reaction, as summarized in Table 6.2.

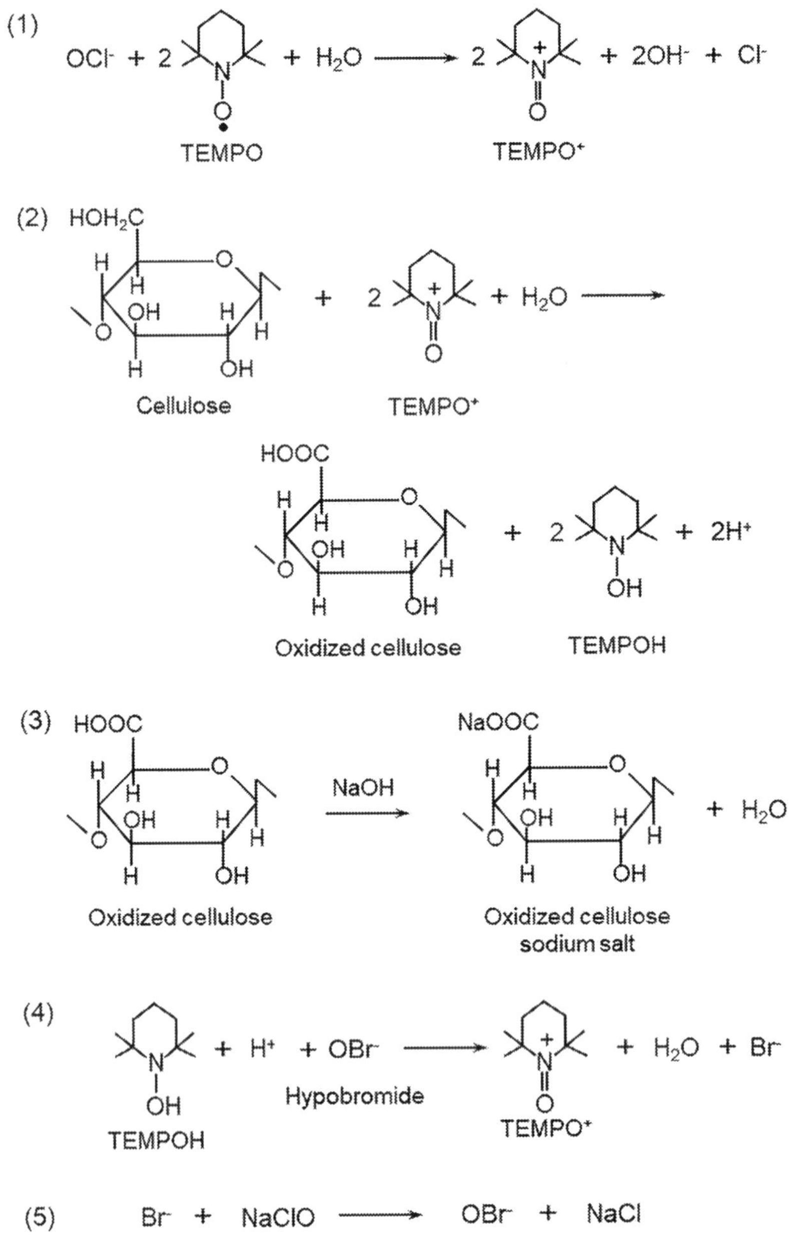

FIGURE 6.1 Representative oxidation mechanism of C6 primary hydroxy of cellulose to C6 carboxylate groups by TEMPO/NaBr/NaClO oxidation system in water at pH 10.

TABLE 6.2
TEMPO-Oxidized Cellulose Nanofibers Preparation from Various Native Celluloses and Optimal Condition for Oxidation

Substrate		Catalytic System				Oxidation Condition			References
		TEMPO	NaBr	NaClO	Water	pH	Temperature	Time	
Potato starch	20 mmol primary alcohol	0.13 mmol	7.8 mmol	44 mmol	500 mL	10.8	2°C	80 min	Nooy et al. (1995)
Amylodextrin						10	20°C	45 min	
Pullulan						10.5	2°C	70 min	
Starch	100 g	0.048 g	0.635 g	4.5–44 g	100 mL	10.75	5°C	unsteady[c]	Kato et al. (2003)
Cellulose cotton linter	5 g	0.0125 g	0.125 g	5–48.35 mmol	375 mL	10.5	RT[b]	24 h	Saito and Isogai (2004)
Cotton linters	0.648 g	0.065 mmol	2 mmol	0.5–2.0 mol	90 mL	10	20°C	unsteady	Montanari et al. (2005)
Sugar beet pulp									
Cotton linters	2 g	0.025 g	0.25 g	4.84 mmol	200 mL	10.5	RT	unsteady	Saito et al. (2006)
Ramie									
Spruce holocelluloses									
Elementally chlorine-free bleached softwood kraft pulp	20 g	0.32 mmol	KBr 4.202 mmol	17 mmol	1,840 mL	9.1	RT	2 h	Dang et al. (2007)
Hardwood bleached kraft pulp	1 g	0.1 mmol	1 mmol	1.3–5.0 mmol	100 mL	10	RT	Unsteady	Saito et al. (2007)
Chitin	1 g	0.1 mmol	1 mmol	5 mmol	100 mL	10	RT	Unsteady	Fan et al. (2008)
Cotton	10 g	0.025 g	0.25 g	40.84 mmol	750 mL	10.5	RT	4 h	Praskalo et al. (2009)
Lyocell									
Softwood bleached kraft pulp	1 g	0.1 mmol	1 mmol	5 mmol	100 mL	10	RT	1.5 h	Fukuzumi et al. (2010)

(Continued)

TABLE 6.2 (Continued)
TEMPO-Oxidized Cellulose Nanofibers Preparation from Various Native Celluloses and Optimal Condition for Oxidation

		Catalytic System				Oxidation Condition			
Substrate		TEMPO	NaBr	NaClO	Water	pH	Temperature	Time	References
Japanese cedar	1 g	0.016 g	0.1 g	2.5–15 mmol	100 mL	10	RT	<2 h	Kuramae et al. (2014)
Ginkgo									
Birch									
Eucalyptus									
Rice straw									
Kenaf									
Birch sawdust pulp	2 g	0.2 mmol	2 mmol	10–20 mmol	100 mL	10.5	RT	Unsteady	Liu et al. (2014)
Waste mushroom bed of shiitake	1 g	0.016 g	0.1 g	20–80 mmol	130 mL	10–10.5	RT	4 h	Konno et al. (2016)
Pine sawdust pulp	15 g	0.24 g	1.5 g	75–225 mmol	1,500 mL	10	RT	Unsteady	Ehman et al. (2016)
Hemp bast	1 g	0.016 g	0.1 g	5–30 mmol	100 mL	10	RT	2 h	Puangsin et al. (2017)
Flax yarn	1 g	0.13 mmol	4.7 mmol	5.65 mmol	100 mL	10	RT	60 s	Fathi et al. (2017)
Oil-palm-trunk Empty fruit bunches of oil palm tree	1 g	0.1 mmol	1 mmol	10 mmol	100 mL	10	RT	>8 h	Sirisansaneeyakul and Chuensangjun (2017)
Bamboo pulp	5 g	0.5 mmol	5 mmol	15–60 mmol	500 mL	10	20°C	6 h	Xiao et al. (2018)

(*Continued*)

TABLE 6.2 (Continued)
TEMPO-Oxidized Cellulose Nanofibers Preparation from Various Native Celluloses and Optimal Condition for Oxidation

		Catalytic System				Oxidation Condition			
Substrate		TEMPO	NaBr	NaClO	Water	pH	Temperature	Time	References
Softwood bleached kraft pulp	1 g	0.1 mmol	1 mmol	10 mmol	100 mL	10	RT	>8 h	Chuensangjun et al. (2019a,b)
Bagasse bleached pulp									
Bleached kraft eucalyptus pulp	30 g	0.06 g	3 g	150–750 mmol	3 L	10	RT	Unsteady	Patiño-Masó et al. (2019)
Cellulose from Kelp	0.5	0.008 g	0.05 g	3.6 mL	500 mL	10	5°C	Unsteady	Wu et al. (2020)
Rubberwood cellulose	1 g	0.06 g	0.1 g	Not reported	100 mL	10	RT	Unsteady	Onkarappa et al. (2020)
Juncus microfibers	20 g	0.32 g	2 g	Not reported	2 L	10	23°C	Unsteady	Kassab et al. (2020)
Furcraea macrophylla	1 g	0.016 g	0.1 g	0.037 mol	100 mL	10.5	RT	Unsteady	Calderón-Vergara et al. (2020)
Spent coffee grounds	1 g	0.020 g	0.1 g	60 mL	130 mL	10-10.5	RT	Unsteady	Kanai et al. (2020)
Elm pulp	45 g	0.72 g	4.5 g	450 mmol	4.5 L	10	RT	Unsteady	Jiménez-López et al. (2020)

[a] Condition for the consumption of sodium hydroxide.
[b] RT means room temperature.
[c] Unsteady means time to maintain the pH until no sodium hydroxide consumption is observed.

a. condition for the consumption of sodium hydroxide;
b. RT means room temperature;
c. unsteady means time to maintain the pH until no sodium hydroxide consumption is observed.

Cellulose substrate was generally suspended in water containing an amount of TEMPO and NaBr. Once TEMPO had fully dissolved, NaClO solution was gradually added to the mixture. The pH of the mixture was adjusted to 10–11 and maintained at pH 10–11 by continuous feeding of NaOH until further NaOH consumption was not observed. The reaction was quenched by adding ethanol and the pH of the mixture was adjusted to 7 (Chuensangjun et al., 2019a). Many researches recently enhanced the efficiency of the conversion of C6 primary hydroxy to carboxy groups exposed on cellulose surfaces by postoxidation treatment process (Fujisawa et al., 2013, Kuramae et al., 2014, Shinoda et al., 2012). Briefly, after TEMPO-mediated oxidation, the recovered TOCNs are postoxidized by adding sodium chlorite ($NaClO_2$) in acetate buffer at pH 4.8 and mixing for 2 days at room temperature. This is done to convert any residual C6-aldehyde to C6-carboxylate, resulting in an increase of carboxylate content up to a high value of ~~1.7 mmol/g (Isogai et al., 2011). After oxidation, the pH of the mixture is adjusted to 7 and washed thoroughly with water to eliminate inorganic impurities (Chuensangjun et al., 2019a). A scheme for the process of TEMPO-oxidized celluloses (TOCs) preparation by the TEMPO/NaBr/NaClO oxidation system in the water at pH 10 is presented in Figure 6.2.

Achievement of TEMPO-mediated oxidation could be presented in terms of change of cellulose structures. Determination of carboxy group content of the oxidized cellulose using a conductometric titration is the tentative procedure to confirm the likelihood of conversion of C6 primary hydroxy to carboxy groups exposed on

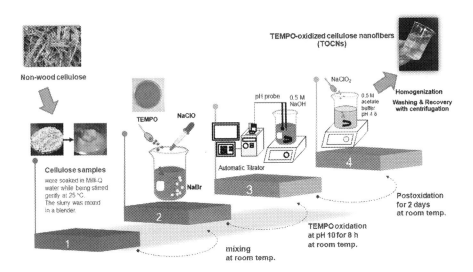

FIGURE 6.2 Process of TEMPO-oxidized wood celluloses preparation by the TEMPO/NaBr/NaClO oxidation system in water at pH 10.

Structure and Surface Modification Techniques

cellulose surfaces (Fan et al., 2008; Perez et al., 2003). The oxidized cellulose was suspended into acid solution (HCl), and after stirring, the suspension was titrated with sodium hydroxide solution (NaOH). The presence of conductivity and pH during titration was plotted against the volume of NaOH. The titration curves showed strong acid corresponding to the excess of HCl and weak acid corresponding to the carboxy content, as shown in Figure 6.3 (Chuensangjun, 2019). The carboxylate content can be calculated using the following equation:

$$\text{Carboxylate content (mmol/g)} = \frac{NV}{W} \text{Carboxylate content (mmol/g)} = \frac{NV}{W} \quad (6.1)$$

where N (M) is the molarity of NaOH solution, V (mL) is the volume of alkali consumed for titration and W (g) is the mass of dried TOCNs sample (Chuensangjun et al., 2019a). The modified C6 primary hydroxy group on cellulose surface to be the aldehyde and carboxy contents is a strong point for the development of functional biocomposites, such as antibacterial film (Huang et al., 2013, Soni et al., 2018), catalytic or medical materials (Tamura et al., 2018; Ranaivoarimanana et al., 2019).

Chemical structure, surface morphology and crystallinity of the TOCNs are basic characteristics for the study of surface modification of cellulose nanofibers by TEMPO-mediated oxidation. Nuclear magnetic resonance (NMR), Fourier transform infrared (FTIR), scanning electron microscopy (SEM), transmission electron microscopy (TEM), atomic force microscopy (AFM) and X-ray diffraction (XRD) are regularly used as analytical implements. For instance, Kato et al. (2003) reported the structure formation of surface-modified cellulose nanofibers measured by ^{1}H and ^{13}C NMR analysis. Their report showed that the spectra of C6 primary hydroxy group was cleaved after the oxidation process, whereas the spectra of C6 carboxy group obviously occurred. The FTIR analysis can be also applied for explaining the chemical structures of surface-modified cellulose nanofibers. The previous reports described that the FTIR absorption bands corresponding to carboxylate group barely occurred in the spectrum of cellulose substrate, but would be increased in intensity after the oxidation process (Chuensangjun et al., 2019a, Fujisawa et al., 2013). Furthermore, the crystallinity of the TOCNs is an important property for surface engineering toward the production of value-added biocomposites, in addition to its morphology. The morphology of TOCNs could be determined by SEM, TEM or AFM analysis (Chuensangjun et al., 2019a, Fujisawa et al., 2013, Puangsin et al., 2017, Puangsin et al., 2013, Saito et al., 2006); meanwhile the crystallinity could be measured by XRD analysis (Kobe et al., 2016, Saito and Isogai, 2004). The application of TEMPO-mediated oxidation for preparation of surface-modified cellulose does not change its crystal structure or the high degree of crystallinity (Chuensangjun et al., 2019a). This property of surface-modified cellulose is a distinctive advantage for improvement of the mechanical and thermal properties of the biocomposites through using the cellulose nanofibers as nucleating agents in the polymer matrix (Fujisawa et al., 2014, Wang and Drzal, 2012).

The TEMPO-mediated oxidation in an aqueous TEMPO/NaBr/NaClO system at pH 10 followed by postoxidation using $NaClO_2$ at pH 4.8 has been recently carried out for producing a high density of carboxylate groups, high crystallinity

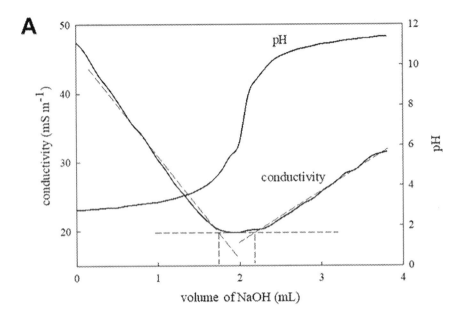

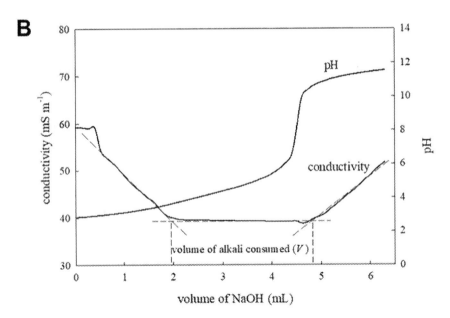

FIGURE 6.3 Conductometric titration curve of (a) original softwood bleached kraft pulp (SBKP) (carboxylate content = 0.03 mmol g^{-1}); (b) TEMPO-oxidized cellulose nanofibers sodium salts (TOCN-COONa) (carboxylate content = 1.68 mmol g^{-1}).

of nanocomposites of poly(lactic acid) grafted on surface-modified TOCNs, or TOCN-*graft*-poly(lactic acid) (Chuensangjun et al., 2019b). The results indicated that using this modification method provided a TOCN product with the highest carboxylate content of 1.68 mmol g^{-1}, which support to the formation of ester bonds between the monomer (L-lactide and PLA-oligomer) and the carboxylate groups on the surfaces of the modified TOCN backbone. Finally, the completed TOCN-*graft*-PLA nanocomposites had a high crystallinity (>71%), a small crystallite size (3.8 nm), good dispersibility in a polar aprotic solvent and enhanced thermal stability.

6.3 SURFACE MODIFIED CELLULOSE TRENDS FOR PRODUCTION OF VALUE-ADDED BIOCOMPOSITES

TOCNs derived from the surface modification of natural fiber by TEMPO-mediated oxidation method not only present high degree of crystallinity but also selective oxidized C6-primary carboxylate groups which can be made other functional groups in chemical reactions, high aspect ratio (length/width) which effect on the mechanical and thermal properties, self-standing light transparent films, gas barrier and adsorption properties (Isogai et al., 2011). Therefore, it has been attractively employed for the production of various value-added biocomposites, as briefly summarized in Figure 6.4.

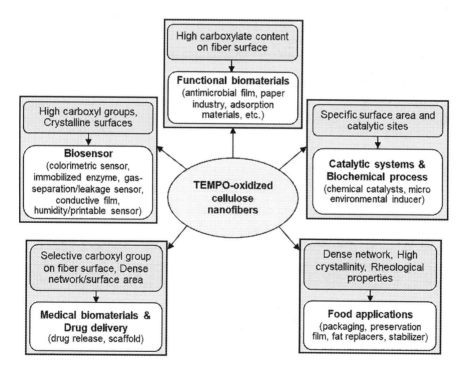

FIGURE 6.4 Remarkable properties of TEMPO-oxidized cellulose nanofiber and its application in various fields.

6.3.1 FUNCTIONAL BIOMATERIALS

Highly selective oxidized C6-primary carboxylate group on the TEMPO-oxidized cellulose surfaces is a good structure for potentially producing functional biocomposites-based celluloses, e.g., antimicrobial film, catalyst support for the paper industry, adsorption materials, etc. Several studies have reported the modification of TEMPO-oxidized cellulose for various functional utilizations, as described below.

The hydrophobic TOCNs films could be prepared by adsorption of a cationic surfactant, namely, n-hexadecyl timethylammonium bromide (cetyltrimethylammonium bromide, CTAB) on the film surfaces (Syverud et al., 2011). The result suggested that one CTAB molecule can be reacted with approximately every 10th glucose monomer of the TOCNs surface which has 0.52 mmol g^{-1} carboxylate content. Because CTAB is an effective antimicrobial agent, the modified film with CTAB layer showed an effective antiseptic property. It has also been expected for a novel utilization of cellulose-based products, such as medical application and food industry. Likewise, Huang et al. (2013) designed a biocomposite film for antibacterial applications by using TEMPO-oxidized cellulose (TOC) as a cellulose-based material. TOC was functionalized with amino acid, i.e., L-phenylalanine (Phe) or L-tryptophan (Try), by using the aqueous solution of N-ethyl-N'-(3-dimethylaminopropyl)carbodiimide hydrochloride (EDAC)/N-hydroxy-succinimide (NHS) as the activator. The amino acid-modified TOC was then mixed with AgNO$_3$ and NaBH$_4$ in ionic liquid [bmim] Cl to form *in situ* silver nanoparticles (AgNPs) in the matrix by reduction reaction before preparing as the nanocomposite films. The amino acid functionalized TOC/AgNPs composite films (i.e., TOC-Phe-AgNPs and TOC-Try-AgNPs) exhibited an excellent *in vitro* antibacterial activity against bacteria, i.e., *Staphylococcus aureus* and *Escherichia coli*.

Grafting technology has been applied for reinforcement biocomposites production for antibacterial packaging. The structural morphology of cellulose isolated from corn cobs bleached pulp was improved by TEMPO-mediated oxidation and then grafting TOCNs with wheat gluten protein (G-CNF), aimed to develop a reinforced PLA bionanocomposite production for active packaging (El-Gendy et al., 2017). Tensile strength and Young's modulus of nanocomposites film increased due to adding of G-CNF. Furthermore, the addition of TiO$_2$ to PLA/G-CNF films exhibited the antibacterial effect against Gram-positive (*Bacillus subtilis* and *Staphylococcus aureus*) and Gram-negative bacteria (*Escherichia coli* and *Pseudomonas aeruginosa*). Besides, Soni et al. (2018) developed the production of antimicrobial and antioxidant films from chitosan/TEMPO-cellulose nanofibers biocomposite. The chitosan/TEMPO-CNFs showed the improved thermal stability and the inhibition against food pathogenic bacteria, i.e. *Salmonella enterica*, *E. coli* O157:H7 and *Listeria monocytogenes*, which can be applied for packaging materials for the food industry.

Gold nanoparticle has also been employed for the production of the functional biocomposites. Azetsu et al. (2013) directly synthesized a gold nanocatalyst on TEMPO-oxidized pulp paper (TOPP). The TEMPO-oxidized pulp fibers with carboxy and aldehyde contents of 1.18 and 0.349 mmol g^{-1}, respectively, were mixed with polyamidepolyamine epichlorohydrin for producing the TOPP by using a papermaking technique. The TOPP containing aldehyde groups showed the simple immersion in an aqueous solution of tetrachloroauric acid (HAuCl$_4$) which resulted in the formation of gold nanoparticles (AuNPs) on the paper surfaces. This novel biocomposite, AuNPs, on TOPP was applied as a catalyst support and demonstrated a great catalytic activity for the chemical reaction.

Adsorption material is another expected functional biocomposites which could be produced from the surface-modified TEMPO-oxidized cellulose. Recently, Pottathara et al. (2020) reported an innovated dye adsorption composite film by incorporation of graphene oxide with TCNs. Basic Violet 1 and Rhodamine 6g dyes were used as the representative dye in adsorption experiment. The results indicated the increase of surface porosity and cationic dye adsorption capacity due to the addition of ammonia-functionalized graphene oxide in TCNF films. Meanwhile, Köse et al. (2020) reported the successful preparation of TOCNs-modified poly(2-hydroxyl ethyl methacyrlate-co-glycidyl methacryalte) (p(HEMA–GMA)) cryogels for adsorption of Fe(II) cations. This novel material has the potential for further use in heavy metal removal for water and soil treatment. The other prototypes of functional biocomposites based on TOCNs have recently been reported, for instance, the antifouling ultrafiltration membranes with self-cleaning property for wastewater treatment (Yang et al., 2021), the nanostructured cellulose sponges for heavy metals removal from seawater (Fiorati et al., 2020a) and removal of anionic organic dye (Riva et al., 2020), a photocatalytic hydrogel for degradation of organic water pollutants (Yue et al., 2020), and the water-repellent and wash-sustainable coatings for textile applications (Ke et al., 2020).

6.3.2 CATALYTIC SYSTEMS AND BIOCHEMICAL PROCESS

TOCN provides the narrowest fiber width (3–5 nm), the largest specific surface area, high-density catalytic sites and surfaces of crystalline due to carboxylate groups, which support to the superior physicochemical properties and bring about to the development of green chemical industries. Kanomata et al. (2018) successfully enhanced the catalytic efficiency in a direct aldol reaction by using cooperative organocatalysis of TOCN and proline. Likewise, Ranaivoarimanana et al. (2019) prepared the TOCN for enhancing the proline-catalyzed Michael additions of ketones to nitroalkenes and, subsequently, discovering the potential of TOCN for the enrichment of enantiomers and enhancing catalytic efficiency in direct asymmetric aldol reactions (Ranaivoarimanana et al., 2020). Furthermore, TOCN could be used as acid catalysts in acetal hydrolysis (Tamura et al., 2018), including to integrate with copper-cystine for catalyzed nitric oxide generation (Darder et al., 2020). In case of biocatalyst (i.e., enzyme), TOCNs can be used as an adsorbent for immobilization of enzymes also, such as the prolonged activity of pyrroloquinoline quinone-dependent glucose dehydrogenase adsorbed on TOCNs (Yamaguchi et al., 2020).

The TOCN has also been applied in microbial fermentation systems for the production of butanol (Hastuti et al., 2019). Briefly, the oil palm empty fruit bunches pulp was used as a substrate for preparation of TOCN, then the TOCN has been added to extractive acetone-butanol-ethanol broth media for butanol production. The increased total butanol yield from the microbial fermentation of *Clostridium saccharoperbutylacetonicum* N1–4 resulted from the induction of better microenvironment by adding TOCN.

6.3.3 BIOSENSOR APPLICATION

The abundant carboxy groups exposed on the oxidized cellulose surface also provide conductivity and barrier properties for a novel biocomposites production, especially sensing materials. For instance, Koga et al. (2013) studied the preparation of transparent conductive films and printable sensors by mixing single-walled carbon nanotubes (CNTs) with TOCNs. Because of the abundant sodium carboxy groups on the crystalline TOCN surfaces, it showed high potential for reinforcing and nanodispersion effects on the CNTs. Moreover, the CNTs/TOCNs nanocomposite dispersion could be applied as printable conductive nanoink because it demonstrated ultrastrong flexible, transparent and conductive properties for various substrates. This research suggests the potential use of cellulose nanofibers for promising humidity or printable sensors and electrical devices applications. The several recent studies also reported the potential of TOCNs for producing the electronic devices, for example, a transparent gel electrolyte actuator based on the composite of TOCN with the ionic liquids and poly(dimethylsiloxane) (Terasawa, 2020), a conductive biocomposite wire (Xu et al., 2020) and Fe_3O_4 nanocomposite aerogel for lithium-ion battery (Liu et al., 2020).

High-performance gas-separation materials and gas-leakage sensors have also been expected to be biocomposites products from TOCN application. In brief, the preparation of metal-organic frameworks (MOFs) on crystalline TOCN with carboxylate groups was conducted by using an ion-exchange reaction between Zn and Na ions in the mixture of TOCN and $Zn(NO_3)_2 \cdot 6H_2O$ (Matsumoto and Kitaoka, 2016). The MOF(ZIF-90)-TOCN film revealed the high flux of CO_2 and high selectivity for separation of CO_2 from a CO_2/CH_4 mixture. Therefore, it is possibly applied for high-performance gas-separation materials including gas-leakage sensors.

Other reports regarding the application of TOCN for biosensor such as colorimetric sensor and urea biosensor application are shortly described. Masruchin et al. (2018) prepared biocomposites of TOCN capable of sensing color change correspond to pH level, by esterification of TOCN with activated bromocresol green, compared to amination approach. This biocomposite showed distinctly the changing color to yellow at the pH below 2 while it became green at pH over 5. It has been expected to possible use in colorimetric applications. The modified cellulose nanocrystal CNC surface was also successfully immobilized with the urease enzyme using carbodiimide coupling method with the support of N-hydroxysuccinimide/N-(3Dimethylaminopropyl)-N'-ethyl-carbodiimide hydrochloride. The immobilized urease on the CNC surface showed the potential use for urea biosensor application (Khalid et al., 2018).

6.3.4 FOOD APPLICATIONS

Although the TOCN presents high crystallinity, dense fibers network and rheological properties, it shows low oxygen permeability and water resistance which might be a drawback for food packaging applications. Shimizu et al. (2016) improved the low oxygen permeability, water-resistance and physical properties of TOCN film through a simple ion-exchange of interfibrillar cross-linkages with multivalent metal ions, aimed for providing a food packaging film and pharmaceutical products. Aqueous metal ions, i.e., $MgCl_2$, $CaCl_2$, $AlCl_3$ and $FeCl_3$ solutions, were reacted with sodium carboxylate groups on TOCN surfaces (TOCN-COONa) for forming the TOCNs-COOM films (M: metal ion) with high Yong's moduli and tensile strength, even at high water contents. In terms of food preservation, Bideau et al. (2017) prepared the bionanocomposite based on TOCN, polypyrrole (PPy) and polyvinyl alcohol (PVA), and characterized its barrier and antioxidant properties for applying in food preservation. The results indicated that the addition of PVA increases the dense network due to hydrogen bonding potential, resulting in water vapor barrier. While a decrease in the oxygen transmission rates of the nanocomposite film was detected due to the PPy particles, which fill the pores of the film. Moreover, a simulation in contact/noncontact with banana was conducted to evaluate a possible application of food packaging. The result demonstrated that the bananas could be retained with no visible brown color and no trace of oxidation, after 5 days of preservation by using TOCN/PVA/PPy film, due to the good barrier property against oxygen of TOCN/PVA/PPy. The TOCN/PVA/PPy film also showed a DPPH (2,2-diphenyl-1-picrylhydrazyl) inhibition capacity and capable degradation under soil burial condition (100 days). In addition, the TOCN was applied as the material base for production of essential oils foam hybrid systems for fresh beef preservation (Zhang et al., 2020).

Besides the application of TOCN for packaging materials, in the recent year, it has been increasingly interesting for applying in food additive systems, for example, fat replacers in foodstuff and food-grade Pickering emulsions. Rheological properties of cellulose nanofibers are most important effects for usability as a food additive. Aaen et al. (2019) reported the rheological properties of TOCN both in pure water and in electrolyte solutions, including rheological interaction with xanthan gum. This report concluded that the addition of high electrolyte effected to a change in flocculation or aggregation which is also reflected in the rheological measurements, such as loss of viscosity. However, the increase in viscosity and storage modulus for TOCN dispersions could be observed after properly adding xanthan. Moreover, Zhang et al. (2019) investigated the improvement of surface hydrophobicity and emulsifying capacity of TEMPO-oxidized bacterial cellulose by adsorption of soy protein isolate nanoparticles through electrostatic attraction. This aimed to apply cellulose nanofibers as a promising Pickering stabilizer in future functional food production.

6.3.5 Medical Biomaterials and Drug Delivery

Novel medical biomaterials and drug delivery are the most attractive high value-added biocomposites application. The surface modification of natural fibers by TEMPO-mediated oxidation method has increasingly been studied for use in the preparation of biomedical materials. For example, Akhlaghi et al. (2013) developed a novel drug delivery system from the modified cellulose nanocrystal (CNC).

TEMPO-mediated oxidation was applied for selectively oxidizing the primary alcohol moieties of CNC surface to carboxy groups, after that chitosan oligosaccharide (CS_{OS}) was mixed with CNC to carry out the carbodiimide reactions using NHS and 1-ethyl-3-(3-dimethylaminopropyl)-carbodiimide as coupling agents. From the reaction, amino groups of CS_{OS} could be reacted with carboxylic acid groups on the oxidized CNC with high conversion (90%) of carboxy groups and the degree of substitution of 0.26, resulting in the formation of CNC-CS_{OS} nanoparticles. The model drug release of procaine hydrochloride (PrHy) loaded to CNC-CS_{OS} nanoparticles with a binding efficiency of 21.5% and drug loading of 14%w/w revealed a fast release of drug in 1 h.

Likewise, due to the TEMPO-mediated oxidation approach could provide the anionic primary C6 hydroxy groups exposed on the surface of cellulose, the anionic TEMPO-oxidized cellulose film has been applied for tissue engineering products such as a 2D scaffolds which be useful for cell viability (Courtenay et al., 2017). Bacterial cellulose film was modified by TEMPO-mediated oxidation to produce anionic cellulose film, thereafter to prepare a scaffold for the study of human osteoblast cancer cells attachment. However, the results described that ionic interactions between cellulose scaffold and phospholipid bilayer of the cell membrane are an important factor in cell attachment. Thus, the better cellulose surfaces should be grafted with a positively charged moiety such as a quaternary ammonium group (GTMAC) to significantly increase cell attachment and spreading. Afterward, the prototype of biomaterials, e.g., bone repair and/or regeneration (Salama et al., 2020), hydrogel scaffolds for diabetic wound healing application (Chao et al., 2020), biomaterial ink for 3D printed lung tissue scaffold (Huang et al., 2020), hydrogel for drug (ibuprofen, IB) release model (Fiorati et al., 2020b), has been developed and recently reported.

6.4 CONCLUDING REMARKS

Surface modification technique of native cellulose has been continuously developed in order to employ in preparation of good properties of biocomposites (good mechanical, thermal and biodegradable properties), which brings about enhancing the feasible application for industrial production processes. Fibers networks with high crystallinity, specific surface area and catalytic sites on the cellulose surfaces are the most important properties for cellulose-reinforced organic matrix or biopolymer production. The previous reports disclosed several pretreatment techniques of structure and surface of natural fibers, namely, physical treatment (e.g., heat treatment, steam explosion, plasma, dielectric discharge, UV treatment, ultrasonication), chemical treatment (e.g., acid hydrolysis, alkaline treatment, acetylation, oxidation), and biological treatment (e.g., microorganisms and enzymatic treatment). These methods promoted; (a) the improvement of cellulose structures such as defibrillation, pore generation and crystallinity; (b) the removal of lignin, hemicellulose and other extracts (oil, wax, hydrophilic pectin) on cellulose surfaces, resulting in better interfacial adhesion; (c) the decrease of hydrophilic nature of fibers, resulting in surface roughness and moisture resistance. Selective oxidation of C6-primary hydroxy groups exposed on cellulose fiber surfaces by using catalytic agent, i.e.,

TEMPO, showed remarkable potential for surface modification under mild condition and low production cost. TOCNs exhibited a high degree of crystallinity, high aspect ratio, light transparent, gas barrier, adsorption property and abundant oxidized C6-primary carboxylate groups on the surfaces. These are significant factors for the invention of the novel value-added biocomposites products, for example, antimicrobial films, adsorption materials, catalysts support, food packaging, sensing materials, drug delivery system and biomedical materials, etc. Many researches which are reviewed and described in this chapter suggest the recent trends and the future opportunity for the biocomposites applications in sustainable manufacturing industries.

ACKNOWLEDGMENTS

This work was continuously supported by the following: Science and Technology Research Institute, King Mongkut's University of Technology North Bangkok; Department of Biotechnology, Faculty of Agro-Industry, Kasetsart University; Laboratory of Bioresources Chemistry, Department of Forest and Forest Products Sciences, Kyushu University. In addition, this work has been funded by the National Research Council of Thailand through the NRCT Senior Research Scholar Program (Contract No. 814–2020).

REFERENCES

Aaen, R., Simon, S., Brodin, F.W. and Syverud, K. (2019), "The potential of TEMPO-oxidized cellulose nanofibrils as rheology modifiers in food systems", *Cellulose*, 26: 5483–5496.

Akhlaghi, S.P., Berry, R.C. and Tam, K.C. (2013), "Surface modification of cellulose nanocrystal with chitosan oligosaccharide for drug delivery applications", *Cellulose*, 20: 1747–1764.

Alsaeed, T., Yousif, B.F. and Ku, H. (2013), "The potential of using date palm fibres as reinforcement for polymeric composites", *Materials and Design*, 43: 177–184.

Amin, F.R., Khalid, H., Zhang, H., Rahman, S., Zhang, R., Liu, G. and Chen, C. (2017), "Pretreatment methods of lignocellulosic biomass for anaerobic digestion", *AMB Express*, 7: 72.

Andlar, M., Rezić, T., Mardetko, N., Kracher, D., Ludwig, R. and Šantek, B. (2018), "Lignocellulose degradation: An overview of fungi and fungal enzymes involved in lignocellulose degradation", *Engineering in Life Sciences*, 18: 768–778.

Auxenfans, T., Crônier, D., Chabbert, B. and Paës, G. (2017), "Understanding the structural and chemical changes of plant biomass following steam explosion pretreatment", *Biotechnology for Biofuels*, 10: 36.

Avérous, L. and Pollet, E. (2012), "Biodegradable polymers", pp. 13–39. In *Environmental Silicate Nano-Biocomposites, Green Energy andTechnology*, Springer-Verlag, London.

Azetsu, A., Koga, H., Yuan, L.-Y. and Kitaoka, T. (2013), "Direct synthesis of gold nanocatalysts on TEMPO-oxidized pulp paper containing aldehyde groups", *Bioresources*, 8(3): 3706–3717.

Béguin, P., Millet, J., Chauvaux, S., Salamitou, S., Tokatlidis, K., Navas, J., Fujino, T., Lemaire, M., Raynaud, O., Daniel, M.K. and Aubert, J.P. (1992), "Bacterial cellulases", *Biochemical Society Transactions*, 20: 42–46.

Bensah, E.C. and Mensah, M. (2013), "Chemical pretreatment methods for the production of cellulosic ethanol: Technologies and innovations", *International Journal of Chemical Engineering*, 2013: 719607.

Bharath, K.N. and Basavarajappa, S. (2016), "Applications of biocomposite materials based on natural fibers from renewable resources: A review", *Science and Engineering of Composite Materials*, 23(2): 123–133.

Bhuiyan, Md.T.R., Hirai, N. and Sobue, N. (2000), "Changes of crystallinity in wood cellulose by heat treatment under dried and moist conditions", *Journal of Wood Science*, 46: 431–436.

Bideau, B., Bras, J., Adoui, N., Loranger, E. and Daneault, C. (2017), "Polypyrrole/nanocellulose composite for food preservation: Barrier and antioxidant characterization", *Food Packaging and Shelf Life*, 12: 1–8.

Bledzki, A.K., Mamun, A.A., Lucka-Gabor, M. and Gutowski, V.S. (2008), "The effects of acetylation on properties of flax fibre and its polypropylene composites", *eXPRESS Polymer Letters*, 2(6): 413–422.

Buschle-diller, G., Fanter, C. and Loth, F. (1999), "Structural changes in hemp fibers as a result of enzymatic hydrolysis with mixed enzyme systems", *Textile Research Journal*, 69(4): 244–251.

Calderón-Vergara, L.A., Ovalle-Serrano, S.A., Blanco-Tirado, C. and Combariza, M.Y. (2020), "Influence of post-oxidation reactions on the physicochemical properties of TEMPO-oxidized cellulose nanofibers before and after amidation", *Cellulose*, 27: 1273–1288.

Calvimontes, A., Mauersberger, P., Nitschke, M., Dutschk, V. and Simon, F. (2011), "Effects of oxygen plasma on cellulose surface", *Cellulose*, 18(3): 803–809.

Camargo, J.S.G., Menezes, A.J., Cruz, N.C., Rangel, E.C. and Delgado-Silva, A.O. (2017), "Morphological and chemical effects of plasma treatment with oxygen (O_2) and sulfur hexafluoride (SF_6) on cellulose surface", *Materials Research*, 20: 842–850.

Chao, F.-C., Wu, M.-H., Chen, L.-C., Lin, H.-L., Liu, D.-Z., Ho, H.-O. and Sheu, M.-T. (2020), "Preparation and characterization of chemically TEMPO-oxidized and mechanically disintegrated sacchachitin nanofibers (SCNF) for enhanced diabetic wound healing", *Carbohydrate Polymers*, 229: 115507.

Chuensangjun, C. (2019), "Biotechnological synthesis of polylactic acid for the production of PLA cellulose compound for injection molding applications", PhD. Thesis, Kasetsart University.

Chuensangjun, C., Kanomata, K., Kitaoka, T., Chisti, Y. and Sirisansaneeyakul, S. (2019a), "Surface-modified cellulose nanofibers-*graft*-poly(lactic acid)s made by ring-opening polymerization of L-lactide", *Journal of Polymers and the Environment*, 27: 847–861.

Chuensangjun, C., Kitaoka, T., Chisti, Y. and Sirisansaneeyakul, S. (2019b), "Chemo-enzymatic preparation and characterization of cellulose nanofibers-*graft*-poly(lactic acid)s", *European Polymer Journal*, 114: 308–318.

Courtenay, J.C., Johns, M.A., Galembeck, F., Deneke, C., Lanzoni, E.M., Costa, C.A., Scott, J.L. and Sharma, R.I. (2017), "Surface modified cellulose scaffolds for tissue engineering", *Cellulose*, 24: 253–267.

Cui, L., Liu, Z., Si, C., Hui, L., Kang, N. and Zhao, T. (2012), "Influence of steam explosion pretreatment on the composition and structure of wheat straw", *Bioresources*, 7(3): 4202–4213.

Dang, Z., Zhang, J. and Ragauskas, A.J. (2007), "Characterizing TEMPO-mediated oxidation of ECF bleached softwood kraft pulps", *Carbohydrate Polymers*, 70: 310–317.

Darder, M., Karan, A., Real, G. and DeCoster, M.A. (2020), "Cellulose-based biomaterials integrated with copper-cystine hybrid structures as catalysts for nitric oxide generation", *Materials Science and Engineering: C*, 108: 110369.

Ehman, N.V., Tarrés, Q., Delgado-Aguilar, M., Vallejos, M.E., Felissia, F., Area, M.C. and Mutjé, P. (2016), "From pine sawdust to cellulose nanofires", *Cellulose Chemistry and Technology*, 50: 361–367.

El-Gendy, A., Abou-Zeid, R.E., Salama, A., Diab, M.A. and El-Sakhawy, M. (2017), "TEMPO-oxidized cellulose nanofibers/polylactic acid/TiO$_2$ as antibacterial bionanocomposite for active packaging", *Egyptian Journal of Chemistry*, 60(6): 1007–1014.

Fan, Y., Saito, T. and Isogai, A. (2008), "Chitin nanocrystals prepared by TEMPO-mediated oxidation of α-Chitin", *Biomacromolecules*, 9: 192–198.

Fathi, B., Harirforoush, M., Foruzanmehr, M., Elkoun, S. and Robert, M. (2017), "Effect of TEMPO oxidation of flax fibers on the grafting efficiency of silane coupling agents", *Journal of Materials Science*, 52: 10624–10636.

Fiorati, A., Grassi, G., Graziano, A., Liberatori, G., Pastori, N., Melone, L., Bonciani, L., Pontorno, L., Punta, C. and Corsi, I. (2020a), "Eco-design of nanostructured cellulose sponges for sea-water decontamination from heavy metal ions", *Journal of Cleaner Production*, 246: 119009.

Fiorati, A., Negrini, N.C., Baschenis, E., Altomare, L., Faré, S., Schieroni, A.G., Piovani, D., Mendichi, R., Ferro, M., Castiglione, F., Mele, A., Punta, C. and Melone, L. (2020b), "TEMPO-nanocellulose/Ca^{2+} hydrogels: Ibuprofen drug diffusion and *in vitro* cytocompatibility", *Materials*, 13(1): 183.

Fiore, V., Scalici, T., Nicoletti, F., Vitale, G., Prestipino, M. and Valenza, A. (2016), "A new eco-friendly chemical treatment of natural fibres: Effect of sodium bicarbonate on properties of sisal fibre and its epoxy composites", *Composites Part B*, 85: 150–160.

Fujisawa, S., Saito, T., Kimura, S., Iwata, T. and Isogai, A. (2013), "Surface engineering of ultrafine cellulose nanofibrils toward polymer nanocomposite materials", *Biomacromolecules*, 14: 541–1546.

Fujisawa, S., Zhang, J., Saito, T., Iwata, T. and Isogai, A. (2014), "Cellulose nanofibrils as templates for the design of poly(L-lactide)-nucleating surfaces", *Polymer*, 55: 2937–2942.

Fukuzumi, H., Saito, T., Okita, Y. and Isogai, A. (2010), "Thermal stabilization of TEMPO-oxidized cellulose", *Polymer Degradation and Stability*, 95: 1502–1508.

García-Cubero, M.T., Coca, M., Bolado, S. and González-Benito, G. (2010), "Chemical oxidation with ozone as pre-treatment of lignocellulosic materials for bioethanol production", *Chemical Engineering Transactions*, 21: 1273–1278.

Gassan, J. and Gutowski, V.S. (2000), "Effects of corona discharge and UV treatment on the properties of jute-fibre epoxy composites", *Composites Science and Technology*, 60: 2857–2863.

Hastuti, N., Darmayanti, R.F., Hardiningtyas, S.D., Kanomata, K., Sonomoto, K., Goto, M. and Kitaoka, T. (2019), "Nanocellulose from oil palm biomass to enhance microbial fermentation of butanol for bioenergy applications", *Bioresources*, 14(3): 6936–6957.

He, Q., Ziegler-Devin, I., Chrusciel, L., Obame, S.N., Hong, L., Lu, X. and Brosse, N. (2020), "Lignin-first integrated steam explosion process for green wood adhesive application", *ACS Sustainable Chemistry & Engineering*, 8: 5380–5392.

Hossain, S.I., Hasan, M., Hasan, Md.N. and Hassan, A. (2013), "Effect of chemical treatment on physical, mechanical and thermal properties of ladies finger natural fiber", *Advances in Materials Science and Engineering*, 2013: 824274.

Huang, K., Zhang, M., Zhang, G., Jiang, X. and Huang, D. (2014), "Acetylation modification of rice straw fiber and its thermal properties", *Cellulose Chemistry and Technology*, 48: 199–207.

Huang, L., Yuan, W., Hong, Y., Fan, S., Yao, X., Ren, T., Song, L., Yang, G. and Zhang, Y. (2020), "3D printed hydrogels with oxidized cellulose nanofibers and silk fibroin for the proliferation of lung epithelial stem cells", *Cellulose*. doi: 10.1007/s10570-020-03526-7.

Huang, M., Chen, F., Jiang, Z. and Li, Y. (2013), "Preparation of TEMPO-oxidized cellulose/amino acid/nanosilver biocomposite film and its antibacterial activity", *International Journal of Biological Macromolecules*, 62: 608–613.

Isogai, A., Saito, T. and Fukuzumi, H. (2011), "TEMPO-oxidized cellulose nanofibers", *Nanoscale*, 3: 71–85.

Jhu, Y.S., Hung, K.C., Xu, J.W. and Wu, J.H. (2019), "Effects of acetylation on the thermal decomposition kinetics of Makino bamboo fibers", *Wood Science and Technology*, 53: 873–887.

Jiménez-López, L., Eugenio, M.E., Ibarra, D., Darder, M., Martín, J.A. and Martín-Sampedro, R. (2020), "Cellulose nanofibers from a dutch elm disease-resistant *Ulmus minor* clone", *Polymers*, 12: 2450.

Kabir, M.M., Wang, H., Lau, K.T. and Cardona, F. (2012), "Chemical treatments on plant-based natural fibre reinforced polymer composites: an overview", *Composites: Part B*, 43: 2883–2892.

Kalia, S. and Vashistha, S. (2012), "Surface modification of sisal fibers (*Agave sisalana*) using bacterial cellulase and methyl methacrylate", *Journal of Polymers and the Environment*, 20: 142–151.

Kan, C.W., Yuen, C.W.M. and Jiang, S.Q. (2007), "Effect of surface treatment on the enzymatic treatment of cellulosic fiber", *Surface Review and Letters*, 14(4): 565–569.

Kanai, N., Honda, T., Yoshihara, N., Oyama, T., Naito, A., Ueda, K. and Kawamura, I. (2020), "Structural characterization of cellulose nanofibers isolated from spent coffee grounds and their composite films with poly(vinyl alcohol): a new non-wood source", *Cellulose*, 27: 5017–5028.

Kanomata, K., Tatebayashi, N., Habaki, X. and Kitaoka, T. (2018), "Cooperative catalysis of cellulose nanofiber and organocatalyst in direct aldol reactions", *Scientific Reports*, 8: 4098.

Kassab, Z., Mansouri, S., Tamraoui, Y., Sehaqui, H., Hannache, H., Qaiss, A.E.K. and Achaby, M.E. (2020), "Identifying *Juncus* plant as viable source for the production of micro- and nano-cellulose fibers: Application for PVA composite materials development", *Industrial Crops & Products*, 144: 112035.

Kato, Y., Matsuo, R. and Isogai, A. (2003), "Oxidation process of water-soluble starch in TEMPO-mediated system", *Carbohydrate Polymers*, 51: 69–75.

Ke, W.-T., Chiu, H.-L. and Liao, Y.-C. (2020), "Multifunctionalized cellulose nanofiber for water-repellent and wash-sustainable coatings on fabrics", *Langmuir*, 36: 8144–8151.

Khalid, W.E.F.W., Heng, L.Y. and Arip, M.N.M. (2018), "Surface modification of cellulose nanomaterial for urea biosensor application", *Sains Malaysiana*, 47(5): 941–949.

Kobe, R., Iwamoto, S., Endo, T., Yoshitani, K. and Teramoto, Y. (2016), "Stretchable composite hydrogels incorporating modified cellulose nanofiber with dispersibility and polymerizability: Mechanical property control and nanofiber orientation", *Polymer*, 97: 480–486.

Koga, H., Saito, T., Kitaoka, T., Nogi, M., Suganuma, K. and Isogai, A. (2013), "Transparent, conductive, and printable composites consisting of TEMPO-Oxidized nanocellulose and carbon nanotube", *Biomacromolecules*, 14: 1160–1165.

Konno, N., Kimura, M., Okuzawa, R., Nakamura, Y., Ike, M., Hayashi, N., Obara, A., Sakamoto, Y. and Habu, N. (2016), "Preparation of cellulose nanofibers from waste mushroom bed of shiitake (*Lentinus edodes*) by TEMPO-mediated oxidation", *Mokuzai Hozon (Wood Protection)*, 42(3): 157–164.

Köse, K., Mavlan, M., Nuruddin, M. and Youngblood, J.P. (2020), "TEMPO-oxidized cellulose nanofiber based polymeric adsorbent for use in iron removal", *Cellulose*, 27: 4623–4635.

Kuramae, R., Saito, T. and Isogai, A. (2014), "TEMPO-oxidized cellulose nanofibrils prepared from various plant holocelluloses", *Reactive & Functional Polymers*, 85: 126–133.

Lopes, T.A., Bufalino, L., Claro, P.I.C., Martins, M.A., Tonoli, G.H.D. and Mendes, L.M. (2018), "The effect of surface modifications with corona discharge in pinus and eucalyptus nanofibril films", *Cellulose*, 25: 5017–5033.

Li, X., Tabil, L.G. and Panigrahi, S. (2007), "Chemical treatments of natural fiber for use in natural fiber-reinforced composites: a review", *Journal of Polymers and the Environment*, 15: 25–33.

Liu, J., Korpinen, R., Mikkonen, K.S., Willfor, S. and Xu, C. (2014), "Nanofibrillated cellulose originated from birch sawdust after sequential extractions: A promising polymeric material from waste to films", *Cellulose*, 21: 2587–2598.

Liu, L., Xiang, Y., Zhang, R., Li, B. and Yu, J. (2019), "Effect of NaClO dosage on the structure of degummed hemp fibers by 2,2,6, 6-tetramethyl-1-piperidinyloxylaccase degumming", *Textile Research Journal*, 89(1): 76–86.

Liu, Y., Chen, J., Liu, Z., Xu, H., Zheng, Y., Zhong, J., Yang, Q., Tian, H., Shi, Z., Yao, J. and Xiong, C. (2020), "Facile fabrication of Fe_3O_4 nanoparticle/carbon nanofiber aerogel from Fe-ion crosslinked cellulose nanofibrils as anode for lithium-ion battery with superhigh capacity", *Journal of Alloys and Compounds*, 829: 154541.

Lucas, M., Hanson, S.K., Wagner, G.L., Kimball, D.B. and Rector, K.D. (2012), "Evidence for room temperature delignification of wood using hydrogen peroxide and manganese acetate as a catalyst", *Bioresource Technology*, 119: 174–180.

Macedo, M.J.P., Silva, G.S., Feitor, M.C., Costa, T.H.C., Ito, E.N. and Melo, J.D.D. (2020), "Surface modification of kapok fibers by cold plasma surface treatment", *Journal of Materials Research and Technology*, 9(2): 2467–2476.

Maki, M., Leung, K.T. and Qin, W. (2009), "The prospects of cellulase-producing bacteria for the bioconversion of lignocellulosic biomass", *International Journal of Biological Sciences*, 5(5): 500–516.

Masruchin, N., Park, B.-D. and Lee, J.M. (2018), "Surface modification of TEMPO-oxidized cellulose nanofibrils for composites to give color change in response to pH level", *Cellulose*, 25: 7079–7090.

Matsumoto, M. and Kitaoka, T. (2016), "Ultraselective gas separation by nanoporous metal-organic frameworks embedded in gas-barrier nanocellulose films", *Advanced Materials*, 28: 1765–1769.

Montanari, S., Roumani, M., Heux, L. and Vignon, M.R. (2005), "Topochemistry of carboxylated cellulose nanocrystals resulting from TEMPO-mediated oxidation", *Macromolecules*, 38: 1665–1671.

Nam, T.H., Ogihara, S., Tung, N.H. and Kobayashi, S. (2011), "Effect of alkali treatment on interfacial and mechanical properties of coir fiber reinforced poly(butylene succinate) biodegradable composites", *Composites: Part B*, 42: 1648–1656.

Neves, R.M., Lopes, K.S., Zimmermann, M.G.V., Poletto, M. and Zattera, A.J. (2020), "Cellulose nanowhiskers extracted from TEMPO-oxidized Curaua fibers", *Journal of Natural Fibers*, 17: 1355–1365.

Nooy, A.E.J., Besemer, A.C. and van Bekkum, H. (1995), "Highly selective nitroxyl radical-mediated oxidation of primary alcohol groups in water-soluble glucans", *Carbohydrate Research*, 269: 89–98.

Onkarappa, H.S., Prakash, G.K., Pujar, G.H., Rajith Kumar, C.R., Latha, M.S. and Betageri, V.S. (2020), "*Hevea brasiliensis* mediated synthesis of nanocellulose: Effect of preparation methods on morphology and properties" *International Journal of Biological Macromolecules*, 160: 1021–1028.

Oushabi, A., Sair, S., Hassani, F.O., Abboud, Y., Tanane, O. and El Bouari, A. (2017), "The effect of alkali treatment on mechanical, morphological and thermal properties of date palm fibers (DPFs): study of the interface of DPF-Polyurethane composite", *South African Journal of Chemical Engineering*, 23: 116–123.

Patiño-Masó, J., Serra-Parareda, F., Tarrés, Q., Mutjé, P., Espinach F.X. and Delgado-Aguilar, M. (2019), "TEMPO-oxidized cellulose nanofibers: A potential bio-based superabsorbent for diaper production", *Nanomaterials*, 9: 1271.

Perez, D.S., Montanari, S. and Vignon, M.R. (2003), "TEMPO-Mediated Oxidation of Cellulose III", *Biomacromolecules*, 4: 1417–1425.

Pielhop, T., Amgarten, J., Rohr, P.R. and Studer, M.H. (2016), "Steam explosion pretreatment of softwood: the effect of the explosive decompression on enzymatic digestibility", *Biotechnology for Biofuels*, 9: 152.

Pottathara, Y.B., Narwade, V.N., Bogle, K.A. and Kokol, V. (2020), "TEMPO-oxidized cellulose nanofibrils-graphene oxide composite films with improved dye adsorption properties", *Polymer Bulletin*, 77: 6175–6189.

Prajapati, B.P., Jana, U.K., Suryawanshi, R.K. and Kango, N. (2020), "Sugarcane bagasse saccharification using *Aspergillus tubingensis* enzymatic cocktail for 2G bio-ethanol production", *Renewable Energy*, 152: 653–663.

Pramanik, K. and Sahu, S. (2017), "Biological treatment of lignocellulosic biomass to bioethanol", *Advances in Biotechnology & Microbiology*, 5(5). ISSN: 2474-7637.

Praskalo, J., Kostic, M., Potthast, A., Popov, G., Pejic, B. and Skundric, P. (2009), "Sorption properties of TEMPO-oxidized natural and man-made cellulose fibers", *Carbohydrate Polymers*, 77: 791–798.

Puangsin, B., Soeta, H., Saito, T. and Isogai, A. (2017), "Characterization of cellulose nanofibrils prepared by direct TEMPO-mediated oxidation of hemp bast", *Cellulose*, 24: 3767–3775.

Puangsin, B., Yang, Q., Saito, T. and Isogai, A. (2013), "Comparative characterization of TEMPO-oxidized cellulose nanofibril films prepared from non-wood resources", *International Journal of Biological Macromolecules*, 59: 208–213.

Ragoubi, M., George, B., Molina, S., Bienaimé, D., Merlin, A., Hiver, J.-M. and Dahoun, A. (2012), "Effect of corona discharge treatment on mechanical and thermal properties of composites based on miscanthus fibres and polylactic acid or polypropylene matrix", *Composites: Part A*, 43: 675–685.

Ranaivoarimanana, N.J., Habaki, X., Uto, T., Kanomata, K., Yui, T. and Kitaoka, T. (2020), "Nanocellulose enriches enantiomers in asymmetric aldol reactions", *RSC Advances*, 10: 37064–37071.

Ranaivoarimanana, N.J., Kanomata, K. and Kitaoka, T. (2019), "Concerted catalysis by nanocellulose and proline in organocatalytic Michael additions", *Molecules*, 24: 1231.

Reddy, K.O., Maheswari, C.U., Shukla, M., Song, J.I. and Rajulu, A.V. (2013), "Tensile and structural characterization of alkali treated Borassus fruit fine fibers", *Composites: Part B*, 44: 433–438.

Reddy, T.R.K., Kim, H.-J. and Park, J.-W. (2016), "Renewable biocomposite properties and their applications", pp. 177–197. In Poletto, M. (eds.) *Composites from Renewable and Sustainable Materials*. IntechOpen, London.

Riva, L., Pastori, N., Panozzo, A., Antonelli, M. and Punta, C. (2020), "Nanostructured cellulose-based sorbent materials for water decontamination from organic dyes", *Nanomaterials*, 10: 570.

Saito, T. and Isogai, A. (2004), "TEMPO-mediated oxidation of native cellulose. The effect of oxidation conditions on chemical and crystal structures of the water-insoluble fractions", *Biomacromolecules*, 5: 1983–1989.

Saito, T., Kimura, S., Nishiyama, Y. and Isogai, A. (2007), "Cellulose nanofibers prepared by TEMPO-mediated oxidation of native cellulose", *Biomacromolecules*, 8: 2485–2491.

Saito, T., Nishiyama, Y., Putaux, J.-L., Vignon, M. and Isogai, A. (2006), "Homogeneous suspensions of individualized microfibrils from TEMPO-catalyzed oxidation of native cellulose", *Biomacromolecules*, 7(6): 1687–1691.

Saito, T., Okita, Y., Nge, T.T., Sugiyama, J. and Isogai, A. (2006), "TEMPO-mediated oxidation of native cellulose: Microscopic analysis of fibrous fractions in the oxidized products", *Carbohydrate Polymers*, 65: 435–440.

Salama, A., Abou-Zeid, R.E., Cruz-Maya, I. and Guarino, V. (2020), "Soy protein hydrolysate grafted cellulose nanofibrils with bioactive signals for bone repair and regeneration", *Carbohydrate Polymers*, 229: 115472.

Saritha, M., Arora, A. and Lata (2012), "Biological pretreatment of lignocellulosic substrates for enhanced delignification and enzymatic digestibility", *Indian Journal of Microbiology*, 52(2): 122–130.

Schneider, W.D.H., Fontana, R.C., Baudel, H.M., Siqueira, F.G., Rencoret, J., Gutiérrez, A., Eugenio, L.I., Prieto, A., Martínez, M.J., Martínez, Á.T. and Dillon, A.J.P. (2020), "Lignin degradation and detoxification of eucalyptus wastes by on-site manufacturing fungal enzymes to enhance second-generation ethanol yield", *Applied Energy*, 262: 114493.

Seghini, M.C., Tirillò, J., Bracciale, M.P., Touchard, F., Chocinski-Arnault, L., Zuorro, A., Lavecchia, R. and Sarasini, F. (2020), "Surface modification of flax yarns by enzymatic treatment and their interfacial adhesion with thermoset matrices", *Applied Sciences*, 10: 2910.

Sekhon, J.K., Maurer, D., Wang, T., Jung, S. and Rosentrater, K.A. (2018), "Ethanol production by soy fiber treatment and simultaneous saccharification and co-fermentation in an integrated corn-soy biorefinery", *Fermentation*, 4: 35.

Sharma, A., Thakur, M., Bhattacharya, M., Mandal, T. and Goswami, S. (2018), "Commercial application of cellulose nano-composites – a review", *Biotechnology Reports*, 21: e00316.

Shimizu, M., Saito, T. and Isogai, A. (2016), "Water-resistant and high oxygen-barrier nanocellulose films with interfibrillar cross-linkages formed through multivalent metal ions", *Journal of Membrane Science*, 500: 1–7.

Shinoda, R., Saito, T., Okita, Y. and Isogai, A. (2012), "Relationship between length and degree of polymerization of TEMPO-oxidized cellulose nanofibrils", *Biomacromolecules*, 13: 842–849.

Sirisansaneeyakul, S. and Chuensangjun, C. (2017), "Cellulose nanofibers from oil-palm-trunk residues and empty fruit bunches of the oil palm tree and the production process", Petty patent no. 1703002127, Thailand.

Soni, B., Mahmoud, B., Chang, S., El-Giar, E.M. and Hassan, E.B. (2018), "Physicochemical, antimicrobial and antioxidant properties of chitosan/TEMPO biocomposite packaging films", *Food Packaging and Shelf Life*, 17: 73–79.

Syverud, K., Xhanari, K., Chinga-Carrasco, G., Yu, Y. and Stenius, P. (2011), "Films made of cellulose nanofibrils: surface modification by adsorption of a cationic surfactant and characterization by computer-assisted electron microscopy", *Journal of Nanoparticle Research*, 13: 773–782.

Tamura, Y., Kanomata, K. and Kitaoka, T. (2018), "Interfacial hydrolysis of acetals on protonated TEMPO-oxidized cellulose nanofibers", *Scientific Reports*, 8: 5021.

Tang, P., Abdul, P.M., Engliman, N.S. and Hassan, O. (2018), "Effects of pretreatment and enzyme cocktail composition on the sugars production from oil palm empty fruit bunch fiber (OPEFBF)", *Cellulose*, 25: 4677–4694.

Terasawa, N. (2020), "Self-standing high-performance transparent actuator based on poly(dimethylsiloxane)/TEMPO-oxidized cellulose nanofibers/ionic liquid gel", *Langmuir*, 36: 6154–6159.

Thapa, S., Mishra, J., Arora, N., Mishra, P., Li, H., O'Hair, J., Bhatti, S. and Zhou, S. (2020), "Microbial cellulolytic enzymes: diversity and biotechnology with reference to lignocellulosic biomass degradation", *Reviews in Environmental Science and Biotechnology*, 19: 621–648.

Tramontina, R., Brenelli, L.B., Sousa, A., Alves, R., Arenas, A.M.Z., Nascimento, V.M., Rabelo, S.C., Freitas, S., Ruller, R. and Squina, F.M. (2020), "Designing a cocktail containing redox enzymes to improve hemicellulosic hydrolysate fermentability by microorganisms", *Enzyme and Microbial Technology*, 135: 09490.

Vizireanu, S., Panaitescu, D.M., Nicolae, C.A., Frone, A.N., Chiulan, I., Ionita, M.D., Satulu, V., Carpen, L.G., Petrescu, S., Birjega, R. and Dinescu, G. (2018), "Cellulose defibrillation and functionalization by plasma in liquid treatment", *Scientific Reports*, 8: 15473.

Wang, K., Yang, H., Wang, W. and Sun, R. (2013), "Structural evaluation and bioethanol production by simultaneous saccharification and fermentation with biodegraded triploid polar", *Biotechnology for Biofuels*, 6: 42.

Wang, T. and Drzal, L.T. (2012), "Cellulose-nanofiber-reinforced poly(lactic acid) composites prepared by a water-based approach", *ACS Applied Materials & Interfaces*, 4: 5079–5085.

Wielen, L.C.V., Östenson, M., Gatenholm, P. and Ragauskas, A.J. (2006), "Surface modification of cellulosic fibers using dielectric-barrier discharge", *Carohydrate Polymers*, 65: 179–184.

Wu, J., Zhu, W., Shi, X., Li, Q., Huang, C., Tian, Y. and Wang, S. (2020), "Acid-free preparation and characterization of kelp (*Laminaria japonica*) nanocelluloses and their application in Pickering emulsions", *Carbohydrate Polymers*, 236: 115999.

Xiao, H., Zhang, W., Wei, Y. and Chen, L. (2018), "Carbon/ZnO nanorods composites templated by TEMPO-oxidized cellulose and photocatalytic activity for dye degradation", *Cellulose*, 25: 1809–1819.

Xu, J., Zhou, Z., Cai, J. and Tian, J. (2020), "Conductive biomass-based composite wires with cross-linked anionic nanocellulose and cationic nanochitin as scaffolds",) *International Journal of Biological Macromolecules*, 156: 1183–1190.

Yamaguchi, A., Nakayama, H., Morita, Y., Sakamoto, H., Kitamura, T., Hashimoto, M. and Suye, S. (2020), "Enhanced and prolonged activity of enzymes adsorbed on TEMPO-oxidized cellulose nanofibers", *ACS Omega*, 5: 18826–18830.

Yang, M., Hadi, P., Yin, X., Yu, J., Huang, X., Ma, H., Walker, H. and Hsiao, B.S. (2021), "Antifouling nanocellulose membranes: How subtle adjustment of surface charge lead to self-cleaning property", *Journal of Membrane Science*, 618: 118739.

Yildizhan, Ş., Çalik, A., Özcanli, M. and Serin, H. (2018), "Bio-composite materials: a short review of recent trends, mechanical and chemical properties, and applications", *European Mechanical Science*, 2(3): 83–91.

Yue, Y., Wang, X., Wu, Q., Han, J. and Jiang, J. (2020), "Highly recyclable and super-tough hydrogel mediated by dual-functional TiO_2 nanoparticles toward efficient photodegradation of organic water pollutants", *Journal of Colloid and Interface Science*, 564, 99–112.

Zhang, X., Liu, Y., Wang, Y., Luo, X, Li, Y., Li, B., Wang, J. and Liu, S. (2019), "Surface modification of cellulose nanofibrils with protein nanoparticles for enhancing the stabilization of O/W pickering emulsions", *Food Hydrocolloids*, 97: 105180.

Zhang, Y., Lu, X., Pizzi, A. and Delmotte, L. (2003), "Wheat straw particleboard bonding improvements by enzyme pretreatment", *Holz als Roh-und Werkstoff*, 61: 49–54.

Zhang, Z., Wang, X., Gao, M., Zhao, Y. and Chen, Y. (2020), "Sustained release of an essential oil by a hybrid cellulose nanofiber foam system", *Cellulose*, 27: 2709–2721.

7 Design and Fabrication Technology in Biocomposite Manufacturing

K. M. Faridul Hasan, Péter György Horváth, Kovács Zsolt, and Tibor Alpár
University of Sopron

CONTENTS

7.1	Introduction	158
7.2	The Term "BC"	159
7.3	Reinforcement/Filler	160
7.4	Biopolymer Matrix	161
7.5	Design of BC Materials	161
7.6	Engineering Design of BCs	163
	7.6.1 Prediction of Strength and Stiffness Features	163
	7.6.2 Failure Prediction of BCs (Unidirectional Lamina)	163
	7.6.2.1 Theory of Maximum Stress	164
	7.6.2.2 Theory of Maximum Strain	166
	7.6.2.3 Theory of Azzi-Tasi-Hill	166
	7.6.2.4 Theory of Tasi-Wu Failure	166
	7.6.3 Failure Prediction of BCs (Un-Notched Lamina)	166
	7.6.4 Failure Prediction of BCs (Randomly Oriented Fiber Laminates)	166
7.7	Modelling of the Materials for BCs	166
	7.7.1 Modeling of Materials in Terms of BCs Manufacturing	167
7.8	Fabrication Techniques of BCs	169
	7.8.1 Polymer Preparation	169
	7.8.2 Calendering Technique	170
	7.8.3 Extrusion Technique	171
	7.8.4 Extruder Die Technique	173
	7.8.5 Injection Molding Technique	173
	7.8.6 Clamping Unit	174
	7.8.7 Injection Unit	175
	7.8.8 Mold Tools	175
	7.8.9 Pressing and Hot-Pressing Techniques	177

DOI: 10.1201/9781003137535-7

7.9   BCs Characterizations 180
       7.9.1   Tensile Characterization of BCs 180
       7.9.2   Flexural Characterization of BCs 180
       7.9.3   Impact Strength Characterization of BCs 181
7.10   Future Prospects 181
7.11   Conclusion 182
References 183

7.1 INTRODUCTION

The focus on sustainable products as well as processes is increasing continuously throughout the world because of the constantly raising environmental issues. The scientists and researchers are relentlessly trying to develop novel technology-based innovative and competitive products that would meet the customer demands without harming the environment. Biocomposite is one of the major sustainable products, which is also getting significant appeal to the engineers for cost-effective and feasible productions [1–4]. The traditional composites were not produced through ensuring sustainable production protocol in so many cases as natural fibers or polymers were not utilized for the fabrication process in vast scale. The use of green plant materials or polymers is a new and interesting concept which is getting tremendous appeal from the sustainability's perspective. However, more research is still needed for optimizing the design of BCs through minimizing the cost. In this regard, the innovative production processes could contribute much for manufacturing competitive BCs which could be widely applied in the field of aerospace, biomedical, constructions, electronics, packaging, and transportations sectors [5–7].

In general, BCs are produced from green, biodegradable, and renewable material sources [8–10]. Various types of plant extracts and natural fibers [11] are the main reinforcing materials used along with relevant and suitable additives and polymers for producing BCs. Materials of biological origin like wood, wood-based natural fibers, recycled woods, byproducts from crops, and regenerated celluloses are some of the most significantly used plant-based materials for BCs manufacturing [6,12–17]. The selection of plant materials for specific BC manufacturing is a challenging task because of the complex relationship between the reinforcements and polymers. The researchers, designers, and scientists are working together for developing successful designs and optimized methods to ensure production of low-cost, durable, and feasible BCs. In this regard, mechanical, thermal, physical, and electrical properties along with recyclability, availability, durability, economical perspective, and after all the manufacturing ability of the materials should be considered [18–22]. Besides, the selection of perfect material is an interdisciplinary task that needs the attention of architectural design, material technology and science, chemistry and physics, industrial design and engineering knowledge [23–27], and after all definitely the expertise on it. The woven or nonwoven textile fabrics [28–34] are very familiar and widely used textile materials. Recently laminated composites from textiles are also getting attentions [35–37].

The design and fabrication are highly important tasks for BCs production. Al-Oqla et al. [22] have divided the engineering design of BCs in three ways: (a) material

Design and Fabrication Technology

selection, (b) geometry identifications, and (c) manufacturing process selections. The appropriate design of BCs could minimize the failure possibilities [38–40] and associated costs from the final products thus enhancing production efficiency and capturing competitive and potential market. There are several manufacturing methods popularly used for BCs productions such as hand lay-up, spray-up, extrusion, pultrusion, vacuum infusion, filament winding, injection moulding, sheet mold compounding, compression moulding, and transfer moulding [41]. Besides, the numerical finite element analysis (FEA) is also employed by the researchers, academics, and industries to investigate stress field and interface between the polymers and reinforcement materials through simulation. Different elastic and geometric properties of reinforcement fiber/wood material and polymer can also be configured by using FEA-based models [42]. Through employing FEA, versatile material modelling, complex structural dimensions, and shape with various boundary conditions can be easily solved [43]. The usage of FEA can solve the experimental problems virtually; the solutions ware easy to analyze and giving initiatives to modify the experimental design further based on the results.

7.2 THE TERM "BC"

With the span of time, advanced technologies and industrial revolutions, the awareness against the generation of nonbiodegradable waste is increasing around the world. So, the necessity of sustainable and environment-friendly products is getting significant attention. The main raw materials of BCs are natural fibers or wood either in virgin or saw dust from industry, see Figure 7.1. In general, biofillers provide higher stiffness and tensile strengths when reinforced with compatible polymer resins which transmits shear stress to the matrix uniformly and protects the produced BCs from external forces/destructions [6,44,45]. However, the most compelling features of BCs are biodegradability and biocompatibility, in view of generating less burden and carbon footprint to the environment [46]. Although natural fibers provide less durability but when used with polymeric resin/synthetic fibers, they exhibit higher strength, stability, and significant performances. The major challenge is to use the suitable polymeric resin for the BCs production. Durability and sustainable feature of a BC should be adjusted to the expected product life cycle [47,48]. Automotive, airplane, and transportation companies are trying to manufacture and use products having fewer negative impacts on the environment. Besides, different countries are also implementing strict rules and regulations for nonbiodegradable and harmful products. Hence, the manufacturers and researchers are giving more efforts to produce lightweight, low-cost, and sustainable BCs continuously throughout the world.

FIGURE 7.1 Normalized scheme for BCs production.

7.3 REINFORCEMENT/FILLER

Natural fillers are promising reinforcement materials for BC productions due to their flexibility, low cost, locally availability, and biodegradability properties. There are various types of bio-based filler materials available in the nature. The most commonly used filler materials for BC productions are categorized as grass, seed, leaf, and bast fibers, and also different types of softwoods and hardwoods [49]. Among them, kenaf, ramie, jute, hemp, coir, rice husk, bamboo, and flax are the examples of some widely used prominent natural fibers used as filler materials [50]. The fibers are collected from different parts of the plants such as leaves, stem, flowers, fruits, grains, and wood as well. Another promising feature of natural fibers is the availability. According to the report of Siakeng at el. [50], worldwide productions of hemp fiber were 215×10^3 tons, abaca fiber 70×10^3 tons, bamboo $10,000 \times 10^3$ tons, jute $2,500 \times 10^3$ tons, coir 100×10^3 tons, cotton $18,500 \times 10^3$ tons, flax 810×10^3 tons, ramie 100×10^3 tons, sugarcane bagasse $75,000 \times 10^3$ tons, kenaf 770×10^3 tons, sisal 380×10^3 tons, wood $1750,000 \times 10^3$ tons, roselle 250×10^3 tons, and banana fibers by 100×10^3 tons. Generally, most of the cellulosic fibers contain cellulose, lignin, hemicellulose, and small proportions of impurities. The strength and stiffness of the fibers are dependent on the amount of cellulose % [51]. The higher the cellulose, the higher is the strength. The degree of polymerization of cellulose, microfibril angle of cellulose, and after all cellulose content influences the resultant mechanical properties [50,52,53]. Besides, cellulosic fiber-reinforced BCs generate the highest renewable energy (6%) savings in contrast to glass-reinforced composites (2.9%) [49]. The detailed explanations of different natural fibers were discussed in our previous study [6]. However, different routes for the BCs are shown in Figure 7.2.

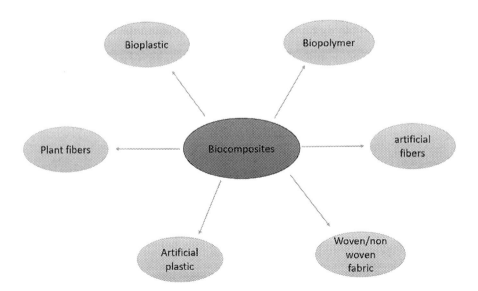

FIGURE 7.2 Different routes of BCs productions.

Design and Fabrication Technology

7.4 BIOPOLYMER MATRIX

In terms of sustainable product development, natural fibers provide enormous benefits thus gaining considerable attention of the research community. As the natural fibers provide attractive mechanical strengths, so there is a big potentiality created for replacing the synthetic fibers. On the other hand, emission of greenhouse gases is making a serious risk for the climate change; so, there is a constant pressure and awareness is rising to replace the petroleum-based resins by using bio-based polymeric resins/additives for BCs manufacturing. The main function of bio-matrix is to transfer the loads uniformly through bonding the fibers together. Some of the major biodegradable polymers are mentioned in Table 7.1. Biodegradable polymers are divided into two major categories: (a) biodegradable polymers of natural origin and (b) synthetic biodegradable polymers. The naturally originated polymers are obtained from agro-based biomass. They are produced by fermentation of biomass or genetically modifying the plants. Biopolymers can also be obtained by chemical modification (like poly(caprolactone) or poly(amides) [41].

Chemical-based biopolymers like aliphatic polyesters in general having convenient life cycle features in contrast to the traditional thermosets/thermoplastic polymers. These kinds of polymers are easily compostable to the nature through enzymatic reactions or by normal micro-organisms activity in atmospheric conditions [54,55].

7.5 DESIGN OF BC MATERIALS

The proper design of BC is highly important to get the expected features and performances. The main purpose of scientific research, innovation, and development

TABLE 7.1
Different Biodegradable Polymers [41]

Naturally Originated Biodegradable Polymers	Synthetic Biodegradable Polymers
Polysaccharides	Polyester
• Chitin	• Poly(caprolactone)
• Cellulose	• Poly(lactic acid)
• Starch	• Poly(glycolic acid)
• Konjac	• Poly(ortho-esters)
• Levan	
Polyester	Poly(anhydrides)
• Polyhydroxyalkanoates	
Lignin	Poly(amides)
Natural rubber	Poly(amide-enamines)
Shellac	Poly(vinyl alcohol)
Proteins	Poly(vinyl acetate)
• Gelatin/collagen	
• Silk, elastin, fibrogen, casein, albumin	
• Protein achieved from grains	
Lignin	Some poly(acrylates)
	Some poly(urethanes)

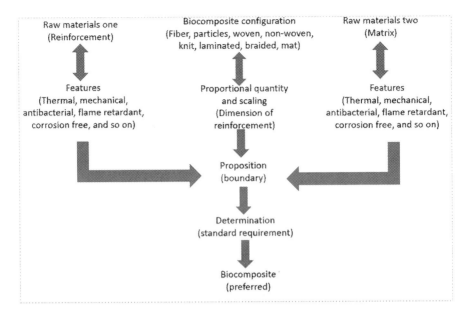

FIGURE 7.3 Different variables used for BCs designing process.

is a part related to the life cycle of products. However, the actual engineering of the products starts with the development of science. The engineered design and fabrication could innovate a novel product or supplement to an available/existing item in the market through enhancing competitive advantage. Engineered BCs exist for multifaceted applications in the automotive, aeronautical, construction, civil, packaging, biomedical, and building sectors. The efficient design of BCs trails numerous challenges along on account of the lack of available information, diverse fiber geometry, orientation of fibers in the composites, variable characteristics of different thermosets and thermoplastic polymers, varied dimensions of fibers (Figure 7.3). There are several factors responsible for fruitful BC design: (a) material selection, (b) fitness to purpose, (c) ease of product manufacturing, (d) product cost, (e) production volume, (f) durability, (g) product quality, (h) maintenance feasibility, (i) operational cost, (j) efficiency, (k) environmental sustainability, and (l) safety aspects [56].

Nowadays, probable experimental results of BCs are also being predicted by FEA. Different software like ANSYS and SolidWorks are used to serve this purpose [57–59]. The ultimate results of the manufactured BCs could be compared with the FEA results, and if needed, the model could be redesigned again according to the achieved output. SolidWorks is an efficient tool for the simulation of mechanical performances of BCs with due respect to stress and strain [60]. So, this software has gained high reputations for predicting accurate results of the BCs. The parameters used for the experimental work could be used to simulate the data in advance through SolidWorks. Rosdi et al. has used SolidWorks for stress and strain analysis to enhance the durability of their produced materials [61]. ANSYS is also used to simulate the mechanical structures through FEA to investigate the strength, temperature

Design and Fabrication Technology

distribution, toughness, elasticity [62,63], and so on. Sometimes assumptions are taken into consideration such as homogeneity and regular orientation of bio-based fibers/wood materials, perfect bonding between the fiber and matrix, and the lack of voids for simplifying the FEA models [64].

7.6 ENGINEERING DESIGN OF BCs

7.6.1 PREDICTION OF STRENGTH AND STIFFNESS FEATURES

The mechanical properties of the BCs on the macro-scale can be derived by using an approach based on homogeneity. The BCs being heterogeneous on the microscopic scale are considered as homogeneous distributions of materials in the composites in macro-scale range [65]. The stress and strains of the composites can be defined using Equation (7.1):

$$\bar{\sigma}_{ij} = \frac{1}{V}\int_V \sigma_{ij} dV, \quad \bar{\varepsilon}_{ij} = \frac{1}{V}\int_V \varepsilon_{ij} dV \tag{7.1}$$

where V is unit cell volume.

The stress-strain relationship could be developed by Equations (7.2) in case of anisotropic materials:

$$\bar{\sigma}_i = E_{ij}\bar{\varepsilon}_j \tag{7.2}$$

where effective stiffness matrix is indicated by E_{ij}. When the global average stress $(\bar{\sigma}_{ij})$ is applied to the boundary surface, the resultant global average strain is $(\bar{\varepsilon}_{ij})$ in case of cuboid unit-cells for FEA, see Equation (7.3)

$$\bar{\sigma}_{ij} = \frac{(P_i)_j}{S_j} \tag{7.3}$$

In Equation (7.3), $(P_i)_j$ is the consequent force and S_j is the area of j-th boundary surface of BCs. The first linear segments of stress-strain curves are used to predict the elastic characteristics of 3-dimensional braided BCs Equation (7.4).

$$E_i = \frac{\bar{\sigma}_i}{\varepsilon_i}, \quad \mu_{ij} = -\frac{\bar{\varepsilon}_i}{\varepsilon_j}, \quad G_{ij} = \frac{\bar{\sigma}_{ij}}{2\varepsilon_{ij}} \tag{7.4}$$

In Equation (7.4)·$\varepsilon_{ij} = \gamma$, the engineering shear strain.

The perceived strength characteristics are determined from the highest stress in the stress-strain curve.

7.6.2 FAILURE PREDICTION OF BCs (UNIDIRECTIONAL LAMINA)

The appropriate design of BCs is highly important to achieve the optimal performance of structural materials. Several methods are used by the scientists and industries to

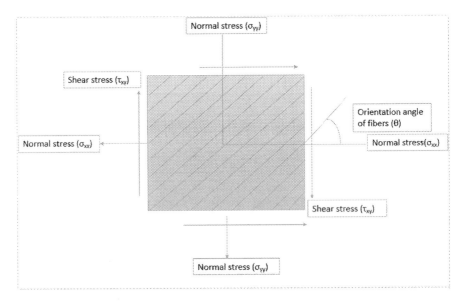

FIGURE 7.4 The stress state of 2D orthotropic lamina.

design and model the structural composites. The failure/breakage/fracture of BCs may happen in different modes of deformations. The loading of reinforcements microstructure (diameter, distribution, and volume fraction of fibers) and matrix polymer is responsible for the kind of deformations [22,66,67]. Besides, the thermal characteristics of fibers also affect the fracture of BCs [68,69]. The natural fiber reinforced polymeric composite will be discussed here for unidirectional loading with fibers oriented at angle θ with respect to the x-axis (Figure 7.4). It is necessary to know four independent elastic constants out of the moduli of elasticities ($E_{d_1d_1}, E_{d_2d_2}$), moduli of rigidity ($G_{d_1d_2}$), and Poisson's ratio ($v_{d_1d_2}$) to predict the failure of BCs in terms of directions 1 (d_1) and directions 2 (d_2). Strength properties are defined with the following terms:

- Tensile strength (longitudinal) = TS_L
- Tensile strength (transverse) = TS_T
- Compressive strength (longitudinal) = CS_L
- Compressive strength (transverse) = CS_T
- Shear strength (in-plane) = SS_{IP}

Stress state in a uni-directional lamina could be predicted according to the theory for planer state of stress.

7.6.2.1 Theory of Maximum Stress

Failure happens when the stress applied in the directions of main material symmetry is equal or exceeds the respective ultimate strengths. In this case, shear stresses ($\tau_{d_1d_2}$) and normal stresses ($\sigma_{d_1d_1}$ and $\sigma_{d_2d_2}$) should follow the bellow limitations (Table 7.2).

TABLE 7.2
Equations on Different Theories of Failure Predictions

Theory of Maximum Stress	Theory of Maximum Strain	Theory of Azzi-Tasi-Hill	Theory of Tasi-Wu Failure
$CS_L < \sigma_{d_1 d_1} < TS_L$	$\varepsilon_{CL} < \varepsilon_{d_1 d_1} < \varepsilon_{SL}$	$\dfrac{\sigma_{d_1 d_2}^2}{TS_L^2} - \dfrac{\sigma_{d_1 d_1} \sigma_{d_2 d_2}}{S_{L-d_1}^2} + \dfrac{\sigma_{d_2 d_2}^2}{TS_T^2} = 1$	$F_{d_1}\sigma_{d_1 d_1} + F_{d_2}\sigma_{d_2 d_2} + F_{d_6}\gamma_{d_1 d_2} + F_{d_1 d_1}\sigma_{d_1 d_1}^2 + F_{d_2 d_2}\sigma_{d_2 d_2}^2 + F_{d_6 d_6}\gamma_{d_1 d_2}^2 + 2F_{d_1 d_2}\sigma_{d_1 d_1}\sigma_{d_2 d_2} = 1$
$CS_T < \sigma_{d_2 d_2} < TS_T$	$\varepsilon_{CT} < \varepsilon_{d_2 d_2} < \varepsilon_{TT}$		$F_{d_1} = \dfrac{1}{TS_L} - \dfrac{1}{CS_L}$
$|\tau_{d_1 d_2}| < |ISS_{IP}|$	$|\gamma_{d_2 d_2}| < |\gamma S_{IP}|$		$F_{d_2} = \dfrac{1}{TS_T} - \dfrac{1}{CS_T}$
			$F_{d_6} = 0,\; F_{d_1 d_1} = \dfrac{1}{TS_L CS_L},\; F_{d_2 d_2} = \dfrac{1}{TS_T \cdot CS_T},\; F_{d_6 d_6} = \dfrac{1}{SS_{IP}^2}$

7.6.2.2 Theory of Maximum Strain

Failure happens when the strain applied in the directions of main material symmetry equals or exceeds the respective ultimate strains. In this case, shear strain ($\gamma_{d_1d_2}$) and normal strain ($\varepsilon_{d_1d_1}$ and $\varepsilon_{d_2d_2}$) should follow the below limitations (Table 7.2).

7.6.2.3 Theory of Azzi-Tasi-Hill

An orthotropic lamina would fail if the equation (Table 7.2) is satisfied, where $\sigma_{d_1d_1}$ and $\sigma_{d_2d_2}$ are assigned for the tensile stresses.

7.6.2.4 Theory of Tasi-Wu Failure

An orthotropic lamina would failed if the equation (Table 7.2) is satisfied, where $F_{d_1}, F_{d_2}, F_{d_3}$, and F_{d_6} are the coefficients of strengths (tensile, compressive, and shear) in MPa unit. Their expressions are also provided in Table 7.2.

7.6.3 FAILURE PREDICTION OF BCs (UN-NOTCHED LAMINA)

The prediction on eventual failure theory could be used for the un-notched lamina with the bellow steps: (a) calculating stress and strain for every particular lamina using lamination theory, (b) using feasible failure theory to predict the first failed lamina, (c) providing decreased strength and stiffness to the failed lamina, (d) recalculating stress and strain for the remaining laminas by using lamination theory, (e) completing steps b and c for the prediction of following laminas, (f) repeating steps b to d until the failure continues to occur.

7.6.4 FAILURE PREDICTION OF BCs (RANDOMLY ORIENTED FIBER LAMINATES)

Hahn's method is one of the popularly used techniques to predict the failure characteristics for randomly arranged discontinuous fiber-based composites. The failure can be predetermined if the maximum tensile stress attains the average strength levels for all the randomly oriented fibers as shown in Equation (7.5), where RS_L is random fiber laminate strength, LS_L is longitudinal strength, and TS_L is tensile strengths of a 0° laminate [70].

$$RS_L = \frac{4}{\pi}\sqrt{LS_L TS_L} \tag{7.5}$$

7.7 MODELLING OF THE MATERIALS FOR BCs

BC materials exhibit inherent inhomogeneity, nonductility, and anisotropic characteristics. The efficient processing of the natural fiber material with thermosetting/thermoplastic polymer is a challenging task to acquire expected performances. Consequently, modelling of the BCs is becoming popular to the researchers and industrialist nowadays for minimizing processing times, cost, and associated efforts through predicting the performances of the designed composites in advance. Modelling could be performed mainly in two ways: (a) simple modelling through implementing traditional analytical modelling and (b) using numerical simulations

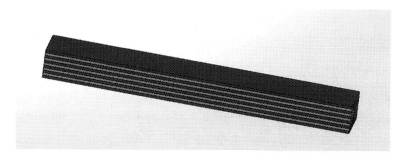

FIGURE 7.5 A designed model based on laminated flax woven fabric reinforced composites by using SolidWorks.

either considering the composites as anisotropic material or focusing on the interactions in the reinforcement-matrix at microscopic level. A simple model for the laminated BC is illustrated in Figure 7.5. Flax woven fabric and polypropylene (PP) sheets were used to design this model, in which the apparent properties of elasticity (isotropic or orthotropic) determined at macroscopic level were assigned to the consecutive laminae of fabric and PP by using SolidWorks software in our lab.

The nonlinear behavior of laminated structural composites is a big challenge. Nonlinear characteristics of unidirectional composites generate confusing analytical and numerical results. To overcome such kind of problems, several studies were reported by the researchers [71,72]. The plane shear test can demonstrate such nonlinear effects by using uniaxial characterization equipment. The failure of BCs entails fiber breakage, layer delamination, and matrix cracking. Although natural fiber reinforced composites exhibit some problems during BCs formations, such problems can be minimized/eliminated through simulating the engineered materials by using FEA. FEA works with specific shape and size of the BCs structure in case of small-scale productions; significant differences can appear in large-scale BCs manufacturing. Any kind of defects in the composite structures could lead to a decline in mechanical performances. By utilizing the capabilities of FEA, consistency in the BC materials can be enhanced. Zhang et al. [65] have reported a simulation of the constituent model and progressive failure/damage of composites, using VUMAT (user-defined material substitute) by an FEA-based software named ABACUS. The level of stress and state of damage could be achieved for all the constituents of every cell. The damage variable can be updated through decline of material properties, when the criterion of failure is satisfied. The entire flow process of numerical analysis is shown in Figure 7.6.

7.7.1 Modeling of Materials in Terms of BCs Manufacturing

BC design is still a challenging task due to the lack of enough information, manufacturing technologies, polymers/matrices, and versatile reinforcement materials. The variation in reinforcing material arrangements/orientation, composite geometry, and matrix types is responsible for the composites' performance. Besides, composites also vary in shape, configurations, and materials used [22,73].

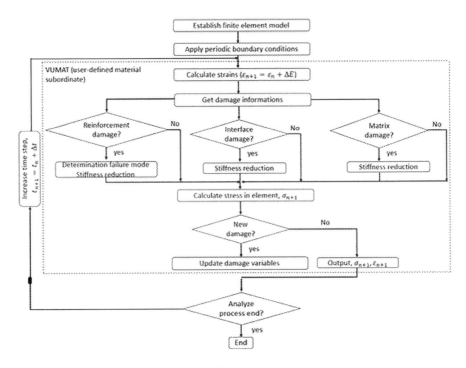

FIGURE 7.6 A flow process regarding FEA. (Adapted with permissions from Zhang et al., A novel interface constitutive model for prediction of stiffness and strength in 3D braided composites. *Composite Structures*, 2017. 163: pp. 32–43. [65] Copyright, Elsevier (2017).)

Different processing parameters like aspect ratio, dispersion, optimum temperature, and orientations, and methods have significant effect on mechanical and functional properties of BCs. Besides, if the fibers are not properly dried to achieve optimum moisture content, it may hamper the bonding between fiber and polymer, the resins of the matrix. Void content is another critical problem for the generation of water during matrix formation, which could decline the mechanical properties of the manufactured BCs. The manufacturing methods of BCs can be divided into two categories: (a) open mold and (b) closed mold. Spray-up and hand-lay-up techniques are most popularly used with open mold methods [74]. However, injection molding, vacuum infusion, extrusion, pultrusion, compression molding, resin transfer molding, press mold compounding, and filament winding (Figure 7.7) are some of the renowned closed mold technology [75–77].

Hand-lay-up is a popular method for BCs manufacturing where the trimmed fibers are spread to the mold according to the predesigned dimensions with several layers. Vacuum bag/Teflon paper is used to wrap the composite while encompassing it by protecting air and to provide barrier against the direct autoclave contact [78,79]. Pultrusion is another prominent method where the fibers from roving strands are pulled out from the creel in a resin bath and then cure with a heated die [80]. This technique has advantages and disadvantages as well. The notable advantages are efficient use of materials, higher throughput achievement, and higher resin content [81].

Design and Fabrication Technology

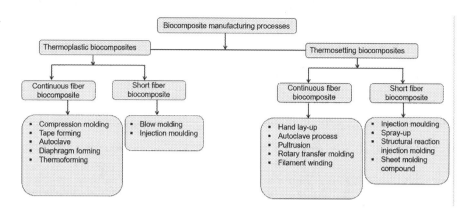

FIGURE 7.7 BCs manufacturing methods depending on thermosets and thermoplastics.

Conversely, some of the major disadvantages for this method are the decrease of BC strength with the higher resin usage and the requirement of uniform cross-section [82]. The fast curing process could reduce the materials strength along with the creation of void if the die is not conformed good enough [83]. In case of extrusion process, pressure and heats are required to produce thermoplastic polymer composites [84,85]. A compatible solvent is required to dissolve nonthermoplastic polymers.

7.8 FABRICATION TECHNIQUES OF BCs

There are several methods available for successful fabrications of BC materials. The most available methods for BC fabrications are compression molding, injection molding, resin transfer molding, pultrusion, hand lay-up, calendering, extrusion, and so on. However, some common fabrication methods with their functioning, processing parameters, and tools are described in this section. The discussions are made based on various books and articles reported by the researchers and scientists earlier [86–92].

7.8.1 Polymer Preparation

The main steps of polymer processing are the mixing of raw materials—polymers and additives, and then pelletizing (if not dry blending), drying of granulates and finally production by extrusion or injection mold or rotational mold, etc. The applied additives can be functional (e.g., reinforcers, plasticizers) and protective (e.g., antidegradation, antioxidants, antistatic). Mixtures can be as follows:

- *Compound*: a mixture designed for a definite purpose (polymers + additives)
- *Blend*: mixture of polymers, which will work as homo- or co-polymers; these should be thermodynamically compatible
- *Alloy*: applied in case of thermodynamically incompatible polymers by the use of compatibilizer additives. In this case, a very intensive mixing is necessary

FIGURE 7.8 Dispersive and distributive mixing. (Drawing by Tibor L. Alpár.)

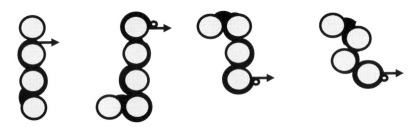

FIGURE 7.9 Calender configurations. (Drawing by Tibor L. Alpár.)

Mixing can be dispersive or distributive (Figure 7.8). The dispersive mixing is an intensive, size reducing process, e.g., mixing of agglomerated pigment powders by a fluid mixer [93,94]. At the contrary, distributive mixing is an extensive mixing without reduction in size, e.g., mixing of polymer melts in a twin screw extruder. The mixed polymer compounds should be pelletized before dosing into an injection molder or into an extruder to process further. Drying might be necessary in case of hygroscopic polymers.

7.8.2 Calendering Technique

Calenders are typically used to produce paper or finishing various textiles like cotton, linen, silk, and other man-made fabrics [95]. Calender is also an important processing machine in the rubber processing industry to produce tires [96], where it is used for the inner and outer fabric layer. To process plastics, the appropriate raw materials used for calendering are amorphous, thermally sensitive, and thermoplastic polymers with a wide melting temperature range; such are hard and soft PVC (poly(vinyl chloride)), PVC co-polymers, hard polystyrene, acrylonitrile-butadiene-styrene (ABS), and cellulose ester. Calenders process molten polymers by their 3–5 parallel, counter rotating, heated cylinders of large diameter (600–800 mm). Typical calendered products are thin panels, foils, or coated fabrics, with a width of 4 m and a thickness as thin as 30–800 µm at a production speed up to 100 m/min. The distances between parallel cylinders can be set precisely, and heated from one end, driven from the other. The cylinders have to bear large loads acting in the narrow gap between them: flexure, torsion, and compression. Therefore, cylinders should have large diameter, hardened (500–550 HB), wear-proof, and smooth surface.

Main calender configurations include type I, type L, type F, and type Z (Figure 7.9). Type I is rare hence difficult to feed. The advantage of type L is easy feeding from the first stage (typical to produce hard PVC). Type F is used to process soft PVC polymers. Type Z is used to coat fabrics among others.

Design and Fabrication Technology

A typical calendering line starts with an extruder to mix and melt the polymer [96]. It results a good homogenous compound which should be buffered and degassed by a rolling mill. It is followed by a filtration and then comes into the calender to produce the foil product. The final stage of calendering is cooling and winding.

7.8.3 Extrusion Technique

The most important and effective technology of polymer processing is as follows: the polymer is plasticized, than homogenized in molten stage, degassed if necessary, than compressed and pressed through a die with fixed cross-section, cooled and a quasi-infinite product is obtained with a constant profile [97,98]. Typical extruded products are foils, panels, threads, pipes, and complex multicavity products.

The main components of an extruder are shown in Figure 7.10. The extruder screw is rotating in a hot barrel. The screw has three zones. The first one is feed zone, also termed as solids conveying. The resin is fed into the extruder through this zone, and the depth of channel is usually remaining the same throughout the entire zone. Second is the melting zone, which is also called as the compression or transition zone. The resins are mostly melted into this section, and the depth of channel gets smaller here progressively. Third is melt conveying or metering zone. This zone, in which channel depth again grows into the same around the zone, melts the last particles and mixes to reach a uniform composition and temperature. In front of the die there is usually a screen pack and a breaker plate. Frequently screw length is also referenced for its diameter as L:D ratio (length : diameter).

Compression can be achieved by increasing the diameter of screw root (root progressive screw) or by decreasing helix angle (angle degressive screw) or by increasing of flight width. The extruder screw geometry is shown in Figure 7.11.

The melt flow describes the volumetric flow rate of the screw (Figure 7.12):

$$\dot{V}t = \dot{V}d - \dot{V}b - \dot{V}l$$

$\dot{V}t$ = total volumetric flow (cm^3/s)
$\dot{V}d$ = drag flow, rate of conveying the melted polymer to the die,

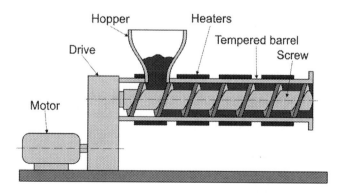

FIGURE 7.10 Main components of an extruder. (Drawing by Péter György Horváth.)

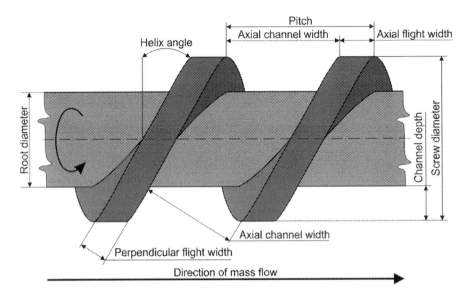

FIGURE 7.11 Extruder screw geometry. (Drawing by Péter György Horváth.)

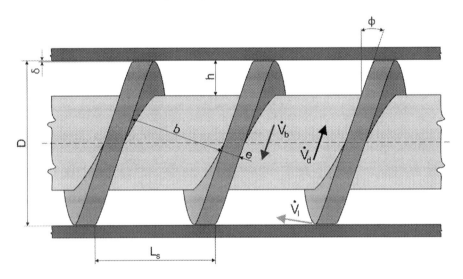

FIGURE 7.12 Flow rate in extruder. (Drawing by Péter György Horváth.)

$\dot{V}b$ = back flow, opposing direction to drag flow,

$\dot{V}l$ = leakage flow, flow through the clearance δ between the flight lands and the barrel wall.

Twin screw extruders are special types of extruders. The double screws result in higher blending efficiency and material flow. Such equipment are often used to make PVC or WPC (wood polymer composite) compounds. The screws may be co-rotating and intermeshing, co-rotating and not intermeshing, counter rotating and intermeshing or counter rotating and not intermeshing. Counter rotating screws result in easier drive, lower shearing in molten polymer; co-rotating screws result in higher shear strain and more extensive compounding. The screws are often configurable and of modular design, so it is possible to gain optimal compounding by different elements like kneaders, helical path mixing, and reverse elements. A special twin screw extruder has conical screws to increase compounding effects especially for WPCs.

7.8.4 Extruder Die Technique

Almost any two-dimensional shape can be a profile for extruded polymer such as pipe, panel, and complex multicavity products. The extrudate is viscoelastic and it has some problems due to material flow effects and swelling after exiting die. During cooling of extrudates "freezing" or inclusions should be omitted. The main sections of die are as follows (Figure 7.13):

- *Transition zone*: leading material flow from circular cross section to near end-shape
- *Shaping zone*: forming of required cross-section
- *Smoothing zone*: stabilizing of profile (smaller cross-section, higher pressure than in shaping zone)
- *Calibrating unit*: ensuring dimensional accuracy, final hardening

The die is followed by a calibrating unit (by vacuum or over pressure), cooling, and optionally special pulling tool for hose production. Vacuum calibration fixes outer dimensions of the product. The viscoelastic extruded pipe enters into a vacuum tank and it is cooled here in water bath between calibrating discs.

7.8.5 Injection Molding Technique

The raw materials for the part—like glasses, metals, elastomers, and most common thermoplastic and thermosetting polymers—are fed into a heated (tempered) barrel,

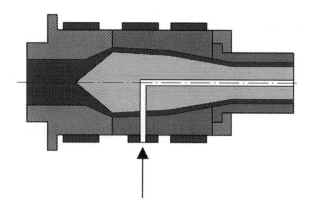

FIGURE 7.13 Pipe extrusion die. (Drawing by Péter György Horváth.)

mixed by screw, and forced into a mold cavity through injection where it is cooled and hardened to the configuration of the cavity to an arbitrary 3D product. It is a discontinuous technology.

Typical polymers for injection molding are:

- *Thermoplastics*: injection in molten phase and cooling in cavity tool: phenylene-ethylene (PE), polypropylene (PP), polystyrene (PS), PVC, poly(methyl methacrylate) (PMMA), ABS, polyoxymethylene (POM), ploy (lactic acid) (PLA), etc.
- *Thermosetting polymers*: reactive injection mould (RIM) from prepolymers, curing in (heated) cavity tool: phenoplasts, aminoplasts, melamine-epoxy combinations, etc.
- *Elastomers*: RIM: rubber, silicone.

The viscosity of injected material should be low, to be able to force into the cavity through a narrow channel ($\varnothing \sim 2$ mm) in a very short time (a few seconds). The main components of injection moulding machines are feeder, injection unit, mold, control, platens, and clamping (Figure 7.14).

7.8.6 Clamping Unit

The pressure of molten polymer is acting on a large specific surface, so the clamping forces, which are keeping the parts of mold together, should be 15%–20% higher than that due to the pressure in cavity caused by the injected polymer. The clamping force may be over 100 t (1,019 kN) even in the case a mid-size injection molding machine. The clamping unit is set up from back platen, moving platen, stationary platen, four robust tie rods, and a clamping mechanism, which can be mechanical or hydraulic.

The maximal F force, exerted by the molten polymer to the closing surface of mold is shown in Equation (7.6),

$$F_{cav} = P_{cav}\left(A_{cav}xn + A_{run}\right) \qquad (7.6)$$

where

F_{cav}: clamping force in N

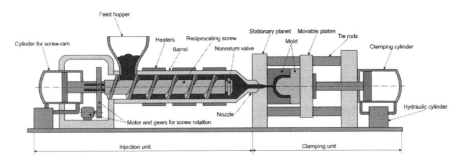

FIGURE 7.14 Scheme on an injection mold machine. (Drawing by Péter György Horváth.)

Design and Fabrication Technology

P_{cav}: average cavity pressure in MPa
A_{cav}: projected surface area of composite product in mm^2
A_{run}: projected surface area of runner system in mm^2
n: number of cavities.

7.8.7 INJECTION UNIT

The main aim of the injection unit is to heat the raw materials to the required temperature until it reaches an optimum viscosity that will let the material to flow into the mold while under force. The main parts of the injection unit are: heated (tempered) barrel nozzle, end cap on the barrel, screw (similar to extruder's screw), heater bands, nonreturn valve, hydraulic cylinder for moving the screw forth and back axially, motor to rotate the screw, hopper, control system for the temperature, operating parameters are time and speed.

The injection of melt into the mold cavity is made of three phases: (a) filling phase, (b) holding phase, and (c) packing phase. Before filling of the cavity, the injection unit moves forward and closes to the mold, so the melt can be forced into the mold on high pressure. After cooling, the injection unit moves back, the nozzle leaves the mold because mold is a cooled unit, and the barrel with the nozzle is heated.

A complete injection cycle is set up of the following steps:

- mold closes
- injection unit closes to the mold
- axial move of the screw, material is forced into the cavity at high speed than the pressure of material is increased to fill completely the cavity
- cooling and parallel moving of the screw back; the material is melted for the next injection (packing)
- injection unit moves back from mold
- old opens and drops the product.

7.8.8 MOLD TOOLS

The mold tool is the vital component used in the injection molding fabrication methods. It provides a passageway for molten plastic materials to travel from the injection unit to mold cavity. It permits the air which would be trapped inside when the mold closes to escape. If the air venting is not possible, the molded component will contain bubbles of air and as a result will provide a poor quality. It cools the molding until it becomes set. The temperature of the mold is regulated by circulated water through a cooler (heat exchanger), because it is imperative that the molding cools at correctly to avoid stresses and distortion. By means of the ejectors the finished moulding is pushed from the mold:

- *Guide pins:* fixed to one half of the mold and aligned with the two halves by entering into the holes in the other half.
- *Gate*: the runner narrows, as it enters into the mold cavity.
- *Runner*: mortise in the mold connecting the cavities to the sprue bush.

- *Sprue bush*: tapered borehole in the center of the mold into which the molten plastic material is initially injected.
- *Locating ring:* ensures the positions of the mold on the fixed platen as if the injection nozzle lines up with the sprue bush when it closes to the mold.
- *Mold cavity*: space in the mold shaped to manufacture the desired product(s).
- *The shot*: total amount of plastic injected into mold.
- *Ejector pins*: push the molding and sprue/runner out of the mold.
- *Sprue*: materials which set in the sprue bush.

The molten material arrives from the barrel—through nozzle—and runs through the sprue bush at high speed into the mold. The sprue bush diameter is within 2.5–5.0 mm and it is conical, so the cooled material can be easily removed from it when ejecting the product. The material proceeds through the runners than through the gates and fills the cavities of products (Figure 7.15).

Runners are needed when multiple products are produced in one tool or in the case of complex product shape when material should be introduced through several runners. The cross section of runners may be round, half-round, trapezoidal, or rectangular. The sum of their starting cross section should be less than the final cross section of sprue bush. In multicavity molds, the runners should be planned so that the polymer should arrive to every gate within the same time and at the same pressure.

The gate is the contact point of runners and tool cavities of the desired products. It increases the flow speed and decreases the viscosity of molten polymer during injection. It makes a weak point enabling the molding to be broken easily or cut from runner.

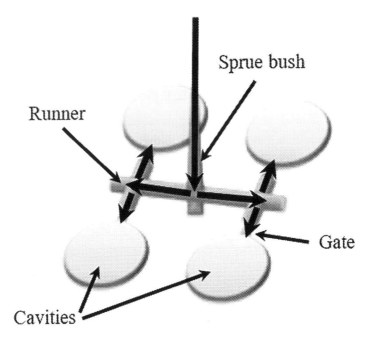

FIGURE 7.15 The route of molten material in the mold. (Drawing by Tibor L. Alpár.)

Design and Fabrication Technology

The following should be designed for a mold:

- The route of melted polymer through runners
- Positions of runners and gates
- The expected orientation of material
- Decrease of melt's pressure in different sections of cavity
- Welding places
- Shrinkage
- Positions of eject needles
- Mechanisms for undercuts
- Mechanisms for releasing molder screw threads
- Strengthenings (ribs and struts) to gain optimal strength parameters
- Cooling method circuit and of the mold
- Programming of injection cycle

7.8.9 Pressing and Hot-Pressing Techniques

Hot pressing has become a very important process for BCs manufacturing, which affects the final product quality and performances significantly [33,99]. The final shape and structure of the particle/fiber composites are given through compressing the mat with or without the application of high temperatures and pressure. The moisture content of the boards is reduced here along with the curation of adhesives. The key technologies of pressings could be:

- prepressing or pressing
- cold or hot pressing
- continuous or discontinuous pressing
- panel pressing or moulding

The prepressing of mat (50%–60%) is easier for introducing the hot pressing. Besides, the time for hot pressing is also reduced along with less possibility of failure if the mat is prepressed. During the first stage of prepressing, air exits from the mat through the sieve belt, where the maximum linear pressure is 30 N/mm. In the second stage, compressive force on the belt is performed by four cylinders, where the maximum linear pressure is 250 N/mm. The belt speed could be adjusted within the range of 1–60 m/min.

In case of hot pressing, ways of heating of the mat can be as follows:

- contact (heating through thermal oil/steam/hot water/electrical)
- microwave
- high frequency (in case of molded products or higher thicknesses)
- steam injections (for higher thicknesses)
- combined technique

In case of fiberboard productions, (a) wet process and (b) dry processes are used. The moisture content of the mat is reduced from the last stage (wet process), whereas mat height could be reduced by applying dry methods.

The hot process can be divided into following categories:

- Continuous
  - Calender press (BISON–MENDE process)
  - Double belt flat-bed system (Siempelkamp, Dieffenbacher, and Küsters)
- Discontinuous/batch [100]
  - Single daylight press
  - Multi daylight press

Continuous common steel belt is used in batch production for transferring pressure and heat to the mat in the prepressing process (Figure 7.16). At the same time, the equipment needs to be able to transfer the mats from continuous to batch pressing. However, continuous process is simpler to produce particleboards. It is possible to manufacture diversified product items with wide range of size and thicknesses as the press is applied on longer piece of panels [100]. Larger boards are needed to produce for getting the maximum benefits of single daylight process. Most of the facilities installed are 25 m long or more [100]. However, a 52 m press was in operation in UK until 2009 [100]. After completion of this process, trims need to cut precisely for avoiding edge effects. In case of multiple daylight pressing, up to 48 panels can be simultaneously. It is important to ensure the same temperature and pressures on all the boards. Generally, 4–8 boards (5–7 m long) are produced with 2.5 m width by using multi-daylight process [100].

The pressing parameters are set in sections such as

- entering section
- high-pressure section
- medium-pressure section
- cooling section

Siempelkamp is a widely used hot press machine. Its typical design includes as follows:

- Flexible press table
- Rigid pressing head
- Upper and lower press plates
- Frame modules
- 2–4 hydraulic cylinders
- The whole volume should be heated above 100°C temperature

The time for pressing depends on the technology. The required temperatures for curing should be reached over the whole cross-sections of the composite panel. Processes as below take place in the course of pressing:

- Heat transport
- Material transport
- Curing of adhesives
- Change of structure of mat

Design and Fabrication Technology

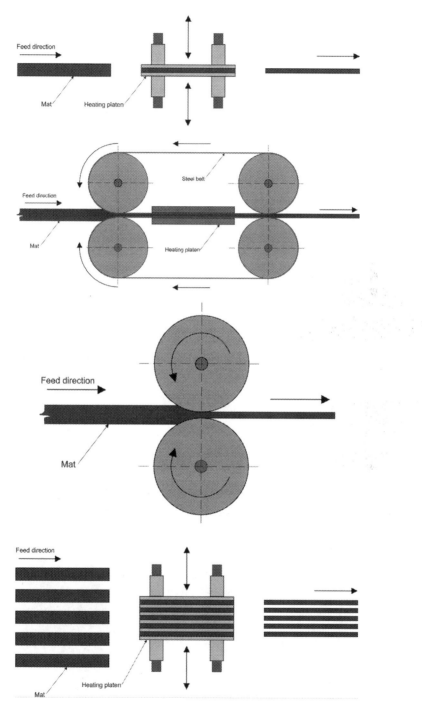

FIGURE 7.16 Different pressing and hot pressing mechanisms. (Drawing by Péter György Horváth.)

7.9 BCs CHARACTERIZATIONS

7.9.1 TENSILE CHARACTERIZATION OF BCs

Tensile properties of BCs are very important parameters, which can be improved with the increase of fiber content but up to a certain level (optimum). After this level, tensile properties start to decline again [101]. Besides, the surface modification of plant fibers also enhances the tensile properties of BCs through ensuring better compatibility between the fiber and polymers [102,103]. It is also found that the presence of noncellulosic fibers is responsible for the poor tensile performance of the BCs. Compatible surface treatments can reduce the impurities in biofibers, thus enhance tensile strength and Young's modulus [104]. This finding was also confirmed by BCs. Ramraji et al. [105] who have reported that the alkaline treatment of flax fiber enhanced the tensile strength by 35% in polymeric BCs. A simple tensile testing model is illustrated in Figure 7.17.

7.9.2 FLEXURAL CHARACTERIZATION OF BCs

Flexural strength and modulus are further significant parameters for assessing the BCs performance. They indicate the capability of the BCs to withstand bending load. The magnitude of reinforcement fiber loading influences the flexural strength. However, different loadings of fibers significantly influence the flexural properties (both the strength and modulus) [106]. Sathish et al. [107] has found that in case of flax and bamboo fiber-reinforced composite with epoxy resin, the flexural properties enhanced with the increase in flax fiber. The highest flexural strength (298.37 MPa) was attained when the proportion of flax and bamboo was 40:0 [107]. When the hemp fiber was bonded with polylactic acid (PLA), the highest flexural strength was

FIGURE 7.17 A tensile testing design for coconut fiber reinforced BC samples using Instron testing machine.

Design and Fabrication Technology

FIGURE 7.18 A flexural testing design for coconut fiber reinforced BC using Instron universal testing equipment.

found at 30% fiber loading [108]. When cellulosic nanofibers were used with epoxy resin, the flexural strength was enhanced by 45 MPa and the bending modulus by 3 GPa [109]. A simple flexural testing model is illustrated in Figure 7.18.

7.9.3 IMPACT STRENGTH CHARACTERIZATION OF BCs

It is challenging for BC products to get the desired impact strength beside the required flexural and tensile strength. Muthuraj et al. reported that various processing parameters such as mixing pressure, speed, temperature, reinforcement dimensions, and quantity play significant role in achieving the ultimate strengths of BCs [110]. Sawpan et al. mentioned that impact, tensile, and flexural properties of the composites increased with the increase in short hemp fiber loading into the matrix [111]. The same study also claimed that the highest impact strength (7.4 kJ/m^2) was obtained by alkaline treatment (35 wt%) of fibers [111]. At the same time, Kim et al. reported that impact strength tends to decrease if the filler fiber/particle size is lower [112]. In this case the biodegradability of the composite was enhanced.

7.10 FUTURE PROSPECTS

As natural fibers are replacing synthetic fibers through generating new routes of the use of green materials with enhanced sustainability, proper design of BC becomes highly important. BCs are showing diverse potentiality to be applicable product for multiple sectors from aeronautical industry to automotives and form packaging to biomedical products (bone plates for fixation of fracture). The interiors of cars and cabins of trucks are made of BCs [113]. The construction and building industries also use BC materials for floor beams, interior paneling, and exterior siding of facades even with attractive appearances. The natural fiber reinforced composites are also used in the electronics sector (circuit boards) [70]. In general, the natural fibers are comparatively cheaper, widely available, and environment-friendly materials. BC materials could be used for housing constructions, building, household materials, flooring, fences, automotive, aeronautical, decking, furniture (chair springs), power industry (housing of transformer), and sport materials [70,113–122]. The worlds BC market is expanding rapidly due to the increased awareness of the end users towards the sustainable products. It is expected to reach the BCs market 36.67 billion USD by 2022 throughout the world with a compound annual growth rate (CAGR) of 14.44% for a 5 years duration starting from 2017 to 2022 [116]. The safer product and recycling capability features are few of the key indicators for this quick expansion of the market.

BC materials still possess some limitations for durability and under-performance issues in contrast to the traditional composites. However, selection of proper biofiller, surface treatment of biofibers, certain nano-treatment, and adequate design could minimize such problems significantly. Besides, different manufacturing methods (compression moulding, injection moulding, etc.) along with the feasible natural fiber materials and design could also create significant improvements of BC performances. The parameters of BC production process (pressure, temperature, pressing time, polymer concentrations) and physics and chemistry of fibers have significant influence for the ultimate BC products too. In future, the efficient design and modelling of BC features before going to production by using different computer aided tools could be a benchmark to produce more cost-effective and user friendly products. Recently, lots of green composite-based products are available in the market. This market has higher potentiality to be increased further significantly for the regulations imposed by the governments/organizations in different countries/regions around the globe.

7.11 CONCLUSION

BCs are getting attentions to ensure the use of green materials in striving to "zero impact on environment" and "sustainability" through using renewable materials. BCs offer a practical solution to the constantly rising environmental concerns. But, at the same time, it is also necessary to produce competitive products which could fulfill the demands of manufacturers and consumers through ensuring perceived performances. In this regard, proper design, fabrication, and technological feasibility are also highly significant. Fabrication of BC plays an important role for achieving the desired properties. If proper design and optimum production procedure are not implemented, dramatical destruction of the BCs' ultimate features will become possible. Selection of materials, processing of biopolymers, and manufacturing technology also influence the BCs' performances. Besides, the interfacial bond and certain reinforcement conditions also govern the perceived mechanical characteristics and failure minimization of BCs. Relevant research works, investigations, manufacturing methods, design and modelling, and development of BCs are also discussed in this report.

REFERENCES

1. Chaitanya, S., I. Singh, and J.I. Song, Recyclability analysis of PLA/sisal fiber biocomposites. *Composites Part B: Engineering*, 2019. **173**: p. 106895.
2. Kargarzadeh, H., N. Johar, and I. Ahmad, Starch biocomposite film reinforced by multiscale rice husk fiber. *Composites Science and Technology*, 2017. **151**: pp. 147–155.
3. Yang, X., F. Berthold, and L.A. Berglund, High-density molded cellulose fibers and transparent biocomposites based on oriented holocellulose. *ACS Applied Materials and Interfaces*, 2019. **11**(10): pp. 10310–10319.
4. Muthuraj, R., et al., Sustainable thermal insulation biocomposites from rice husk, wheat husk, wood fibers and textile waste fibers: Elaboration and performances evaluation. *Industrial Crops and Products*, 2019. **135**: pp. 238–245.
5. Chauhan, V., T. Kärki, and J. Varis, Review of natural fiber-reinforced engineering plastic composites, their applications in the transportation sector and processing techniques. *Journal of Thermoplastic Composite Materials*, 2019. p. 1–41.
6. Hasan, K.M.F., P.G. Horváth, and T. Alpár, Potential natural fiber polymeric nanobiocomposites: A review. *Polymers*, 2020. **12**(5): p. 1072.
7. Rajak, D.K., et al., Fiber-reinforced polymer composites: Manufacturing, properties, and applications. *Polymers*, 2019. **11**(10): p. 1667.
8. Laaziz, S.A., et al., Bio-composites based on polylactic acid and argan nut shell: Production and properties. *International Journal of Biological Macromolecules*, 2017. **104**: pp. 30–42.
9. Totaro, G., et al., A new route of valorization of rice endosperm by-product: Production of polymeric biocomposites. *Composites Part B: Engineering*, 2018. **139**: pp. 195–202.
10. Hasan, K.M.F., P.G. Horváth, and T. Alpár, Effects of alkaline treatments on coconut fiber reinforced biocomposites. *In 9th Interdisciplinary Doctoral Conference*, 2020, Pecs, Hungary: Doctoral Student Association of the University of Pécs.
11. Hasan, K., P.G. Horváth, and T. Alpár, Potential natural fiber polymeric nanobiocomposites: A review. *Polymers*, 2020. **12**(5): p. 1072.
12. Versino, F., M. Urriza, and M.A. García, Cassava-based biocomposites as fertilizer controlled-release systems for plant growth improvement. *Industrial Crops and Products*, 2020. **144**: p. 112062.
13. Tekinalp, H.L., et al., High modulus biocomposites via additive manufacturing: Cellulose nanofibril networks as "microsponges". *Composites Part B: Engineering*, 2019. **173**: p. 106817.
14. Leao, A.L., et al., Agro-based biocomposites for industrial applications. *Molecular Crystals and Liquid Crystals*, 2010. **522**(1): pp. 18/[318]–27/[327].
15. Kalita, D. and A. Netravali, Thermoset resin based fiber reinforced biocomposites. *Textile Finish*, 2017. **425**. pp. 423–484.
16. John, M.J. and S. Thomas, Biofibres and biocomposites. *Carbohydrate Polymers*, 2008. **71**(3): pp. 343–364.
17. Hasan, K., P.G. Horváth, and T. Alpár, Thermo-mechanical behavior of MDI bonded flax/glass woven fabric reinforced laminated composites. *ACS Omega*, 2020. **6**: pp. 6124–6133.
18. Santos, S.F.D., et al., Non-conventional cement-based composites reinforced with vegetable fibers: A review of strategies to improve durability. *Materiales de Construcción*, 2015. **65**(317): p. 041.
19. Satyanarayana, K., J. Guimarães, and F. Wypych, Studies on lignocellulosic fibers of Brazil. Part I: Source, production, morphology, properties and applications. *Composites Part A: Applied Science and Manufacturing*, 2007. **38**(7): pp. 1694–1709.
20. Athira, G., A. Bahurudeen, and S. Appari, Sustainable alternatives to carbon intensive paddy field burning in India: A framework for cleaner production in agriculture, energy, and construction industries. *Journal of Cleaner Production*, 2019. **236**: p. 117598.

21. Queiroz, A.U. and F.P. Collares-Queiroz, Innovation and industrial trends in bioplastics. *Journal of Macromolecular Science®, Part C: Polymer Reviews*, 2009. **49**(2): pp. 65–78.
22. AL-Oqla, F.M., A. Almagableh, and M.A. Omari, Design and fabrication of green biocomposites, in O.Y. Alothman, M. Jawaid, S. M. Sapuan (Eds.), *Green Biocomposites*. 2017, Springer: Berlin, pp. 45–67.
23. Sanchez, C., H. Arribart, and M.M.G. Guille, Biomimetism and bioinspiration as tools for the design of innovative materials and systems. *Nature Materials*, 2005. **4**(4): pp. 277–288.
24. Bensaude-Vincent, B., The construction of a discipline: Materials science in the United States. *Historical Studies in the Physical & Biological Sciences*, 2001. **31**(2): pp. 223–248.
25. McEvoy, M.A. and N. Correll, Materials that couple sensing, actuation, computation, and communication. *Science*, 2015. **347**(6228): p. 1261689.
26. Addington, D.M. and D.L. Schodek, *Smart Materials and New Technologies: For the Architecture and Design Professions*. 2005, Routledge: London.
27. Bechthold, M. and J.C. Weaver, Materials science and architecture. *Nature Reviews Materials*, 2017. **2**(12): pp. 1–19.
28. Hasan, K.F., et al., Wool functionalization through AgNPs: Coloration, antibacterial, and wastewater treatment. *Surface Innovations*, 2020. **9**(1): pp. 25–36.
29. K.M. Faridul Hasan, H. Wang, S. Mahmud, A. Jahid, M. Islam, and C.G. Wangbao Jin, Colorful and antibacterial nylon fabric via in-situ biosynthesis of chitosan mediated nanosilver. *Journal of Materials Research and Technology*, 2020. **9**(6): pp. 16135–16145.
30. Hasan, K.F., et al., Coloration of aramid fabric via in-situ biosynthesis of silver nanoparticles with enhanced antibacterial effect. *Inorganic Chemistry Communications*, 2020. **119**: pp. 108115.
31. Hasan, K., et al., A novel coloration of polyester fabric through green silver nanoparticles (G-AgNPs@PET). *Nanomaterials*, 2019. **9**(4): p. 569.
32. Mahmud, S., et al., In situ synthesis of green AgNPs on ramie fabric with functional and catalytic properties. *Emerging Materials Research*, 2019: pp. 1–11.
33. Xu, X., et al., Microwave curing of carbon fiber/bismaleimide composite laminates: Material characterization and hot pressing pretreatment. *Materials and Design*, 2016. **97**: pp. 316–323.
34. Hasan, K.F., et al., Colorful and antibacterial nylon fabric via in-situ biosynthesis of chitosan mediated nanosilver. *Journal of Materials Research and Technology*, 2020. **9**(6): pp. 16135–16145.
35. Seretis, G., et al., On the graphene nanoplatelets reinforcement of hand lay-up glass fabric/epoxy laminated composites. *Composites Part B: Engineering*, 2017. **118**: pp. 26–32.
36. Karabulut, N., M. Aktaş, and H.E. Balcıoğlu, Surface modification effects on the mechanical properties of woven jute fabric reinforced laminated composites. *Journal of Natural Fibers*, 2019. **16**(5): pp. 629–643.
37. Hasan, K.M.F., P.G. Horváth, G. Markó, and T. Alpár, Thermo-mechanical characteristics of flax woven fabric reinforced PLA and PP biocomposites. *Green Materials*, 2020.
38. Mirmiran, A. and M. Shahawy, Behavior of concrete columns confined by fiber composites. *Journal of Structural Engineering*, 1997. **123**(5): pp. 583–590.
39. Cox, B.N. and G. Flannagan, *Handbook of Analytical Method of Textile Composites*. 1997: National Aeronautics and Space Administration Langley Research Center: Hampton, VA.
40. Fiore, V., et al., Evolution of the bearing failure map of pinned flax composite laminates aged in marine environment. *Composites Part B: Engineering*, 2020. **187**: p. 107864.
41. Jawaid, M., M.S. Salit, and O.Y. Alothman, *Green Biocomposites: Design and Applications*. 2017, Springer: Berlin, Germany.
42. Kern, W.T., et al., Finite element analysis and microscopy of natural fiber composites containing microcellular voids. *Materials & Design*, 2016. **106**: pp. 285–294.

43. Alhijazi, M., et al., Finite element analysis of natural fibers composites: A review. *Nanotechnology Reviews*, 2020. **9**(1): pp. 853–875.
44. Peltola, H., et al., Wood based PLA and PP composites: Effect of fibre type and matrix polymer on fibre morphology, dispersion and composite properties. *Composites Part A: Applied Science and Manufacturing*, 2014. **61**: pp. 13–22.
45. Dhakal, H.N. and M. Sain, Enhancement of mechanical properties of flax-epoxy composite with carbon fibre hybridisation for lightweight applications. *Materials*, 2020. **13**(1): p. 109.
46. Correa, J.P., J.M. Montalvo-Navarrete, and M.A. Hidalgo-Salazar, Carbon footprint considerations for biocomposite materials for sustainable products: A review. *Journal of Cleaner Production*, 2019. **208**: pp. 785–794.
47. Ita-Nagy, D., et al., Life cycle assessment of bagasse fiber reinforced biocomposites. *Science of the Total Environment*, 2020. **720**: p. 137586.
48. Rodriguez, L.J., et al., A literature review on life cycle tools fostering holistic sustainability assessment: An application in biocomposite materials. *Journal of Environmental Management*, 2020. **262**: p. 110308.
49. Agarwal, J., et al., Progress of novel techniques for lightweight automobile applications through innovative eco-friendly composite materials: a review. *Journal of Thermoplastic Composite Materials*, 2020. **33**(7): pp. 978–1013.
50. Siakeng, R., et al., Natural fiber reinforced polylactic acid composites: A review. *Polymer Composites*, 2019. **40**(2): pp. 446–463.
51. Ketola, A.E., et al., Effect of micro-and nanofibrillated cellulose on the drying shrinkage, extensibility, and strength of fibre networks. *BioResources*, 2018. **13**(3): pp. 5319–5342.
52. Thakur, V., A. Singha, and M. Thakur, In-air graft copolymerization of ethyl acrylate onto natural cellulosic polymers. *International Journal of Polymer Analysis and Characterization*, 2012. **17**(1): pp. 48–60.
53. de Souza Fonseca, A., et al., Improving cellulose nanofibrillation of non-wood fiber using alkaline and bleaching pre-treatments. *Industrial Crops and Products*, 2019. **131**: pp. 203–212.
54. Rudnik, E., *Compostable Polymer Materials*. 2019, Newnes: London.
55. Bastioli, C., *Handbook of Biodegradable Polymers*. 2020, Walter de Gruyter GmbH & Co KG: Berlin.
56. Badia, J., et al., Relevant factors for the eco-design of polylactide/sisal biocomposites to control biodegradation in soil in an end-of-life scenario. *Polymer Degradation and Stability*, 2017. **143**: pp. 9–19.
57. Bouakba, M., et al., Cactus fibre/polyester biocomposites: Manufacturing, quasi-static mechanical and fatigue characterisation. *Composites Science and Technology*, 2013. **74**: pp. 150–159.
58. Kashan, J.S. and S.M. Ali, Modeling and simulation for mechanical behavior of modified biocomposite for scaffold application. *Ingeniería e Investigación*, 2019. **39**(1): pp. 63–75.
59. Song, P., et al., Novel 3D porous biocomposite scaffolds fabricated by fused deposition modeling and gas foaming combined technology. *Composites Part B: Engineering*, 2018. **152**: pp. 151–159.
60. Mazlan, M., et al., Experimental and numerical approach to study the effect of biocomposite material to enhance durability of wood based composite material. *International Journal of Engineering & Technology*, 2018. **7**(3.28): pp. 157–162.
61. Rosdi, N., et al., Effect of biocomposite materials to enhance durability of selected wood species (intsia palembanica miq, neobalanocarpus heimii, shorea plagata) in Malaysia. *ARPN Journal of Engineering and Applied Sciences*, 2015. **10**(1): pp. 313–320.
62. Salehnejad, M.A., et al., Cracking failure analysis of an engine exhaust manifold at high temperatures based on critical fracture toughness and FE simulation approach. *Engineering Fracture Mechanics*, 2019. **211**: pp. 125–136.

63. Suman, S. and P. Biswas, Thermo-mechanical study of single and multi-pass welding of CSEF steel for residual stresses and deformations considering solid state phase transformation. *Materials Today: Proceedings*, 2020. **28**(2): pp. 789–795.
64. Potluri, R., Mechanical properties evaluation of T800 carbon fiber reinforced hybrid composite embedded with silicon carbide microparticles. *Multidiscipline Modeling in Materials and Structures*, 2018. **14**(3): pp. 589–608.
65. Zhang, C., J. Curiel-Sosa, and T.Q. Bui, A novel interface constitutive model for prediction of stiffness and strength in 3D braided composites. *Composite Structures*, 2017. **163**: pp. 32–43.
66. Yan, L., et al., Effect of alkali treatment on microstructure and mechanical properties of coir fibres, coir fibre reinforced-polymer composites and reinforced-cementitious composites. *Construction and Building Materials*, 2016. **112**: pp. 168–182.
67. Xue, G., et al., Influence of fiber reinforcement on mechanical behavior and microstructural properties of cemented tailings backfill. *Construction and Building Materials*, 2019. **213**: pp. 275–285.
68. Panyasart, K., et al., Effect of surface treatment on the properties of pineapple leaf fibers reinforced polyamide 6 composites. *Energy Procedia*, 2014. **56**: pp. 406–413.
69. Alamri, H. and I.M. Low, Microstructural, mechanical, and thermal characteristics of recycled cellulose fiber-halloysite-epoxy hybrid nanocomposites. *Polymer Composites*, 2012. **33**(4): pp. 589–600.
70. Mallick, P.K., *Fiber-Reinforced Composites: Materials, Manufacturing, and Design*. 2007, CRC Press: Boca Raton, FL.
71. Melro, A., et al., Micromechanical analysis of polymer composites reinforced by unidirectional fibres: Part I–Constitutive modelling. *International Journal of Solids and Structures*, 2013. **50**(11–12): pp. 1897–1905.
72. Giannadakis, K. and J. Varna, Analysis of nonlinear shear stress–strain response of unidirectional GF/EP composite. *Composites Part A: Applied Science and Manufacturing*, 2014. **62**: pp. 67–76.
73. Cardon, L., et al., Design and fabrication methods for biocomposites, in L. Ambrosio (Ed.), *Biomedical Composites*. 2017, Elsevier: Amsterdam, pp. 17–36.
74. Xiao, B., et al., Hybrid laminated composites molded by spray lay-up process. *Fibers and Polymers*, 2015. **16**(8): pp. 1759–1765.
75. Rubio-López, A., et al., Manufacture of compression moulded PLA based biocomposites: A parametric study. *Composite Structures*, 2015. **131**: pp. 995–1000.
76. Salit, M.S., Manufacturing techniques of tropical natural fibre composites, in *Tropical Natural Fibre Composites*. 2014, Springer: Berlin, Germany, pp. 103–118.
77. Saba, N., et al., Manufacturing and processing of kenaf fibre-reinforced epoxy composites via different methods, in S.M. Sapuan, M. Jawaid, E. Hoque, and N.B. Yusoff (Eds), *Manufacturing of Natural Fibre Reinforced Polymer Composites*. 2015, Springer: Berlin, Germany, pp. 101–124.
78. Nguyen-Dinh, N., C. Bouvet, and R. Zitoune, Influence of machining damage generated during trimming of CFRP composite on the compressive strength. *Journal of Composite Materials*, 2020. **54**(11): pp. 1413–1430.
79. Alzebdeh, K.I., M.M. Nassar, and R. Arunachalam, Effect of fabrication parameters on strength of natural fiber polypropylene composites: Statistical assessment. *Measurement*, 2019. **146**: pp. 195–207.
80. Kim, J.K. and K. Pal, *Recent Advances in the Processing of Wood-Plastic Composites*, vol. 32. 2010, Springer Science & Business Media: Berlin, Germany.
81. Schäfer, J. and T. Gries, Braiding pultrusion of thermoplastic composites, in Y. Kyosev (Ed.), *Advances in Braiding Technology*. 2016, Elsevier: Amsterdam, pp. 405–428.
82. McIlhagger, A., E. Archer, and R. McIlhagger, Manufacturing processes for composite materials and components for aerospace applications, in C. Soutis (Ed.), *Polymer Composites in the Aerospace Industry*. 2020, Elsevier: Amsterdam, pp. 59–81.

83. Goh, G.D., et al., Recent progress in additive manufacturing of fiber reinforced polymer composite. *Advanced Materials Technologies*, 2019. **4**(1): p. 1800271.
84. Bharimalla, A., et al., Nanocellulose-polymer composites for applications in food packaging: Current status, future prospects and challenges. *Polymer-Plastics Technology and Engineering*, 2017. **56**(8): pp. 805–823.
85. Fallon, J.J., S.H. McKnight, and M.J. Bortner, Highly loaded fiber filled polymers for material extrusion: A review of current understanding. *Additive Manufacturing*, 2019. **30**: p. 100810.
86. Baird, D.G. and D.I. Collias, *Polymer Processing: Principles and Design*. 2014, John Wiley & Sons: Hoboken, NJ.
87. Chokshi, R. and H. Zia, Hot-Melt Extrusion Technique: A Review. Iranian Journal of Pharmaceutical Research, 2004. **3**: pp. 3–16.
88. Franck A., H.B., Ruse H., Schulz G., *Kunststoff-Kompendium*. 2011, Vogel Publishing House: Germany.
89. Morton-Jones, D., *Polymer Processing*. 1989, Chapman and Hall: London.
90. Ebeling, F.-W.H., H.; Schirber, H.; Schlör, N.; Schwarz, O., *Kunststoffkunde (Plastics)*. 2007, Vogel Business Media: Germany.
91. Tadmor, Z. and C.G. Gogos, *Principles of Polymer Processing*. 2013, John Wiley & Sons: Hoboken, NJ.
92. Czvikovszky, T., P. Nagy, and J. Gaal, A polimertechnika alapjai. *Műegyetemi Kiadó, Budapest*, 2000: p. 453.
93. Jézéquel, P.-H. and V. Collin, Mixing of concrete or mortars: dispersive aspects. *Cement and Concrete Research*, 2007. **37**(9): pp. 1321–1333.
94. Song, J. and J. Evans, The assessment of dispersion of fine ceramic powders for injection moulding and related processes. *Journal of the European Ceramic Society*, 1993. **12**(6): pp. 467–478.
95. Gupta, N. and N. Kanth, Analysis of nip mechanics model for rolling calender used in textile industry. *Journal of the Serbian Society for Computational Mechanics*, 2018. **12**(2): pp. 39–52.
96. Ciesielski, A., *An Introduction to Rubber Technology*. 1999, iSmithers Rapra Publishing: Shawbury, pp. 3–169.
97. Azaiez, J., et al., State-of-the-art on numerical simulation of fiber-reinforced thermoplastic forming processes. *Archives of Computational Methods in Engineering*, 2002. **9**(2): pp. 141–198.
98. Plackett, D., et al., Biodegradable composites based on L-polylactide and jute fibres. *Composites Science and Technology*, 2003. **63**(9): pp. 1287–1296.
99. Fages, E., et al., Use of wet-laid techniques to form flax-polypropylene nonwovens as base substrates for eco-friendly composites by using hot-press molding. *Polymer Composites*, 2012. **33**(2): pp. 253–261.
100. Barbu, M.I.A.M.C., Wood-based panel technology, in H. Thoemen, M. Irle, and M. Sernek (Eds), *Wood-Based Panels, an Introduction for Specialists*. 2010, Brunel University Press: London, England, pp. 1–90.
101. Ku, H., et al., A review on the tensile properties of natural fiber reinforced polymer composites. *Composites Part B: Engineering*, 2011. **42**(4): pp. 856–873.
102. Dányádi, L., J. Móczó, and B. Pukánszky, Effect of various surface modifications of wood flour on the properties of PP/wood composites. *Composites Part A: Applied Science and Manufacturing*, 2010. **41**(2): pp. 199–206.
103. Mahjoub, R., et al., Tensile properties of Kenaf fiber due to various conditions of chemical fiber surface modifications. *Construction and Building Materials*, 2014. **55**: pp. 103–113.
104. Latif, R., et al., Surface treatments of plant fibers and their effects on mechanical properties of fiber-reinforced composites: A review. *Journal of Reinforced Plastics and Composites*, 2019. **38**(1): pp. 15–30.

105. Ramraji, K., K. Rajkumar, and P. Sabarinathan, Tailoring of tensile and dynamic thermomechanical properties of interleaved chemical-treated fine almond shell particulate flax fiber stacked vinyl ester polymeric composites. *Proceedings of the Institution of Mechanical Engineers, Part L: Journal of Materials: Design and Applications*, 2019. **233**(11): pp. 2311–2322.
106. Khan, M.Z., S.K. Srivastava, and M. Gupta, Tensile and flexural properties of natural fiber reinforced polymer composites: A review. *Journal of Reinforced Plastics and Composites*, 2018. **37**(24): pp. 1435–1455.
107. Sathish, S., et al., Experimental investigation on volume fraction of mechanical and physical properties of flax and bamboo fibers reinforced hybrid epoxy composites. *Polymers and Polymer Composites*, 2017. **25**(3): pp. 229–236.
108. Durante, M., et al., Creep behaviour of polylactic acid reinforced by woven hemp fabric. *Composites Part B: Engineering*, 2017. **124**: pp. 16–22.
109. Saba, N., et al., Mechanical, morphological and structural properties of cellulose nanofibers reinforced epoxy compofsites. *International Journal of Biological Macromolecules*, 2017. **97**: pp. 190–200.
110. Muthuraj, R., et al., Influence of processing parameters on the impact strength of biocomposites: A statistical approach. *Composites Part A: Applied Science and Manufacturing*, 2016. **83**: pp. 120–129.
111. Sawpan, M.A., K.L. Pickering, and A. Fernyhough, Improvement of mechanical performance of industrial hemp fibre reinforced polylactide biocomposites. *Composites Part A: Applied Science and Manufacturing*, 2011. **42**(3): pp. 310–319.
112. Kim, H.S., H.S. Yang, and H.J. Kim, Biodegradability and mechanical properties of agro-flour–filled polybutylene succinate biocomposites. *Journal of Applied Polymer Science*, 2005. **97**(4): pp. 1513–1521.
113. Balaji, A., B. Karthikeyan, and C.S. Raj, Bagasse fiber–the future biocomposite material: A review. *International Journal of Cemtech Research*, 2014. **7**(1): pp. 223–233.
114. Li, M., et al., Recent advancements of plant-based natural fiber–reinforced composites and their applications. *Composites Part B: Engineering*, 2020. **200**: p. 108254.
115. Shamsuyeva, M., et al., Surface modification of flax fibers for manufacture of engineering thermoplastic biocomposites. *Journal of Composites Science*, 2020. **4**(2): p. 64.
116. Hasan, K., P.G. Horváth, B. Miklós and T. Alpár, A state-of-the-art review on coir fiber-reinforced biocomposites. *RSC Advances*, 2021. **11**(18): pp. 10548–10571.
117. Hasan, K., P.G. Horváth, and T. Alpár, Development of lignocellulosic fiber reinforced cement composite panels using semi-dry technology. *Cellulose*, 2021. **28**(6): pp. 3631–3645.
118. Hasan, K., P.G. Horváth, and T. Alpár, Lignocellulosic Fiber Cement Compatibility: A State of the Art Review. *Journal of Natural Fibers*, 2021. pp. 1–26.
119. Hasan, K., P.G. Horváth, K. Zsófia and T. Alpár, Development of lignocellulosic fiber reinforced cement composite panels using semi-dry technology. *Scientific Reports*, 2021. **11**(1): pp. 1–13.
120. Hasan, K., P.G. Horváth, and T. Alpár, Thermomechanical Behavior of Methylene Diphenyl Diisocyanate-Bonded Flax/Glass Woven Fabric Reinforced Laminated Composites. *ACS Omega*, 2021. **6**(9): pp. 6124–6133.
121. Hasan, K., P.G. Horváth, and T. Alpár, Potential Fabric Reinforced Composites: A Comprehensive Review. *Journal of Material Science*, 2021. DOI: 10.1007/s10853-021-06177-6.
122. Biocomposites Market 2020, 29th October, 2020; Available from: https://www.market-sadmarkets.com/Market-Reports/biocomposite-market-258097936.html.

8 Progress in Development of Biorefining Process

Toward Platform Chemical-Derived Polymeric Materials

Malinee Sriariyanun
King Mongkut's University of Technology
North Bangkok (KMUTNB)

Prapakorn Tantayotai
Srinakarinwirot University

Yu-Shen Cheng
National Yunlin University of Science and Technology

Peerapong Pornwongthong, Santi Chuetor, and Kraipat Cheenkachorn
King Mongkut's University of Technology
North Bangkok (KMUTNB)

CONTENTS

8.1	Introduction	190
8.2	Lignocellulosic Biorefinery for Production of Platform Chemicals	193
8.3	Biocomposite from Platform Chemicals	199
8.4	Trends and Applications of Platform Chemical-Derived Polymers	203
	8.4.1 1,4-Diacid-Based Platform	204
	8.4.2 5-HMF-Based Platform	205
	8.4.3 Aspartic Acid-Based Platform	206
	8.4.4 Itaconic Acid-Based Platform	206
	8.4.5 Sorbitol-Based Platform	206
	8.4.6 Lactic Acid-Based Platform	207
8.5	Summary	208
Acknowledgments		208
References		208

DOI: 10.1201/9781003137535-8

8.1 INTRODUCTION

Polymeric biomaterials have been included in attractive fields of research and industries due to the continuous demands of users in different downstream industries. Regarding to prompt awareness worldwide to response in mitigation of polluted environments, United Nations policy sets the sustainable development goals to establish collaborations to prevent poverty and increase world security in different aspects, namely, Bioeconomy, Circular economy and Green economy (BCG economy). Therefore, this motivation drives the shift from using petroleum-based materials to biomaterials. In nature, three main categories of fibers are available for formulations and productions to various types of polymers, including (a) plant (bast, leaf, seed, fruit, grass and wood), (b) animal (wool, hair and silk) and (c) mineral (asbestos, glass, mineral wool, ceramic, silicate and carbon). There is an economic estimation of world market share of biomaterials to grow from 3.5 million tons in 2011 to 12 million tons by 2020 (Aeschelmann and Carus 2015). However, several limitations have been mentioned, such as availability of feedstocks, consistency in quality of natural material, competitions in material consumption for foods or polymers, and conflict in ethic and religion, etc. Among these natural fibers, lignocellulose is one of the raw materials used in making biomaterials due to its high availability with approximate annual global production at 75–150 billion tons (de Souza and Borsali 2004, Klemm et al. 2005, Cao et al. 2009). The predicted market demand of cellulose will increase up to 300 billion USD with CAGR of 4.2% between 2019 and 2026 (O'Dea 2015).

Although the natural biomaterials could be obtained from natural resources, however, when considered in large-scale production, it is not simple to find such secured supplies and economical feasible stocks of these materials. This concern leads to the development of technology and process to utilize renewable and sustainable resources, which can promptly integrate with current infrastructure and existing process for polymer synthesis from petroleum derivatives. Biorefinery has been developed during the late 1990s in parallel to petroleum refinery with the analogous concept to use refinery process to convert raw materials to an array of chemicals, powers and fuels. The petroleum refinery uses crude oil or petroleum, whereas biorefinery uses biomass as raw materials.

Biorefineries are classified differently based on various criteria (Figure 8.1). There are three technological phases applied in biorefineries, from phase one to three (Kamm and Kamm 2004, Clack and Deswarte 2008). Phase 1 and Phase 2 are similar that they utilize only one type of biomass feedstock. Technology availability for Phase 1 is developed to produce just one type of product, such as production of biodiesel from vegetable oil or bioethanol from sugarcane molasse. Phase 2 process is equipped with technology to produce more varieties of products, such as production of different oligosaccharides and bioethanol from sugarcane molasse. Phase 3 technology can utilize different types of feedstocks to produce a wide range of products, for example, lignocellulose biorefinery, starch biorefinery or whole crop refinery for productions of fuels and chemicals in the same facilities and plants (Kamm and Kamm 2004, Clack and Deswarte 2008). Based on types of conversion processes from biomass to products, there are four main groups, including mechanical/physical, biochemical, chemical and thermochemical conversion. Furthermore, types

Progress in Development of Biorefining Process

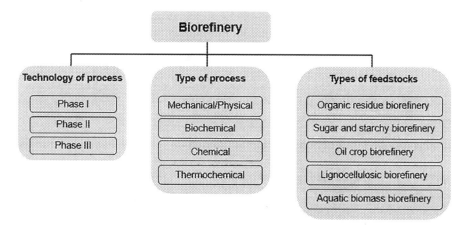

FIGURE 8.1 Classifications of biorefineries based on different criteria.

of feedstocks could be used for classification of biorefinery to be based on organic residue, sugar and starch, oil crop, lignocellulose and aquatic biomass (Werpy and Peterson 2004, Cheng et al. 2020) (Figure 8.1). According to all supported technologies, processes and feedstocks, biorefining process is theoretically not restricted to a specific raw material because there are possibilities to develop processes to convert any types of biomass to target chemicals, fuels and energy. In fact, small numbers of industrial facilities are success in market share and economic aspects, and most of them located in West Europe, North America, China and Japan. The major players in the current market are BASF (USA), Cargill (USA), PTT Global Chemicals (Thailand) and Champlor (France), and most of them located in North America. On the other hand, SME entrepreneurs for biorefinery business are currently growing in West Europe by using co-op collaborations to overcome the bottleneck in logistics and transportations.

Among different products produced from biorefinery, platform chemicals are gaining interesting in modern research and industries. Platform chemicals are building block chemicals or small molecule chemicals, which are readily converted or reacted to chemical intermediates to form final biochemical products. The values of biochemicals were reported to be around 6.8 billions USD in 2015. It is estimated that the market share of these renewable chemicals will increase to 113 million tons by 2050, which is about one-third of total market share of organic chemicals (Jong et al. 2012). Some examples of platform chemicals are ethanol, furfutal, hydroxymethylfurfural (HMF), 2, 5-furandicarboxylic acid (FDCA), glycerol, succinic acid, lactic acid, levulinic acid, 3-hydroxypropionic acid (3-HP), 3-hydroxypropionaldehyde (3-HPA), sorbitol, xylitol, and isoprene (Cheng et al. 2020). These platform chemicals are structural compatible to existing monomeric chemicals obtained from petroleum refinery. Additionally, these platform chemicals have been studies about the catalytic conversion reactions or polymerizations to target final products, and some of their feasibilities in terms of supplied quantities, technical barriers and production costs have been analyzed to determine their potential positions in the market. Based on

National Renewable Energy Laboratory (NREL) analysis in 2004, twelve platform chemicals were considered to be candidates for further development to commercialization, and those candidates included 1,4 diacids, FDCA, 3-HPA, aspartic acid, glucaric acid, glutamic acid, itaconic acid, levulinic acid, 3-hydroxybutyrolactone, glycerol, sorbitol, xylitol/arabinitol (Werpy and Peterson 2004). This nomination of potential platform chemicals was decided based on several parameters, such as technological barricades to existing infrastructure and technology, supplies for raw materials, and replacement of similar petroleum-derived products.

In response to BCG economy concept to minimize release of waste to environment, various types of waste generated from household and industries have been introduced to biorefinery, such as used cooking oil, animal by-products, lignocellulose biomass and mixed household food waste (Show and Sriariyanun 2021, Sriariyanun and Kitsubthawee 2020) (Figure 8.2). Used cooking oil is mostly composed of triacylglycerides, fatty acids and minor nitrogen-containing compounds, which currently is used as raw materials for biodiesel productions. Animal by-products from slaughterhouse, including blood, bone and fat from unused parts of meats or leftover residues from cutting, have high nitrogen contents, which currently are processed and mixed as animal feeds, and fats are used for biodiesel production. Lignocellulose biomass wastes contain high polysaccharide contents and currently they are mixed in compost and animal feeds. Lastly, mixed household wastes, such as cheese, milk, sugars, and breads, are mostly disposed in landfill or mixed in compost. The examples of waste utilizations mentioned here are simple processes making them become a mainstream for waste management.

In fact, these wastes were obtained from community households and industries, which produced in daily life with large quantities. Unfortunately, biodiesels, animal feeds and fertilizing compost are relatively low-value products and there are production routes from other low-cost resources for these products to compete the market prices and shares. Also, the petroleum-derived products are still competitive in pricing and availability of market stocks. Furthermore, these wastes have been continuously produced in vast amounts worldwide, and the situations are getting worse in

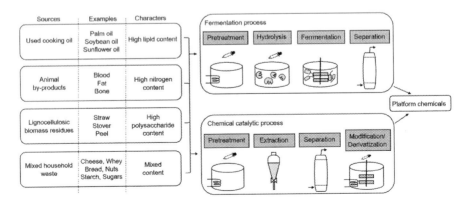

FIGURE 8.2 Utilization of household and industrial wastes for productions of platform chemicals via fermentation and chemical conversion processes.

waste management driving many countries to export wastes to other countries due to the over-capacity amounts of wastes for them to handle. Therefore, conversions of these wastes to platform chemicals have become the focus of R&D sectors and industry sectors to boost up the price of wastes and their by-products, and also provide the options to utilize the over-production amounts of wastes.

Currently, two types of processes, via fermentation and chemical catalytic conversion, for conversion of platform chemicals from these wastes are conducted in laboratory scales and are up-scaled for pilot and industrials scales, such as HMF, succinic, and itaconic acid, etc. (Sriariyanun et al. 2019, Rachmontree et al. 2020) (Figure 8.2). Briefly, for fermentation, wastes, which are composing of biochemical molecules, such as lipids, proteins and carbohydrates, are pretreated by chemical, biological or physical means to prepare waste biomass become more vulnerable for enzymatic hydrolysis reaction. Then, the pretreated biomass is hydrolyzed to smaller-sized biochemical molecules and monomers, such as oligosaccharide, polypeptide, monosaccharide, amino acid and fatty acid that microorganisms can absorb and convert to targeted platform chemicals via fermentations. Lastly, the fermented products are separated and purified to obtain the targeted platform chemicals with expected purity. In chemical catalytic process, waste biomass is pretreated to remove impurities and proceeded to extraction to obtain targeted biochemicals. Then, the targeted biochemicals are separated to even more remove by-products and finally are catalytic converted to platform chemicals by using chemical reactions or thermo-chemical reactions (Cheng et al. 2020). Based on these two concepts of biorefining processes for production of platform chemicals, various types of biomass could be supplied as raw materials of the processes, and these platform chemicals are subsequently used in production of biocomposite materials.

8.2 LIGNOCELLULOSIC BIOREFINERY FOR PRODUCTION OF PLATFORM CHEMICALS

Lignocellulose biomass is one of the most abundance biomass produced in nature because it is the main component in plant cell walls. It is composed of three major compounds, cellulose, hemicellulose and lignin. Within plant cell wall, ratios of cellulose, hemicellulose and lignin are in ranges of 40%–60%, 20%–30% and 15%–20%, respectively, depending on plant species, climate, temperature, nutrients, growth stage, pathogen and environment (Table 8.1) (Rodiahwati and Sriariyanun 2016, Akkharasinphonrat et al. 2017). Cellulose and hemicellulose are polysaccharides which contain sugar monomers linked to each other in a polymeric chain. Cellulose is made of glucose sugars (hexose, namely, β-D-glucopyranose) linked to each molecule by β-1,4-glycosidic bonds. The length of linear polymeric cellulose chain ranges in 100–10,000 monomeric units causing different properties of plant fibers. Cellulose is naturally synthesized by various living organisms, such as plants, algae and some bacteria. Cellulose fibrils are packed together as bundles with size of 5–50 nm in diameter and each fibril is packed by intramolecular hydrogen bonds and van der Waals forces (occur where proximity of cellulose chain stacking). The bonding strength between each fibril and intramolecular hydrogen bonds influences in physical, mechanical and physical properties for fibrils, for example, stability,

TABLE 8.1
Composition Analysis of Lignocellulose Biomass Obtained from Different Plant Species

Type	Cellulose	Hemicellulose	Lignin	Analysis Methods	References
Napier grass straw	41.18	30.15	6.68	Van Soest Method	Amnuaycheewa et al. (2017)
Cassava pulp	84.93	0.5	10.83	Van Soest Method	Runajak et al. (2020)
Sugarcane baggasse	34.78	25.04	24.64	NREL standard method	Fan et al. (2020)
Rice straw	34.63	29.74	15.34	Van Soest Method	Amnuaycheewa et al. (2016)
Alamo switchgrass	35.8	25.6	28.8	NREL standard method	Xu et al. (2020)
Durian peel	25.7	18.5	15.9	Van Soest Method	Siwina and Leesing (2020)
Corn cob	40	41.4	5.8	AOAC standard method	Kaliyan and Morey (2010)
Switchgrass	43.8	28.8	9.2	Near infrared reflectance spectroscopy	Kaliyan et al. (2009)
Corn stover	49.4	26.2	8.8	Near infrared reflectance spectroscopy	Kaliyan et al. (2009)
Switchgrass	34.71	25.08	12.53	Van Soest Method	Jian et al. (2020)
Cotton liner	89.7	1	2.7	FRD® by solvent extraction	Dorez et al. (2014)
Flax	80	13	2	FRD® by solvent extraction	Dorez et al. (2014)
Hemp	74.1	7.6	2.2	FRD® by solvent extraction	Dorez et al. (2014)
Sugarcane	51.8	27.6	10.7	FRD® by solvent extraction	Dorez et al. (2014)
Bamboo	54.6	11.4	21.7	FRD® by solvent extraction	Dorez et al. (2014)
Coconut	51.3	11.7	30.7	FRD® by solvent extraction	Dorez et al. (2014)
Rice straw	38.14	31.2	26.35	N.D.	Lo et al. (2017)
Rice husk	30.42	28.03	36.02	N.D.	Lo et al. (2017)
Corn stover	43.97	28.94	21.82	N.D.	Lo et al. (2017)
Sugarcane baggasse	46.55	27.4	20.61	N.D.	Lo et al. (2017)
Sugarcane peel	41.11	26.4	24.31	N.D.	Lo et al. (2017)
Waste coffee grounds	33.1	30.03	24.52	N.D.	Lo et al. (2017)

(Continued)

TABLE 8.1 (*Continued*)
Composition Analysis of Lignocellulose Biomass Obtained from Different Plant Species

Type	Compositions (%)			Analysis Methods	References
	Cellulose	Hemicellulose	Lignin		
Bamboo leaves	34.14	25.22	35.03	N.D.	Lo et al. (2017)
Pine wood sawdust	48.6	10.5	25.3	A standard method (GBT-20805-2006)	Shi and Wang (2014)
Pistachio shell	42.7	29.9	13.5	Seifert method (PN-P-50092:1992 standard)	Salasinska et al. (2016)
Sunflower husk	37.3	35	22.9	Seifert method (PN-P-50092:1992 standard)	Salasinska et al. (2016)
Arundo donax	35.83	31.12	18.86	NREL standard method	Yang et al. (2020)
Garlic skin	25.03	30.94	19.53	NREL standard method	Ji et al. (2020)
Grass waste	31.2	19.9	20	NREL standard method	Yan et al. (2020)

N.D, Not determined.

inertness, dissolution, gelation, and plasticization, glass transition temperature (T_g), appearance, heat resistance and elasticity (Klemm et al. 2005).

Hemicellulose is heteropolymeric compounds of hexose (i.e., glucose, mannose, galactose and rhamnose) and pentose (xylose, arabinose) sugars linked together in linear and branch chain manner. The polymerization degrees of hemicelluloses are in the range of 50–200 units. Hemicelluloses have variations in compositions, such as glucuronoxylan, galactomannan, glucomannan, xyloglucan and arabinoxylan, depending on plant species and parts (Pauly et al. 2013). Lignin is polymeric phenolic compounds synthesized from aromatic monolignol precursors (such as *p*-coumaryl, coniferyl and sinapyl alcohol). It is mostly found in secondary plant cell walls located in fully grown parts of plants, especially barks and peels. Hemicellulose fibers cross-link with cellulose fibers by ether bonds and they link with lignin molecules with ester bond or ether bond. Therefore, these three compounds arrange in bundles by having cellulose packed as base and covered with hemicellulose and lignin (Figure 8.3).

Since the 2000s, due to awareness in insecure situation of petroleum's market price fluctuations and concerns in greenhouse effects worldwide, many researchers have been focused on finding the alternative platform chemicals for substituting the use of petroleum-derived chemicals. As one of the most abundant natural biomass, lignocellulose biomass were proposed from many studies to be processed in biorefinery for productions of many types of platform chemicals, including cellulose

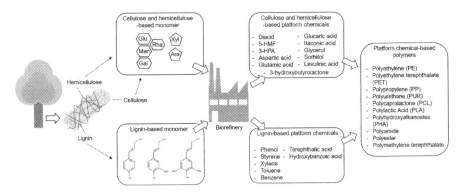

FIGURE 8.3 Biorefining process of lignocellulose biomass for production of polymeric biocomposites.

and hemicellulose derivatives (diacid, furfural and derivatives, 3-HPA, aspartic acid, glutamic acid, glucaric acid, itaconica acid, glycerol, sorbitol, levulinic acid, 3-hydroxybutyrolactone) and lignin derivatives (phenol, styrene, xylene, toluene, benzene, terephthalic acid, hydroxybenzoic acid) (Figure 8.3). These platform chemicals are also demonstrated to be converted to platform chemical-based polymers, such as polyethylene, polyethylene terephthalate (PET), polypropylene (PP), polyurethane (PUR), polycaprolactone (PCL), polylactic acid (PLA), polyhydroxyalkanoates (PHA), polyamide, polyester and polymethylene terephthalate, which are conventional polymers used in various purposes worldwide (Isikgor and Becer 2015).

Base on the fact that various types of plant biomass obtained from wastes and by-products produced from households and agriculture-processing industries are composed of cellulose, hemicellulose and lignin, with different proportions. Thus, theoretically, they all have same raw materials in common, so we can take different sources of lignocelluloses to convert to the target products using the same process or technological platform or facility. This concept fits in the sustainability in supply chain of raw materials to feed the process. Also, it provides the opportunity for business matching, based on BCG concept, that the lignocellulose wastes from different factories could be used as raw materials for lignocellulose biorefinery plants. Altogether, based on these reasons, lignocellulose has become one of raw material candidates for biorefining process for productions of platform chemicals.

In general, biorefining process of lignocellulosic biomass is composed of four steps, including pretreatment, hydrolysis, fermentation and product recovery/purification (Cheng et al. 2020) (Figure 8.4). Among these steps of biorefining, pretreatment and hydrolysis steps require more budget and processing time to invest; therefore, many new techniques are being researched and developed to reduce the time and cost in these steps. Of course, each pretreatment method has its own merits; it is unlikely to develop a universal approach to fit all types of biomass with different lignocellulose structural and compositional properties. A desirable pretreatment technology should be able to increase the enzymatic hydrolysis ability of biomass by simultaneously removing lignin, decreasing polymerization degree of holocellulose, and increasing the porosity of biomass, while also meet the techno-economic

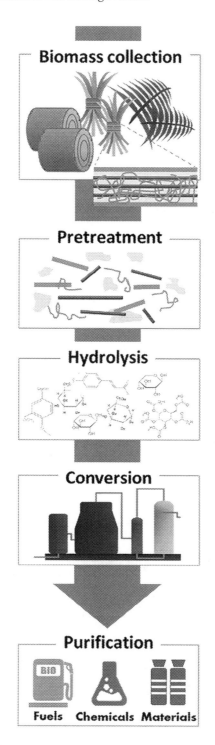

FIGURE 8.4 Steps of biorefining process of lignocellulose biomass to products.

and environmental requirements. Although the conventional methods are commonly used for biomass pretreatment, there are still some disadvantages that could be potentially relieved by employing other new developed methods. Thus, the overall process is considered to be complex as it is a multistep process and every step is operated under conditions that are possibly affected by the previous step. For example, if acid or alkaline chemicals are used for pretreatment process, which bring the pH to extremely high or low value. Before proceeding to enzymatic hydrolysis step, the pH of biomass is needed to be adjusted to pH 4.5–5.0, the optimal condition of cellulase enzyme. In another scenario, during hydrolysis and pretreatment with thermo-physical method, some inhibitory compounds are generated such as HMF and furfural, and they could interfere with the functions of microorganisms in fermentation processes (Rachmontree et al. 2020, Boontum et al. 2019). In this case, detoxification process is needed to be installed between hydrolysis and fermentation. However, to set up and operate the industrial process and make it feasible, smooth work flow, continuous operation and less numbers of process units are necessary. Consolidated processing concept is developed to be one solution by combining steps of conventional biorefinery from four to one reactor (Figure 8.5).

Consolidated processing is proposed with several benefits: (a) simplification of overall operation, (b) numbers of vessels or units are reduced, (c) contamination rate is reduced, (d) capital investment is reduced and (e) less energy consumption (Hasunuma and Kondo 2012). The design of consolidated processing could be done in different combinations, such as pretreatment and hydrolysis, hydrolysis and fermentation, fermentation and product recovery (Figure 8.4). To achieve this design, it was suggested by many research that genetic engineered microorganisms are necessary, since they can tolerate and function in extreme operational conditions, such as high temperature, high acidity, high pressure, or immunity to feedback repression of metabolic pathways. Furthermore, genetic engineered microorganisms could be

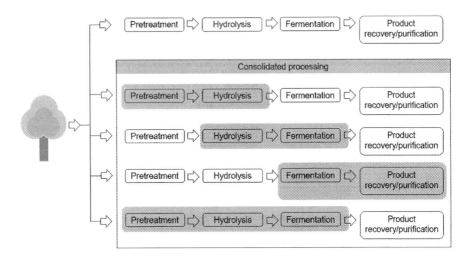

FIGURE 8.5 Biorefining process of lignocellulosic biomass and consolidated processing concept.

manipulated to contain multiple genes, which encoded a set of enzymes that catalyze exogenous metabolic pathways (Devarapalli and Atiyeh 2015). To prove the concept of consolidated processing, Amoah et al. (2017) constructed a recombinant yeast, *S. cerevisiae* MT8-1, expressing five cellulase genes (BGL, XYNII, EGII, CBHI and CBHII), which was subsequently used for simultaneous saccharification and fermentation (SSF) for ethanol production from ionic liquid pretreated bagasse and unbleached kraft pulp. This recombinant *S. cerevisiae* MT8-1 produced ethanol from pretreated bagasse and unbleached kraft pulp with higher yield for 4 and 2 folds, respectively, compared to the wild type yeast (Amoah et al. 2017). Similarly, another recombinant *S. cerevisiae* containing hemicellulase genes (BGL, XYLA, XYN) was constructed and used in SSF process to hydrolyze beechwood xylan and hemicellulosic liquor obtained from hydrothermal treatment of corncob and to convert to ethanol with the ethanol titer of 11.1 g/L and 32.8% conversion (Cunha et al. 2020). These studies demonstrated the benefits of recombinant microorganism and consolidated processing to improve process efficiency to gain product titers and to reduce the numbers of process steps and operational cost.

8.3 BIOCOMPOSITE FROM PLATFORM CHEMICALS

During the early 20th century, the major products of biorefining process have been primarily focused on biofuels, especially ethanol, biodiesel and biogas due to the market competition with crude oil. During the Gulf war in 2008, the supply of crude oil was decreased and the price was raised to 140 USD/Barrel, leading to motivation to find alternative biofuel to supply world demands. After continuous drops in crude oil price until now, the current crude oil price in 2020 is about 30 USD/Barrel. In 2008, the bioethanol price producing from primary fermentation of corn starch or sugarcane molasse was set at 60–62 USD/Barrel, which was a big motivation for biorefining process to beat with the crude oil market. Since then, parallel to the crude oil trend, the current market price of bioethanol was gradually dropped to 32–34 USD/Barrel in 2020 (www.bloomberg.com). Due to the slim gap of prices between crude oil and ethanol, as well as the world concerns about carbon footprints and environmental impacts, the target products of biorefinery deliberately shift from primary biofuels to other value-added products, especially platform chemicals.

Starting from 2010, the records of published research in ScienceDirect (https://www.sciencedirect.com/) and PubMed (https://pubmed.ncbi.nlm.nih.gov/) database related to "Platform chemicals" have been increased (Figure 8.6). Concurrently, among the numbers of published papers of Platform chemicals in each year, clearly more than 60% of them are related to polymers and composite. These emerging evidences suggested the trends of platform chemicals for production of polymers and composites. The motivations of using platform chemicals for polymers and composites are the growing needs for biopolymers and biocomposites to reduce the use of petroleum polymers to achieve BCG concept and reduce the environmental impacts. In 2020, the global market values of fiber types of biocomposites were doubly increased from 2014 (about 2.5 Billion USD). Also the total biocomposite market size was estimated to be 14–16 billion USD in 2016 (Fowler and Hughes 2006). Currently, the biggest sector of biocomposites is made of natural fiber composites,

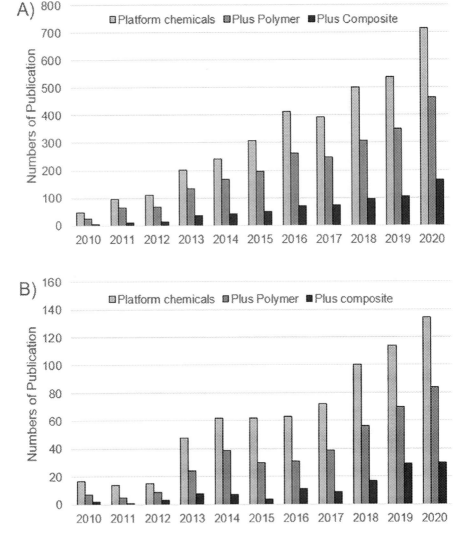

FIGURE 8.6 Published scientific research in platform chemicals, polymer and composites in with (a) ScieneDirect database and (b) PubMed database.

especially lignocellulose fiber, the manufacturing and processing have been continuously developed to the production of biocomposite in industrial scale with the purpose to modify the properties and characteristics of composites. Variations in formulations and fabrications of fibers as matrix or filler making differentiation in physical and mechanical properties, such as viscosity (m.Pa.s), gel temperature (°C), gel time (min), time to peak isotherm (min), temperature to peak isotherm (°C), cure cycle (days), tensile strength (psi), tensile modulus (psi), flexural strength (psi), flexural modulus (psi), elongation at break (%), glass transition temperature (°C), Heat deflection temperature (°C), water absorption (mg/sample), stiffness (GPa), etc.

Progress in Development of Biorefining Process

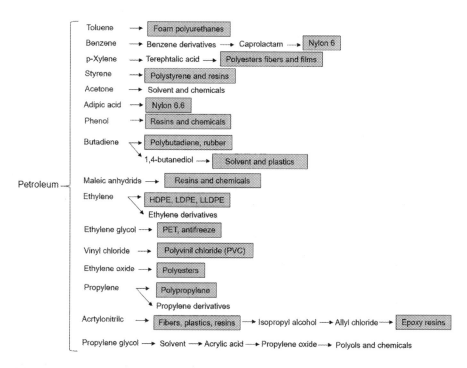

FIGURE 8.7 Platform chemicals produced from crude oil by petro-refinery plants and their pathways to be utilized in polymer and composite industries.

In addition to fiber polymeric composites, de-novo synthesized composite is developed to fulfill the need of the markets and customers because fiber type composites are still unable to function in the same manner as petroleum derived composites. Conventionally, numerous platform chemicals are refined and obtained from crude oil though processed in petrorefinery plants. The primary platform chemicals from petrorefinery include BTX (Benzene, Toluene and Xylene), acetone, butadiene, ethylene, ethylene glycol, ethylene oxide, propylene and propylene glycol (Takkellapati et al. 2018) (Figure 8.7). These chemicals are practically fractionated from crude oil and synthesized by catalytic conversion in petrorefinery plants to different classes of chemicals. Some BTX derivatives were used for productions of polymers and plastic beads, such as nylon, polyester and PUR. Numerous types of petro-derived polymeric compounds are polybutadiene, polyethylene (HDPE, LDPE, LLDPE, PET), polyvinyl chloride (PVC), PP and epoxy resins, and they are consumed in conventional markets for different downstream industries (Dwi Prasetyo et al. 2020).

Analogously, biomass, especially lignocellulose biomass, is disintegrated and processed by fractionation and catalytic conversion (via chemical, physical, thermochemical and biological reactions), similar to petrorefining process. Nowadays, several classes of platform chemicals obtained from biorefining processes are 1,4-diacid, furan, polyol, ABE (acetone, butanol and ethanol), HPA (hydroxylprionic acid), gluconic acid and lactic acid (Figure 8.8). Starting from few decades ago, glycerol, ethanol and butanol are produced as major products worldwide since they are all

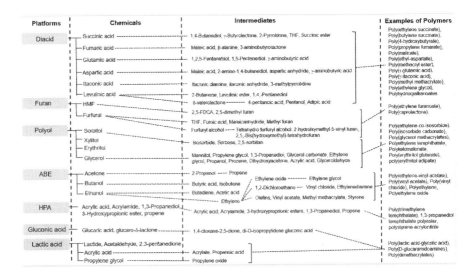

FIGURE 8.8 Platform biochemicals obtained from biomass and their pathways to be utilized in polymer and composite industries.

engaged in fuel production industries. Glycerol is a major byproduct from biodiesel production with various applications, such as solvent, sweetener, humectant, food preservative, and it is an important component in cosmetic, pharmaceutical and homecare products. Unfortunately, glycerol's purity obtained from biodiesel production is not qualified for other downstream processing, thus the recovery and purification processes are necessary, which are not yet economic feasible in several cases (Lari et al. 2018, D'Angelo et al. 2018). Therefore, glycerol obtained from biodiesel has been used as raw materials for microbial fermentations to produce various biochemical products, such as succinic acid, levulinic acid, itaconic acid, sorbitol, xylitol, erythritol, ethanol, butanol and bioplastic, because certain species of microorganisms have specific metabolic pathway to convert glycerol to other metabolites in glycolysis pathway. Additionally, these selected microorganisms also tolerate to the impurities of glycerol obtained from biodiesel production.

Ethanol and butanol are also blended to gasoline and jet fuel as alternative to benzene transportation engine. More than 90% of ethanol produced to date are obtained from fermentations of byproducts of agriculture-processing industries, such as sugarcane molasse, corn starch and cassava starch. After glycerol, ethanol and butanol, other platform chemicals have been attracted attentions from R&D sections including succinic acid, lactic acid, HMF and polyols because there are downside trends of fuels but upside trends of other businesses, especially bioplastic, biocomposite, pharmaceutical and cosmetic (Cheng et al. 2020). Due to advancement of R&D, diacid compounds could be catalytic converted to ester intermediate compounds and subsequently polymerized to different groups of polymers, such as polyethylene succinate, polymethacryl ester, polymethyl methacrylate and PHA (Figure 8.8). Due to functional groups and chemical structures of HMF and furfural, their derivatives are used in condensation reactions to form biopolymers (Rachmontree et al. 2020).

Polyols, especially glycerol, are coupled with ethylene glycol and propylene glycol to form polyglycerol polymers (Figure 8.8). Some other classes of platform chemical-derived polymers are also synthesized in different research and many of them are currently available as products in the market such as polytetramethylene ether glycol (PTMEG)-derived polymers ((Elastane (Lycra), Arnitel (DSM), Hytrel (Dupont), Pibiflex (SOFTER)), polycaprolactone ((Perstorp (CAPA), Dow chemical (Tone), Daicel (Celgreen)), polyacrylic acid ((Acrylsol), Lubrizol (Carbopol), BASF (Sokalan)) (Isikgor and Becer 2015).

According to the progress of R&D activities and manufacturing technology, as mentioned, utilization of platform chemicals to commercialization is more possible and practical. However, the major bottleneck of commercialization of biorefinery-derived platform chemicals is a competition to the petro-refinery. The costs per GJ for energy production of heat obtained from petroleum and biomass are 3.66 and 4.88 USD, respectively. For transportation fuel, the costs per GJ for using petroleum and biomass are 9.76 and 12.20 USD, respectively. For bulk chemical fuels, the costs from these two sources are 36.59 and 91.48 USD, respectively. On the other hand, PHA synthesis industry for production of bioplastic used raw materials including lignocellulose biomass (food crops, sugarcane, etc.) and vegetable oil. Unfortunately, the current production cost of PHA from waste biomass is 5–10 times higher than that of petroleum-derived chemicals (Tsang et al. 2019). It could be observed from these products that the price gaps between petroleum-derived products and biorefinery products are slimmer for the simplex products, such as heat and power, while for complex products, including platform chemicals, this gap is getting bigger. Difference in price gaps is due to several factors, such as economic of production scale, technological availability and logistics and transportation of raw materials. Therefore, the analysis of Strengths, Weakness, Opportunity, and Threats (SWOT) for commercialization of biorefinery-derived platform chemicals is conducted as in Table 8.2.

8.4 TRENDS AND APPLICATIONS OF PLATFORM CHEMICAL-DERIVED POLYMERS

The production of many types of platform chemicals from biorefining processes has currently reached maturation in technology for industrial-scale production, especially, succinic acid, lactic acid, itaconic acid, levulinic acid and 5-HMF. These platform chemicals are produced as bulk chemicals by number of chemical industries, especially BASF, Dupont, Merck, Evonik, etc. (Menon and Rao 2012). However, several platform chemicals, nowadays, such as succinic acid, lactic acid, itacomic acid, are also produced by petroleum-derived feedstocks and chemical catalytic conversion resulting in the competition in terms of market price. In such case, those platform chemicals have security in supply stocks for further end product development, and consequently, the main focus of R&D and industrial manufacturing shift to development and improvement in polymerization reaction, formulation and fabrication of platform chemical-derived polymers. Here, attempts on commercially available platform chemical-derived polymers are explained as follows.

TABLE 8.2
Analysis of Strengths, Weakness, Opportunity and Threats (SWOT) on Biorefinery in Commercialization

Strengths	Weakness
Conventional biorefinery products are common in current market. The customer perception and acceptance are possible	The high potential platform chemicals in downstream industries are not nominated due to early stage of development
Conversions of waste biomass to biorefining products are solution for waste management to reduce environmental impact and increase profit to industries	The value chains of industrial processing are not clear, such as supplied chain of biomass, and targeted downstream industrial users
Biorefinery products are also sensational welcome to specific sectors of customers by using bio-derived products, or ethical/religion restriction	Inconsistency in quantity and quality of supplied biomass
Biorefinery products motivate social movement as they are potentially fix CO_2 released to environment	Less in scale of production compared to petroleum biorefinery
Biorefinery products are the option for processing to avoid contacts to carcinogen chemicals used in petroleum refinery	Some sectors of biorefining process need further development in technology because they cannot readily fit in infrastructures and technologies for petroleum refinery

Opportunities	Threats
Biorefinery fits well with BCG economy concept, which could get support from social and government support worldwide	Inconsistency in governmental policy and support
Consolidating of the high-end market position of various downstream sectors, such as pharmaceutical, cosmetic and medical use	Fluctuation of petroleum price
	Insecurity in biomass supplies in terms of quantity and quality
International cooperations and fundings in research and technological developments to promote biorefining process	Creditability for financial investment as the biorefinery is young-age business
Some infrastructure and technology for petroleum refinery are readily to be integrated with biorefinery	Slow transition and implementation to downstream industries and customers
Several plarform chemical-derived bioplastic have been proved to have comparable properties to traditional chemical resins	

8.4.1 1,4-Diacid-Based Platform

Currently, production of succinic acid, also known as amber acid, using microorganisms via fermentation has become commercialized as bulk chemicals by several companies, such as BASF, Myriant and Mitsubishi Chemical. The global production of succinic acid in 2014 was estimated at 16,000–30,000 tons (~110 millions USD) with CAGR rate at 9% during 2019–2024 (https://www.mordorintelligence.com). Polymeric products of succinic acid and its derivatives, such as 1,4-butanediol, γ-butyrolactone, tetrahydrofuran, turn to be a major target for business focus. Various heteropolymeric products, including polyester, polyamide, polyester amide, could be formulated through condensation reactions. The available types of these heteropolymers are polybutylene succinate (PBS), polyethylene succinate (PES), polypropylene succinate (PPS), polyalkylene succinate (PAS), PTMEG, and currently, productions of PBS and PES are commercially available in industries (Xu and Guo 2010). PBS is a biodegradable thermoplastic polyester that is produced via condensations of succinic acid and 1,4-butanediol. PBS is used as the greener packagings, shopping bags, kitchen utensils, due to its high crystallinity structures, which allows for the high volume and heat-resistant products. PES is also a biodegradable polyester derived from condensations of succinic acid and ethylene glycol, and it is used in wrap film and membranes (Gigli et al. 2016). PTMEG is a type of polyester thermoplastic elastomers. Due to its low flexibility and hydrolytic stability, it is mainly commercialized as synthetic fibers such as TERATHANE® PTMEG which is liquid polymer at room temperature and is applicable in production of cable jacketing, medical tubing, adhesive and sealants (https://terathane.com/). Other diacids, including fumaric and malic acid, are produced by commercialized fermentation technology and are also used in manufacturing for polymeric products. Their polyesters could be formulated as homopolymeric or heteropolymeric compounds, such as polypropylene fumarate, which is applicable in tissue scaffolds and orthopedic implants (Cai et al. 2019).

8.4.2 5-HMF-Based Platform

5-HMF is one of the important renewable platform chemicals derived from deconstruction and dehydration of hemicellulose in lignocellulosic biomass. The major producers of 5-HMF are Nowa Pharmaceutical, NBB company, AVA Biochem and Everlight Chemical (https://www.marketwatch.com). The molecule of 5-HMF consists of furan ring and alcohol and aldehyde functional groups. Therefore, its molecule is readily for derivatizations, and the most important derivatives are 2,5-furandicarboxylic acid, 2,5-dimethylfuran, gamma-hydroxyvaleric acid and gamma-valerolactone (Rachmontree et al. 2020). The important commercialized products of 5-HMF-based platform are polyethylene furanoate (PEF) and PCL. PEF is a renewable polyester bioplastic as a replacement product for fossil-based PET to be used as bottle in conventional market. PEF is synthesized by two-step polymerization of 2,5-furandicarboxylic acid and ethylene glycol. PEF has several properties with a potential to substitute the use of PET, including liquid/gas barrier performance, thermal resistant, high glass transition temperature, low melting temperature, light weight, high modulus, recyclable, long shelf-life and cost-competitive. Currently, PEFs are produced in the forms of bottles, films and fibers, making their uses in beverage bottles, apparels, diapers, home furnishing, gardening and industrial fibers. Due to concern to the environmental impact, Coca Cola with collaborations of Verent, Gevo and Avantium targeted to produce bio-based PlantBottle™ and use them in their product's containers (Loos et al. 2020, Rosenboom et al. 2018). Similar to PEF, PCL is also a biodegradable polyester with a low melting point, good resistance to water, oil, solvent and chlorine. PCL is normally used as additives for resin formulations to improve a specific property, such as impact resistance. Its estimated market value of PCL in 2018 was 530 million USD. PCL is presently used often as the filament for 3D printers and prototype preparation, and it is approved from FDA for application in medical sector for human organ transplants (Guarino et al. 2017, Lyu et al. 2019).

8.4.3 ASPARTIC ACID-BASED PLATFORM

Aspartic acid is an amino acid produced by micro-oraganisms via fermentations of sugars obtained from lignocellulose. Aspartic acid, due to its active functional group, can convert to various metabolic compounds and it can be used as substrates for different types of polymers, such as polyaspartic acid (PASA) and thermal polyaspartate. PASA is a biodegradable water soluble polymer, so it was often used as hydrogel (Adelnia et al. 2019). Additionally, it is applied as water-swelling material in sanitary products, such as feminine pads and baby diaper, food packaging and coating (Thombre and Sarwade 2007, Yavvari et al. 2019). The market of PASA coating including bridge, flooring material, water-proofing tops, sealants and pipelines in the petroleum refinery industry is estimated to be 329 million in 2020, and it is competitive with PUR. The main players in PASA coating are BASF SE (Germany), AkzoNobel (Netherlands) and Sherwin-Williams Company (US). The chemical production of PASA is produced by dehydration of L-aspartic acid to polysuccinimide and then is subsequently hydrolyzed to PASA. Alternatively, a specific bacterium, *Bacillus subtilis*, possesses the protease to hydrolyze diethyl L-aspartate to form polyethyl L-aspartate; however, this route is not yet suitable for large-scale production (Thombre and Sarwade 2007).

8.4.4 ITACONIC ACID-BASED PLATFORM

Due to the current progress on technology, itaconic acid could be produced by chemical reaction and fermentation method. In chemical reaction method, citric acid is used as substrates and it thermolyzes to generate itaconic acid. While in biological conversion methods, different microorganisms such as *Aspergillus niger* could consume various types of biomass substrates ranging from sugarcane molasses, cassava flour and lignocellulose hydrolysate to generate itaconic acid. Itaconic acid has the potential to convert to different derivatives resulting in variation of polymeric products used in different industries, such as absorbent, acrylic glass, sanitary membrane product, coating, construction sheet materials (Teleky and Vodnar 2019, Sriariyanun et al. 2019). The market value of itaconic acid was estimated to be 81 million USD in 2016, which are relatively less than other platform chemicals, so it was considered in an economical aspect to be the niche market product. However, the future market of itaconic acid product is predicted to be focused in Asia-Pacific regions. One candidate of itaconic-derived polymers is a latex made of methyl methacrylate (MMA), which is an unsaturated polyester resins. MMA is also mixed in styrene butadiene rubber and MMA could also be produced from petroleum feedstocks as well, which make the supplies sufficient for the market need (Sano et al. 2020, Teleky and Vodnar 2019). Itaconic acid could also form homopolymer, polyitaconic acid by radical polymerization, which is potentially alternative option for polyacrylic acid (Robert and Friebel 2016).

8.4.5 SORBITOL-BASED PLATFORM

Sorbitol is a member of polyols and sugar alcohol. Its application is broad range of uses, for example, sweetener, bulky food thickener, personal care, cosmetic, toothpaste and losange. The biggest producers of sorbitol in the world are Cargill, Pfizer and SPI Pharma with the estimated total values of 1,245 million USD in 2018. Sorbitol is produced by both direct hydrolysis of cellulose using heterogenous catalyst to sorbitol or by fermentation of microorganisms (Silveira and Jonas 2002, Rice et al. 2020). Due to the functional group in sorbitol molecule making it as an active chemical to convert to different derivatives and forms different polymers, such as polyethylene-co-isosorbide terephthalate (PEIT) by coupled with other chemical monomers, such as other glycols. PEIT has high transition glass temperature so it is suitable to be used as biodegradable hot-filled bottle instead of PET. PEIT is synthesized by multistep reaction starting from dehydration reaction of sorbitol to form ring structure isosorbide, which is subsequently ring opening by base catalytic reaction and coupled with ethylene oxide to form polysorbates. Isosorbide can also ring opened and coupled with ethylene glycol and terephthalic acid in condensation reaction to form PEIT (Feng et al. 2015). Isosorbide can also in demand of Asia-Pacific region by having NOVAPHENE, Mitsubishi chemical, as major business players since it could be converted to different derivatives, such as diesters, a plasticizer in making PVC (Pasupuleti and Madras 2011, Wang et al. 2018).

8.4.6 Lactic Acid-Based Platform

Lactic acid or 2-hydroxypropionic acid is polymerized to PLA by two main reactions: (a) direct condensation of lactic acid (LA) and (b) ring-opening polymerization of the cyclic lactide. Due to strong equilibriums of PLA, and LA, the direct condensation is a more difficult process to control forward reaction, especially during synthesis of large polymeric molecules; therefore, the ring opening of lactide monomer is the preferred pathway. The polymerizations of lactide intermediates are homogeneous catalyzed in the absence of water (Inkinen et al. 2011). Lactic acid is also industrialized produced by fermentation of cheap cost raw materials, such as sugarcane molasses, glycerol, starch and lignocellulose hydrolysates by microorganisms, and the technology already reached the maturation phase (Juturu and Wu 2016, Zhang et al. 2018). Similar to other platform chemicals, polymeric products of lactic acids are synthesized by two methods: polycondensation of lactic acid and ring opening reaction of its cyclic dimer or lactide (Inkinen et al. 2011). Based on the history of development of lactic acid, the catalog of polymeric lactic acids, such as lactates, polyalkylene oxalates, polyethers, 2,3-pentanedione, acrylates and polyacetaldehyde-recorcinol, is demonstrated. Also, due to the availability of technology and feedstocks, the price of PLA is quite relatively lower than other bioplastic resin making its demands to grow to be over 4,800 million USD before 2019. PLA has superior weather-resistant properties, high tensile strength and fracture resistance; therefore, it has the potential to be used in textile, packaging, electronics segments and as an alternative option for PET, PS and PC derived from petroleum feedstocks (Castro-Aguirre et al. 2016, Tawakkal et al. 2014). Another important biodegradable polymer derived from lactic acid is polylactic-co-glycolic acid (PLGA), which plays role in drug delivery in human uses. PLGA gains attention from R&D segments especially with the application of controllable release of active compounds in digestive tract because the structural stability of PLGA is sensitive to several factors, especially change in pH. Therefore, the

PLGA could be formulated to carry bioactive compounds and this bioactive particle could be designed to release the active compounds at the specific targeted organ for absorption (Kapoor et al. 2015, Danhier et al. 2012).

In additional to these five platform chemical-based polymers, other chemical platforms as listed in Table 8.2 are also developed in research scale and industrial scale. However, their market values and market sectors are differentiated, with different levels of technology and social readiness to survive in market competition with petroleum derived polymers. Their feasibilities are determined by availability of technology to convert feedstock to specific product with high conversion rate and less by-products. The downstream formulation and fabrication process are also critical to determine the destination of uses and customer sections. Although the supplies of petroleum feedstock are negative factors to lessen the opportunity of biorefinery-derived platform chemicals; however, it helps to secure the quantitative supplies when petro- and bio-derived chemicals sharing the manufacturing process, which will be positive factor for manufacturer and business players. Together with the society motivation of moving forward to greener processes, platform chemical-derived polymeric products are still in rising trends of demands and supplies.

8.5 SUMMARY

Cellulose biomass is an abundant and renewable resource for production of biocomposite because it offers various routes to produce wide varieties of platform chemicals via biorefining process. To make the biorefining process become economic feasible and practical, state-of-the-art in planning and design of processes are required to reduce complexity and enhance efficiency in production conversion. During early era of biorefining process, the major products have been focused on biofuel sectors to substitute the demands of fossil fuels. Currently, the market sectors of platform chemicals are shifted to biochemicals and polymers sectors with higher market value per unit and to achieve some specific properties that petroleum-derived products do not possess. Due to the available technology of catalysis conversion and fermentation, different platform chemicals are converted to their derivatives with analogical functions of petroleum-derived building blocks, which are compatible to polymerization reactions to form polymeric products. Still, it needs state-of-the-art in selection of conversion routes and collaborations to downstream industries to transform these polymer products to final market products in various sectors.

ACKNOWLEDGMENTS

The authors would like to thank King Mongkut's University of Technology, North Bangkok (Research University Grant No. KMUTNB-BasicR-64-37) and Srinakarinwirot University (Research University Grant No. 671/2563) for financial support of this work.

REFERENCES

Adelnia H, Blakey I, Little PJ and Ta HT., Hydrogels based on poly(aspartic acid): Synthesis and applications. *Front. Chem.* 2019, 7(2019), 755.

Aeschelmann F and Carus M. Bio-based building blocks and polymers in the world: Capacities, production and applications: Status quo and trends toward. *Ind. Biotechnol.* 2015, 11, 154–159.

Akkharasinphonrat R, Douzou T and Sriariyanun M. Development of ionic liquid utilization in biorefinery process of lignocellulosic biomass. *KMUTNB Int. J. Appl. Sci. Technol.* 2017, 10(2), 89–96.

Amnuaycheewa P, Hengaroonprasan R, Rattanaporn K, Kirdponpattara S, Cheenkachorn K and Sriariyanun M. Enhancing enzymatic hydrolysis and biogas production from rice straw by pretreatment with organic acids. *Ind. Crops Prod.* 2016, 87, 247–254.

Amnuaycheewa P, Rodiahwati W, Sanvarinda P, Cheenkachorn K, Tawai A and Sriariyanun M. Effect of organic acid pretreatment on Napier grass (Pennisetum purpureum) straw biomass conversion. *KMUTNB Int. J. Appl. Sci. Technol.* 2017, 10(2), 107–117.

Amoah J, Ishizue N, Ishizaki M, Yasuda M, Takahashi K, Ninomiya K, Yamada R, Kondo A and Ogino C. Development and evaluation of consolidated bioprocessing yeast for ethanol production from ionic liquid-pretreated bagasse. *Bioresour. Technol.* 2017, 245, 1413–1420.

Boontum A, Phetsom J, Rodiawati W, Kitsubthawee K and Kuntothom T. Characterization of diluted acid pretreatment of water hyacinth. *Appl. Sci. Eng. Prog.* 2019, 12(4), 253–263. DOI: 10.14416/j.asep.2019.09.003.

Cai Z, Wan Y, Becker ML, Long YZ and Dean D. Poly(propylene fumarate)-based materials: Synthesis, functionalization, properties, device fabrication and biomedical applications, *Biomaterials*, 2019, 208, 45–71.

Cao Y, Wu J, Zhang J, Li HQ, Zhang Y and He JS. Room temperature ionic liquids (RTILs): A new and versatile platform for cellulose processing and derivatization. *Chem. Eng. J.* 2009, 147, 13.

Castro-Aguirre E, Iñiguez-Franco F, Samsudin H, Fang X and Auras R. Poly(lactic acid)-mass production, processing, industrial applications, and end of life. *Adv. Drug Delivery Rev.* 2016, 15(107), 333–366

Cheng YS, Mutrakulcharoen P, Chuetor S, Cheenkachorn K, Tantayotai P, Panakkal EJ and Sriariyanun M. Recent situation and progress in biorefining process of lignocellulosic biomass: Toward green economy. *Appl. Sci. Eng. Prog.* 2020, 13(4), 299–311.

Clark JH and Deswarte FEI (Eds). The biorefinery concept: An integrated approach. In *Introduction to Chemicals from Biomass*. John Wiley & Sons: Hoboken, NJ, 2008.

Cunha JT, Romaní A, Inokuma K, Johansson B, Hasunuma T, Kondo A and Domingues L. Consolidated bioprocessing of corn cob-derived hemicellulose: Engineered industrial Saccharomyces cerevisiae as efficient whole cell biocatalysts. *Biotechnol. Biofuels* 2020, 13, 1–15.

D'Angelo SC, Dall'ARA A, Mondelli C, Perez-Raminez J and Papadokonstantakis S., Techno-economic analysis of a glycerol biorefinery. *ACS Sustainable Chem. Eng.* 2018, 6(12), 16563–16572.

Danhier F, Ansorena E, Silva JM, Coco R, Breton AL and Préat V. PLGA-based nanoparticles: An overview of biomedical applications. *J. Controlled Release* 2012, 161(2), 505–522.

de Souza Lima MM and Borsali R. Rodlike cellulose microcrystals: Structure, properties, and applications, Macromol. *Rapid Commun.* 2004, 25(7), 771–787.

Devarapalli M and Atiyeh HK. A review of conversion processes for bioethanol production with a focus on syngas fermentation. *Biofuel Res. J.* 2015, 2, 268–280.

Dorez G, Ferry L, Sonnier R, Taguet A and Lopez-Cuesta JM. Effect of cellulose, hemicellulose and lignin contents on pyrolysis and combustion of natural fibers. *J. Anal. Appl. Pyrolysis* 2014, 107, 323–331.

Dwi Prasetyo W, Putra ZA, Bilad MR, Mahlia TMI, Wibisono Y, Nordin NAH and Wirzal MDH. Insight into the sustainable integration of bio- and petroleum refineries for the production of fuels and chemicals. *Polymers (Basel)* 2020, 12(5), 1091.

Fan Z, Lin J, Wu J, Zhang L, Lyu X, Xiao W, Gong Y, Xu Y and Liu Z. Vacuum-assisted black liquor-recycling enhances the sugar yield of sugarcane bagasse and decreases water and alkali consumption. *Bioresour. Technol.* 2020, 309, 123349.

Feng L, Zhu W, Zhou W, Li C, Zhang D, Xiao Y and Zheng L. A designed synthetic strategy toward poly(isosorbide terephthalate) copolymers: A combination of temporary modification, transesterification, cyclization and polycondensation. *Polym. Chem.* 2015, 6, 7470–7479.

Fowler PA and Hughes JM. Biocomposites: Technology, environmental credentials and market forces. *J. Sci. Food Agric.* 2006, 86(12), 1781–1789.

Gigli M, Fabbri M, Lotti N, Gamberini R, Rimini B and Munari A., Poly(butylene succinate)-based polyesters for biomedical applications: A review, *Eur. Polym. J.* 2016, 75, 431–460.

Guarino V, Gentile G, Sorrentino L and Ambrosio L. Polycaprolactone: Synthesis, properties, and applications. *Encycl. Polym. Sci. Technol.* 2017. doi: 10.1002/0471440264.pst658.

Hasunuma T. and Kondo A. Consolidated bioprocessing and simultaneous saccharification and fermentation of lignocellulose to ethanol with thermotolerant yeast strains. *Process. Biochem.* 2012, 47, 1287–1294.

Inkinen S, Hakkarainen M, Albertsson AC and Södergård A. From lactic acid to poly(lactic acid) (PLA): Characterization and analysis of PLA and its precursors. *Biomacromolecules* 2011, 12(3), 523–532.

Isikgor FH and Becer CR. Lignocellulosic biomass: A sustainable platform for the production of bio-based chemicals and polymers. *Polym. Chem. Polym. Chem.* 2015, 6, 4497–4559.

Ji Q, Yu X, Yagoub AE, Chen L and Zhou C. Efficient removal of lignin from vegetable wastes by ultrasonic and microwave-assisted treatment with ternary deep eutectic solvent. *Ind. Crops Prod.* 2020, 149, 112357.

Jian, S, JiYu, Z, Wen, W, Guang Quing, L and Chang C. Assessment of pretreatment effects on anaerobic digestion of switchgrass: Economics-energy-environment (3E) analysis. *Ind. Crops Prod.* 2020, 145, 111957.

Jong E, Higson A, Walsh P and Wellisch M. Product developments in the bio-based chemicals arena. *Biofuels, Biofuel Bioprod. Biorefin.* 2012, 6(6), 606–624.

Juturu V and Wu JC. Microbial production of lactic acid: the latest development. *Crit Rev. Biotechnol.* 2016, 36(6), 967–977.

Kaliyan N and Morey RV. Densification characteristics of corn cobs. *Fuel Process. Technol.* 2010, 91(5), 559–565.

Kaliyan N, Morey RV, White MD and Doering A. Roll press briquetting and pelleting of corn stover and switchgrass. *T ASABE.* 2009, 52(2), 543–555.

Kamm B and Kamm M. Principles of biorefineries. *Appl. Microbiol. Biotechnol.* 2004, 64(2), 137–145.

Kapoor DN, Bhatia A, Kaur R, Sharma R, Kaur G and Dhawan S. PLGA: A unique polymer for drug delivery. *Ther. Delivery* 2015, 6(1), 41–58.

Klemm D, Heublein B, Fink HP and Bohn A. Cellulose: Fascinating biopolymer and sustainable raw material. *Angew. Chem. Int. Ed. Engl.* 2005, 44(22), 3358–3393.

Lari GM, Pastore G, Haus M, Ding Y, Papadokonstantakis S, Mondelli C and Perez-Raminez J. Environmental and economical perspectives of a glycerol biorefinery. *Energy Environ. Sci.* 2018, 11, 1012–1029.

Lo SL, Huang YF, Chiueh PT and Kuan WH. Microwave pyrolysis of lignocellulosic biomass. *Energy Procedia* 2017, 105, 41–46.

Loos K, Zhang R, Pereira I, Agostinho B, Hu H, Maniar D, Sbirrazzuoli N, Silvestre AJD, Guigo N and Sousa AF. A perspective on PEF synthesis, properties, and end-life. *Front Chem.* 2020, 8, 585.

Lyu JS, Lee JS, and Han J. Development of a biodegradable polycaprolactone film incorporated with an antimicrobial agent via an extrusion process. *Sci. Rep.* 2019, 9, 20236.

Menon V. and Rao M. Trends in bioconversion of lignocellulose: Biofuels, platform chemicals & biorefinery concept. *Prog. Energy Combust. Sci.* 2012, 38(4), 522–550.

O'Dea N. Emerging innovation trends in composites. *Paper Presented at Composites Engineering Show*, Lucintel, Inc; NEC Birmingham, UK, 2015.

Pasupuleti S and Madras G. Synthesis and degradation of sorbitol-based polymers. *J. Appl. Polym. Sci.* 2011, 121(5), 2861–2869.

Pauly M, Gille S, Liu L, Mansoori N, de Souza A, Schultink A and Xiong G. Hemicellulose biosynthesis. *Planta.* 2013, 238(4), 627–642.

Rachamontree P, Sriariyanun M, Tepaamorndech S and Somboonwatthanakul I. Optimization of oil production from cassava pulp and sugarcane bagasse using oleaginous yeast. *Orient. J. Chem.* 2019, 35(2), 668–677.

Rachmontree P, Douzou T, Cheenkachorn K and Sriariyanun M. Furfural: A sustainable platform chemical and fuel. *Appl. Sci. Eng. Prog.* 2020, 13(1), 3–10.

Rice T, Zannini E, Arendt E and Coffey A. A review of polyols: Biotechnological production, food applications, regulation, labeling and health effects. *Crit. Rev. Food Sci. Nutr.* 2020, 60(12), 2034–2051.

Robert T, Friebel S., Itaconic acid: A versatile building block for renewable polyesters with enhanced functionality. *Green Chem.* 2016, 18, 2922–2934.

Rodiahwati W and Sriariyanun M. Lignocellulosic biomass to biofuel production: Integration of chemical and extrusion (screw press) pretreatment. *KMUTNB Int. J. Appl. Sci. Technol.* 2016, 9(4), 289–298.

Rosenboom JG, Hohl DK, Fleckenstein P, Storti G and and Morbidelli M. Bottle-grade polyethylene furanoate from ring-opening polymerisation of cyclic oligomers. *Nat. Commun.* 2018, 9, 2701.

Runajak R, Chuetor S, Rodiahwati W, Sriariyanun M, Tantayotai P and Phornphisutthimas S. Analysis of microbial consortia with high cellulolytic activities for cassava pulp degradation. *E3S Web Conf.* 2020, 141, 03005. doi:10.1051/e3sconf/202014103005.

Salasinska K, Polka M, Gloc M and Ryszkowska J. Natural fiber composites: The effect of the kind and content of filler on the dimensional and fire stability of polyolefin-based composites. *Polimery* 2016, 61, 255–265.

Sano M, Tanaka T, Ohara H and Aso Y., Itaconic acid derivatives: Structure, function, biosynthesis, and perspectives. *Appl. Microbiol. Biotechnol.* 2020, 104(21), 9041–9051.

Shi X and Wang JA. comparative investigation into the formation behaviors of char, liquids and gases during pyrolysis of pinewood and lignocellulosic components. *Bioresour Technol.* 2014, 170, 262–269.

Show P.L and Sriariyanun M. Prospect of liquid biphasic system in microalgae research. *Appl. Sci. Eng. Prog.* 2021, 14(3). doi:10.14416/j.asep.2020.12.001.

Silveira MM and Jonas R. The biotechnological production of sorbitol. *Appl. Microbiol. Biotechnol.* 2002, 59(4–5), 400–408.

Siwina S and Leesing R. Bioconversion of durian (Durio zibethinus Murr.) peel hydrolysate into biodiesel by newly isolated oleaginous yeast Rhodotorula mucilaginosa KKUSY14. *Renew Energy.* 2020, 163, 237–245.

Sriariyanun M, Heitz JH, Yasurin P, Asavasanti S and Tantayotai P. Itaconic acid: A promising and sustainable platform chemical? *Appl. Sci. Eng. Prog.* 2019, 12(2), 75–82.

Sriariyanun M and Kitsubthawee K. Trends in lignocellulosic biorefinery for production of value-added biochemicals. *Appl. Sci. Eng. Prog.* 2020, 13(4), 283–284.

Takkellapati S, Li T and Gonzalez MA. An overview of biorefinery derived platform chemicals from a cellulose and hemicellulose biorefinery. *Clean Technol. Environ. Policy.* 2018, 20(7), 1615–1630.

Tawakkal IS, Cran MJ, Miltz J and Bigger SW. A review of poly(lactic acid)-based materials for antimicrobial packaging. *J. Food Sci.* 2014, 79(8), R1477–90.

Teleky BE and Vodnar DC. Biomass-derived production of itaconic acid as a building block in specialty polymers. *Polymers (Basel).* 2019, 11(6), 1035.

Thombre SM and Sarwade BD. Synthesis and biodegradability of polyaspartic acid: A critical review. *J. Macromol. Sci. A.* 2007, 42(9): 1299–1315.

Tsang YF, Kumar V, Samadar P, Yang Y, Lee J, Ok YS, Song H, Kim KH, Kwon EE, and Jeon YJ. Production of bioplastic through food waste valorization. *Environ. Int.* 2019, 127, 625–644.

Wang YF, Xu BH, Du YR and Zhang SJ. Heterogeneous cyclization of sorbitol to isosorbide catalyzed by a novel basic porous polymer-supported ionic liquid. *Mol. Catal.* 2018, 457(2018), 59–66.

Werpy T and Peterson G. Top value added chemicals from biomass. Results of screening for potential candidates from sugars and synthesis gas. The National Renewable Energy Laboratory (NREL) Report. 2004, 1.

Xu J and Guo BH. Poly(butylene succinate) and its copolymers: Research, development and industrialization. *Biotechnol. J.* 2010, 5(11), 1149–1163.

Xu K, Shi Z, Lyu J, Zhang Q, Zhong T, Du G and Wang S. Effects of hydrothermal pretreatment on nano-mechanical property of switchgrass cell wall and on energy consumption of isolated lignin-coated cellulose nanofibrils by mechanical grinding. *Ind. Crops. Prod.* 2020, 149, 112317.

Yan X, Cheng JR, Wang YT and Zhu MJ. Enhanced lignin removal and enzymolysis efficiency of grass waste by hydrogen peroxide synergized dilute alkali pretreatment. *Bioresour. Technol.* 2020, 301, 122756.

Yang J, Wang X, Shen B, Hu Z, Xu L and Yang S. Lignin from energy plant (Arundo donax): Pyrolysis kinetics, mechanism and pathway evaluation. *Renew Energy* 2020, 161, 963–971.

Yavvari PS, Awasthi AK, Sharma A, Bajaj A and Srivastava A., Emerging biomedical applications of polyaspartic acid-derived biodegradable polyelectrolytes and polyelectrolyte complexes. *J. Mater. Chem. B.* 2019, 7, 2102–2122.

Zhang Y, Yoshida M and Vadlani PV. Biosynthesis of D-lactic acid from lignocellulosic biomass. *Biotechnol. Lett.* 2018, 40(8), 1167–1179.

9 Biocomposite Production from Ionic Liquids (IL)-Assisted Processes Using Biodegradable Biomass

Marttin Paulraj Gundupalli
King Mongkut's University of Technology North Bangkok

Kittipong Rattanaporn
Kasetsart University

Santi Chuetor
King Mongkut's University of Technology North Bangkok

Wawat Rodiahwati
University of New England

Malinee Sriariyanun
King Mongkut's University of Technology North Bangkok

CONTENTS

Abbreviations	214
9.1 Introduction	214
9.2 Fractionation of Lignocellulosic Biomass	218
9.3 Ionic Liquids	220
9.3.1 Properties of Ionic Liquid	221
9.3.2 Ionic Liquids and Their Applications in Different Industries	225
9.4 Fabrication of Biocomposites from Lignocellulosic Biomass Using Ionic Liquid	226
9.4.1 Biopolymer Films	227
9.4.2 Aerogels	231
9.4.3 Hydrogel	233
9.4.4 All Cellulose Composites	235

DOI: 10.1201/9781003137535-9

9.5 Ionic Liquid-Integrated Composites in Other Biotechnological
    Applications 238
9.6 Summary 239
Acknowledgments 239
References 239

ABBREVIATIONS

[Amim][Cl]:	1-allyl-3-methylimidazolium chloride
[Bmim][BF$_4$]:	1-butyl-3-methyl-imidazolium tetrafluoroborate
[Bmim][Cl]:	1-butyl-3-methylimidazolium chloride
[Bmim][OAc]:	1-butyl-3-methylimidazolium acetate
[Emim][(MeO)(H)PO$_2$]:	1-ethyl-3-methylimidazolium methylphosphonate
[Emim][BF$_4$]:	1-ethyl-3-methylimidazolium tetrafluoroborate
[Emim][DEP]:	1-ethyl-3-methylimidazolium diethyl phosphate
[Emim][OAc]:	1-ethyl-3-methylimidazolium acetate
[Emim][(CF$_3$SO$_2$)2N]:	1-ethyl-3-methylimidazolium bis(trifluoromethanesulfonyl)imide
[EOHMIM]:	1-hydroxyethyl-3-methylimidazolium
[Hmim][HSO$_4$]:	1-methylimidazolium hydrogen sulfate
[HOEmim][Cl]:	1-(2-hydroxyethyl)-3-methylimidazolium chloride
[Mim][OAc]:	1-methylimidazolium acetate
[MOYI][Cl]:	1-methyl-3-(oxiran-2-ylmethyl)-1H-imidazol-3-ium chloride (IL-oxiran)
[Smim][HSO$_4$]:	1-sulfobutyl-3-methylimidazoliumhydrogen sulfate
AIL:	acidic ionic liquid
ASA:	amidosulfonic acid
BASF:	Badische Anilin und Soda Fabrik
BASIL:	biphasic acid scavenging utilizing ionic liquids
BIL:	basic ionic liquid
CMC:	carboxymethylcellulose
DP:	degree of polymerization
NMMO:	N-methylmorpholine N-oxide
PEI:	polyethylenimine hydrogel
PET:	polyethylene terephthalate
PVA:	polyvinyl alcohol
TGA:	thermo-gravimetric analysis
TMCS:	trimethylchlorosilane
VOC:	volatile organic compounds

9.1 INTRODUCTION

In recent years, lignocellulosic biomass has been considered an important source of biomaterial worldwide. Traditionally, lignocellulosic biomass are used to produce bioethanol, energy, and heat (Gupta and Verma, 2015). Lignocellulosic biomass includes dedicated energy crops, agricultural residues, grasses and forestry residues, etc. Lignocelluloses include woody substrates that fall under the category of softwood, hardwood, and grasses. It is abundantly available and is currently utilized for the production of the energy and food sector (Ge et al., 2014). Lignocellulosic biomass comprises cellulose, hemicellulose, and lignin in varying ranges at 33%–51%, 19%–34%, and 20%–30%, respectively (van Maris et al., 2006). Lignocellulosic biomass is a potential source of predominantly second-generation bioethanol. Currently, several industrial plants for ethanol production are in operation, especially in Europe and North America (Sindhu et al., 2016). It is considered as one of the most promising alternative raw materials for producing composites as a matrix or a filler (Mahmood et al., 2015). In nature, a large numbers of natural fibers are produced and lignocellulosic biomass is the main component of the plant cell wall, which could be used to enhance the properties of many composite materials (Gurunathan et al., 2015). Sustainability concept and global awareness have influenced biocomposite's design and engineering through the utilization of lignocellulosic biomass. Lignocellulosic polymeric composites are mixtures of polymers (natural/petroleum resources) and lignocellulose components (cellulose, hemicellulose, and lignin). These materials act as reinforcements to provide the desired characteristics in the end products of composite materials (Fernandes et al., 2017). Lignocellulosic biomass is considered to be abundant and renewable, consisting of polysaccharides, such as hemicellulose and cellulose. It is also rich in an aromatic polymer such as lignin (Zoghlami and Paës, 2019) (Figure 9.1). These plant components constitute a complex assembly of polymers that are inherently resistant to enzymatic conversion in biological processes. On the other hand, due to their rigid structures and complex interactions between

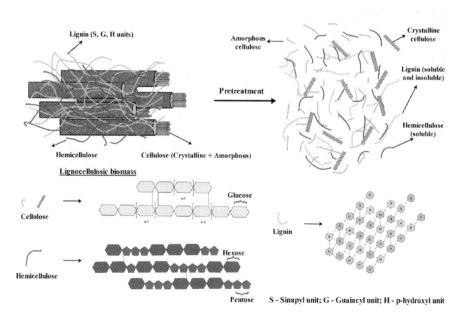

FIGURE 9.1 Structural arrangements of cellulose, hemicellulose and lignin in lignocellulose biomass.

hemicellulose, cellulose, and lignin, these plant component's utilizations face challenges for downstream processing (Luo et al., 2019).

Cellulose is an abundant biopolymer on earth with a yield of 1.5×10^{12} tons/year. It is the major polymer in lignocellulosic biomass, equivalent to 40–60 wt% (Sharma et al., 2017). It is a fibrous, tough, and water-insoluble polysaccharide that plays an indispensable role in maintaining the strength of plant cell wall structure (Bringmann et al., 2012). Cellulose comprises more than two β-D-glucose molecules covalently linked via β-(1,4)-glycosidic bonds (Chami Khazraji and Robert, 2013). These glycosidic bonds exist between cellobiose (a dimer of β-D-glucose), the basic repeating unit of the cellulose chain. These chains consist of 500–1,400 D-glucose units linearly arranged to form microfibrils (Robak and Balcerek, 2018). A total of 36 microfibrils packed together within the plant's primary cell wall form long cellulose fibrils (Kubicki et al., 2018). Lignin and hemicellulose are associated with cellulose microfibrils (Eriksson and Bermek, 2009). Cellulose structure includes a crystalline and amorphous region, resulting in the variation in fiber properties, in chemical and physical aspects. When cellulose was hydrolyzed at 320°C and 25 MPa, changes of crystalline region of cellulose into an amorphous region occurred (Limayem and Ricke, 2012). Cotton, cellulose, and flux are the purest cellulose sources (up to 90% Cellulose), and wood contains about 50% cellulose (Hemmati et al., 2019). In recent years, cellulose's mechanical properties have become the focus of research on preparing eco-friendly composite materials using cellulosic fibers (Nishino and Arimoto, 2007). Cellulose is a significant material in many industries due to its diverse physical and chemical properties. The physical and chemical properties shown in Figure 9.2 illustrate cellulose's importance (Heinze, 2015; Khattab et al., 2017; Wade and Welch, 1965).

Cellulose-derived materials such as cloth, tissues, pharmaceuticals, paper, coatings and polymer composites have been used worldwide since the 19th century (Zhang et al., 2017). Cellulose is used as a potential material due to its abundance, lightweight, low cost, biodegradability, low abrasiveness, and reproducibility (Bledzki et al., 1996). Some application studies of cellulose in fibers, films, beads, and cellulose derivatives have been reported in recent years (Zhang et al., 2009). However, cellulose application is limited due to the cohesion between the cellulose molecules with extended intermolecular and intramolecular hydrogen bonds (Pang et al., 2015). Intramolecular hydrogen bonds provide chain rigidity. On the other hand, intermolecular hydrogen bonds make linear macro-polymeric molecules assembling into sheet-like structures (Medronho and Lindman, 2015). Therefore, cellulose is almost insoluble in all common inorganic and organic solvents. To utilize the cellulose present in biomass, it is necessary to separate or fractionate it from other plant components (e.g., hemicellulose and lignin).

Hemicellulose is the second most abundant polysaccharides in the cell wall of plants after cellulose. They account for 15–30 wt% of lignocellulosic biomass and are considered a promising material for producing value-added biochemical products (Luo et al., 2019). Compared with cellulose, hemicellulose is composed of straight and branched short polymer chains depending on plant's types. These polymers are usually composed of C_5 sugars (arabinose, xylose) and C_6 sugars (galactose, glucose, and mannose) making them have numerous combinations in compositions, namely, xylan, xyloglucan, β-glucan, mannan, and galactoglucomanan (Ji et al., 2011). The degree of polymerization (DP) of hemicellulose is usually in the range

PHYSICAL	CHEMICAL
- Non-toxic - Biodegradable - Homo-biopolymer - White appearance - 1.44 X 10^6 to 1.8 X 10^6 g (molecular mass) - Density - 1.52 - 1.54 g/cm^3 (20°C) - High compressive strength - High tensile - Intermolecular and intramolecular hydrogen bonds gives maximum strength	- Degree of polymerization (DP) of 8000 - 10000 - Insoluble in water - Soluble in organic solvents - Reaction with concentrated acids (anhydrous) - Reacts with hydrocarbons, ketones, alcohols, acids, amides, esters, halogenated hydrocarbons, etc - Higher affinitiy in solubility to alkaline base in order LiOH > NaOH > KOH > RbOH > CsOH - High compressive strength - High tensile - Intermolecular and intramolecular hydrogen bonds gives maximum strength

FIGURE 9.2 Physical and chemical properties of cellulose. (From Heinze, T., 2015. Cellulose: Structure and properties. In: *Advances in Polymer Science*. pp. 1–52; Khattab, M.M., Abdel-Hady, N.A., Dahman, Y., 2017. Cellulose nanocomposites. In: *Cellulose-Reinforced Nanofiber Composites*. Elsevier, pp. 483–516; Wade, R.H., Welch, C.M., 1965. The role of alkali metal hydroxides in the benzylation of cotton cellulose. *Text. Res. J.* 35, 930–934.)

of 100–200 units (Zeng et al., 2017) and is hydrolyzed into monomers and other products during the fractionation of lignocellulosic biomass. Due to this property, the removal of hemicellulose by various pretreatment methods, such as steam and acid, can enhance the enzyme's catalytic activity in the saccharification process (de Oliveira Santos et al., 2018). The interactions between hemicellulose and cellulose in plant cell walls (primary and secondary) play an important role in providing tensile properties to wood materials (Mikkonen, 2013).

Lignin is also considered the second most abundant plant component with different chemical structures and compositions (Khan et al., 2019; Nasrullah et al., 2017). The lignin content varies greatly upon species, growth phase, parts of plants, and environments. In softwood and hardwood, the lignin content ranges from 25% to 35% and 15% and 30%, respectively (Lourenço and Pereira, 2018). Lignin is an aromatic polymer (heterogeneous) with three precursors (monolignols) such as (a) p-coumaryl alcohol, (b) coniferyl alcohol, and (c) sinapyl alcohol. These monolignols are from different lignin units, such as (a) p-hydroxyphenyl (H), guaiacyl (G), and syringyl (S) (del Río et al., 2015). Due to lignin's complex structure with linear and branched chain of the lignin polymeric units, the plant cell wall exhibits recalcitrance to the hydrolysis process and physical rigidity. Also, lignin inhibits the hydrolysis of cellulose during the enzymatic saccharification process. Therefore, to enhance the enzyme catalytic activity, it is necessary to hydrolyze or modify lignin's structure. Lignin has been used in the design of a biocomposite material (Cicala et al., 2017). Furthermore, lignin can be blended with protein, cellulose, and starch to reinforce a biocomposite material with a special property (Doherty et al., 2011).

9.2 FRACTIONATION OF LIGNOCELLULOSIC BIOMASS

Cellulose, hemicellulose, and lignin can be disintegrated and fractionated by applying various pretreatment methods. Also, the effectiveness of enzymatic saccharification is increased by various pretreatment methods (Mosier et al., 2005). The compact structures of plant components are disrupted during pretreatment. This process is carried out to overcome the resistance caused by cellulose and lignin present in the cell wall (Singh et al., 2015). There are many pretreatment methods, which are divided into physical, chemical, and biological methods. Some of the pretreatment methods are summarized and shown in Figure 9.3 (Kumar, 2018). The pretreatment's mechanisms to promote lignocellulose disintegration have been demonstrated to associate with fiber swelling, increased porosity, increased contact area, reduced hydrophobicity, and removed inhibitors.

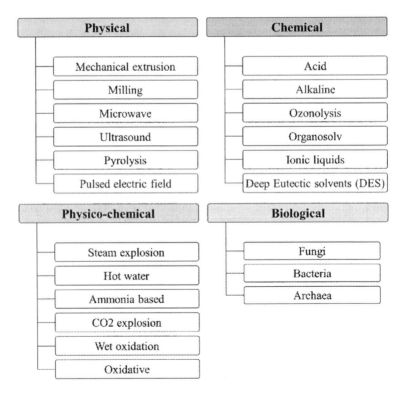

FIGURE 9.3 Various pretreatment methods utilized for partial and complete solubilization of hemicellulose, lignin, and cellulose components of the plant cell (Kumar, 2018).

Though there are different methods to pretreat lignocellulosic biomass, some of the pretreatments under harsh conditions convert hemicellulose and cellulose to other byproducts. Some of the byproducts include acetic acid, formic acid, levulinic acid, furfural, 5-hydroxymethylfurfural cause a reduction in yields and recovery rates of targeted products (Jönsson et al., 2013). Therefore, milder methods, such as ionic liquid and organosolv pretreatments, were demonstrated to cause partial or complete solubilization of hemicellulose and lignin with less byproducts. Also, the ionic liquid can dissolute cellulose presented in the lignocellulosic biomass during pretreatment. For fractionation purposes, the solubilized cellulose can be simply regenerated by additions of antisolvents such as alcohol and water causing by interfering with hydrogen bond arrangements. Some of the advantages and disadvantages of these methods are summarized and shown in Table 9.1. The derivatization of regenerated cellulose

TABLE 9.1
Advantage and Disadvantage of Alkaline, Ionic Liquid and Organosolv Pretreatment Method

Pretreatment Methods	Advantages	Disadvantages	References
Alkaline	• Partial or complete hydrolysis of hemicellulose • Solubilization and recovery of lignin • Decrease in the degree of crystallinity • Retention of cellulose in biomass	• High and low solubility in hardwood and softwood, respectively • large amount of wastewater generated • Longer reaction time during pretreatment required • Recovery of chemical increases the process cost	Singh et al. (2016)
Ionic liquid	• Decrease in crystallinity of regenerated cellulose • Increase in the external and internal surface of cellulose due to swelling • Alteration or modification of hemicellulose and lignin • Solubilization and recovery of lignin	• High cost of Ionic liquids • Additional process required to recover monosaccharides from the hydrolysate • Toxic nature of ionic liquids • Separations of sugars from ionic liquid are difficult	Bhatt and Shilpa (2014), Sun et al. (2016)
Organosolv	• Hydrolysis of hemicellulose and lignin • Recovery of solubilized lignin • Decrease in the crystallinity of cellulose	• High cost of solvents • High-pressure reactor required • Highly volatile results in loss of solvent • Recycling and reuse of solvents required therefore not feasible for industrial application	Bhatt and Shilpa (2014)

can be carried out by mixing it with another material, which leads to the formation of various composites (Huber et al., 2012b). In recent years, the application of ionic liquid in the processing of cellulose has attracted attention due to its high affinity towards cleavage of hydrogen bonds (Mahmood et al., 2017b). In this chapter, ionic liquids (IL) and its application for synthesis of different biocomposites have been discussed in detail.

9.3 IONIC LIQUIDS

Ionic liquids are generally defined as nonmolecular compounds composed entirely of ions (cations and anions) with a melting point below 100°C (Gorke et al., 2010). It is also referred to as the molten salts because they are mostly liquid phases at room temperature. However, native molten salts are usually referred to as liquid compounds with high viscosity, high melting point, and high corrosive properties (Zhao, 2003). Alternatively, ionic liquid possesses several properties to make it considered to be environmentally friendly and green solvents as an alternative to volatile organic solvents, such as low volatility, low corrosive, low viscosity, and recyclable (Brandt-Talbot et al., 2017). Anion charge delocalization and cation structure are the two main factors that affect physical, chemical, and biological characteristics of ionic liquid (Kumari and Singh, 2018). The length and branch of the alkyl chain connected to the cation can be adjusted to design the ionic liquid according to requirements and performance. Therefore, the ionic liquid is also called designed solvent (Mai et al., 2014). Due to the different characteristics of ionic liquids, they are considered to be used for many purposes such as medicine, environmental chemistry, medicine, biocomposites materials, etc. The ability to recover and reuse ionic liquid improves the economic efficiency of its application (Usmani et al., 2020).

Ionic liquid can be divided into three different main types on the basis of the physical and chemical properties. Ionic liquid is divided as neutral, acid, and basic based on the properties like conductivity, solubility, viscosity, acidity, water miscibility, and basicity (Hajipour and Rafiee, 2015). Neutral ionic liquid (NIL) possesses properties like good thermal stability, electrochemical stability, low melting points, low viscous, and can be used as an inert solvent. These properties exist due to the weak electrostatic interaction between cations and anions. Some of the NIL anions are thiocyanate, hexafluorophosphate, p-toluenesulfonate (tosylate), bis(trifluoromethanesulfonyl)amide, and tricyanomethide, and their structures are shown in Figure 9.4 (Hajipour et al., 2010).

Acidic ionic liquid (AIL) is composed of acids which represent cation or anions. AIL is classified as Lewis AIL and Bronsted AIL. In general, Brønsted acids are known for proton (H$^+$) donor, and Lewis acid is known for electron acceptor. The proton donor and electron acceptor function is usually localized in anion or cation of AIL. Due to its acidic nature, the properties possessed are water solubilities and enhanced acidic nature, thus, making AIL as a potential green catalyst with enhanced catalytic activities (Olivier-Bourbigou et al., 2010). Some of the common AILs are shown in Figure 9.5.

On the other hand, ionic liquids also open up new opportunities for the development of environmentally friendly base catalysts. Thus, creating a way through the

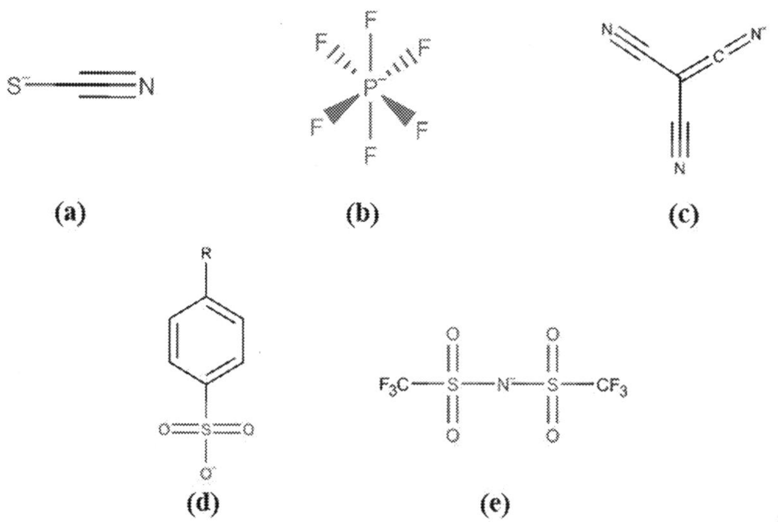

FIGURE 9.4 Neutral ionic liquids (NIL) with different anions. (a) Thiocyanate. (b) Hexafluorophosphate. (c) Tricyanomethide. (d) p-toluenesulfonate. (e) Bis(trifluoromethanesulfonyl)amide.

advantages of combining inorganic bases and ionic liquid is offered. Some of the properties exhibited by basic ionic liquid (BIL) are nonvolatile, flexible, immiscible to many organic solvents, and noncorrosive (Olivier-Bourbigou et al., 2010). These ionic liquids have potentials to replace the traditional alkaline chemicals such as NaOH, KOH, NaHCO$_3$, K$_2$CO$_3$, triethylamine, and NaOAc. BIL has been used in the process, such as (a) Markovnikov addition, (b) Michael addition, (c) Knoevenagel condensation, (d) Mannich reaction, (e) Henry reaction, (f) Feist-Benary reaction, (g) oximation, etc. (Hajipour and Rafiee, 2009). Some of the BILs, along with their structures, are shown in Figure 9.6.

9.3.1 Properties of Ionic Liquid

Ionic liquids are defined as salts that have a melting point lower than the boiling point of water (Welton, 1999). An important criterion for assessing ionic liquid and its application is its melting point. There are several ionic liquids with different anions and cations, and only a few ionic liquids with low melting points have been reported. The derivatization of melting points for various ionic liquids has been reported to aid in the molecular design of novel compounds (Trohalaki and Pachter, 2005). The melting point depends on the type of anion and cation used during the synthesis of the ionic liquid. In another study, attempts were made to determine the melting points by correlating it with the chemical structure of imidazolium bromides and benzimidazolium bromides using molecular descriptors generated by the CODESSA program. It was reported that the melting point was influenced due to the intermolecular and intramolecular interactions between the anions and cations (Katritzky

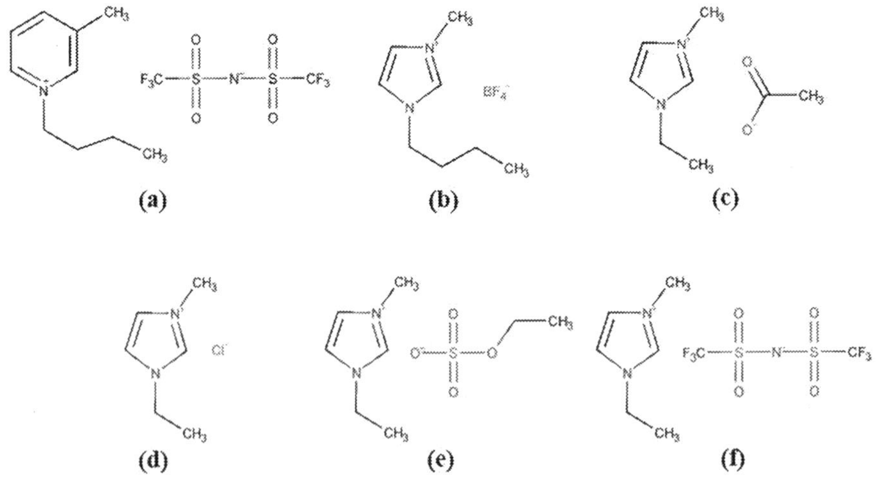

FIGURE 9.5 Common aprotic ionic liquids (AIL). (a) 1-Butyl-3-methylimidazolium bis(trifluoromethylsulfonyl)imide;.(b) 1-butyl-3-methylimidazolium tetrafluoroborate. (c) 1-ethyl-3-methylimidazolium acetate. (d) 1-ethyl-3-methylimidazolium chloride. (e) 1-ethyl-3-methylimidazolium ethyl sulfate. (f) 1-ethyl-3-methylimidazolium bis(trifluoromethanesulfonyl)amide.

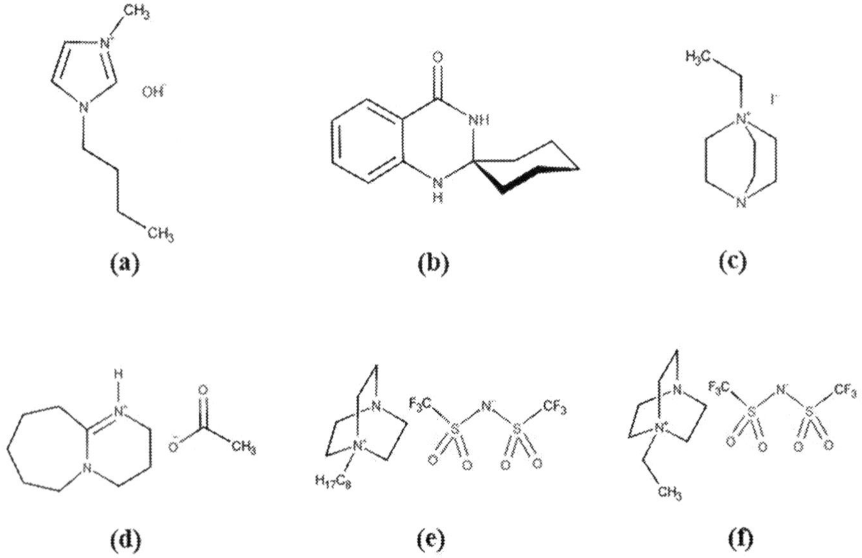

FIGURE 9.6 Basic Ionic liquid (BIL) examples. (a) 1-Methyl-3-butylimidazolium hydroxide. (b) 1-Propyl-3-alkylimidazole hydroxide. (c) 1-Ethyl-4-aza-1-azoniabicyclo [2.2.2] octane iodide. (d) 1,8-Diazabicyclo[5.4.0]-undec-7-en-8-ium Acetate. (e) 1-Ethyl-4-aza-1-azoniabicyclo[2.2.2] octane bis(trifluoromethane sulfonyl)amide. (f) 1-Octyl-4-aza-1-azonia-bicyclo[2.2.2]octane-bis-(trifluoromethanesulfonyl)amide.

et al., 2002). Furthermore, it has been reported that inorganic anion makes a significant contribution to decrease the melting point of the ionic liquid. It is stated that the increase in the size of the anion with similar cation has a distinct effect on increasing or decreasing the melting point (Zhao, 2003).

The vapor pressures of ionic liquids at ambient temperature are negligible, suggesting the low volatility property and its possibility to be used as green solvents (Akbari et al., 2018). The boiling temperatures of some ionic liquids cannot be determined because the vapor pressure is related to 1 atm, and such ionic liquids start to decompose even at lower temperatures (Ahrenberg et al., 2016). This is one of the most important reasons for the emergence of ionic liquid function as a new alternative solvent to volatile organic compounds (VOC) solvents (Passos et al., 2014). Since ionic liquids are low-volatile solvents, they are applied to be used in the azeotrope distillation process to reduce the release of solvent to the environment, considering to be an eco-friendly process (Mallakpour and Dinari, 2012). Consequently, with low volatility of IL, it is also classified as low toxicity solvent to users (Montalbán et al., 2015; Zhao et al., 2007).

Thermal stability is the unique property of ionic liquid. Most of the ionic liquids are thermally stable when the operating temperature is greater than 350°C. However, the stability of the ionic liquid depends upon the strength of heteroatom-hydrogen and heteroatom-carbon bonds between anions and cations (Wasserscheid and Keim, 2000). The thermal stability of the ionic liquid can be experimentally determined by monitoring of the thermal decomposition obtained by thermo-gravimetric analysis (TGA) under vacuum conditions. It was reported that ionic liquids with imidazolium cations such as [Bmim][BF_4], [Emim][BF_4], and 1,2-dimethyl-3-propy limidazolium bis(trifl uorosulfonyl)imide are stable up to temperatures of 423°C, 445°C, and 457°C, respectively (Welton, 2004). Similarly, [Emim][$(CF_3SO_2)_2N$] with a melting point of −3°C was reported to be stable when the temperature was more than 400°C (Bonhôte et al., 1996). The thermal stability of the ionic liquid is also dependent on the residence time of the reaction. It was reported that loss of mass was observed for 1-decyl-3-methylimidazolium triflate [$C_{15}MIM$][triflate] and 1-alkyl-3-methylimidazolium hexafluorophosphates [RMIM][PF_6] at a temperature of 200°C after 10h. In addition, the thermal decomposition of different cations decreases as the anions' hydrophilicity increases (Kosmulski et al., 2004).

Viscosity describes the intrinsic flow resistance of a fluid across the certain medium causing by molecular interactions between liquid molecules, including van der Waals interactions, repulsions, hydrogen bonds, and electrostatic forces. Many types of ionic liquids have relatively higher viscosity compared to organic solvents, which can cause problems in industrial processing. These include (a) adverse effects on power requirements; (b) dissolution, extraction, and filtration; (c) reduction of mass and heat transfer rate in the reaction; and (d) separation process (Aparicio et al., 2010). Therefore, viscosity must be considered in unit operation calculation, analysis, and design since high viscosity can be the major issue for an industrial application. It is necessary to understand the relationship between the viscosity and the structure of ions (anions and cations) for the design of ionic liquid with low viscosity (Zhang et al., 2015). In addition to viscosity, ionic liquid density influences the process parameters, such as (a) mass transfer, (b) immiscible phase's separation,

(c) substrates and products diffusion, (d) cost of pumping, (e) active centers accessibility, and (f) apparatuses adhesion (Ochędzan-Siodłak et al., 2013). Theoretically, density could be altered by temperature. It was reported that the density of the ionic liquid, pyridinium, and imidazolium species decreased as the temperature increased from 20°C to 55°C (Ochędzan-Siodłak et al., 2013). Furthermore, the molecular size of the cation inversely determines the ionic liquid density as the size of the cation increases; the density decreases (Laali, 2003).

Ionic liquid also plays a role as electrolytes in electrochemical reactions due to its conductivity (Shahzad et al., 2019). Theoretically, the conductivity of the solution depends on the charge carriers and other properties such as viscosity, ion size, density, ion mobility, delocalization, anion charge, and aggregation (Yuan et al., 2018). Since ionic liquids (highly viscous) are composed of ions (cations and anions), they have high ionic conductivity. Electrical conductivity is inversely proportional to viscosity. The temperature has an effect on the viscosity of ionic liquid that is during the temperature increases, the viscosity decreases. Therefore, as the viscosity decreases, the conductivity of the ionic liquid increases with an increase in mobility of ions (Yusuf et al., 2017). Among different ionic liquids, imidazolium species of ionic liquids have higher conductivity. The conductivity of ionic liquids ranges from 0.1 to 20 ms/cm (Olivier-Bourbigou et al., 2010).

The dielectric constant is a critical factor for understanding the solvation potential of an ionic liquid and also an important input parameter for many solvent models (Huang et al., 2011). The dielectric constant is used in measuring the polarity of the molecular solvents. In order to evaluate the solvation of ionic liquid, it is necessary to characterize and understand its static dielectric constant (Weingärtner, 2006). In one study, the dielectric constant of ionic liquids based on imidazolium was monitored to be in the range of 8.8–15.2 at 25°C (Wakai et al., 2005). A decrease in the dielectric constant is associated with the length of the alkyl chain on the cation of the imidazolium ionic liquid. Therefore, it is a polar solvent (Weingärtner et al., 2007). Most ionic liquids have dielectric constants below 20.5, which are the values of acetone and the average dielectric constant of the ionic liquid is 15.5 (Rybinska-Fryca et al., 2018). Most ionic liquids at room temperature have dielectric constant above six, and they are easily mixed with solvents such as water, ethanol, acetone, and dimethylformide (Berthod and Cara-Broch, 2004). Another property related to dielectric constant is the polarity that is important for designing ionic liquids (Hallett and Welton, 2011). Polarity is the sum of all nonspecific and specific intermolecular interactions that exist between the solvent and the solute (Reichardt and Welton, 2010). Polarity and polarizability of a solvent are the critical impacts on chemical reactions (Zhao, 2003). The polarity of several ionic liquids is determined using solvatochromic dye, fluorescent dye, Reichardt's dye, and Nile Red dye (Dzyuba and Bartsch, 2002). Ionic liquids show high polarity due to their ionic properties. Despite their high polarity, many ionic liquids are inherently hydrophobic (Weuster-Botz, 2007). In a study, 1-alkyl-3-methylimidazolium cation was probed with Nile red to determine the polarity and found that it had the polarity in comparable level to the polarity of short-chain alcohols (Carmichael and Seddon, 2000). The nucleophilicity of the anion was tested with the solvatochromic dye showed that the polarities of [Tf$_2$N] and [PF$_6$] were much lower than that of alcohols (Infante and Huszagh, 1987).

9.3.2 Ionic Liquids and Their Applications in Different Industries

For the past few years, ionic liquids are considered interesting chemicals due to their unique properties that are suitable for many industrial applications. Ionic liquids are members of salts, but they are different from other salts, such as sodium chloride. Due to its strong ionic interaction and tightly packed structure, sodium chloride only melts at very high temperatures (>800°C). However, the melting point of ionic liquids is lower than 100°C, due to the asymmetric structure and close-packed structure of the constituent ions (Earle and Seddon, 2000). Many ionic liquids have high thermal stability and low flammability due to the negligible vapor pressure at ambient temperatures. It is believed that these properties can improve the safety of solvents at high temperatures used in batteries or in outer space (Nancarrow and Mohammed, 2017). It should be noted, however, that many ionic liquids can still be distilled under extreme conditions.

Ionic liquids have high electrical and thermal conductivity making them have potential applications as heat transfer fluids and electrolytes. Due to different aspects of their properties, as described in Section 9.3.2, they are designed and applied in different fields based on their compositions and characteristics. Over the years, various ionic liquids have been synthesized to reduce cost, improve recoverability, and reusability. Although ionic liquids are currently not widely used, several chemical companies have begun to commercialize them for use in a variety of applications. In 1988, the French Petroleum Institute commercialized an ionic liquid for the synthesis of polybutene through the Difasol process. The product is considered very important for manufacture rubber, plastics, and other similar materials (Martins et al., 2017). Ionic liquids display potential use in different industries that are related to solvents, catalysis, additives, chemistry, electrochemistry, and analysis (Figure 9.7) (Greer et al., 2020).

Ionic liquids are utilized as solvents for acid scavenging, separation, dissolution, thermal fluids, and catalyst recovery. In 2003, the application of the ionic liquid was commercialized on a large scale, in line with the launch of the BASF's BASIL process. The BASF Company uses imidazolium-based ionic liquids to remove acids in the process. It has been reported that the ionic liquid can be easily removed from the reaction. In addition, this process was awarded the prestigious innovation award in 2004. The process is currently being carried out on a multi-ton scale (Ruiz-Angel and Berthod, 2006). The BASIL process scavenges acid from the process using an ionic liquid during the production of alkyl phenyl phosphines. This compound is an important precursor to products, such as photoinitiators for curing by using UV. Photoinitiators are used in printing inks and coatings (Gutowski, 2018). Similarly, in the 1980s, the lyocell process was introduced to the market with advantages over viscose production (Sharma et al., 2019). The lyocell process involved the use of ionic liquids to dissolve cellulose to produce lyocell fibers for the textile industry. In the viscose manufacturing process, NaOH is used for mercerizing cellulose pulp, and carbon disulfide is used for xanthanion (Zhang et al., 2018). The mercerized pulp sheet is soaked in 18% NaOH to produce alkaline cellulose. However, this process is not environmentally friendly and produces more wastewater. Lyocell process is usually carried out using ionic liquid or N-methylmorpholine N-oxide for the production

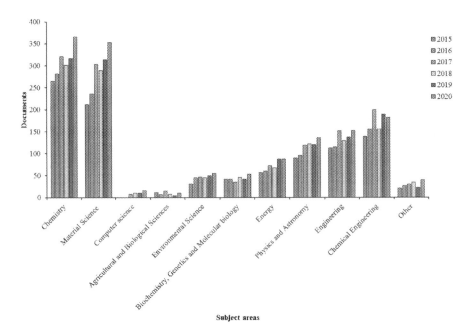

FIGURE 9.7 Numbers of publications during 2015–2020 available in ScienceDirect database (www.sciencedirect.com) with the keywords ionic liquids and different areas of research.

of lyocell fiber. In this process, ionic liquid cations, such as imidazolium, ammonium, and pyridinium, are utilized, while anions like bromide, chloride, hexafluorophosphate, bis(trifluoro-methylsulfonyl)imide, and trifluoromethyl sulfonate are utilized in this process (Sun and Rogers, 2010).

Ionic liquids are used in other areas such as electrochemistry, organic synthesis, liquid-phase extraction, polymerization processes, and clean technology catalysis (Duchet et al., 2010). Separation technology has received great attention in a variety of applications. Imidazole-based ionic liquids have potential use as stationary phases in gas chromatography columns (Yao and Anderson, 2009). Ionic liquids exhibit an extraordinary selectivity with dual nature that separates both polar and nonpolar compounds. Therefore, ionic liquids are considered a multimodal medium for chromatographic separation (Han and Row, 2010). In addition, over the past few years, ionic liquids are applied in the field of extraction and separation of the bioactive compound, such as antioxidants, alkaloids, saponins, isoflavones, anthraquinones, polyphenolic, and phenolic compounds (Capela et al., 2018).

9.4 FABRICATION OF BIOCOMPOSITES FROM LIGNOCELLULOSIC BIOMASS USING IONIC LIQUID

The use of lignocellulosic biomass to replace petroleum-derived polymers for the production of green materials has been increasing worldwide in recent years (Mahmood et al., 2017b). Fractionation of lignocellulosic biomass into hemicellulose, cellulose, and lignin is carried out using different pretreatment methods as described in Section 9.2. Compared to various pretreatment methods, ionic liquid pretreatment is considered very beneficial for the production of biocomposites materials (Mahmood et al., 2015). An ionic liquid in the production of biocomposites has a number of advantages, such as (a) nonvolatile, (b) noncorrosive, (c) high thermal stability, and (d) dissolution of polymers under mild conditions (Mahmood et al., 2017a). As discussed in Section 9.1, cellulose is a polymer that contains intermolecular and intramolecular hydrogen bonds, making it becomes tough and rigid biopolymer (Medronho and Lindman, 2015; Pang et al., 2015). In general, cellulose in lignocellulosic biomass is present in their native form as cellulose I. During the pretreatment of lignocellulosic biomass with ionic liquid, cellulose I dissolves and is regenerated by using antisolvents, such as water. The regenerated cellulose loses its natural form and turns into either amorphous form, or cellulose II, or in its intermediate form (Zhou et al., 2013). Therefore, cellulose can be converted to a form with lower crystallinity and blended with other biopolymer materials to synthesize a new type of biopolymer or biocomposites. These blending or formulations improve chemical, physical, mechanical, and thermal properties of biocomposites to fit with targeted purposes of applications (Stanton et al., 2018). In addition, ionic liquids increase the flexibility of the fabricated composite materials.

The synthesis and manufacture of biocomposites by IL-mediated process need to be designed upon the types of cations and anions of the ionic liquid. Previously, the influence of three types of ionic liquids with different cations was demonstrated by blending them with chitosan and nanocrystal cellulose for biocomposite production. This study reported that the type of cation had a significant effect on the dispersity and particle size of the synthesized biocomposites (Grząbka-Zasadzińska et al., 2019). Up to now, most of the common cations used in ionic liquids to dissolve the cellulose present in lignocellulosic biomass are imidazolium. Imidazolium ionic liquids are the most studied in the dissolution of cellulose for the manufacture of different biocomposites. Some examples of the studies related to biocomposite production by the imidazolium-mediated process are displayed in Table 9.2. Based on Table 9.2, it could be seen that ionic liquids were applied as a solvent for different biological macromolecules or biological polymers, such as proteins (silk, keratin, soy protein, collagen, elastin, corn zein, reflectin) and polysaccharides (starch, chitin, chitosan, alginate). Ionic liquids were demonstrated to assist fabrications of polymeric composites based on different effects as reviewed in Gough et al. (2020), including induction of coagulation of materials and also ionic liquids was integrated into composites to modify fiber structure (such as changing from alpha-helix structure to beta-pleat sheet of cellulose) and ion conductivity (Gough et al., 2020). It is noted that with different anions for similar cations, different biocomposites material can be synthesized. Manufactures of different biocomposites from cellulose have been discussed in further details.

9.4.1 Biopolymer Films

Natural biopolymers (such as polysaccharides, lipids, and proteins) are common materials for the synthesis of edible and/or biodegradable films. Polysaccharides

TABLE 9.2
Application of Ionic Liquid with Imidazolium Cation for the Synthesis of Different Biocomposites

Type of Ionic Liquid	Categories	Materials	Loading Ratio	Conditions	Fabrication Techniques	Applications	References
[Bmim][Cl]	All-cellulose	Hemp fabric	–	100°C and 30 min	Dissolution/hot-press process (self-welding) in the sandwich scheme	Fabric	K. Chen et al. (2020)
[Amim][Cl]	Polysaccharide	Thai silk/AVICEL Microcrystalline cellulose	10 wt%	90°C, and 60 rpm	Solution casting	Film	DeFrates et al. (2017)
[Bmim][OAc]	Polysaccharide	Chitosan/wood pulp	6 wt%	85°C–95°C, 3–4 days	Solution casting	Film	Stefanescu et al. (2012)
[Bmim][BF$_4$]	Polysaccharide/enzyme	Chitosan/glucose oxidase	0.5% CS solution containing 20 mM BQ, 5.0 mg mL^{-1} GOD and 1% (v/v) BMIM·BF$_4$	–	Solution casting	Biosensor	Zeng et al. (2009)
[Amim][Cl] and [Smim][HSO$_4$]	Polysaccharide	Oxycellulose/chitosan	2 wt%	room temperature	Solution casting	Film	Zhou et al. (2014)
[EOHMIM] + [TEA]	oil/polymer	Castor oil/Polyurethane	1:2 mole ratio of Castor oil and TEA+15% [Bmim][Cl]	2 h, 150°C	Solution casting	Foam for CO$_2$ capture	Fernández Rojas et al. (2017)
[Emim][OAc]	Polysaccharide	Chitosan/CMC	0.585 wt%	90°C, 5 h.	Dissolution/Electrospinning	Film	Shamshina et al. (2018)
[Emim][OAc]	Polysaccharide	Chitin/carbon nanotube	1.5 % et of chitin in 5 g of [Emim][OAc]	130°C, 2 h	Solution casting	Tissue engineering	N. Singh et al. (2016)

(*Continued*)

TABLE 9.2 (Continued)
Application of Ionic Liquid with Imidazolium Cation for the Synthesis of Different Biocomposites

Type of Ionic Liquid	Categories	Materials	Loading Ratio	Conditions	Fabrication Techniques	Applications	References
[Hmim][HSO$_4$]	Polysaccharide	Chitosan/chondroitin sulfate hydrogels	4.0 wt%/vol.% CHT and CS-solutions	100°C, 10 min	Solution casting	Tissue engineering	Nunes et al. (2017)
[MOYI][Cl]	Polysaccharide	Chitosan/methoxy polyethylenglycol-aldehyde (mPEG-aldehyde)	1 g of microcrystalline cellulose in 2-(Chloromethyl)oxirane (50 mmol) and 1-methylimidazole (50 mmol)	60°C, 24 h	Polymerization	Drug delivery	Rahimi et al. (2018)
[Emim][OAc]	Polysaccharide	Chitin/PLA+Nano-Hydroxyapatite (nHAP)	0.5 wt% nHAp in IL	90°C	Solution casting	Scaffold-tissue engineering	Chakravarty et al. (2020)
[Mim][Ac]	Polysaccharide	Collagen/Na-alginate	20–40% w/v Na-alginate in IL	40°C	Solution casting	Tissue engineering	Iqbal et al. (2017)
[Bmim][Cl]	Polysaccharide	Collagen/microcrystalline cellulose	6 wt% collagen/cellulose/[C4mim]Cl solution	100°C	Solution casting	Hydrogel for heavy metal adsorbent	Wang et al. (2013)
[Bmim][Cl]	Polysaccharide	Zein protein/Starch	15–30 wt%	160°C, 2 min	Melt-processing	Plasticizer	Leroy et al. (2012)
[Amim][Cl]	All-cellulose	Agave microfibrils	4 wt% (cellulose microfibrils to AmimCl)	80°C	Solution casting	Film	Reddy et al. (2014)

(such as cellulose, chitosan, and chitin) can be formulated into different biocomposites due to their different mechanical, physical, chemical, and biological properties (Zimmermann et al., 2010). Biopolymers (degradable) can be obtained from a variety of natural sources, including microorganisms, animal, and plant (Vieira et al., 2011). Ionic liquid pretreatment facilitates the conversion of cellulose to its soluble form by dissolution. Therefore, the soluble form of cellulose can be formulated with other components to form biopolymer films that are applied as packaging materials. In recent years, biodegradable biopolymer films are being utilized as packaging materials. This initial work was carried out to replace the synthetic petroleum-based packing materials which are not eco-friendly and nondegradable. Blending cellulose with a polymer is one of the simplest ways to develop new biocomposite materials with different physicochemical characteristics (Bajpai et al., 2008). In one study, the absorption capacity of biofilms was studied by blending cellulose with various polymers. Cellulose was dissolved by using [Emim][OAc] and blended with chitosan and silk. The evaluations of absorptivity to methyl orange, crystal violet, and lysozyme were reported for these composite biofilms. The absorption capacity of methyl orange by cellulose/chitosan film was 9.1 times higher than that of lysozyme, which was 7.9 times higher suggesting the good characters in absorptivity in this blended biofilm (Park et al., 2020).

In another study, biocomposite films were derived from cellulose, starch, and lignin. The composite biofilm was prepared by using [Amim][Cl] as a solvent and coagulated by using a nonsolvent method. Biocomposite films synthesized from different sources of polysaccharides and lignin have been demonstrated to have different mechanical properties. The selected formulation of biofilm showed high gas barrier capacity and good thermal stability under tested condition. In addition, the $CO_2:O_2$ permeability ratio of biofilms was close to one (Wu et al., 2009). On the other hand, other ionic liquid such as [Amim][Cl], [Bmim][Cl], [Emim][Cl], and [Emim][OAc] was used in the synthesis of cellulose films through the regeneration of soluble cellulose after pretreatment. Cellulose membranes regenerated from [Emim][Cl]-mediated process have been reported to have higher tensile strength (119 MPa). The properties of the cellulose film have shown favorable application for agricultural purposes and in biodegradable packaging (Pang et al., 2014). Similarly, imidazolium-based ionic liquids were used to regenerate microcrystalline cellulose to blend with *Bombyx mori* (silk). This study was conducted to understand the effects of intermolecular and intramolecular hydrogen bond interactions on the thermal properties and morphologies of biocomposite materials. This study reported that ionic liquids had a strong effect on the morphology of regenerated cellulose/silk composite. In addition, it was reported that the effects of intermolecular interactions on anions of ionic liquids and biofilms are directly related by impacting on macromolecular arrangements of cellulose in the β-sheet forms. On the other hand, it has been reported that cations of ionic liquids do not influence on thermal properties of biofilms (Stanton et al., 2018).

Niroomand (2016) made and studied the properties of biocomposite films synthesized by mixing cellulose and nano-chitosan. Cellulose was dissolved in ionic liquid [Bmim][Cl], regenerated by addition of anti-solvent, and cast to form a biofilm with and without chitosan. The study revealed changes in mechanical, physical, and optical properties of this biocomposite film. The transparency of the cellulose/nano-chitosan film has been improved compared to biofilm consisting of pure cellulose. However, it was also reported that increasing the nano-chitosan content did not improve the tensile strength and transparency of the biofilm, and the observations were similar to that of pure cellulose biofilm. This was caused due to the formation of nano-chitosan aggregates (Niroomand et al., 2016). Another study selected a hemicellulose member, arabinoglucuronoxylan, to be the additive in the fabrication of cellulose/xylan composite films. The film was prepared by dissolving wood polymers in [Emim][OAc]. The polymers were regenerated and cast into films by using ethanol as a coagulant. The films were optically transparent, and their mechanical properties were improved. The study reports that the mechanical properties are independent of the compositions of the polysaccharides. However, an increase in Young's modulus was observed at low humidity, with a 20% arabinoglucuronoxylan content in the biofilm (Sundberg et al., 2015). The higher flexibility and thermoplasticity cellulose composite film were fabricated by using [Bmim][Cl]-mediated process. The dissolved cellulose was regenerated by gelation. Interestingly, it was found that the retained residues of [Bmim][Cl] enhance the mechanical properties (Haq et al., 2019). Similarly, it was reported that for [Emim][Cl], the regeneration of cellulose was difficult compared to regeneration by using other ionic liquids, [Amim][Cl] and [Bmim][Cl]. [Amim][Cl] showed stronger intermolecular interactions during cellulose regeneration and more ordered molecular arrangement. The biofilm made by using [Amim][Cl] had higher tensile strength, crystallinity, and transparency (Zheng et al., 2019). [Amim][Cl] ionic liquids have the ability to dissolve and fractionate lignocellulosic biomass for a variety of applications such as fibers, films, and cellulose derivatives (Wang et al., 2011). The regenerated celluloses were obtained from Borassus fruit fibers by using [Amim][Cl] and water as an antisolvent. Changes in mechanical properties were reported, and the average tensile strength, elongation, and modulus of the biofilm were found to be 111 ± 19 MPa, $3.1\% \pm 0.8\%$, and 6149 ± 603 MPa, respectively, which are suitable for a variety of industrial applications (Reddy et al., 2017).

9.4.2 Aerogels

Aerogel is an ultra-lightweight porous material created by replacing the liquid present in a gel with air without changing the three-dimensional (3D) framework of the gel (Pierre and Pajonk, 2002). Aerogel is used in a variety of applications due to its large specific surface area, low density, and high porosity of its internal structure (Bheekhun et al., 2013). Some applications of aerogels are in the fields of cosmic dust capture, pesticides, insulation, high-energy physics, electronics, aero- capacitors, catalysts, molds, imaging equipment, etc. (Hrubesh, 1998). Among the various types of aerogels, silica aerogels exhibit excellent physical properties such as thermal conductivity, low density, permittivity, and refractive index (Gurav et al., 2010). It has been noted that most silica aerogels are not environmentally friendly and not biodegradable, but are biocompatible due to its inertness property (Stergar and Maver, 2016). Aerogels are usually prepared by using the sol-gel process. When these liquids are replaced with air using a supercritical drying process or freeze-drying method, the gel is referred to as aerogel. Supercritical drying technology is the most widely used and expensive drying method in the aerogel manufacturing process (Yu et al., 2016).

In recent years, ionic liquids have been used as an effective solvent in the synthesis of aerogels without the need for supercritical drying techniques. Synthesis of aerogels by using the ionic liquids-mediated process is demonstrated to produce stable products with enhanced properties. Ionic liquids can dissolve cellulose to form a gel with various shapes such as films, fibers, and beads. These aerogel shapes are depended on liquid-filled cellulose structures, which determined by coagulation of cellulose (caused by water and alcohol filling the liquid volumes between these solid structures of celluloses) (Chang and Zhang, 2011). Among the various ionic liquids, imidazolium-based ionic liquids, such as [Amim][Cl], [Emim][Cl], and [Emim][Cl], are widely used in the production of aerogels from cellulose (Zhang et al., 2017). Recently, a cellulose aerogel was prepared by IL-mediated process to be used in cleaning up oil spills. Two types of ionic liquids, [Emim][OAc] and [Bmim][Cl], were selected to solubilize celluloses and the results showed that an aerogel prepared by using [Bmim][Cl] exhibited improved properties such as higher thermal stability compared to [Emim][OAc]-based aerogel. [Bmim][Cl]-based aerogel showed an 84% oil spill removal efficiency with a 12% increase over an [Emim][OAc]-based aerogel. In addition, for [Bmim][Cl], an increase in the porous structure was observed. The enhanced properties are explained by the presence of halide anions with strong hydrogen bonds and the hydrophobicity of cations (Warsi Khan et al., 2020). Similarly, the cellulose extracted from the corn stalks was fabricated into aerogels by using different ionic liquids ([Emim][Cl], [Pmim][Cl], [Amim][Cl], [Bmim][Cl], [Hmim][Cl], and [Omim][Cl] with combination of amido-sulfonic acid (ASA). Lignin from corn stalks was removed or modified using ASA during IL pretreatment. Among the various ionic liquids, [Amim][Cl] can assist in the formation of a composite aerogel that composed of 88.1% cellulose. This [Amim][Cl]-based aerogel was highly porous and showed good thermal stability. The aerogel finds use in the dye industry for absorbing dyes due to its highly porous structure. These synthesized aerogels were used to study the absorption of dyes. The gel showed an absorbency of 549 mg/g for Congo red and 302 mg/g for Coomassie brilliant blue (Li et al., 2019). Therefore, aerogel has the potential to make suitable material for absorption of dyes. Other types of raw materials of cellulose fibrils were investigated for IL-mediated extraction and used for aerogel synthesis. Celluloses recovered from waste newspapers were blended in composite to form aerogel for applications in the absorption of oil and chloroform. The cellulose aerogel was prepared from newspaper by dissolving the cellulose in [Amim][Cl], and the liquid between the cellulose solids in wet aerogel was removed by using the freeze-drying method. The aerogel was treated with trimethylchlorosilane to enhance the absorption capacity. The silane-coated aerogel showed increase absorption capacity for oil and organic solvent (chloroform). Absorption of solvent increased the weight of aerogel ranging from 11 to 22 times. The study was concluded that coating of cellulose aerogel increased the absorption capacity and had potential application in sewage purification (Jin et al., 2015). On the other hand, cellulose aerogels were also studied for drug loading capacity of impregnating phytol (pharmaceutical drug) within the aerogel as reported by Lopes (2017). The cellulose aerogel was prepared by using different ionic liquids, such as [Amim][Cl], [Bmim][Cl], [Emim][OAc], [Emim][DEP], and [HOEmim][Cl]. The phytol drug was impregnated using the supercritical CO_2 method at 100 bar, 40°C with phytol to aerogel ratio at 10:1. The aerogel with 2% cellulose was prepared by dissolving in [Emim] [DEP], and it showed the higher drug loading capacity. The prepared aerogel had the capacity to hold 50% w/w of the drug, thus making it a suitable material for application in medicinal field (Lopes et al., 2017).

Modifications of IL-assisted processes were demonstrated to determine the macrostructure of composite aerogels. Mixtures of [Amim][Cl] and dimethyl sulfoxide (DMSO) were formulated to reduce the cost and viscosity of IL and used for extraction of cellulose in the cotton stalk. Two drying processes of cellulose composite aerogel were compared, including (a) cyclic conventional freezing (CCF) at −20°C followed by slow thawing and (b) cyclic freezing with liquid nitrogen (CFN) at −196°C followed by slow thawing at 20°C. The porosities of aerogels prepared by these two methods were compared by using Brunauer-Emmett-Teller (BET) analysis. The CCF aerogel showed a hierarchical 3D open porous "web-like" structure with a small specific surface area and a large pore diameter. On the other hand, the CFN aerogel showed a "film-like" porous structure with a large specific surface area and a relatively small pore diameter. It was observed that different drying methods resulted in different porous structures of aerogel (Mussana et al., 2018). The modifications of composite aerogel were done in a different way to make a certain desired property. A synthetic cellulose/ZnO_2 aerogel is currently prepared by using acid and alkaline treatments. Thus the process has been reported to be harmful to the environment. The ionic liquid-mediated process was applied for the synthesis of this type of aerogel. ZnO_2 nanoparticles were incorporated on solubilized cellulose in [Amim][Cl] and diethylene glycol (polyhydric alcohol). It was reported that the aerogels with surface area, lightweight, and strength of $267 \pm 4\,m^2/g$, $0.0463\,g/cm^3$ and $51.53\,N/cm^2$, respectively. This cellulose/ ZnO_2 aerogel was applied due to its catalytic activity for PET degradation, which could reduce the negative impact on the environment (X. Li et al., 2018). Therefore, composite aerogels have several applications in different fields due to their tailor-made properties.

9.4.3 Hydrogel

The hydrogel is a type of hydrophilic material with three-dimensional polymeric networks that are highly swollen due to liquid present in between the network structures (Klouda and Mikos, 2008). Hydrogels are hydrophilic due to the presence of hydrophilic groups, such as -SO_3H, -$CONH_2$, - CONH -, -OH, -NH_2, and -COOH (Bahram et al., 2016). Over the decades, hydrogels are being manufactured using acrylamide or (meth)acrylic compounds. These compounds are not eco-friendly and are toxic to human health (Bashir et al., 2020). In recent years, studies are carried out to replace the synthetic polymers for the production of the hydrogel by using biodegradable and biocompatible polymers. Celluloses with hydroxyl and acetamide groups exhibit hydrophilic properties, making cellulose a potential source for fabrication of hydrogels (Chang and Zhang, 2011). Hydrogels are considered to be an intelligent material due to its swelling property and its potential for response to environmental factors. Some of these environmental factors are pH, temperature, electric fields, and solvent composition. Due to these properties, hydrogels have gained attraction in different industrial and pharmaceutical applications (Liang et al., 2015).

Hydrogels could be synthesized by physical or chemical cross-linking process of soluble cellulose. In fact, cellulose is immiscible with water due to its crystallinity, and ionic liquid could assist in the improvement of cellulose solubilization in aqueous solution. Therefore, hydrogels can be made from cellulose dissolved using ionic liquids. In addition, celluloses are blended with other materials to make hydrogels with enhancing physical, chemical, and mechanical properties. A study reported that the hydrogel prepared with varying concentration of cellulose with DMSO and [Bmim][Ac] solution showed varying compression modulus (0.2–3.1 MPa). This study suggested the importance of the relationship between cellulose content and hydrogel compression modulus property (Satani et al., 2020). Furthermore, cellulose was blended in [Bmim][Cl] to cast a hydrogel membrane, and it was demonstrated that cellulose content also determines the transparency and mechanical properties of composite hydrogel (Peng et al., 2018). Also, cellulose was blended with petroleum-derived polymer, PVA to form hydrogel membrane by immersion-precipitation phase transformation using [Bmim][Ac] as a solvent. It was observed that the compact dual network structures between the cellulose and PVA were formed. It was noted that the tensile strength of the membrane increased compared to pure cellulose hydrogel. The absorption capacity of water increased thus increasing the swelling of the hydrogel. Therefore, it is understood that blending hydrogel with other materials can enhance the mechanical properties and has potential to apply in the field of biomedical engineering (Peng et al., 2017).

Kimura (2015) fabricated the hydrogel to validate and investigate the shape-persistent and toughness of the cellulose-based hydrogel. In this approach, the hydrogel was fabricated from cellulose dissolved in [C2mim][(MeO)(H)PO$_2$], and the hydrogels were exposed to methanol. It was reported that the shape of the hydrogel was unchanged though the pH, temperature and solvents were varied. However, methanol addition exhibited enhanced mechanical properties of hydrogels (Kimura et al., 2015). Similarly, another study evaluated the mechanical strength of pure cellulose hydrogel and composite cellulose/graphene hydrogel. Reduced graphene oxide and cellulose from wood were dissolved in [Bmim][Cl] solution. The mixture was coagulated by the addition of anti-solvent or coagulant, e.g., water. Cellulose/graphene composite hydrogel showed a 4-fold increase in Young's modulus when adding 0.5% reduced graphene oxide in hydrogel (Xu et al., 2015). In another study, the properties of cellulose/graphene oxide composite hydrogel were studied. The cellulose was recovered from the tea residue by using ionic liquid ([Amim][Cl]) extraction. The effect of graphene in cellulose hydrogel was investigated through a study on adsorption of methylene blue dye. It is hypothesized that graphene enhances the absorption capacity of the hydrogel. The cellulose/graphene oxide composite hydrogel showed increased thermal stability and textural property. The practicability, hardness, and gumminess for hydrogel were reported as 4.1, 12.7, and 17.8 times higher than the pure cellulose hydrogel. The absorption capacity and absorption ratio of 466.35 mg/g and 92.7%, respectively, were reported for the composite hydrogel. Therefore, this composite hydrogel has the potential application in dye industry or the absorption process (Liu et al., 2017). Another study to test the application of composite hydrogel synthesized by using IL-mediated process for treatment of dye-contaminated waste was conducted by preparation of cellulose/magnetic diatomite hydrogel. Cellulose was extracted from pineapple peel by using [Bmim][Cl], and then magnetic diatomite particles were blended into composite and cast to a hydrogel. The absorption capacity of composite hydrogel was tested with methylene blue dye and evaluated based on the Langmuir isotherm model. The maximum absorption capacity for pure cellulose hydrogel and hybrid hydrogel was 75.87 mg/mg and 101.94 mg/mg, respectively. Furthermore, it was reported that the composite hydrogel exhibited enhanced thermal stability and reusability for methylene blue absorption. Due to the presence of magnetic diatomite material, the swelling ability increased, resulting in increased absorption of the dye. This study was concluded with the possibilities of using the hybrid hydrogel in dye industries for absorption (Dai et al., 2019).

Aforementioned, many studies demonstrated that composite hydrogels have potential uses in the dye industry. Hydrogels are also applied in the field of biotechnology related to enzyme capture/immobilization. Cellulose hydrogel microspheres were prepared using an oil emulsion-ionic liquid ([Emim][OAc]) method. Hydrogels with microspheres were prepared using the sol-gel transition, followed by the freeze-drying method. Production parameters, such as ionic liquid to oil ratio, agitation rate, and surfactant concentration, were optimized for effective production of microspheres. The hydrogel was used to immobilize the lipase enzyme extracted from *Candida rugose*. In this study, the activity of the immobilized enzyme on hydrogel was compared with the free enzyme in suspension. The specific activity of the immobilized enzyme showed a 1.4-fold increase in activity compared to the free enzyme in suspension. Microspheres in hydrogels showed increased loading capacity, specificity constant, and yield. In addition, Fe_3O_4 was added along with the enzyme to recover the enzyme in the microsphere after use. Therefore, hydrogels and magnetic particles can be combined to reuse the enzyme without losing enzyme activity (Jo et al., 2019). Similarly, studies on the capture of cellulase enzymes in cellulose hydrogel microspheres to promote *in situ* enzymatic saccharification of bagasse were reported. This study was conducted to investigate enzyme resistance to interference during the saccharification process. By comparison, the activity of immobilized cellulase in hydrogel and free cellulase enzyme in suspension in ionic liquid [Emim][OAc] was 40.9% and 5%, respectively. An increase in the enzymatic activity of cellulase immobilized on hydrogel microspheres was reported. Furthermore, it was reported that the enzyme activity efficiency of immobilized cellulase was maintained up to 85.2% in the sixth cycle. Therefore, by immobilizing the enzyme in hydrogel microspheres, this method has the potential to reduce the costs incurred during the saccharification process (Zhou et al., 2019).

9.4.4 All Cellulose Composites

All Cellulose Composites (ACC) are single-component cellulosic-based composites. The matrix of ACC is a generated cellulose and is composed of high strength cellulose fibers (Li et al., 2018). In recent years, ACC has received a great deal of attention from researchers due to its unique features such as biocompatibility, excellent interfacial compatibility, and biodegradability. This composite exhibits superior mechanical, barrier, and optical properties. Capiati and Porter (1975) introduced the concept of one-component polymer composites by using all polyethylene related polymers

(Capiati and Porter, 1975). Inspired by the polythene composite Nishino (2004) propose ACC by using cellulose as the main component (Nishino et al., 2004). The ACC composites showed good interfacial interactions and high strength in cellulose-rich phases. ACC is usually prepared by two different processes. The first process partially dissolves and regenerates cellulose to form a matrix of composite, and this matrix is surrounded by undissolved portions of cellulose. In the second process, the cellulose is completely dissolved, and the cellulose is regenerated to form a matrix with the addition of the cellulose reinforcing materials. Compared to all-composite polyethylene, ACC is eco-friendly and nontoxic to humans due to their inertness and degradability. ACC has a great potential in applications of packaging materials, biomedical engineering, structural and photoelectric devices (Kalka et al., 2014). The example of the synthesis process for ACC is shown in Figure 9.8.

Ionic liquid has been reported to be used in the manufacturing of ACC as targeting to be eco-friendly process due to liquid ionic properties. In one study, cellulose was dissolved using [Emim][OAc] and blended with rayon fiber to prepare ACC. After mixing the high-strength rayon fibers with an ionic liquid solution containing cellulose, the ionic liquid was removed. Mechanical properties of this ACC, such as tensile strength, impact testing and bending, were investigated and compared to thermoplastic fiberglass reinforced with plastics. Tensile strength, Young's modulus, and elongation of ACC increased to 45–75 MPa, 1.5–2.3 GPa, and 4%–17%, respectively. The flexural modulus and Charpy impact strength of ACC compared to thermoplastics increased to 5.3–7.6 GPa and 50–115 kJ/m^3, respectively. In addition, it was reported that the use of cellulose in the production of ACC made it possible to recycle composite materials (up to four recycling productions) (Spörl et al., 2018). In addition

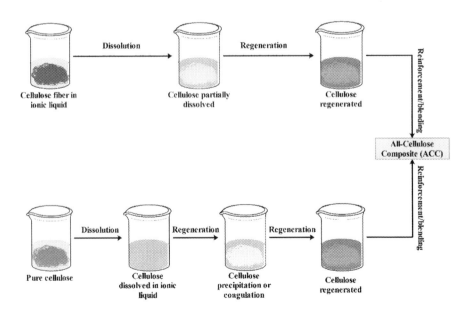

FIGURE 9.8 Process flow diagram for synthesis/fabrication of All-Cellulose Composite (ACC).

to [Emim][OAc], [Bmim][Cl] was used to dissolve the cellulose present in wood to produce ACC in a similar manner. This type of ACC was made by a process with a shorter period of time and a lower concentration of the ionic liquid. The composite was examined for its mechanical properties. The mechanical strength of the ACC improved after the removal of the ionic liquid. The study demonstrated that ACC that was synthesized at a lower temperature (25°C) showed greater transparency and reduced crystallinity. In addition, the recovery and reusability of ionic liquids were also studied to find the possibility to reduce the production cost and environmental impact (Thirion, 2015).

Although the ionic liquid process is considered to be a green process due to its properties, especially low toxicity and recyclable, the cellulose recovery process by using liquid process consumes significant energies in the form of heat. An attempt was made to synthesize ACC from hemp cellulose by using lower energy consumption strategy. The fiber in the hemp was dissolved in the [Bmim][Cl] at a lower temperature before proceeding with the hot pressing process to cast the ACC composite. It was reported that the ACC showed a tensile strength and modulus of 81.1 MPa and 1.50 GPa, respectively. It was observed that the mechanical properties of ACC increased when the hot pressing temperature reached to 120°C. Hemp-based ACC illustrated the potential to be utilized in the industries (Chen et al., 2020). Similarly, Chen (2020) dissolved the flax-based cellulose in the ionic liquid ([Emim][OAc]). The synthesis process was carried by using the facile room temperature for impregnation method. The ACC showed the highest strength and Young's modulus of 151.3 MPa, and 10.1 GPa, respectively with a crystallinity index of 43% and 20% w/w of cellulose II (Chen et al., 2020).

Practically, during the dissolution and regeneration of cellulose in ionic liquid, cellulose I could be converted to cellulose II. Huber (2012) produced continuous reinforced fiber for preparation of ACC laminates from a rayon textile by dissolving it in [Bmim][Ac]. The ACC laminates were prepared via solvent infusion process. In this process, the outer surface of the cellulose fiber is partially dissolved due to the penetration of the solvent through the dry cellulose fiber. The composite produced was reported as dimensionally stable and thick with high volume fractions of continuous fibers greater than 70 vol%. This character of ACC was the results of (a) homogenous partial dissolution of cellulose fiber and (b) high-pressure application during regeneration and drying of the ACC laminates. This study suggested the advantage for the production of ACC in the laminate forms compared to the thin film ACC (Huber et al., 2012a). Nanocrystal cellulose was also produced and used as reinforced fibers for ACC to fill in the [Amim][Cl]-dissolved cellulose matrix to prepare ACNC. First, the cellulose was partially dissolved and regenerated as a matrix by controlling the time and temperature during the dissolution of cellulose. The nanocomposite ACNC film was transparent in visible light, compact, and isotropic, and it exhibited enhanced mechanical properties, in terms of the elastic modulus and tensile strength compared to the microcrystalline ACC films (Zhang et al., 2016). Therefore, it is understood that materials with cellulose reinforced into the cellulose solution can further enhance the mechanical properties. ACC has the potential to be used in the textile industry, medical fields, biomedical engineering, and more.

9.5 IONIC LIQUID-INTEGRATED COMPOSITES IN OTHER BIOTECHNOLOGICAL APPLICATIONS

Aforementioned, ionic liquid has been demonstrated in many studies to play important roles in fractionation and recovery of cellulose for productions of degradable composites. Ionic liquids when applied in the fabrication process, it modified the structures and arrangements of cellulose fibers resulting in modification of mechanical and physical properties of polymeric composites. However, due to the various characters of ionic liquids, they could function in different purpose further than solvent. The ionic liquid was recently impregnated on the structure of cyclodextrin nanosponge (CDNS-IL), polymeric carbohydrates, and blended with graphene oxide and chitosan. This mixed biocomposite was used as a medium for immobilization of catalysts, including Pd and Fe_3O_4 nanoparticles, for hydrogenation of nitroarenes in aqueous media. This composite carrying catalyst expressed magnetic property, which makes the recovery of catalysts is possible and enhances the reaction to operate under mild conditions (Sadjadi et al., 2019). In another study, polymeric ionic liquids (PILs) were fabricated in biocomposites and applied for separation of protein molecules in whole human blood. PILs were selected as they are good polarity and dispersants of the mixtures. This composite was prepared by using imidazolium-based PILs and 1-vinyl-3-ethylimidazolium bromide, $P(ViEtIm^+Br^-)$, to mix with graphene oxide (GO) to be PIL-G composites. Then, PIL-G composites were deposited on the surface of SiO_2 nanoparticles via electrostatic interactions. This $PIL(Br)$-G/SiO_2 nanocomposite was tested as an adsorbent for whole human blood for separation of human serum albumin (BSA), haemoglobin (Hb), and cytochrome C (Cyt-C) proteins. This $PIL(Br)$-G/SiO_2 nanocomposite had maximum adsorption efficiency at 98% for BSA, pI at 4.9 when the pH of tested media was adjusted to pH 5.0, while the specific adsorption of Hb and Cyt-C were significantly lower. Due to difference in pI values of these three proteins, when the pH of adsorption media was adjusted the specific adsorption of each protein was modified and had high selectivity to a certain protein. This finding suggested the possibility of using ionic liquid-composite in the separation of biological materials for medical purposes (Liu et al., 2018).

Ionic liquid-derived composites were also demonstrated to be used as sensors for different targeted molecules (Joshi et al., 2017). Boobphahom and others (2019) developed a paper-based analytical device (PAD) with an electrode for the detection of creatinine molecule. Creatinine is a biological molecule produced from human muscles during the breakdown of the creatine, and creatinine is normally removed from the human body by the function of the kidney in the form of urine. Therefore, detection of creatine could be used to be an indicator of the functional status of the kidney. This PAD equipped with the electrochemical sensor is made of fabrications of copper oxide (CuO) and ionic liquid and deposited on printed graphene. The CuO-IL-graphene embedded in PAD exhibited a linear range of creatinine detection suggested the possibility of using this device for medical cares (Boobphahom et al., 2019). A similar concept of application in sensor, Wang et al. (2016) prepared a microporous silk carbon-ionic liquid (Silk C-IL) composite for sensing of dopamine, a human neurotransmitter molecule. For the production process, the natural silk cocoons were activated by carbonization and KOH to prepare silk carbon to generate a microporous carbon base with the large surface area. The ionic liquid was then integrated into Silk C by non-covalent bonding under grinding conditions. This Silk C-IL sensor was able to detect dopamine with low false-positive signals from other interferences such as ascorbic acid, glucose and uric acid (Wang et al., 2016).

9.6 SUMMARY

Ionic liquids are molten salts made from combinations of cations and anions, making them available for designing and tailor-making based on different targets of uses. Nowadays, commercial ionic liquids are available in current markets, and they are used in different industries. In this chapter, ionic liquids have been discussed as an eco-friendly solvent for solubilization of natural biomass to extract specific biopolymers, especially cellulose for further application in the production of biocomposites or degradable polymeric materials. Due to their variations in formulations between cations and anions, when applied ionic liquids in a process, it is necessary to conduct process optimization to understand their properties, such as viscosity, polarity, melting point, recyclability, to evaluate whether they are suitable to the purposes of users. Ionic liquids have been demonstrated to be excellent green solvents with high selectivity for biopolymeric molecules, especially cellulose, which are subsequently used in fabrication and manufacturer of biocomposites. Due to their functions as electrolytes, ionic liquids have been impregnated into biocomposites for production of smart biomaterials, such as sensors for medical purposes.

ACKNOWLEDGMENTS

The authors would like to thank King Mongkut's University of Technology, North Bangkok (Research University Grant No. KMUTNB-BasicR-64-37, KMUTNB-Post-64-05) for financial support of this work.

REFERENCES

Ahrenberg, M., Beck, M., Neise, C., Keßler, O., Kragl, U., Verevkin, S.P., Schick, C., 2016. Vapor pressure of ionic liquids at low temperatures from AC-chip-calorimetry. *Phys. Chem. Chem. Phys.* 18, 21381–21390. doi:10.1039/C6CP01948J.

Akbari, F., Alavianmehr, M.M., Behjatmanesh Ardakani, R., Mohammad-Aghaie, D., 2018. Thermophysical properties of ionic liquids and their mixtures from a new equation of state. *Ionics (Kiel).* 24, 1357–1369. doi:10.1007/s11581-017-2310-8.

Aparicio, S., Atilhan, M., Karadas, F., 2010. Thermophysical properties of pure ionic liquids: Review of present situation. *Ind. Eng. Chem. Res.* 49, 9580–9595. doi:10.1021/ie101441s.

Bahram, M., Mohseni, N., Moghtader, M., 2016. An introduction to hydrogels and some recent applications. In: *Emerging Concepts in Analysis and Applications of Hydrogels.* InTech. doi:10.5772/64301.

Bajpai, A.K., Shukla, S.K., Bhanu, S., Kankane, S., 2008. Responsive polymers in controlled drug delivery. *Prog. Polym. Sci.* 33, 1088–1118. doi:10.1016/j.progpolymsci.2008.07.005.

Bashir, S., Hina, M., Iqbal, J., Rajpar, A.H., Mujtaba, M.A., Alghamdi, N.A., Wageh, S., Ramesh, K., Ramesh, S., 2020. Fundamental concepts of hydrogels: Synthesis, properties, and their applications. *Polymers (Basel).* 12, 2702. doi:10.3390/polym12112702.

Berthod, A., Cara-Broch, S., 2004. Uses of Ionic Liquids in Analytical Chemistry. *Ann. Marie Curie Fellowsh. Assoc.* 3, 1–6.

Bhatt, S.M., Shilpa, 2014. Lignocellulosic feedstock conversion, inhibitor detoxification and cellulosic hydrolysis – A review. *Biofuels* 5, 633–649. doi:10.1080/17597269.2014.1003702.

Bheekhun, N., Abu Talib, A.R., Hassan, M.R., 2013. Aerogels in aerospace: An overview. *Adv. Mater. Sci. Eng.* 2013, 1–18. doi:10.1155/2013/406065.

Bledzki, A.K., Reihmane, S., Gassan, J., 1996. Properties and modification methods for vegetable fibers for natural fiber composites. *J. Appl. Polym. Sci.* 59, 1329–1336. doi:10.1002/(SICI)1097-4628(19960222)59:8<1329::AID-APP17>3.3.CO;2-5.

Bonhôte, P., Dias, A.-P., Papageorgiou, N., Kalyanasundaram, K., Grätzel, M., 1996. Hydrophobic, highly conductive ambient-temperature molten salts. *Inorg. Chem.* 35, 1168–1178. doi:10.1021/ic951325x.

Boobphahom, S., Ruecha, N., Rodthongkum, N., Chailapakul, O., Remcho, V.T., 2019. A copper oxide-ionic liquid/reduced graphene oxide composite sensor enabled by digital dispensing: Non-enzymatic paper-based microfluidic determination of creatinine in human blood serum. *Anal. Chim. Acta* 1083, 110–118. doi:10.1016/j.aca.2019.07.029.

Brandt-Talbot, A., Gschwend, F.J.V., Fennell, P.S., Lammens, T.M., Tan, B., Weale, J., Hallett, J.P., 2017. An economically viable ionic liquid for the fractionation of lignocellulosic biomass. *Green Chem.* doi:10.1039/c7gc00705a.

Bringmann, M., Landrein, B., Schudoma, C., Hamant, O., Hauser, M.-T., Persson, S., 2012. Cracking the elusive alignment hypothesis: The microtubule–cellulose synthase nexus unraveled. *Trends Plant Sci.* 17, 666–674. doi:10.1016/j.tplants.2012.06.003.

Capela, E.V., Coutinho, J.A.P., Freire, M.G., 2018. Application of ionic liquids in separation and fractionation processes. In: Meyers, R. (eds.) *Encyclopedia of Sustainability Science and Technology*. Springer, New York, pp. 1–29. doi:10.1007/978-1-4939-2493-6_1005-1.

Capiati, N.J., Porter, R.S., 1975. The concept of one polymer composites modelled with high density polyethylene. *J. Mater. Sci.* 10, 1671–1677. doi:10.1007/BF00554928.

Carmichael, A.J., Seddon, K.R., 2000. Polarity study of some 1-alkyl-3-methylimidazolium ambient-temperature ionic liquids with the solvatochromic dye, Nile Red. *J. Phys. Org. Chem.* 13, 591–595. doi:10.1002/1099-1395(200010)13:10<591::AID-POC305>3.0.CO;2-2.

Chakravarty, J., Rabbi, M.F., Chalivendra, V., Ferreira, T., Brigham, C.J., 2020. Mechanical and biological properties of chitin/polylactide (PLA)/hydroxyapatite (HAP) composites cast using ionic liquid solutions. *Int. J. Biol. Macromol.* 151, 1213–1223. doi:10.1016/j.ijbiomac.2019.10.168.

Chami Khazraji, A., Robert, S., 2013. Interaction effects between cellulose and water in nanocrystalline and amorphous regions: A novel approach using molecular modeling. *J. Nanomater.* 2013, 1–10. doi:10.1155/2013/409676.

Chang, C., Zhang, L., 2011. Cellulose-based hydrogels: Present status and application prospects. *Carbohydr. Polym.* 84, 40–53. doi:10.1016/j.carbpol.2010.12.023.

Chen, F., Sawada, D., Hummel, M., Sixta, H., Budtova, T., 2020. Unidirectional all-cellulose composites from flax via controlled impregnation with ionic liquid. *Polymers (Basel)*. 12, 1010. doi:10.3390/polym12051010.

Chen, K., Xu, W., Ding, Y., Xue, P., Sheng, P., Qiao, H., He, J., 2020. Hemp-based all-cellulose composites through ionic liquid promoted controllable dissolution and structural control. *Carbohydr. Polym.* 235, 116027. doi:10.1016/j.carbpol.2020.116027.

Cicala, G., Tosto, C., Latteri, A., La Rosa, A., Blanco, I., Elsabbagh, A., Russo, P., Ziegmann, G., 2017. Green composites based on blends of polypropylene with liquid wood reinforced with hemp fibers: Thermomechanical properties and the effect of recycling cycles. *Materials (Basel)*. 10, 998. doi:10.3390/ma10090998.

Dai, H., Huang, Y., Zhang, Y., Zhang, H., Huang, H., 2019. Green and facile fabrication of pineapple peel cellulose/magnetic diatomite hydrogels in ionic liquid for methylene blue adsorption. *Cellulose*. doi:10.1007/s10570-019-02283-6.

DeFrates, K., Markiewicz, T., Callaway, K., Xue, Y., Stanton, J., Salas-de la Cruz, D., Hu, X., 2017. Structure–property relationships of Thai silk–microcrystalline cellulose biocomposite materials fabricated from ionic liquid. *Int. J. Biol. Macromol.* 104, 919–928. doi:10.1016/j.ijbiomac.2017.06.103.

de Oliveira Santos, V.T., Siqueira, G., Milagres, A.M.F., Ferraz, A., 2018. Role of hemicellulose removal during dilute acid pretreatment on the cellulose accessibility and enzymatic hydrolysis of compositionally diverse sugarcane hybrids. *Ind. Crops Prod.* 111, 722–730. doi:10.1016/j.indcrop.2017.11.053.

del Río, J.C., Lino, A.G., Colodette, J.L., Lima, C.F., Gutiérrez, A., Martínez, Á.T., Lu, F., Ralph, J., Rencoret, J., 2015. Differences in the chemical structure of the lignins from sugarcane bagasse and straw. *Biomass Bioenergy* 81, 322–338. doi:10.1016/j.biombioe.2015.07.006.

Doherty, W.O.S., Mousavioun, P., Fellows, C.M., 2011. Value-adding to cellulosic ethanol: Lignin polymers. *Ind. Crops Prod.* 33, 259–276. doi:10.1016/j.indcrop.2010.10.022.

Duchet, L., Legeay, J.C., Carrié, D., Paquin, L., Vanden Eynde, J.J., Bazureau, J.P., 2010. Synthesis of 3,5-disubstituted 1,2,4-oxadiazoles using ionic liquid-phase organic synthesis (IoLiPOS) methodology. *Tetrahedron* 66, 986–994. doi:10.1016/j.tet.2009.11.079.

Dzyuba, S.V., Bartsch, R.A., 2002. Expanding the polarity range of ionic liquids. *Tetrahedron Lett.* 43, 4657–4659. doi:10.1016/S0040-4039(02)00858-4.

Earle, M.J., Seddon, K.R., 2000. Ionic liquids. Green solvents for the future. *Pure Appl. Chem.* 72, 1391–1398. doi:10.1351/pac200072071391.

Eriksson, K.-E.L., Bermek, H., 2009. Lignin, lignocellulose, ligninase. In: *Encyclopedia of Microbiology*. Elsevier, pp. 373–384. doi:10.1016/B978-012373944-5.00152-8.

Fernandes, E.M., Pires, R.A., Reis, R.L., 2017. Cork biomass biocomposites. In: *Lignocellulosic Fiber and Biomass-Based Composite Materials*. Elsevier, pp. 365–385. doi:10.1016/B978-0-08-100959-8.00017-2.

Fernández Rojas, M., Pacheco Miranda, L., Martinez Ramirez, A., Pradilla Quintero, K., Bernard, F., Einloft, S., Carreño Díaz, L.A., 2017. New biocomposites based on castor oil polyurethane foams and ionic liquids for CO_2 capture. *Fluid Phase Equilib.* 452, 103–112. doi:10.1016/j.fluid.2017.08.026.

Ge, X., Yang, L., Sheets, J.P., Yu, Z., Li, Y., 2014. Biological conversion of methane to liquid fuels: Status and opportunities. *Biotechnol. Adv.* 32, 1460–1475. doi:10.1016/j.biotechadv.2014.09.004.

Gorke, J., Srienc, F., Kazlauskas, R., 2010. Toward advanced ionic liquids. Polar, enzyme-friendly solvents for biocatalysis. *Biotechnol. Bioprocess Eng.* 15, 40–53. doi:10.1007/s12257-009-3079-z.

Gough, C.R., Rivera-Galletti, A., Cowan, D.A., Salas-de la Cruz, D., Hu, X., 2020. Protein and polysaccharide-based fiber materials generated from ionic liquids: A review. *Molecules* 25, 3362. doi:10.3390/molecules25153362.

Greer, A.J., Jacquemin, J., Hardacre, C., 2020. Industrial applications of ionic liquids. *Molecules* 25, 5207. doi:10.3390/molecules25215207.

Grząbka-Zasadzińska, A., Skrzypczak, A., Borysiak, S., 2019. The influence of the cation type of ionic liquid on the production of nanocrystalline cellulose and mechanical properties of chitosan-based biocomposites. *Cellulose* 26, 4827–4840. doi:10.1007/s10570-019-02412-1.

Gupta, A., Verma, J.P., 2015. Sustainable bio-ethanol production from agro-residues: A review. *Renew. Sustain. Energy Rev.* 41, 550–567. doi:10.1016/j.rser.2014.08.032.

Gurav, J.L., Jung, I.-K., Park, H.-H., Kang, E.S., Nadargi, D.Y., 2010. Silica aerogel: Synthesis and applications. *J. Nanomater.* 2010, 1–11. doi:10.1155/2010/409310.

Gurunathan, T., Mohanty, S., Nayak, S.K., 2015. A review of the recent developments in biocomposites based on natural fibers and their application perspectives. *Compos. Part A Appl. Sci. Manuf.* 77, 1–25. doi:10.1016/j.compositesa.2015.06.007.

Gutowski, K.E., 2018. Industrial uses and applications of ionic liquids. *Phys. Sci. Rev.* 3. doi:10.1515/psr-2017-0191.
Hajipour, A.R., Rafiee, F., 2015. Recent progress in ionic liquids and their applications in organic synthesis. *Org. Prep. Proced. Int.* 47, 249–308. doi:10.1080/00304948.2015.1052317.
Hajipour, A.R., Rafiee, F., 2009. Basic ionic liquids. A short review. *J. Iran. Chem. Soc.* 6, 647–678. doi:10.1007/BF03246155.
Hajipour, A.R., Rafiee, F., Ruoho, A.E., 2010. A rapid and convenient method for the synthesis of aldoximes under microwave irradiation using in situ generated ionic liquids. *J. Iran. Chem. Soc.* 7, 114–118. doi:10.1007/BF03245867.
Hallett, J.P., Welton, T., 2011. Room-temperature ionic liquids: Solvents for synthesis and catalysis. *Chem. Rev.* 111, 3508–3576. doi:10.1021/cr1003248.
Han, D., Row, K.H., 2010. Recent applications of ionic liquids in separation technology. *Molecules* 15, 2405–2426. doi:10.3390/molecules15042405.
Haq, M.A., Habu, Y., Yamamoto, K., Takada, A., Kadokawa, J., 2019. Ionic liquid induces flexibility and thermoplasticity in cellulose film. *Carbohydr. Polym.* 223, 115058. doi:10.1016/j.carbpol.2019.115058.
Heinze, T., 2015. Cellulose: Structure and properties. In: *Advances in Polymer Science*. pp. 1–52. doi:10.1007/12_2015_319.
Hemmati, F., Jafari, S.M., Taheri, R.A., 2019. Optimization of homogenization-sonication technique for the production of cellulose nanocrystals from cotton linter. *Int. J. Biol. Macromol.* 137, 374–381. doi:10.1016/j.ijbiomac.2019.06.241.
Hrubesh, L.W., 1998. Aerogel applications. *J. Non. Cryst. Solids* 225, 335–342. doi:10.1016/S0022-3093(98)00135-5.
Huang, M.-M., Jiang, Y., Sasisanker, P., Driver, G.W., Weingärtner, H., 2011. Static relative dielectric permittivities of ionic liquids at 25°C. *J. Chem. Eng. Data* 56, 1494–1499. doi:10.1021/je101184s.
Huber, T., Bickerton, S., Müssig, J., Pang, S., Staiger, M.P., 2012a. Solvent infusion processing of all-cellulose composite materials. *Carbohydr. Polym.* 90, 730–733. doi:10.1016/j.carbpol.2012.05.047.
Huber, T., Müssig, J., Curnow, O., Pang, S., Bickerton, S., Staiger, M.P., 2012b. A critical review of all-cellulose composites. *J. Mater. Sci.* doi:10.1007/s10853-011-5774-3.
Infante, J.P., Huszagh, V.A., 1987. Is there a new biosynthetic pathway for lung surfactant phosphatidylcholine? *Trends Biochem. Sci.* 12, 131–133. doi:10.1016/0968-0004(87)90062-4.
Iqbal, B., Muhammad, N., Jamal, A., Ahmad, P., Khan, Z.U.H., Rahim, A., Khan, A.S., Gonfa, G., Iqbal, J., Rehman, I.U., 2017. An application of ionic liquid for preparation of homogeneous collagen and alginate hydrogels for skin dressing. *J. Mol. Liq.* 243, 720–725. doi:10.1016/j.molliq.2017.08.101.
Ji, X.-J., Huang, H., Nie, Z.-K., Qu, L., Xu, Q., Tsao, G.T., 2011. Fuels and chemicals from hemicellulose sugars. In: *Advances in Biochemical Engineering/Biotechnology*. pp. 199–224. doi:10.1007/10_2011_124.
Jin, C., Han, S., Li, J., Sun, Q., 2015. Fabrication of cellulose-based aerogels from waste newspaper without any pretreatment and their use for absorbents. *Carbohydr. Polym.* 123, 150–156. doi:10.1016/j.carbpol.2015.01.056.
Jo, S., Park, S., Oh, Y., Hong, J., Kim, H.J., Kim, K.J., Oh, K.K., Lee, S.H., 2019. Development of cellulose hydrogel microspheres for lipase immobilization. *Biotechnol. Bioprocess Eng.* 24, 145–154. doi:10.1007/s12257-018-0335-0.
Jönsson, L.J., Alriksson, B., Nilvebrant, N.O., 2013. Bioconversion of lignocellulose: Inhibitors and detoxification. *Biotechnol. Biofuels.* doi:10.1186/1754-6834-6-16.
Joshi, V.S., Kreth, J., Koley, D., 2017. Pt-Decorated MWCNTs–ionic liquid composite-based hydrogen peroxide sensor to study microbial metabolism using scanning electrochemical microscopy. *Anal. Chem.* 89, 7709–7718. doi:10.1021/acs.analchem.7b01677.

Kalka, S., Huber, T., Steinberg, J., Baronian, K., Müssig, J., Staiger, M.P., 2014. Biodegradability of all-cellulose composite laminates. *Compos. Part A Appl. Sci. Manuf.* 59, 37–44. doi:10.1016/j.compositesa.2013.12.012.

Katritzky, A.R., Jain, R., Lomaka, A., Petrukhin, R., Karelson, M., Visser, A.E., Rogers, R.D., 2002. Correlation of the melting points of potential ionic liquids (Imidazolium Bromides and Benzimidazolium Bromides) using the CODESSA program. *J. Chem. Inf. Comput. Sci.* 42, 225–231. doi:10.1021/ci0100494.

Khan, T.A., Lee, J.-H., Kim, H.-J., 2019. Lignin-based adhesives and coatings. In: *Lignocellulose for Future Bioeconomy*. Elsevier, pp. 153–206. doi:10.1016/B978-0-12-816354-2.00009-8.

Khattab, M.M., Abdel-Hady, N.A., Dahman, Y., 2017. Cellulose nanocomposites. In: *Cellulose-Reinforced Nanofiber Composites*. Elsevier, pp. 483–516. doi:10.1016/B978-0-08-100957-4.00021-8.

Kimura, M., Shinohara, Y., Takizawa, J., Ren, S., Sagisaka, K., Lin, Y., Hattori, Y., Hinestroza, J.P., 2015. Versatile molding process for tough cellulose hydrogel materials. *Sci. Rep.* 5, 16266. doi:10.1038/srep16266.

Klouda, L., Mikos, A.G., 2008. Thermoresponsive hydrogels in biomedical applications. *Eur. J. Pharm. Biopharm.* 68, 34–45. doi:10.1016/j.ejpb.2007.02.025.

Kosmulski, M., Gustafsson, J., Rosenholm, J.B., 2004. Thermal stability of low temperature ionic liquids revisited. *Thermochim. Acta* 412, 47–53. doi:10.1016/j.tca.2003.08.022.

Kubicki, J.D., Yang, H., Sawada, D., O'Neill, H., Oehme, D., Cosgrove, D., 2018. The shape of native plant cellulose microfibrils. *Sci. Rep.* 8, 13983. doi:10.1038/s41598-018-32211-w.

Kumar, A., 2018. Assessment of different pretreatment technologies for efficient bioconversion of lignocellulose to ethanol. *Front. Biosci.* 10, 521. doi:10.2741/s521.

Kumari, D., Singh, R., 2018. Pretreatment of lignocellulosic wastes for biofuel production: A critical review. *Renew. Sustain. Energy Rev.* doi:10.1016/j.rser.2018.03.111.

Laali, K.K., 2003. Ionic liquids in synthesis. *Synthesis (Stuttg).* 2003, 1752–1752. doi:10.1055/s-2003-40869.

Leroy, E., Jacquet, P., Coativy, G., Reguerre, A.L., Lourdin, D., 2012. Compatibilization of starch–zein melt processed blends by an ionic liquid used as plasticizer. *Carbohydr. Polym.* 89, 955–963. doi:10.1016/j.carbpol.2012.04.044.

Li, J., Nawaz, H., Wu, J., Zhang, Jinming, Wan, J., Mi, Q., Yu, J., Zhang, J., 2018. All-cellulose composites based on the self-reinforced effect. *Compos. Commun.* 9, 42–53. doi:10.1016/j.coco.2018.04.008.

Li, X., Lu, X., Yang, J., Ju, Z., Kang, Y., Xu, J., Zhang, S., 2019. A facile ionic liquid approach to prepare cellulose-rich aerogels directly from corn stalks. *Green Chem.* 21, 2699–2708. doi:10.1039/C9GC00282K.

Li, X., Zhang, J., Ju, Z., Li, Y., Xu, J., Xin, J., Lu, X., Zhang, S., 2018. Facile synthesis of cellulose/ZnO aerogel with uniform and tunable nanoparticles based on ionic liquid and polyhydric alcohol. *ACS Sustain. Chem. Eng.* 6, 16248–16254. doi:10.1021/acssuschemeng.8b03106.

Liang, X., Qu, B., Li, J., Xiao, H., He, B., Qian, L., 2015. Preparation of cellulose-based conductive hydrogels with ionic liquid. *React. Funct. Polym.* 86, 1–6. doi:10.1016/j.reactfunctpolym.2014.11.002.

Limayem, A., Ricke, S.C., 2012. Lignocellulosic biomass for bioethanol production: Current perspectives, potential issues and future prospects. *Prog. Energy Combust. Sci.* doi:10.1016/j.pecs.2012.03.002.

Liu, J., Liang, Y., Shen, J., Bai, Q., 2018. Polymeric ionic liquid-assembled graphene-immobilized silica composite for selective isolation of human serum albumin from human whole blood. *Anal. Bioanal. Chem.* 410, 573–584. doi:10.1007/s00216-017-0758-z.

Liu, Z., Li, D., Dai, H., Huang, H., 2017. Enhanced properties of tea residue cellulose hydrogels by addition of graphene oxide. *J. Mol. Liq.* 244, 110–116. doi:10.1016/j.molliq.2017.08.106.

Lopes, J.M., Mustapa, A.N., Pantić, M., Bermejo, M.D., Martín, Á., Novak, Z., Knez, Ž., Cocero, M.J., 2017. Preparation of cellulose aerogels from ionic liquid solutions for supercritical impregnation of phytol. *J. Supercrit. Fluids* 130, 17–22. doi:10.1016/j.supflu.2017.07.018.

Lourenço, A., Pereira, H., 2018. Compositional variability of lignin in biomass. In: *Lignin - Trends and Applications*. InTech. doi:10.5772/intechopen.71208.

Luo, Y., Li, Z., Li, X., Liu, X., Fan, J., Clark, J.H., Hu, C., 2019. The production of furfural directly from hemicellulose in lignocellulosic biomass: A review. *Catal. Today* 319, 14–24. doi:10.1016/j.cattod.2018.06.042.

Mahmood, H., Moniruzzaman, M., Yusup, S., Akil, H.M., 2017a. Green composites from ionic liquid-assisted processing of sustainable resources: A brief overview. In: *Progress and Developments in Ionic Liquids*. InTech. doi:10.5772/65796.

Mahmood, H., Moniruzzaman, M., Yusup, S., Akil, H.M., 2015. Comparison of some biocomposite board properties fabricated from lignocellulosic biomass before and after Ionic liquid pretreatment. *Chem. Eng. Trans.* doi:10.3303/CET1545119.

Mahmood, H., Moniruzzaman, M., Yusup, S., Welton, T., 2017b. Ionic liquids assisted processing of renewable resources for the fabrication of biodegradable composite materials. *Green Chem.* 19, 2051–2075. doi:10.1039/C7GC00318H.

Mai, N.L., Ha, S.H., Koo, Y.-M., 2014. Efficient pretreatment of lignocellulose in ionic liquids/co-solvent for enzymatic hydrolysis enhancement into fermentable sugars. *Process Biochem.* 49, 1144–1151. doi:10.1016/j.procbio.2014.03.024.

Mallakpour, S., Dinari, M., 2012. Ionic liquids as green solvents: Progress and prospects. In: *Green Solvents II*. Springer, Dordrecht, pp. 1–32. doi:10.1007/978-94-007-2891-2_1.

Martins, P.L.G., Braga, A.R., de Rosso, V.V., 2017. Can ionic liquid solvents be applied in the food industry? *Trends Food Sci. Technol.* 66, 117–124. doi:10.1016/j.tifs.2017.06.002.

Medronho, B., Lindman, B., 2015. Brief overview on cellulose dissolution/regeneration interactions and mechanisms. *Adv. Colloid Interface Sci.* 222, 502–508. doi:10.1016/j.cis.2014.05.004.

Mikkonen, K.S., 2013. Recent studies on hemicellulose-based blends, composites and nanocomposites. In: *Advanced Structured Materials*. pp. 313–336. doi:10.1007/978-3-642-20940-6_9.

Montalbán, M.G., Bolívar, C.L., Díaz Baños, F.G., Víllora, G., 2015. Effect of Temperature, Anion, and Alkyl Chain Length on the Density and Refractive Index of 1-Alkyl-3-methylimidazolium-Based Ionic Liquids. *J. Chem. Eng. Data* 60, 1986–1996. doi:10.1021/je501091q.

Mosier, N., Wyman, C., Dale, B., Elander, R., Lee, Y.Y., Holtzapple, M., Ladisch, M., 2005. Features of promising technologies for pretreatment of lignocellulosic biomass. *Bioresour. Technol.* 96, 673–686. doi:10.1016/j.biortech.2004.06.025.

Mussana, H., Yang, X., Tessima, M., Han, F., Iqbal, N., Liu, L., 2018. Preparation of lignocellulose aerogels from cotton stalks in the ionic liquid-based co-solvent system. *Ind. Crops Prod.* 113, 225–233. doi:10.1016/j.indcrop.2018.01.025.

Nancarrow, P., Mohammed, H., 2017. Ionic liquids in space technology - current and future trends. *ChemBioEng Rev.* 4, 106–119. doi:10.1002/cben.201600021.

Nasrullah, A., Bhat, A.H., Sada Khan, A., Ajab, H., 2017. Comprehensive approach on the structure, production, processing, and application of lignin. In: *Lignocellulosic Fiber and Biomass-Based Composite Materials*. Elsevier, pp. 165–178. doi:10.1016/B978-0-08-100959-8.00009-3.

Niroomand, F., Khosravani, A., Younesi, H., 2016. Fabrication and properties of cellulose-nanochitosan biocomposite film using ionic liquid. *Cellulose* 23, 1311–1324. doi:10.1007/s10570-016-0872-7.

Nishino, T., Arimoto, N., 2007. All-cellulose composite prepared by selective dissolving of fiber surface. *Biomacromolecules* 8, 2712–2716. doi:10.1021/bm0703416.

Nishino, T., Matsuda, I., Hirao, K., 2004. All-cellulose composite. *Macromolecules* 37, 7683–7687. doi:10.1021/ma049300h.

Nunes, C.S., Rufato, K.B., Souza, P.R., de Almeida, E.A.M.S., da Silva, M.J.V., Scariot, D.B., Nakamura, C.V., Rosa, F.A., Martins, A.F., Muniz, E.C., 2017. Chitosan/chondroitin sulfate hydrogels prepared in [Hmim][HSO$_4$] ionic liquid. *Carbohydr. Polym.* 170, 99–106. doi:10.1016/j.carbpol.2017.04.073.

Ochędzan-Siodłak, W., Dziubek, K., Siodłak, D., 2013. Densities and viscosities of imidazolium and pyridinium chloroaluminate ionic liquids. *J. Mol. Liq.* 177, 85–93. doi:10.1016/j.molliq.2012.10.001.

Olivier-Bourbigou, H., Magna, L., Morvan, D., 2010. Ionic liquids and catalysis: Recent progress from knowledge to applications. *Appl. Catal. A Gen.* doi:10.1016/j.apcata.2009.10.008.

Pang, J.-H., Liu, X., Wu, M., Wu, Y.-Y., Zhang, X.-M., Sun, R.-C., 2014. Fabrication and characterization of regenerated cellulose films using different ionic liquids. *J. Spectrosc.* 2014, 1–8. doi:10.1155/2014/214057.

Pang, J.-H., Wu, M., Zhang, Q., Tan, X., Xu, F., Zhang, X., Sun, R., 2015. Comparison of physical properties of regenerated cellulose films fabricated with different cellulose feedstocks in ionic liquid. *Carbohydr. Polym.* 121, 71–78. doi:10.1016/j.carbpol.2014.11.067.

Park, S., Oh, Y., Yun, J., Yoo, E., Jung, D., Park, K.S., Oh, K.K., Lee, S.H., 2020. Characterization of blended cellulose/biopolymer films prepared using ionic liquid. *Cellulose* 27, 5101–5119. doi:10.1007/s10570-020-03152-3.

Passos, H., Freire, M.G., Coutinho, J.A.P., 2014. Ionic liquid solutions as extractive solvents for value-added compounds from biomass. *Green Chem.* 16, 4786–4815. doi:10.1039/C4GC00236A.

Peng, H., Wang, S., Xu, H., Dai, G., 2018. Preparations, properties, and formation mechanism of novel cellulose hydrogel membrane based on ionic liquid. *J. Appl. Polym. Sci.* 135, 45488. doi:10.1002/app.45488.

Peng, H., Wang, S., Xu, H., Hao, X., 2017. Preparation, properties and formation mechanism of cellulose/polyvinyl alcohol bio-composite hydrogel membranes. *New J. Chem.* 41, 6564–6573. doi:10.1039/C7NJ00845G.

Pierre, A.C., Pajonk, G.M., 2002. Chemistry of aerogels and their applications. *Chem. Rev.* 102, 4243–4266. doi:10.1021/cr0101306.

Rahimi, M., Shafiei-Irannejad, V., Safa, K.D., Salehi, R., 2018. Multi-branched ionic liquid-chitosan as a smart and biocompatible nano-vehicle for combination chemotherapy with stealth and targeted properties. *Carbohydr. Polym.* 196, 299–312. doi:10.1016/j.carbpol.2018.05.059.

Reddy, K.O., Maheswari, C.U., Dhlamini, M.S., Mothudi, B.M., Zhang, Jinming, Zhang, Jun, Nagarajan, R., Rajulu, A.V., 2017. Preparation and characterization of regenerated cellulose films using borassus fruit fibers and an ionic liquid. *Carbohydr. Polym.* 160, 203–211. doi:10.1016/j.carbpol.2016.12.051.

Reddy, K.O., Zhang, J., Zhang, J., Rajulu, A.V., 2014. Preparation and properties of self-reinforced cellulose composite films from Agave microfibrils using an ionic liquid. *Carbohydr. Polym.* 114, 537–545. doi:10.1016/j.carbpol.2014.08.054.

Reichardt, C., Welton, T., 2010. *Solvents and Solvent Effects in Organic Chemistry, Solvents and Solvent Effects in Organic Chemistry*: Fourth Edition. Wiley-VCH Verlag GmbH & Co. KGaA, Weinheim. doi:10.1002/9783527632220.

Robak, K., Balcerek, M., 2018. Review of second-generation bioethanol production from residual biomass. *Food Technol. Biotechnol.* 56. doi:10.17113/ftb.56.02.18.5428.

Ruiz-Angel, M.J., Berthod, A., 2006. Reversed phase liquid chromatography of alkyl-imidazolium ionic liquids. *J. Chromatogr. A* 1113, 101–108. doi:10.1016/j.chroma.2006.01.124.

Rybinska-Fryca, A., Sosnowska, A., Puzyn, T., 2018. Prediction of dielectric constant of ionic liquids. *J. Mol. Liq.* 260, 57–64. doi:10.1016/j.molliq.2018.03.080.

Sadjadi, S., Heravi, M.M., Raja, M., 2019. Composite of ionic liquid decorated cyclodextrin nanosponge, graphene oxide and chitosan: A novel catalyst support. *Int. J. Biol. Macromol.* 122, 228–237. doi:10.1016/j.ijbiomac.2018.10.160.

Satani, H., Kuwata, M., Shimizu, A., 2020. Simple and environmentally friendly preparation of cellulose hydrogels using an ionic liquid. *Carbohydr. Res.* 494, 108054. doi:10.1016/j.carres.2020.108054.

Shahzad, S., Shah, A., Kowsari, E., Iftikhar, F.J., Nawab, A., Piro, B., Akhter, M.S., Rana, U.A., Zou, Y., 2019. Ionic liquids as environmentally benign electrolytes for high-performance supercapacitors. *Glob. Challenges* 3, 1800023. doi:10.1002/gch2.201800023.

Shamshina, J.L., Zavgorodnya, O., Choudhary, H., Frye, B., Newbury, N., Rogers, R.D., 2018. In search of stronger/cheaper chitin nanofibers through electrospinning of chitin-cellulose composites using an ionic liquid platform. *ACS Sustain. Chem. Eng.* doi:10.1021/acssuschemeng.8b03269.

Sharma, A., Sen, D., Thakre, S., Kumaraswamy, G., 2019. characterizing microvoids in regenerated cellulose fibers obtained from viscose and lyocell processes. *Macromolecules* 52, 3987–3994. doi:10.1021/acs.macromol.9b00487.

Sharma, H.K., Xu, C., Qin, W., 2017. Biological pretreatment of lignocellulosic biomass for biofuels and bioproducts: An overview. *Waste Biomass Valorization* 10, 1–17. doi:10.1007/s12649-017-0059-y.

Sindhu, R., Binod, P., Pandey, A., 2016. Biological pretreatment of lignocellulosic biomass – An overview. *Bioresour. Technol.* 199, 76–82. doi:10.1016/j.biortech.2015.08.030.

Singh, N., Chen, J., Koziol, K.K., Hallam, K.R., Janas, D., Patil, A.J., Strachan, A., Hanley, J.G., Rahatekar, S.S., 2016. Chitin and carbon nanotube composites as biocompatible scaffolds for neuron growth. *Nanoscale* 8, 8288–8299. doi:10.1039/C5NR06595J.

Singh, R., Krishna, B.B., Kumar, J., Bhaskar, T., 2016. Opportunities for utilization of non-conventional energy sources for biomass pretreatment. *Bioresour. Technol.* 199, 398–407. doi:10.1016/j.biortech.2015.08.117.

Singh, S., Cheng, G., Sathitsuksanoh, N., Wu, D., Varanasi, P., George, A., Balan, V., Gao, X., Kumar, R., Dale, B.E., Wyman, C.E., Simmons, B.A., 2015. Comparison of different biomass pretreatment techniques and their impact on chemistry and structure. *Front. Energy Res.* 2. doi:10.3389/fenrg.2014.00062.

Spörl, J.M., Batti, F., Vocht, M.-P., Raab, R., Müller, A., Hermanutz, F., Buchmeiser, M.R., 2018. Ionic liquid approach toward manufacture and full recycling of all-cellulose composites. *Macromol. Mater. Eng.* 303, 1700335. doi:10.1002/mame.201700335.

Stanton, J., Xue, Y., Pandher, P., Malek, L., Brown, T., Hu, X., Salas-de la Cruz, D., 2018. Impact of ionic liquid type on the structure, morphology and properties of silk-cellulose biocomposite materials. *Int. J. Biol. Macromol.* 108, 333–341. doi:10.1016/j.ijbiomac.2017.11.137.

Stefanescu, C., Daly, W.H., Negulescu, I.I., 2012. Biocomposite films prepared from ionic liquid solutions of chitosan and cellulose. *Carbohydr. Polym.* 87, 435–443. doi:10.1016/j.carbpol.2011.08.003.

Stergar, J., Maver, U., 2016. Review of aerogel-based materials in biomedical applications. *J. Sol-Gel Sci. Technol.* 77, 738–752. doi:10.1007/s10971-016-3968-5.

Sun, N., Rogers, D.R.D., 2010. Dissolution and processing of cellulosic materials with ionic liquids: fundamentals and applications. Chemistry (Easton). 208.Sun, S., Sun, S., Cao,

X., Sun, R., 2016. The role of pretreatment in improving the enzymatic hydrolysis of lignocellulosic materials. *Bioresour. Technol.* doi:10.1016/j.biortech.2015.08.061.

Sundberg, J., Toriz, G., Gatenholm, P., 2015. Effect of xylan content on mechanical properties in regenerated cellulose/xylan blend films from ionic liquid. *Cellulose* 22, 1943–1953. doi:10.1007/s10570-015-0606-2.

Thirion, C., 2015. All-cellulose composite production using an ionic liquid: dissolution of wood derived high purity cellulose with BmimCl.

Trohalaki, S., Pachter, R., 2005. Prediction of melting points for ionic liquids. *QSAR Comb. Sci.* 24, 485–490. doi:10.1002/qsar.200430927.

Usmani, Z., Sharma, M., Gupta, P., Karpichev, Y., Gathergood, N., Bhat, R., Gupta, V.K., 2020. Ionic liquid based pretreatment of lignocellulosic biomass for enhanced bioconversion. *Bioresour. Technol.* 304, 123003. doi:10.1016/j.biortech.2020.123003.

van Maris, A.J.A., Abbott, D.A., Bellissimi, E., van den Brink, J., Kuyper, M., Luttik, M.A.H., Wisselink, H.W., Scheffers, W.A., van Dijken, J.P., Pronk, J.T., 2006. Alcoholic fermentation of carbon sources in biomass hydrolysates by Saccharomyces cerevisiae: Current status. *Antonie Van Leeuwenhoek* 90, 391–418. doi:10.1007/s10482-006-9085-7.

Vieira, M.G.A., da Silva, M.A., dos Santos, L.O., Beppu, M.M., 2011. Natural-based plasticizers and biopolymer films: A review. *Eur. Polym. J.* 47, 254–263. doi:10.1016/j.eurpolymj.2010.12.011.

Wade, R.H., Welch, C.M., 1965. The role of alkali metal hydroxides in the benzylation of cotton cellulose. *Text. Res. J.* 35, 930–934. doi:10.1177/004051756503501010.

Wakai, C., Oleinikova, A., Ott, M., Weingärtner, H., 2005. How polar are ionic liquids? Determination of the static dielectric constant of an imidazolium-based ionic liquid by microwave dielectric spectroscopy. *J. Phys. Chem. B* 109, 17028–30. doi:10.1021/jp053946+.

Wang, J., Wei, L., Ma, Y., Li, K., Li, M., Yu, Y., Wang, L., Qiu, H., 2013. Collagen/cellulose hydrogel beads reconstituted from ionic liquid solution for Cu(II) adsorption. *Carbohydr. Polym.* 98, 736–743. doi:10.1016/j.carbpol.2013.06.001.

Wang, M., Bai, L., Zhang, L., Sun, G., Zhang, X., Dong, S., 2016. A microporous silk carbon–ionic liquid composite for the electrochemical sensing of dopamine. *Analyst* 141, 2447–2453. doi:10.1039/C6AN00016A.

Wang, X., Li, H., Cao, Y., Tang, Q., 2011. Cellulose extraction from wood chip in an ionic liquid 1-allyl-3-methylimidazolium chloride (AmimCl). *Bioresour. Technol.* 102, 7959–7965. doi:10.1016/j.biortech.2011.05.064.

Warsi Khan, H., Moniruzzaman, M., Mahmoud Elsayed Nasef, M., Azmi Bustam@Khalil, M., 2020. Ionic liquid assisted cellulose aerogels for cleaning an oil spill. *Mater. Today Proc.* 31, 217–220. doi:10.1016/j.matpr.2020.05.139.

Wasserscheid, P., Keim, W., 2000. Ionic liquids—New "Solutions" for transition metal catalysis. *Angew. Chemie* 39, 3772–3789. doi:10.1002/1521-3773(20001103)39:21<3772::AID-ANIE3772>3.0.CO;2-5.

Weingärtner, H., 2006. The static dielectric constant of ionic liquids. *Zeitschrift für Phys. Chemie* 220, 1395–1405. doi:10.1524/zpch.2006.220.10.1395.

Weingärtner, H., Sasisanker, P., Daguenet, C., Dyson, P.J., Krossing, I., Slattery, J.M., Schubert, T., 2007. The dielectric response of room-temperature ionic liquids: Effect of cation variation. *J. Phys. Chem. B* 111, 4775–4780. doi:10.1021/jp0671188.

Welton, T., 2004. Ionic liquids in catalysis. *Coord. Chem. Rev.* 248, 2459–2477. doi:10.1016/j.ccr.2004.04.015.

Welton, T., 1999. Room-temperature ionic liquids. Solvents for Synthesis and Catalysis. *Chem. Rev.* 99, 2071–2084. doi:10.1021/cr980032t.

Weuster-Botz, D., 2007. Process intensification of whole-cell biocatalysis with ionic liquids. *Chem. Rec.* 7, 334–340. doi:10.1002/tcr.20130.

Wu, R.-L., Wang, X.-L., Li, F., Li, H.-Z., Wang, Y.-Z., 2009. Green composite films prepared from cellulose, starch and lignin in room-temperature ionic liquid. *Bioresour. Technol.* 100, 2569–2574. doi:10.1016/j.biortech.2008.11.044.

Xu, M., Huang, Q., Wang, X., Sun, R., 2015. Highly tough cellulose/graphene composite hydrogels prepared from ionic liquids. *Ind. Crops Prod.* 70, 56–63. doi:10.1016/j.indcrop.2015.03.004.

Yao, C., Anderson, J.L., 2009. Retention characteristics of organic compounds on molten salt and ionic liquid-based gas chromatography stationary phases. *J. Chromatogr. A* 1216, 1658–1712. doi:10.1016/j.chroma.2008.12.001.

Yu, M., Li, J., Wang, L., 2016. Preparation and characterization of magnetic carbon aerogel from pyrolysis of sodium carboxymethyl cellulose aerogel crosslinked by iron trichloride. *J. Porous Mater.* 23, 997–1003. doi:10.1007/s10934-016-0157-4.

Yuan, W.-L., Yang, X., He, L., Xue, Y., Qin, S., Tao, G.-H., 2018. Viscosity, conductivity, and electrochemical property of dicyanamide ionic liquids. *Front. Chem.* 6. doi:10.3389/fchem.2018.00059.

Yusuf, S.N.F., Yahya, R., Arof, A.K., 2017. Ionic liquid enhancement of polymer electrolyte conductivity and their effects on the performance of electrochemical devices. In: *Progress and Developments in Ionic Liquids*. InTech. doi:10.5772/65752.

Zeng, X., Li, X., Xing, L., Liu, X., Luo, S., Wei, W., Kong, B., Li, Y., 2009. Electrodeposition of chitosan–ionic liquid–glucose oxidase biocomposite onto nano-gold electrode for amperometric glucose sensing. *Biosens. Bioelectron.* 24, 2898–2903. doi:10.1016/j.bios.2009.02.027.

Zeng, Y., Himmel, M.E., Ding, S.Y., 2017. Visualizing chemical functionality in plant cell walls Mike Himmel. *Biotechnol. Biofuels.* doi:10.1186/s13068-017-0953-3.

Zhang, J., Luo, N., Zhang, X., Xu, L., Wu, J., Yu, J., He, J., Zhang, J., 2016. All-Cellulose nanocomposites reinforced with in situ retained cellulose nanocrystals during selective dissolution of cellulose in an ionic liquid. *ACS Sustain. Chem. Eng.* 4, 4417–4423. doi:10.1021/acssuschemeng.6b01034.

Zhang, J., Wu, J., Cao, Y., Sang, S., Zhang, J., He, J., 2009. Synthesis of cellulose benzoates under homogeneous conditions in an ionic liquid. *Cellulose*. doi:10.1007/s10570-008-9260-2.

Zhang, J., Wu, J., Yu, J., Zhang, X., He, J., Zhang, J., 2017. Application of ionic liquids for dissolving cellulose and fabricating cellulose-based materials: State of the art and future trends. *Mater. Chem. Front.* 1, 1273–1290. doi:10.1039/C6QM00348F.

Zhang, S., Chen, C., Duan, C., Hu, H., Li, H., Li, J., Liu, Y., Ma, X., Stavik, J., Ni, Y., 2018. Regenerated cellulose by the lyocell process, a brief review of the process and properties. *BioResources* 13, 4577–4592. doi:10.15376/biores.13.2.Zhang.

Zhang, X., Huo, F., Liu, X., Dong, K., He, H., Yao, X., Zhang, S., 2015. Influence of microstructure and interaction on viscosity of ionic liquids. *Ind. Eng. Chem. Res.* 54, 3505–3514. doi:10.1021/acs.iecr.5b00415.

Zhao, D., Liao, Y., Zhang, Z.D., 2007. Toxicity of ionic liquids. *Clean - Soil, Air, Water* 35, 42–48. doi:10.1002/clen.200600015.

Zhao, H., 2003. Review: Current studies on some physical properties of ionic liquids. *Phys. Chem. Liq.* 41, 545–557. doi:10.1080/003191031000117319.

Zheng, X., Huang, F., Chen, L., Huang, L., Cao, S., Ma, X., 2019. Preparation of transparent film via cellulose regeneration: Correlations between ionic liquid and film properties. *Carbohydr. Polym.* 203, 214–218. doi:10.1016/j.carbpol.2018.09.060.

Zhou, L., Wang, Q., Wen, J., Chen, X., Shao, Z., 2013. Preparation and characterization of transparent silk fibroin/cellulose blend films. *Polymer (Guildf)*. 54, 5035–5042. doi:10.1016/j.polymer.2013.07.002.

Zhou, Y., Fan, M., Luo, X., Huang, L., Chen, L., 2014. Acidic ionic liquid catalyzed crosslinking of oxycellulose with chitosan for advanced biocomposites. *Carbohydr. Polym.* 113, 108–114. doi:10.1016/j.carbpol.2014.06.081.

Zhou, Z., Ju, X., Zhou, M., Xu, X., Fu, J., Li, L., 2019. An enhanced ionic liquid-tolerant immobilized cellulase system via hydrogel microsphere for improving in situ saccharification of biomass. *Bioresour. Technol.* 294, 122146. doi:10.1016/j.biortech.2019.122146.

Zimmermann, T., Bordeanu, N., Strub, E., 2010. Properties of nanofibrillated cellulose from different raw materials and its reinforcement potential. *Carbohydr. Polym.* 79, 1086–1093. doi:10.1016/j.carbpol.2009.10.045.

Zoghlami, A., Paës, G., 2019. Lignocellulosic biomass: Understanding recalcitrance and predicting hydrolysis. *Front. Chem.* 7. doi:10.3389/fchem.2019.00874.

10 Deep Eutectic Solvent-Mediated Process for Productions of Sustainable Polymeric Biomaterials

Elizabeth Jayex Panakkal
King Mongkut's University of Technology North Bangkok

Yu-Shen Cheng
National Yunlin University of Science and Technology

*Theerawut Phusantisampan and
Malinee Sriariyanun*
King Mongkut's University of Technology North Bangkok

CONTENTS

10.1	Introduction	252
10.2	Lignocellulose	253
10.3	Deep Eutectic Solvent	256
10.4	Dissolution and Pretreatment of Lignocellulose by Deep Eutectic Solvent	260
10.5	Deep Eutectic Solvent in Polymeric Composite Synthesis and Fabrication	267
	10.5.1 Application of Deep Eutectic Solvent in Polymerization and Polymer Extraction	268
	10.5.2 Application of Deep Eutectic Solvent as Plasticizer	273
	10.5.3 Application of Deep Eutectic Solvent in Lignocellulose and Natural Biopolymeric Composite	275
10.6	Conclusion	278
Acknowledgments		278
References		278

DOI: 10.1201/9781003137535-10

10.1 INTRODUCTION

Growth in the global economy has always depended on the natural resources causing worldwide depletion. Even though global economic growth is progressing, it has had a rising negative impact on our planet and on living organism's habitats. Over the last century, the economic development has endangered our natural resources and disturbed environmental equilibrium. Global climatic change, biodiversity loss, ozone depletion and greenhouse effect are some examples of impacts that we have done to our nature. Hence sustainable development by protecting our natural resources and environment has become a challenge (Vilaplana et al. 2010). Even though implementation of Life Cycle Assessment studies in industries has been effective to reduce environmental impacts at all stages of productions, a prompt action needs to be taken for sustainable industrial development with reduced impact on environment. In this scenario, BCG economy, including bioeconomy, circular economy and green economy, can be a promising concept. BCG economy, which includes bioeconomy, circular economy and green economy, is a concept in industrialization to produce value-added products from waste materials in a cost-effective manner and thereby reducing the waste generation leading to less impact on environment (Ngammuangtueng et al. 2020, Sriariyanun and Kitsubthawee 2020, Cheng et al. 2020).

Based on this concept, some of the industries have stepped forward to utilize waste or byproduct as a raw material to produce value-added products. This has helped in waste reduction and maximization of resource utilization. Polymeric industries are conventionally using petroleum-based materials to produce composites used in packaging, paint, medicines, automobiles, etc. However, due to depletion of fossil fuels and problems relating to product disposal after the service life of product, there is motivation to shift from petroleum-based materials to renewable and degradable sources. The industries have attempted to overcome the issues of conventional non-biodegradable, petroleum-based materials by replacing them with biocomposites.

Biocomposites are degradable composites, comprising matrix made of biopolymer and natural fiber as a reinforcing phase. The first known use of biocomposite was dated 3400 BC, where it was used to make plywood. Later engineers, builders, artisans and manufacturers in ancient time had used biocomposite using straw, bones, horns, etc. Over the years, more research and knowledge on crude oil chemistry led to the usage of petrochemical polymers. However, again when safe-guarding environment raised to be a necessity, more researches were carried out to produce new composites using raw materials from renewable resources (Bledzki et al. 2012). Natural fibers have been used as an alternative for artificial fiber to synthesis composite materials with properties of low cost, renewability and environment friendliness. As natural fibers are available abundantly, at low cost, it is preferred over synthetic fiber to make composites lighter. The most commonly used natural fibers derived either from plants or animals. Natural fibers utilized from plants are jute, coir, hemp, banana, etc. Silk, wool and hair are the common type of natural fibers used from animal sources. Yet plant-based natural fibers are of more demanded owing to their availability and renewability. Cellulose, present in the lignocellulose of plant, is the most abundant and inexhaustible resource used in biocomposite. The properties like biocompatibility, high tensile strength, low density and high thermal stability also make cellulose the most widely used raw material for biocomposite (Baghaei and Skrifvars 2020, Miao and Hamad 2013). Agriculture waste is a potential source for lignocellulose. A huge amount of agro waste is generated from postharvest activities, and these wastes are mostly combusted, which can lead to pollution and serious health concerns. Instead of combusting, utilizing these agro wastes for the production of value-added products is a promising step toward BCG economy. Hence, these agro wastes, a.k.a. lignocellulose biomass, are being used in polymeric industry as a source of lignocellulose, instead of petroleum-based materials for production of biocomposite.

10.2 LIGNOCELLULOSE

Lignocellulose is the most abundant biomass on earth as it exists in plant cell walls. It is a natural biopolymer composed of cellulose, hemicellulose and lignin, arranged in a complex manner, providing rigid nature to the plant cell. Lignocellulosic biomass has been identified as a promising alternative for fossil fuel usage in this growing world with increasing energy needs. Lignocellulose also has high potential to be used in the production of biofuels and bio-based chemicals in an eco-friendly manner (Zoghlami and Paës 2019, Sriariyanun and Kitsubthawee 2020, Cheng et al. 2020). However, the recalcitrant nature of the lignocellulosic biomass makes it impossible to be used in biorefining process without undergoing prior steps of treatments. These steps include collection of biomass, pretreatment, hydrolysis, conversion, recovery and purification of product (Figure 10.1) (Rodiahwati and Sriariyanun 2016).

Once the lignocellulose biomass is collected from agriculture fields or from agro-processing industries, it has to be pretreated. The necessity of pretreatment

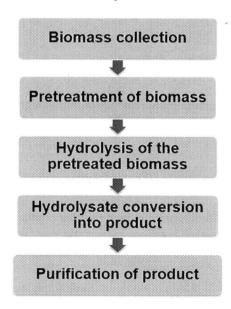

FIGURE 10.1 Steps of biorefining process of lignocellulose biomass.

arises due to the recalcitrant nature of lignocellulose. The complex arrangement of cellulose, hemicellulose and lignin in the plant cell makes it more rigid in nature. Cellulose is the primary structural component in plant cell wall. Cellulose is the polymer of D-glucose linked through β-1,4-glycosidic linkage. Long polymeric chains of these glucose units with their intermolecular and intramolecular hydrogen bonds are arranged as parallel fibrils in the matrix of lignin and hemicellulose (Bhat et al. 2017). Hence, enzymatic hydrolysis of cellulose is hindered by the presence of lignin and hemicellulose. Moreover, intramolecular hydrogen bonds of cellulose between the hydroxyl groups and oxygen of nearby molecules give extra protection for cellulose. Along with the hydrogen bond linkages of cellulose, hemicellulose is also covalently linked with lignin. Lignin is another polymer acting as a binding agent between cellulose and hemicellulose; consequently, it provides a rigid structure to plant cells (Oh et al. 2020). This complexity in the structure of lignocellulosic biomass makes pretreatment an essential step in biomass processing. Pretreatment can enhance the enzyme accessibility to improvise the usability of biomass. Pretreatment also breaks down the large, bulky biomass into smaller particles, and increases surface area so that biomass will be more vulnerable for hydrolysis and enzyme activity (Rodiahwati and Sriariyanun 2016, Akkharasinphonrat et al. 2017). Hydrolysis reaction will convert the polymers from pretreated lignocellulosic biomass into monomers. These monomers are later transformed to bio-based products and platform chemicals after fermentation or chemical reactions (Galbe and Wallberg 2019, Sriariyanun et al. 2019, Rachmontree et al. 2020). These products are then recovered, purified and marketed in different downstream industries, such as foods, feeds, cosmetics, pharmaceuticals, chemicals, agriculturals and polymers.

Among these steps of biorefining, pretreatment and hydrolysis steps require more budget and processing time to invest (Baruah et al. 2018); therefore, many new techniques are being researched and developed to reduce the time and cost in these steps. The conventional methods used for pretreatment and their drawbacks are given in Table 10.1. Even though conventional methods are used for pretreatment, each technology has its own insufficiency at pretreatment. Physical pretreatment cannot expose most of the biomass for enzymatic hydrolysis; however, it is necessary to breakdown the large biomass to smaller size enough for further handling in the downstream process. Chemical pretreatment can help in releasing the sugars and removes lignin or hemicellulose from their complex structure, but it has other impacts on pretreatment. Corrosion caused to reactor while using acid for large-scale pretreatment purpose is a concern in acid pretreatment. Moreover, the acid has to be neutralized before proceeding to the next steps in biomass processing. This can increase the cost and risk involved in pretreatment. Alkaline pretreatment, on the other hand, uses noncorrosive reagents and does not require specialized equipment for pretreatment. However, it produces lot of waste water during the operation and this can cause environment and water pollution, which is a more serious problem. Ionic liquids being more eco-friendlier technique than acid and alkaline pretreatment, it uses harsh operational conditions, along with complex synthesis process involved in the synthesis of ionic liquid. Even though recovery of ionic liquid is possible, it requires more energy as well (Cheenkachorn et al. 2016, Akkharasinphonrat et al. 2017). This can increase the risk and cost of the process when it is operated in

TABLE 10.1
Conventional Methods of Pretreatment and Their Disadvantages

Type of Pretreatment	Different Methods Used	Characteristics	Disadvantages	References
Physical	Milling, Microwave assisted, Ultrasonication, Extrusion	No solvent usage, less smoke and waste, pre-requisite for all treatments for size reduction	Mild, insufficient to disintegrate lignin matrix	Baruah et al. (2018), Aguilar-Reynosa et al. (2017)
Chemical	Acid hydrolysis, Alkaline hydrolysis, Organic solvents and salts, Ionic liquids	Can easily free bound sugars	Inhibitor formation, high cost involved, requirement of energy intensive process to recover solvents, complexity in synthesis of ionic liquids	Baruah et al. (2018), Guo et al. (2018)
Physiochemical	Steam explosion, CO_2 explosion, Ammonia fibre explosion, Liquid hot water	Limited usage of chemicals, minimum environmental damage	Thermal energy consumption, cost involved, inhibitor formation, harsh conditions	Baruah et al. (2018)
Biological	Bacterial, Fungal, Enzymatic	Fungus is the most preferred, low capital cost, more environment friendly	Long reaction time, contamination, consumption of some amount of sugars for microbial growth	Baruah et al. (2018), Rastogi and Srivastava (2017)

large-scale process. Biological treatment requires no chemicals for pretreating the biomass and is environment-friendly. Despite using less energy and mild conditions for operation, it takes long time for pretreatment. Microbes involved in this type of pretreatment may use some amount of sugar in biomass for their growth as well causing reduction in yields of target products and this limits the usage of biological pretreatment (Lin et al. 2020).

Due to these limitations, a need for much more advanced technique had raised. In the past few decades, researches more focused toward ionic liquids which were considered to be the better solvent than conventional techniques and currently the trend has changed. A more user-friendly technique has been adopted recently for pretreatment with less cost and time involved in the process (Procentese et al. 2018). This technique uses Deep Eutectic Solvent (DES) which requires less energy for pretreatment and thereby reduces the cost of pretreatment. Researches have started focusing on DESs pretreatment as it is more environmental friendly and better recyclable than ionic liquids. DESs have properties very similar to ionic liquids, with less energy consumption and low cost involved in the synthesis of solvent which attracting attention from researchers worldwide. Figure 10.2 represents the numbers

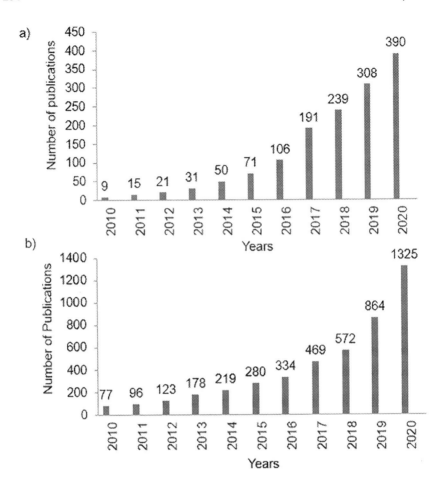

FIGURE 10.2 Trends of publications with deep eutectic solvents during 2010–2020 in (a) PubMed database and (b) Science Direct database.

of publications on DESs in PubMed and Science Direct database during 2010–2020. The graphs in Figure 10.2 clearly show the rising trend in research on DESs. DES usage is in its infancy stage and more research is necessary to understand more about it in different applications. Due to the increasing research and efficiency of DES in biomass processing, this chapter will discuss more about DES, its effect on biomass processing, recycling and reuse of DES and its various application including production of polymeric biomaterial.

10.3 DEEP EUTECTIC SOLVENT

DES is a mixture of two or three compounds consisting of at least one hydrogen bond donor and hydrogen bond acceptor, interacting to each other by self-association and forming a eutectic mixture with low melting point than each of its constituents. DES can be synthesized from mixing of wide ranges of chemical compounds, which can accept or donate protons or electrons. It was initially reported by Abbott et al. in (2003) as eutectic mixture between choline chloride and urea at 1:2 ratios, which was liquid in ambient temperature with unusual solvent properties. Also, this work reported that the freezing point for this eutectic mixture was much lower than the freezing point of its individual compounds. This unusual property of solvent was described to be deeply influenced by hydrogen bonding between each compound (Abbott et al. 2003). The charge delocalization through the hydrogen bonding between the halide anion and amine is pointed out to be the reason for formation of eutectic mixture. Considering the fact that this solvent was very similar to ionic liquid, it was named as DES, in order to differentiate it from ionic liquids (Abbott et al. 2004). The term DES also reflects the physical property to have low freezing point than each of its constituents (Kohli 2019). These solvents are generally liquid below 100°C (Baruah et al. 2018). Even though DES is very similar to ionic liquids, it is suggested to be an alternative option to ionic liquids in these recent years. A comparison of properties and characters of ionic liquids and DES is given in Table 10.2. Due to the low cost, low toxicity and biodegradability, DES is desirable for industrial scale biomass processing.

Smith et al., in 2014, has come forth with a general formula to describe the synthesis of DES (Smith et al. 2014). The formula is given as Cat^+X^-zY, where Cat^+ is ammonium, phosphonium or sulfonium cation, X denotes Lewis base anion, Y is Lewis or Bronsted acid and z is the number of Y molecules interacting with anion. There are four different types of DESs based on its constituents (Table 10.3 and Figure 10.3). Among the four different types of DESs, Type 3 is most commonly used in biomass processing owing to its availability and environmental friendliness of its constituents (Tan et al. 2020).

DES has been used in many industrial fields for different purposes (Figure 10.4). In spite of using it in biomass processing, it has also been used in medical field to

TABLE 10.2

Comparison between Ionic Liquid and Deep Eutectic Solvent (Kohli 2019, Zdanowicz et al. 2018)

Ionic Liquid	Deep Eutectic Solvent
Low melting point	Low melting point
Low vapor pressure	Low vapor pressure
Noninflammable	Noninflammable
High thermal stability	Less thermal stability than ionic liquids
High dissolution ability	High dissolution ability
Moderate to high solution conductivity	High conductivity
Highly viscous	Lower viscosity
Can be toxic	Low or no toxicity
Not environment friendly	Often biodegradable
Complex synthesis process and purification required	Simple synthesis process via mixing and heating and no purification required
Recycling is expensive	Recycling is cheaper

TABLE 10.3
Types of Deep Eutectic Solvent Classified Based on Compositions (Smith et al. 2014, Longo et al. 2018)

Type of Deep Eutectic Solvent	Composition	Solvent 1	mp°C	Solvent 2	mp°C	Deep Eutectic Solvent	T_m °C
Type 1	Quaternary ammonium salt + Metal halides	Choline chloride	303	Zinc chloride	293	Choline chloride/Zinc chloride (1:2)	24
Type 2	Quaternary ammonium salt + Hydrated metal halides	Choline chloride	303	$MgCl_2.6H_2O$	116	Choline chloride/ $MgCl_2.6H_2O$ (1:1)	16
Type 3	Quaternary ammonium salt + Hydrogen bond donor	Choline chloride	303	Urea	134	Choline chloride/urea (1:2)	12
Type 4	Metal salt + Hydrogen bond donor	Zinc chloride	293	Urea	134	Zinc chloride/urea	9

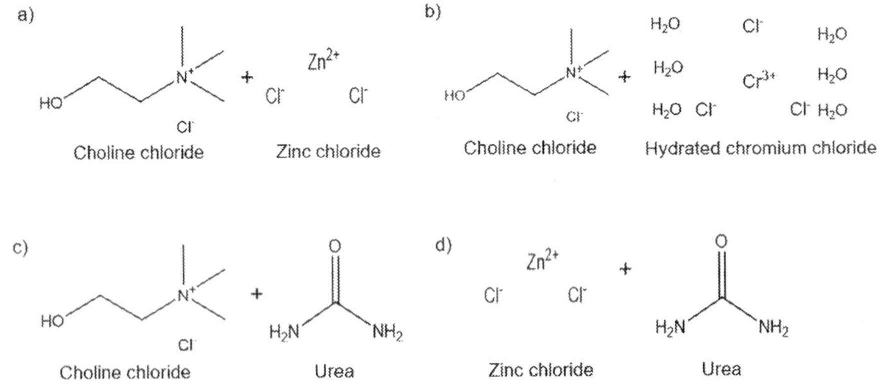

FIGURE 10.3 Example of different types of deep eutectic solvent: (a) Type 1, (b) Type 2, (c) Type c, (d) Type 4.

solubilize many drugs and it aids in drug delivery (Emami and Shayanfar 2020). The DES has also found its application in metal processing. Earlier, electrofinishing industry used water to prepare highly conducting solution. However, due to the narrow window of water's electrical potential, deposition of metals was affected and the technology was not efficient enough to provide the desired performance. DES such as choline chloride:ethylene glycol in zinc electro deposition, choline chloride:oxalic

Deep Eutectic Solvent Mediated Process

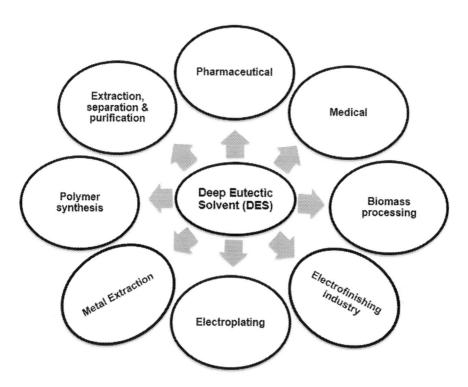

FIGURE 10.4 Various applications of deep eutectic solvents in different industries.

acid in copper plating, with high conductivity and the ability to dissolve metal salt were identified as potential candidates for metal processing, metal dissolution and metal deposition (Smith et al. 2014). The wider potential applications of DESs were used in metal processing. Using water as a solvent for electroplating has many disadvantages, although water is nontoxic. Due to inefficiency of water as the solvent, the electroplating technique requires to use additives, brighteners or even plasma and chemical vapor technique leading to increased operation cost. Sometimes, highly hazardous complexing agents are also required to be dissolved in aqueous solvents. DES with its unique properties can overcome these limitations. Electroplating using hazardous chemicals like chromium, nickel, zinc and aluminum has also been done by DES (Smith et al. 2014). Aqueous phosphoric acid or sulfuric acid along with additives is used for electropolishing commercially, but these solutions are toxic and very corrosive in nature and they release toxic gas during the process operation. These undesirable outcomes of electropolishing can be tackled by using DES. DES has proved to be noncorrosive, nontoxic and has also reduced gas release during the process, making it suitable for electropolishing. DES is also used to extract metals from different matrices and also to process formation of the metal oxides. The ability of DES to dissolve a wide variety metal oxide has helped in this application. Copper oxide and fluorides have been removed from postetching using DESs, containing choline chloride:urea and choline chloride:malonic acid (Taubert and Raghavan 2014). It has also helped in removing organic sulfides from fuels (Li et al. 2013).

Additionally, DES is also applied as a suitable media for synthesizing nanoparticles, because it helps to control the shape of nanoparticles and helps to reduce the usage of surfactants or seeds. By adjusting the water content of the DES, it can help in synthesizing nanoparticles with different particle sizes (Smith et al. 2014). It is also used in synthesizing chemicals, for instance, synthesis of nitriles from aldehydes (Patil et al. 2013), peptides (Maugeri et al. 2013), xanthanes and tetraketones (Smith et al. 2014). Furthermore, DES, comprising choline chloride:lactic acid, is demonstrated in gas adsorption study by challenging with carbon dioxide (Casal et al. 2013). Similarly, sulfur dioxide released during burning of fossil fuels can be captured by DES containing choline chloride and glycerol (Yang et al. 2013). Biotransformation is another application of DES. Generally, biotransformation is carried out by using a polar organic solvent, but DES helps in dissolving the substrate without deactivating enzyme (Smith et al. 2014). In biotransformation, DES can be used to dissolve nonpolar substrate in aqueous solution along with water and it can also be used as an alternative to nonaqueous solvent. The DES has been used to extract various chemical compounds from different sources. For example, it was selected based on its polarity to separate residual glycerol from raw biodiesel (Mbous et al. 2017) and as mobile phase in chromatographic and electrochemical analyses (Shishov et al. 2020).

10.4 DISSOLUTION AND PRETREATMENT OF LIGNOCELLULOSE BY DEEP EUTECTIC SOLVENT

The plant cell wall is primarily composed of lignocellulose biomass, which is the mixture of cellulose, hemicellulose and lignin. The content of cellulose, hemicellulose and lignin varies among different types of plants; additionally this mixture ratio depends on growth phase, environment and cultivation condition. Generally, lignocellulosic biomass contains 35%–50% of cellulose, 20%–35% hemicellulose and 10%–25% lignin (Isikgor and Remzi 2015), as shown in Table 10.4. These lignocellulosic biomass when treated with DES could yield each component during fractionation process.

Solubilizing capability of DES to solubilize lignocellulosic biomass was initially demonstrated by Francisco et al. in 2012 (Francisco et al. 2012). Several types of DESs are used for disintegration and delignification of lignocellulosic biomass and their efficiencies are depended on the ionic strength, molar ratio of acid, and strength of hydrogen bond acceptors (Zhao et al. 2018). Numbers of the current studies point to the fact that DES has good solubility for lignin than cellulose (Kumar et al. 2016, Lynam et al. 2017); however, relatively less numbers of research are focused on the dissolution of cellulose to understand the mechanism of cellulose dissolution (Zhang et al. 2020a). This scenario may be due to the fact that pretreatment using DES is still in its early stage of development and extensive researches are being carried out to explore more facts about it.

A study conducted by Alvarez-Vasco et al., in 2016, explained that the polarity and basicity of DES can affect delignification efficiency. DES has the capability to cleave ether bonds of lignin phenyl propane units selectively, without interfering C-C linkages in cellulose and thus plays a vital role in lignin extraction (Alvarez-Vasco

TABLE 10.4
Compositions of Different Types of Natural Lignocellulosic Biomass

Biomass	Cellulose	Hemicellulose	Lignin	References
Wheat straw	42.49	20.27	18.08	Tsegaye et al. (2017)
Rice straw	32.15	28.00	19.64	Shawky et al. (2011)
Corn stalk	29.80	33.30	16.65	Shawky et al. (2011)
Corn cob	32.56	38.42	15.59	Shawky et al. (2011)
Corn leaf	33.56	25.00	14.35	Shawky et al. (2011)
Sugarcane bagasse	47	16	27	de Souza Moretti et al. (2016)
Sugarcane straw	43	15	23	de Souza Moretti et al. (2016)
Napier grass	46.58	34.14	2.25	Kamarullah et al. (2015)
Durian peel	25.7	18.5	15.9	Siwina and Leesing (2020)
Cotton liner	89.7	1	2.7	Dorez et al. (2014)
Bamboo	54.6	11.4	21.7	Dorez et al. (2014)
Hemp	74.1	7.6	2.2	Dorez et al. (2014)
Grass waste	31.2	19.9	20	Yan et al. (2020)
Waste coffee grounds	33.1	30.03	24.52	Lo et al. (2017)
Rice husk	30.42	28.03	36.02	Lo et al. (2017)
Bamboo leaves	34.14	25.22	35.03	Lo et al. (2017)
Garlic skin	25.03	30.94	19.53	Ji et al. (2020)
Sunflower husk	37.3	35	22.9	Salasinska et al. (2016)
Sorghum leaves	28.56	29.18	3.94	Rorke and Kana (2016)
Water hyacinth	16.4	32.7	5.7	Rezania et al. (2019)
Switch grass	39	26	23	Arora et al. (2010)

et al. 2016). On the other hand, dissolution of cellulose can be affected by two factors: (i) strong hydrogen bond network between cellulose and DES and (ii) strong cohesive energy of cellulose fibrils (Vigier et al. 2015). Dissolution of cellulose requires a more thermodynamically stable system to be formed after the dissociation and reorganization of hydrogen bonds in cellulose and DES. The cellulose dissolution in DES depends upon the number and strength of hydrogen bonds formed by free ions and cellulose. An increase in total bond energy corresponded to high solubility of cellulose. The DES competed with the hydrogen bonds of cellulose, instead of weakening the interaction between cations and anions in hydrogen bonds of cellulose. Along with the internal hydrogen bonds, the remaining ions of the DES formed hydrogen bond between the cellulose and helped in its dissolution (Zhang et al. 2020a). Alternatively, mechanism of hemicellulose dissolution in DES is not yet clear. The studies up to date have not tried to explore the mechanism of hemicellulose dissolution in DES. Comparing with the cellulose, hemicellulose dissolution and extraction have been considered to be a secondary approach in biorefining process (Morais et al. 2020).

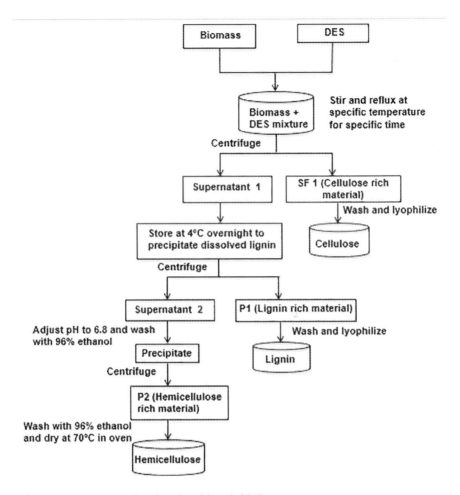

FIGURE 10.5 Biomass fractionation (Li et al. 2018).

Due to the properties of DESs in solubilization of lignocellulose components, the DES-mediated fractionation process of lignocellulose was developed. A schematic diagram of biomass fractionation undertaken is given in Figure 10.5 (Li et al. 2018). Briefly, choline chloride and lactic acid were the selected DES for the rice straw, which was mixed in different ratio, by keeping 10 wt% water content. Powdered rice straw was mixed with DES and stirred for a specific time at a specified temperature. The cellulose-rich material was recovered from solid fraction 1, after centrifugation of pretreated biomass. The recovered cellulose-rich material was washed with ethanol and water and later stored after lyophilization. The supernatant after collecting the cellulose-rich material, along with the condensed washings of cellulose-rich material, was added with water and stored at 4°C overnight for lignin precipitation. The precipitated lignin was collected by centrifugation and was washed with water and stored as lignin-rich material after lyophilization. The supernatant after collecting the lignin-rich material along with the condensed washings of precipitated lignin was used to fractionate hemicellulose-rich material. The pH was adjusted to 6.8 followed by precipitation using 96% ethanol with continuous stirring for 24 h. The hemicellulose-rich material was obtained as precipitate and was washed with 96% ethanol and oven-dried at 70°C. Thus, the biomass was fractionated by centrifugation and fractions were collected as precipitates. Also, this study proved that the addition of water in DES system helped in removing 30%–35% of lignin from rice straw (Li et al. 2018). This study demonstrated that addition of water in DESs system or aqueous DESs improved the ability to remove lignin. Also it is reported that addition of small amount of water into the mixture can reduce the preparation time, temperature and viscosity (Li et al. 2018).

With the differential capability for biomass dissolution by DES, there are several studies being conducted using DES for pretreatment process of lignocellulosic biomass. Using a series of solvent with hydrogen bond acceptors as lactic acid, malic acid and oxalic acid in combination with different hydrogen bond donors, in varying ratios, showed that the solubility of lignin and cellulose depended on hydrogen bond donor and the molar ratio of solvent (Francisco et al. 2012). In the case of solvents with malic acid as hydrogen bond donor, the lignin and cellulose solubility varied from 0.00 to 14.90 wt% and 0.00 to 0.78 wt%, respectively. Similarly, when oxalic acid was used as hydrogen bond acceptor, the solubility of lignin and cellulose varied from 0.00 to 3.62 wt% and 0.00 to 0.25 wt%, respectively. Whereas, in the lactic acid series, the lignin solubility varied from 4.55 to 12.03 wt% and no solubility for cellulose (Francisco et al. 2012). The results showed that when molar ratio of DES mixtures were varied, the cellulose and lignin dissolution were changed accordingly. Another study in 2014 depicted variations of cellulose solubilities at 1.03, 1.79 and 2.83 wt% using urea:acetamide (1:2), caprolactam:acetamide (1:1), and urea:caprolactam (1:3) as DES, respectively (Zhou and Liu 2014). These studies point to the fact that dissolution of the biomass depends upon the formulations of the DESs and pretreatment conditions.

The biomass dissolution by DES could be influenced by the pH of DES. In 2018, six types of the acidic, basic and neutral DESs were comparatively applied to delignification and biomass fractionation of oil palm empty fruit bunch (EFB). Among all tests in this work, the acidic choline chloride:lactic acid DES showed the best performance with 100% hemicellulose recovery, 88% delignification and 50% lignin extraction from EFB (Yt et al. 2018). This work was further extended to understand the effect of functional group in acidic component of DES on lignin extraction. Nine types of carboxylic acids with different functional groups as hydrogen bond donors in DES system were tested for fractionation process of EFB. The study found out that choline chloride:lactic acid (1:15) and choline chloride:formic acid (1:2) can extract more than 60 wt% of lignin. It could be concluded that biomass fractionation and lignin extraction could be improved with the presence of hydroxyl group and short alkyl chain (Yt et al. 2019). An overview of similar studies conducted using DES for lignocellulose pretreatments and fractionations has been mentioned in Table 10.5.

There are several factors that determine the sugar yield obtained from biomass processing. Studies conducted by Liu et al. revealed that the molar ratio, temperature and time of pretreatment can affect wheat straw processing. As the pretreatment time increased, lignin removal and xylan loss had increased but the cellulose

TABLE 10.5
Lignocellulose Pretreatment with Optimal Conditions and Yields by Different Types of Deep Eutectic Solvents

Biomass	Deep Eutectic Solvent Solvent 1	Solvent 2	Molar ratio	Temperature	Time	Yield	References
Eucalyptus camaldulensis	Choline chloride	Lactic acid	1:10	110°C	6 h	Glucose: 94.3%	Shen et al. (2019)
Bamboo residue	Choline chloride	Lactic acid	1:4	130°C	90 min	76.9%	Lin et al. (2020)
Corn cob	Ethylamine chloride	Lactic acid	1:1	150°C	0.5 h	Total sugar: 58.8g/L	Xu et al. (2020)
Oil palm empty fruity bunch	Choline chloride	Lactic acid	1:2	120°C	3 h	20.7%	Thi and Lee (2019)
	Choline chloride	Glycerol	1:2	120°C	3 h	20.0%	
	Choline chloride	Urea	1:2	120°C	3 h	16.9	
Sago waste	Choline chloride	Urea	1:2	110°C	3 h	Glucose yield: 5.2 mg/ml	Wan and Mun (2018)
Corn stover	Choline chloride	Glycerol	1:2	180°C	2 hours	Approx glucose: 80%; Approx xylose: 40%	Xu et al. (2018)
Lettuce leaves	Choline chloride	Glycerol	1:2	150°C	16 h	Glucose yield: 94.9% Xylose yield: 75.0%	Procentese et al. (2017)
Wheat straw	Choline chloride	Urea	1:2	80°C	24 h	94.93	Jablonský et al. (2015)
	Choline chloride	Malonic acid	1:1	60°C	24 h	90.57	
	Choline chloride	Lactic acid	1:9	60°C	24 h	85.28	
	Choline chloride	Malic acid	1:1	80°C	24 h	80.74	
	Choline chloride	Lactic acid	1:10	60°C	24 h	83.49	
	Choline chloride	Oxalic acid x 2H$_2$O	1:1	60°C	24 h	59.07	
Rice straw	Choline chloride	Lactic acid	5:1	60°C	12 h	Maximum reducing sugar yield: approx. 333 mg/g	Kumar et al. (2016)

(Continued)

TABLE 10.5 (Continued)
Lignocellulose Pretreatment with Optimal Conditions and Yields by Different Types of Deep Eutectic Solvents

Biomass	Deep Eutectic Solvent Solvent 1	Solvent 2	Molar ratio	Temperature	Time	Yield	References
Corn cob	Betaine	Lysine-W87	1:1	60°C	5 h	Glucose yield:60%	Liang et al. (2020)
	Betaine	Arginine-W82	1:1	60°C	5 h	Glucose yield: 52.2%	
Rice husk	Choline chloride	Glycerol	1:2	115°C	3 h	Glucose production enhancement: 183.3%	Gunny et al. (2015)
	Choline chloride	Ethylene glycol	1:2	115°C	3 h	Glucose production enhancement: 200%	
Corn cob	Lactic acid	Choline chloride	5:1	90°C	24 h	Glucose yield: 83.5%	Zhang et al. (2016a)
	Malonic acid	Choline chloride	1:1	90°C	24 h	Glucose yield: 61.5%	
	Ethylene glycol	Choline chloride	2:1	90°C	24 h	Glucose yield: 85.3%	
	Glycerol	Choline chloride	2:1	90°C	24 h	Glucose yield: 96.4%	
Switch grass	Acidified, aqueous Choline chloride	Glycerol	1:2	120°C	1 h	Glucose yield:88.9% Xylose yield: 98.3%	Chen et al. (2018b)
Potato peel	Choline Chloride	Glycerol	1:2	115°C	3 h	Glucose yield: 41%	Procentese et al. (2018)
Apple residue	Choline Chloride	Glycerol	1:2	115°C	3 h	Glucose yield:76%	
Coffee silver skin	Choline Chloride	Glycerol	1:2	115°C	3 h	Glucose yield:29%	
Brewer's spent grain	Choline Chloride	Glycerol	1:2	115°C	3 h	Glucose yield: 34%	
Rice Straw	90% Lactic acid	Choline chloride-water	3:1	90°C	6 h	Cellulose rich material: 56.3%; Lignose rich material: 9.1%; Hemicellulose rich material: 0.5%; Glucose: 60%; Xylose: 22%	Li et al. (2018)
Corn stover	Choline chloride	Formic acid	1:2	130°C	3 h	Glucose yield: 99%	Xu et al. (2016a)
Corn cob	Choline chloride	Imidazole	3:7	80°C	15 h	Glucose: 86%; Xylose: 63%	Procentese et al. (2015)

recovery decreased. Compared to pretreated straw, the untreated wheat straw had less digestions of cellulose and xylan due to the presence of lignin and xylan coating which hindered the hydrolysis. The increasing temperature can promote DES to break the bonds between cellulose, xylan and lignin and thereby assist the removal of lignin (Liu et al. 2019). Difference in molar ratio of DES, using choline chloride and lactic acid, in bamboo pretreatment also demonstrated that increased ratios of lactic acid in DES gained hemicellulose destruction in biomass and enhanced the subsequent enzymatic hydrolysis. It is hypothesized that the increase in lactic acid content caused a change in the pH of solvent, thereby leading to instability in the structure of hemicellulose (Lin et al. 2020). Nevertheless, in the case of a study conducted on EFB, as the molar ratio of choline chloride:lactic acid increased, the yield of reducing sugar decreased. The high acidity condition applied during the pretreatment caused the EFB to clump together, thereby reducing the surface area for enzymatic hydrolysis. This has led to reduction in sugar yield when high molar ratio of choline chloride:lactic acid was used (Thi and Lee 2019). According to the study conducted by Mamilla et al. on rice straw, it was estimated that there could be breakage between bridging bonds of hydrogen bond donor and hydrogen bond acceptor at higher temperature, since they are loosely bound chemical bonds. This helps DES to interact with –OH group of cellulose and lignin (49). Similarly, the impact of temperature was observed in pretreatment of *Eucalyptus camaldulensis* to increase lignin yield when temperature was increased suggesting the breakage of hydrogen bonds of lignocellulose fibrils at high temperatures (Shen et al. 2019). Effect of pretreatment time on biomass was also studied by Mamilla et al. and found out that as the pretreatment time increased, the amount of biomass dissolved also increased (Mamilla et al. 2019). These studies revealed that biomass dissolution depend on pretreatment time, temperature and molar ratio of DES. More research and studies are conducted on these factors to establish a clear relation between them and pretreatment efficiency to maximize the yield recovery from the process.

Besides these factors and their relationships with the pretreatment efficiency, recyclability and reusability of DES has also paved the way for its wide acceptance in biomass processing. This property can help in making biorefinery more cost-effective and economic feasible. Taking this into consideration, there are studies being carried out on recyclability of DES. Recycled DES was used by Shen et al. for four cycles in the study. The pretreatment capacity and the chemical structure of the DES were maintained even after four cycles of recycling. The study also reused the recycled DES, without any further treatment to estimate the enzymatic saccharification. Using the fourth recycled DES, a saccharification ratio of 73.8% was still attained, which was more than the untreated sample (Shen et al. 2019). This can prove the maximum utilization and efficiency of recycled DES. Since the regeneration of DES does not involve many chemical reaction, rather than disruption of hydrogen bonds between hydrogen bond acceptor and hydrogen bond donor, the recycling of DES is simple. In a study conducted on Switch grass, the DES filtrate was collected after pretreatment from biomass slurry and it was vacuum evaporated to remove the acetone, which was used along with water to wash the pretreatment slurry. The lignin, which precipitated as a result, was removed, and the supernatant was again vacuum evaporated to remove water residue and was reused for next pretreatment cycle (Chen et al. 2018b).

This recycle and reuse can help in reducing the waste generation and maximum utilization of the DES compounds. The recycled DES performed well for the first two cycles, but the performance decreased in the third cycle. This decreased efficiency is possibly due to the cumulative acid consumption during the previous pretreatment cycles, which was confirmed by the increase in pH. Hence, addition of acid into the recycled DES to correct the pH as the original solvent was demonstrated to improve the efficiency of biomass fractionation by recycled DES (Chen et al. 2018b). Another study on rice straw pretreatment applied vacuum evaporation to remove residues of ethanol and water remaining in the pretreated slurry after biomass washing, and consequently, the DES was recovered. The pretreatment by using recycled DES in this work showed 69% recovery of cellulose after five cycle reuse (Li et al. 2018). In spite of using vacuum evaporation, another group of researchers used membrane based technologies, ultrafiltration and electrodialysis to separate and recover DES after biomass fractionation (Liang et al. 2019). These studies suggested the possibility of biomass fractionation with DES-mediated process to be a cost-effective design with reduced waste generation.

Even though DES has gained much interests from researchers due to its positive effects, it also has few noticeable drawbacks. One such drawback is the viscosity of DES. High viscosity of DES hinders its usage in industries and also for many intended applications. Addition of water can reduce the viscosity, but there is limitation for the amount of water that can be added without affecting the bonding arrangements between hydrogen bond donor and hydrogen bond acceptor. Selection of hydrogen bond donor and hydrogen bond acceptor is yet another drawback in using DES. Different combination of hydrogen bond donor and hydrogen bond acceptor can lead to fractionation of different compounds. Hence, selection of hydrogen bond donor and hydrogen bond acceptor should be done carefully. A deeper knowledge on interaction of each compound of DES on the biomass is necessary to fractionate a specific compound from biomass. Since the usage of DES is in its infancy, more studies are required for the industrial-scale usage of this solvent.

10.5 DEEP EUTECTIC SOLVENT IN POLYMERIC COMPOSITE SYNTHESIS AND FABRICATION

Being biodegradable and nontoxic solvent, DES has been utilized in polymer industry as well, along with its application in many other fields. The research trend of using DES in polymer science is steadily increased during 2010–2020 based on numbers of publications (Figure 10.6). To enhance the application of DES in polymer studies, it not only requires deeper knowledge about the solvent but also should understand the interaction of DES with polymers. When a polymer is used along with a solvent, the polarity needs to be considered. The polarity of the solvent can affect the polymerization, polymer constitution, structure and its behavior (Jablonský et al. 2019). Hence, detailed study on the properties of polymer and DES is required.

DES usually gets involved in polymeric composite synthesis in various roles such as a solvent, which is not directly involved in the polymerization of monomer to polymer, but it influences in the kinetics of polymerization. DES can alter the properties

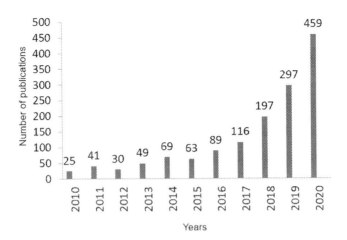

FIGURE 10.6 Publications on deep eutectic solvents and polymer science from year 2010 to 2020.

of polymer or even can get polymerized on itself (Jablonský et al. 2019). The first article on eutectic mixture being used for polymerization was published in 1985 (Genies and Tsintavis 1985). Polyaniline produced by the application of eutectic mixture at that time had better nucleation and polymerization. Many other studies were conducted since then for producing synthetic and natural polymers, but only in 2011, DES was used for polymerization studies (Mota-Morales et al. 2011). Table 10.6 has representations of few studies in polymer using DES. DES adopts different roles in polymer science as represented in Figure 10.7.

10.5.1 Application of Deep Eutectic Solvent in Polymerization and Polymer Extraction

The polymerization can be enhanced by applications of DES based on four different functions, including (a) by reduction of components for reaction, (b) through action as a substitute solution to organic solvents with enhanced performance, (c) by minimization of the waste production and (d) by recycling the residual compounds (Roda et al. 2019). Itaconic acid, obtained from biorefining process of lignocellulosic biomass, is one of the most basic building blocks to be used as a hydrogen bond donor along with DES for polymerization reactions (Jablonský et al. 2019). Table 10.7 shows examples of research conducted on polymerization of monomers using DES. DES was also used in biopolymer dissolution and mesoporous templating to synthesize chondroitin sulfate mesoporous material (Ferreira et al. 2019). The researchers suggested that using DES in the biopolymer composite synthesis helped to attain mesoporosity and they were able to maintain stable structure of polymers even after removing DES. Study on polymerization of 3-octylthiophene (P3OT) in DES portrayed the use of DES as a solvent in polymerization. The study was carried out to compare the yield of polymerization of 3-octylthiophene when prepared in a 1-butyl-3-methylimidazolium hexafluoroantimonate (ionic liquid), chloroform

TABLE 10.6
Deep Eutectic Solvent Applications in Synthesis of Polymeric Composites

Deep Eutectic Solvent with Molar ratio	Function of Deep Eutectic Solvent	Study Purpose	References
Choline chloride:urea (1:2)	Solvent	Comparison of molecular solvation of poly (vinyl pyrrolidone) in water versus deep eutectic solvent	Sapir et al. (2016)
Choline chloride:itaconic acid	Functional monomer for polymerization	Composing of solid extraction matrices	Xu et al. (2016b)
Choline chloride:ethylene glycol (1:2) Choline chloride:urea (1:2) Choline chloride:glycerol (1:2)	Solvent	Electrochemical characterization of poly(3,4-ethylenedioxythiophene) prepared from deep eutectic solvent for sensing biomarkers	Prathish et al. (2016)
Choline chloride:urea (1:2)	Solvent	Porous carbon xerogels production	Chen et al. 2018a)
Choline chloride:glycerol (1:2)	Modifier of molecular imprinted polymer	Chlorogenic acid purification from honeysuckle	Li et al. (2016)
Choline chloride:acrylic acid (1:2)	Functional monomer	Polymeric matrix Production for biomolecule recognition	Fu et al. (2019)
Choline chloride:glycerol (1:2)	Reaction media	Anionic polymerization of olefin	Sánchez-Condado et al. (2019)
Choline chloride:glycerol (1:2)	Modifier	Separation of compound cleistanthol from Phyllanthus flexuosus	Gan et al. (2016)
Choline chloride:urea (1:2) Choline chloride:glycerol (1:2)	Chitosan methylation solvent	Production of methylated chitosan	Bangde et al. (2016)
Boric acid:glycidyl trimethylammonium chloride (1:3)	Derivatization	Treatment of waste waster	Vuoti et al. (2018)
Zinc chloride:urea (3:10)	Solvent and reactant	Processing of lignin	Roda et al. (2019)
Choline chloride:mandelic acid (1:2)	Delivery agent	Production of drug delivery system	Mano et al. (2017)
Choline chloride:urea (1:2) Choline chloride:glycerol (1:2) Imidazole:glycerol (1:1)	Surface modifier	Thermoplastic starch/wood biocomposite processing	Grylewicz et al. (2019)
Choline chloride:Glycerol –hexadecyltrimethylammonium bromide	Compatibilizing agent	HDPE/agar biocomposites compatibilization	Shamsuri et al. (2014)

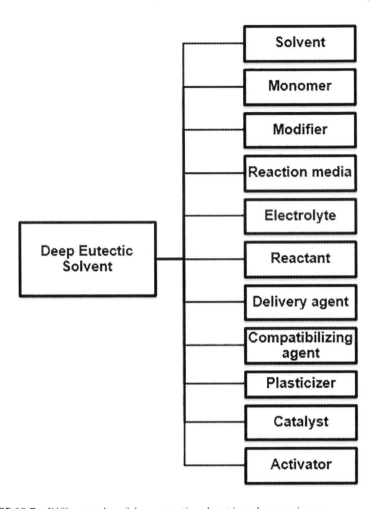

FIGURE 10.7 Different roles of deep eutectic solvent in polymer science.

(conventional polymerization) and various DES. The result revealed that after 10 h of reaction, there was 100% yield when the DES, choline chloride (Chcl) and urea were used. Whereas the polymerization using chloroform and ionic liquid could yield 87% and 99% after a long reaction time of 48 h. Hence, the study could prove that DES can enhance the polymerization yield in short duration (Park and Lee 2017).

DES has been used for production of molecular imprinted polymers. Molecular imprinted polymers (MIP) are polymer molds for setting template molecules and can be used for recognition of template molecule. DES was demonstrated to be used as a monomer for polymerization of molecular imprinted polymer. The results showed that the polymeric product was stable in shape and reusable along with characters like high imprinting factor, fast binding kinetics, and high adsorption capacity (Li et al. 2016). In the preparation of molecular imprinted polymer methodology, DES can be used as a medium, functional monomer, modifier or template. These are termed as

TABLE 10.7
Deep Eutectic Solvents Involved in Polymerization of Monomers

Deep Eutectic Solvent	Molar Ratio	Monomer	Agents Used	Polymer	References
Chcl:ethylene glycol	1:2	2-hydroxy ethyl methacrylate	Poly(ethylene glycol)diacrylate as cross-linker	Poly(2-hydroxy ethyl methacrylate	Qin and Panzer (2017)
Chcl:acrylic acid	1:2	Choline chloride:acrylic acid	N,N-methylenebisacrylamide as crosslinker	Deep eutectic solvent-molecular imprinted polymer	Sánchez-Leija et al. (2016)
Chcl: 2zinc chloride	1:2	10,16-Dihydroxyhexadecanoic acid	*Candida antartica* lipase B	Aliphatic polyester of 10,16-Dihydroxyhexadecanoic acid	Gómez-Patiño et al. (2015)
Acetamide:urea	3:1	methyl methacrylate	$FeBr_2$ as catalyst	Poly(methyl methacrylate)	Wang et al. (2017)
Caprolactam:acetamide	3:1	methyl methacrylate	$FeBr_2$ as catalyst	Poly(methyl methacrylate)	Wang et al. (2017)
Chcl:fructose	2:1	2-hydroxyethyl-methacrylate	Indomethacin as drug	Poly(2-hydroxyethyl-methacrylate)	Mukesh et al. (2016b)
Chcl:ethylene glycol	1:2	1-butyl-3-methyl-imidazolium tetrafluoro-borate	Ethylene glycol dimethacrylate as cross linker	Polymer monolith	Zhang et al. (2016b)
Chcl:ethylene glycol	1:2	Methylene blue		Poly(methylene blue)	Hosu et al. (2017a,b)
Chcl:orcinol	1:1.5	2-Hydroxyethyl-methacrylate		Polymerized 2-hydroxyethyl-methacrylate	Mukesh et al. (2016a)

TABLE 10.8
Deep Eutectic Solvent in Molecular Imprinted Polymers

Deep Eutectic Solvent	Molar Ratio	Substrate	Compound of Interest	References
Choline chloride:ethylene glycol	1:2	Urine	Chlorprenaline, bambuterol	Liang et al. (2017)
Betaine:ethylene glycol:water	1:2:1	Green bean extract	Levofloxacin	Li and Row (2017)
Choline chloride:glycerol	1:2	Milk	Chloromycetin and thiamphenicol	Li et al. (2017)
Choline chloride:acrylic acid	1:2	Human serum	Transferrin	Zhang et al. (2018)
Betain:ethylene glycol:water	1:2:1	Millet extraction	Levofloxacin and contamination	Xiaoxia et al. (2017)
Methanol +Choline chloride:glycerol	1:2	Hawthorn	Caffeic acid	Guizhen et al. (2015)
Methanol+ Choline chloride:urea				
Choline chloride:ethylene glycol	1:2	Rat urine	Aristolochic acid I,II	Yan-Hui et al. (2019)
Choline chloride:glycerol	1:2	Rat urine	Aristolochic acid I,II	
Choline chloride:urea	1:2	Rat urine	Aristolochic acid I,II	
Choline chloride:formic acid	1:2	Marine kelp	Laminarin, fucoidan	Guizhen et al. (2019)
Choline chloride:acetic acid	1:2	Marine kelp	Laminarin, fucoidan	
Choline chloride:propionic acid	1:2	Marine kelp	Laminarin, fucoidan	
Choline chloride:urea	1:2	Marine kelp	Laminarin, fucoidan	

DES-MIP (DES-molecular imprinted polymer) (Jablonský et al. 2020) (Table 10.8). DESs when used as monomer in molecular imprinted polymer, it was able to specifically interact to target substrates due to the presence of functional groups in them. Also, the higher rigidity in DES-MIP has reduced the shrinkage and swelling of biopolymers (Jablonský et al. 2020).

In addition to polymerization, DES is also used to extract polymers from raw materials other than cellulose, hemicellulose and lignin. It has been used to extract chitin, agar, pectin and many other similar biological compounds (Table 10.9). Studies are progressing on the usage of DES for extraction of polymers. The extraction of polymer depends on various factors such as acidity, viscosity of DES and the solubility of the selected polymer in the DES (Roda et al. 2019). The main principle of extraction might be the capacity of DES to diffuse into the polymer containing

TABLE 10.9
Extraction of Biological Polymers Using Deep Eutectic Solvent

Deep Eutectic Solvent	Molar Ratio	Source	Extracted Biopolymer	References
Choline chloride:oxalic acid	1:2	Rabbit hair	Keratin	Wang et al. (2018)
Choline chloride:malic acid	1:1	Shrimp shell	Chitin	Huang et al. (2018)
Choline chloride:lactic acid	1:9	Wood	Lignin	Esmaeili et al. (2018)
Choline chloride:oxalic acid	1:2	Wool	Keratin	Wang and Tang (2018)
Choline chloride:glycerol	1:2	Brown sea weed	Alginate and fucoidan	Sivagnanam et al. (2018)
Lactic acid:glucose:water	6:1:6	Pomelo	Pectin	Liew et al. (2018)
Choline chloride:oxalic acid	1:2	*Poria cocus*	Polysaccharides	Zhang et al. (2020b)
Choline chloride:lactic acid	1:5	Rice straw	Lignin	Dan et al. (2018)

matrix, and rupturing the chemical bonds of matrix, leading to diffusion of solvent. Along with this, hydrogen and electrostatic interactions between the DES and polymer of interest can be helpful in its extraction (Roda et al. 2019).

10.5.2 Application of Deep Eutectic Solvent as Plasticizer

Plasticizers are nonvolatile organic substances which are inert and have low molecular weight and low vapor pressure. These plasticizers can help in decreasing the glass transition temperature of polymers. It can also provide high thermostability and elasticity to the polymer (Roda et al. 2019). The plasticization using DES relies on aspects like composition of DES, ratio and interaction between polymer and DES and the type of polymer (Sokolova et al. 2018). Table 10.10 represents few DESs used to plasticize polymer. Sousa et al., in 2014, has conducted a study to produce agar films with more strength and elasticity using DES and low concentration of polymer (Sousa et al. 2014). The agar film was synthesized through a three-stage approach where initially the polymer was solubilized in DES. The DES used was choline chloride:glycerol and choline chloride:urea at 1:2 molar ratio. The mixture of agar and DES mixture was thermocompressed using hydraulic press after being molded into a square mold, covered with aluminum foil, between two Teflon plates. The films thus obtained were immersed in ethanol and dried in air. Upon storage, the film produced using choline chloride:glycerol produced unpleasant odor and changed its color, whereas the film produced using choline chloride:urea was stable and easy to handle. Overall, the study showed that agar film synthesized using cholin chloride:urea had good mechanical resistance increased elasticity and enhanced hydrophilicity proving the potentiality of DES and polymer mixture as a good plasticizing agent. Plasticizing effect of DES was compared with the conventional plasticizers by Sirvio et al. (2018). The conventional plasticizers used in the study were glycerol, propylene carbonate, and ethylene carbonate; the DESs involved in the study were choline chloride in combination with glycerol, glucose, urea, citric acid and tetrabutylammonium bromide:propylene carbonate and tetrabutylammonium bromide:ethylene carbonate. Their effect was

TABLE 10.10
Polymer Plasticized Using Deep Eutectic Solvent

Deep Eutectic Solvent	Molar Ratio	Plasticized Compound	References
Choline chloride:urea	1:2	Agar films	Sousa et al. (2014)
Choline chloride:lactic acid	1:1	Chitosan films	Almeida et al. (2018)
Choline chloride:glycerol	1:2	Cellulose films	Sirviö et al. (2018)
Choline chloride:glucose	1:2	Cellulose films	
Choline chloride:malic acid	1:1	Chitosan films	Andrea et al. (2018)
Choline chloride:lactic acid	1:1	Chitosan films	
Choline chloride:citric acid	1:1	Chitosan films	
Choline chloride:glycine	1:2	Chitosan films	
Xylitol:choline chloride	2:1	Thermoplastic starch	Magdalena et al.
Xylitol:glycerol	1:2	Thermoplastic starch	(2019)
Starch:choline chloride	2:1	Thermoplastic starch	
Starch:choline chloride	1:2	Thermoplastic starch	
Choline chloride:urea	1:2	Cellulose film	Wang et al. (2014)
Choline chloride:glycerol	1.5:1	Bioactive polysaccharide of *Momordica charantia*	Hakimin et al. (2018)
Choline chloride:glycerol	1:1	Bioactive polysaccharide of *Momordica charantia*	
Choline chloride:glycerol	1:1.5	Bioactive polysaccharide of *Momordica charantia*	
Choline chloride:glycerol	1:2	Bioactive polysaccharide of *Momordica charantia*	
Choline chloride:glycerol	1:3	Bioactive polysaccharide of *Momordica charantia*	

studied on composite films prepared from softwood cellulose fibers and hydroxythyl cellulose. The tensile strength, elongation at break and effect of humidity on composites were compared between conventional plasticizers and DES. Among the conventional plasticizers used, glycerol could increase the elongation at break by 53% whereas among the DESs used, choline chloride based DESs exhibited increase in elongation at break up to 81%. When compared between conventional plasticizers and DES, DES was proved to be a better plasticizer with the capability to improve the conductivity and thermos-formability of the composite. Another study by Wang et al., in 2015, also focused on using eco-friendly DES as a plasticizer for regenerated cellulose films (Wang et al. 2014). The group used choline chloride:urea (1:2) as plasticizer. The significant increase in the elongation from 25.92% to 34.88% confirms the efficient plasticizing effect of choline chloride:urea. Even though the there was a decrease in tensile strength with increasing content of choline chloride:urea, 31.34 MPa tensile strength could be attained with 25.92% elongation, reaffirming the ability to use DES as plasticizer for cellulose films.

In addition to cellulose films, chitosan films were also studied using choline chloride:lactic as DES and curcumin (Almeida et al. 2018). The study focused on the effect of DES and curcumin on physical properties of chitosan films and compared it with pure chitosan films. The chitosan films were produced by adding different concentration of DES and curcumin on chitosan solution. The addition of DES into chitosan has increased the flexibility of the films without compromising its transparency. Addition of curcumin has increased the tensile strength, elongation at break and water vapor permeability. When 10% DES was added alone the tensile strength and elongation at break were 20.5 Mpa and 3.6%, respectively. These values increased to 29.3 Mpa tensile strength and 5.7% elongation at break by addition of 0.5% curcumin along with 10% DES. These chitosan films can be used in food packaging. A study conducted on bioactive polysaccharides of *Momordica charantia* to produce bio-based plastics, used DES, choline chloride:glycerol in varying molar ratio as plasticizer. Tensile strain, antimicrobial and antioxidant activity were studied for the synthesized bioplastics. The highest tensile strain was displayed by the bioactive polysaccharide used along with choline chloride:glycerol (1.5:1) and lowest was displayed by bioactive polysaccharide in combination with choline chloride:glycerol (1:2). Antioxidant activity was more for polysaccharide formulated with choline chloride:glycerol (1:3). Composites contained bioactive polysaccharide with choline chloride:glycerol (1.5:1) also displayed high antimicrobial activity against few Gram-positive and Gram-negative bacteria. This study could prove that bioplastics when prepared using DES as plasticizer have a wide range of applications from food industry to nutraceuticals (Hakimin et al. 2018).

10.5.3 Application of Deep Eutectic Solvent in Lignocellulose and Natural Biopolymeric Composite

DES is used in synthesis of cellulose nanofibrils. Cellulose nanofibrils are delicate fibrils of cellulose obtained by applying mechanical process on plants, woods, algae, etc. However, isolating cellulose nanofibrils requires a huge amount of energy and it is two-step process. Initially, the biomass needs to be pulped or bleached using various methods. This will ensure that only intact fibers are isolated and also help in partial removal of noncellulosic material. The second step involves chemical or biological treatment along with mechanical treatment. This allows the isolated fibrils to be fragmented into cellulose nanofibrils. However, the biological treatment is the time-consuming and chemical treatments are costly and toxic. Hence, DESs are used to reduce the cost of production. A study conducted on ramie fiber for nanocellulose synthesis showed that choline chloride:oxalic acid dihydrate could induce morphological changes in ramie fiber during pretreatment from amorphous to crystalline structure (Wang et al. 2020). DES containing choline chloride:urea (1:2) has been used as a media for pretreatment of birch biomass at 100°C, without hydrolysis for nanofibrillation of birch cellulose pulp by using micro fluidizer (Antti et al. 2015). In addition to this, DES could also be used to synthesize high strength cellulose nanofibers. The pretreatment using choline chloride:imidazole had resulted in swelling of cellulose fibers, causing an increase in diameter from 18.1 to 18.9 μm under mild conditions of 60°C for 15 min. DES could induce swelling, even before the disintegration of cellulose into cellulose nanofibers (Antti et al. 2015). Another study has used imidazole and glycerol at 1:1 molar ratio for manufacturing thermoplastic starch/wood fiber composite. In this work, the imidazole/glycerol could act as a strong wood fiber surface modifier as well as starch plasticizer (Grylewicz et al. 2019). Study by Adamus et al. (2018) on composite material has used DES as plasticizer with urea intercalated montmorillonite to synthesize a composite by extrusion and thermocompression techniques. Choline chloride:urea (1:2) and choline chloride:imidazole (3:7) were used as DES. This composite could be used to make films, foams or injection molded goods (Jakub et al. 2018, Antti et al. 2020).

When most of the research concentrated on removal of lignin and hemicellulose from biomass and utilization of cellulose, Akond and Lynam, in 2020, focused on using lignin in biopolymer synthesis (Rahman and Lynam 2020). In this work, lignin, as a byproduct from cellulose separation, was applied in polymer synthesis as plasticizer. Initially, lignin was extracted from waste biomass such as coffee chaff and sugarcane bagasse using formic acid:choline chloride (2:1) as DES at condition of 152°C for 2 h and 15 min with 16.67% yield of lignin recovery from biomass. A cement paste of Ordinary Portland cement mixed with salt water was prepared and the extracted lignin was added to the cement paste. Salt water was added instead of tap water to prepare cement paste as this composite was aimed to be used in locations nearby ocean where availability of portable water is limited. This lignin-cement composite paste was then tested for its plasticity, porosity and compressive strength. The result demonstrated that addition of 0.3% lignin to cement paste has increased fluidity of cement. The slump area testing using different composite cement batches was carried out. The 2% and 4% carboxymethylated cement could spread only to above 20 cm^2 approximately, whereas the cements prepared from sugarcane bagasse lignin and coffee chaff lignin were above 70 and 60 cm^2, respectively. This implies that the lignin has plasticizing properties. This increased plasticity of the cement by addition of lignin could be due to the large molecular weight of lignin. The viscosity of the cement was decreased by the addition of lignin to the paste resulting in increased plasticity. High compressive strength of biocomposites at approximately 35 MPa and 30 MPa was produced by mixing with sugarcane bagasse lignin and coffee chaff lignin, respectively, along with low porosity was yet another obtained property by addition of lignin in cement. When compared with sugarcane bagasse lignin and coffee chaff lignin, sugarcane bagasse lignin is more promising plasticizer than coffee lignin, because of its low porosity. Low porosity implies to reduced degradation. Hence their studies could show the potential of lignin application as plasticizer (Rahman and Lynam 2020).

DES has also been used to as plasticizer in starch based films by casting from aqueous solutions (Magdalena and Johansson 2016). The study has been conducted on native potato starch and hydroxypropylated and oxidized potato starch. In this work, various types of DES were synthesized to act as plasticizing agent and cross-linking agent. Different DESs containing choline chloride, urea, glycerol and choline dihydrogen citrate were used at different combinations and ratios. Starch-based films were synthesized using different DESs and their mechanical barrier properties, water vapor barrier properties and tensile strength were also tested. Gelatinized starch film without plasticizer exhibited maximum tensile strength of 46.4 MPa, which dropped to 5.0 MPa by addition of glycerol into starch material and elongation at break increased to approximately 95%. Even though gelatinized starch film had high tensile strength, it was stiff and uneven. The film prepared from native potato starch when used in combination with choline chloride:glycerol had maximum tensile strength of 6.1 MPa which was twice more than film produced by starch used in combination with choline chloride:urea. When the choline chloride was changed with choline dihydrogen citrate the mechanical properties of film had improved. For instance, the film produced using choline dihydrogen citrate and glycerol at higher ratio (1:8) had more elasticity (31.0%) and reduced tensile strength (6 MPa). Also, the film produced from choline chloride:urea had more elongation at break at about 73.5% and reduced tensile strength of about 3.0 MPa implying that the starch based films which was plasticized by DES, along with their cross-linking activity can be used in films, coatings, paper industry, etc. (Magdalena and Johansson 2016). A DES containing choline chloride, glycerol and polyethylene glycol was used to synthesize a lignin containing rigid polyurethane foam (Bailiang et al. 2020). The study revealed that the hydrogen bonds between lignin and ternary DES resulted in dissolution of lignin in DES. Also the synthesized lignin-containing polyurethane foam exhibited good mechanical properties and had a lignin replacement rate up to 58.6%, which was the maximum amount of lignin replacement rate for lignin-containing rigid polyurethane foam.

Despite being hydrophilic in nature with weak mechanical and barrier properties, starch has been used in pharmaceutical and many other industries. Starch is thermoplasticized before its being utilized by using water, glycerol, polyalcohols, amines, amides, polycarboxylic acid or ionic liquids. Yet, these plasticizers have their own drawbacks. They make the starch brittle, stiff and recrystallizable, which are unwanted characteristics (Magdalena et al. 2016). Hence, DES has been used as plasticizer. Several research tested different types of DESs such as choline chloride:urea, choline chloride:glycerol, choline chloride:citric acid as plasticizer. The starch that was plasticized using choline chloride:urea was demonstrated to have mechanical properties very similar to low density polyethylene. The ability of plasticization or dissolution of starch was attributed to the presence of water content in the polymerization system (Abbott et al. 2012). A study conducted on starch plasticization using DESs, including choline chloride:imidazole, glycerol:imidazole, citric acid:imidazole, malic acid:imidazole and choline chloride:urea at different ratios of mixtures. Mixed starch and these DESs were used plasticizing test in biocomposite synthesis. The results demonstrated that the starch in the deep eutectic system containing imidazole and choline chloride or glycerol can be easily transformed into thermoplasticized product. Whereas, the DES containing imidazole and citric acid or malic acid displayed very little dissolution of starch (Magdalena et al. 2016). The deep eutectic system can cause the breakage of internal hydrogen bonding making biocomposite become less crystalline in texture. Choline chloride and urea were taken as plasticizing agent and salt was the modifier for biocomposite production as demonstrated by another study using corn starch as matrix. This thermoplastic corn starch was recyclable and compostable. This makes it eco-friendlier and prevents toxicity to the environment (Abbott et al. 2012). Similarly, instead of starch, chitosan is also used to make biofilm composites by thermos-compression molding using DES (Galvis-Sánchez et al. 2016). The DES used here consists of choline chloride and citric acid (1:1). The chitosan sample was prepared by mixing with DES. The fabricated film was then casted by compression molding technique. Using of DES in the preparation of biofilms has helped it in enhancing the mechanical, water resistance, microstructure and thermal properties of the biofilm produced. Addition of choline chloride reduced the brittleness of chitosan-citric acid biofilm and also improved the elongation. Being environmental friendlier and cheaper, the DES has the added advantage of adjusting the nature and content of its hydrogen bond donor and making it more potential candidate in utilizing it for producing biofilms, biocomposites and much more (Eric et al. 2012).

10.6 CONCLUSION

DES, with its unique properties, has already been used in many fields for different purposes. They have attracted researches towards them, with their two main properties: low cost and environment friendliness. The ease of DES synthesis is yet another property which promotes their usage in different fields. Still, usage of DES is in its infancy and awaits much more innovative applications for them. Mixing of different hydrogen bond donors and hydrogen bond acceptors may provide new insight into unique properties that can be utilized in wide variety of fields. In spite of its application in extraction and polymerization, more detailed study of DES can provide new ideas on modifying existing polymers with DES. Being labeled as a "green solvent," more study needs to be concentrated on recyclability and reusability of DES in order to promote its usage in industrial scale. There are only limited number of studies focusing on the aspect of recycling and reusing DES. If DES proves to be recycled and reused for several cycles by minimizing wastage, it would bring a dramatic increase in its usage for industrial applications as it will reduce the cost of production again and can contribute to the profit.

ACKNOWLEDGMENTS

The authors would like to thank King Mongkut's University of Technology, North Bangkok (Research University Grant No. KMUTNB-BasicR-64-37, KMUTNB-PHD-64-02) and Srinakarinwirot University (Research University Grant No. 671/2563) for financial support of this work.

REFERENCES

Abbott, A.P., Ballantyne A.D., Conde J.P., Ryder K.S., Wise W.R. 2012. Salt modified starch: Sustainable, recyclable plastics. *Green Chem.* 14 (5), 1302–1307. https://doi.org/10.1039/C2GC16568F.

Abbott A.P., Boothby D., Capper G., Davies D.L., Rasheed R.K. 2004. Deep eutectic solvents formed between choline chloride and carboxylic acids: versatile alternatives to ionic liquids. *J. Am. Chem. Soc.* 126, 9142–9147. https://doi.org/10.1021/ja048266j.

Abbott A.P., Capper G., Davies D.L., Rasheed R.K., Tambyrajah V. 2003. Novel solvent properties of choline chloride/urea mixtures. *Chem. Commun.* 70–71. https://doi.org/10.1039/B210714G.

Aguilar-Reynosa A., Romaní A., Rodríguez-Jasso R.Ma., Aguilar C.N., Garrote G., Ruiz H.A. 2017. Microwave heating processing as alternative of pretreatment in second- generation bio refinery: An overview. *Ener. Conver.Manag.* 136, 50–65. https://doi.org/10.1016/j.enconman.2017.01.004

Akkharasinphonrat R., Douzou T., Sriariyanun M. 2017. Development of ionic liquid utilization in biorefinery process of lignocellulosic biomass. *KMUTNB Int. J. Appl. Sci. Technol.* 10(2), 89–96.

Almeida C.M.R., Magalhães, J.M.C.S., Souza H.K.S., Gonçalves M.P. 2018. The role of choline chloride- based deep eutectic solvent and curcumin on chitosan films properties. *Food Hydrocolloids.* 18, 456–466.

Alvarez-Vasco C., Ma R., Quintero M., Guo M., Geleynse S., Ramasamy K.K., Wolcott M., Zhang X. 2016. Unique low-molecular-weight lignin with high purity extracted from wood by deep eutectic solvents (DES): A source of lignin for valorization. *Green Chem.* 18, 5133–5141. https://doi.org/10.1039/C6GC01007E.

Andrea G., Castro M.C., Biernacki K., Gonçalves M., Souza H. 2018. Natural deep eutectic solvents as green plasticizers for chitosan thermoplastic production with controlled/desired mechanical and barrier properties. *Food Hydrocolloids* 8 (April). https://doi.org/10.1016/j.foodhyd.2018.04.026.

Antti S.J., Visanko M., Liimatainen H. 2015. Deep eutectic solvent system based on choline chloride- urea as a pre-treatment for nanofibrillation of wood cellulose. *Green Chem.* 17(6), 3401–3406. https://doi.org/10.1039/C5GC00398A.

Antti S.J., Hyypiö K., Asaadi S., Junka K., Liimatainen H. 2020. High-strength cellulose nanofibers produced via swelling pretreatment based on a choline chloride–imidazole deep eutectic solvent. *Green Chem.* 22(5), 1763–1775. https://doi.org/10.1039/C9GC04119B.

Arora R., Manisseri C., Li C., Ong M.D., Scheller H.V., Vogel K., Simmons B.A., Singh S. 2010. Monitoring and analyzing process streams towards understanding ionic liquid pretreatment of switchgrass (Panicum virgatum L.). *BioEner. Res.* 3, 134–145. https://doi.org/10.1007/s12155-010-9087-1.

Bangde P.S., Jain R., Dandekar P. 2016. Alternative approach to synthesize methylated chitosan using deep eutectic solvents, biocatalyst and "green" methylating agents. *ACS Sustainable Chem. Eng.* 4, 3552–3557. https://doi.org/10.1021/acssuschemeng.6b00653.

Baghaei B. and Skrifvars M. 2020. All-cellulose composites: A review of recent studies on structure, properties and applications. *Molecules* 25. https://doi.org/10.3390/molecules25122836.

Bailiang X., Yang Y., Tang R., Xue D., Sun Y., Li X. 2020. Efficient dissolution of lignin in novel ternary deep eutectic solvents and its application in polyurethane. *Int. J. Biol. Macromol.* 164 (December), 480–488. https://doi.org/10.1016/j.ijbiomac.2020.07.153.

Baruah J., Nath B.K., Sharma R., Kumar S., Deka R.C., Baruah D.C., Kalita E. 2018. Recent trends in the pretreatment of lignocellulosic biomass for value-added products. *Front. Energy Res.* 6. https://doi.org/10.3389/fenrg.2018.00141.

Bhat A.H., Dasan Y., Khan I., Jawaid M. 2017. Cellulosic biocomposites: Potential materials for future, in: *Green Energy and Technology.* pp. 69–100. https://doi.org/10.1007/978-3-319-49382-4_4.

Bledzki A, Jaszkiewicz A, Urbaniak M, Walczak D. 2012. Biocomposites in the Past and in the Future. Fibres and Textiles in Eastern Europe 96, 15–22.

Casal M.F., Bruinhorst A.V.D., Zubeir L.F., Peters C.J., Kroon M.C. 2013. A new low transition temperature mixture (LTTM) formed by choline chloride + lactic acid: characterization as solvent for CO_2 capture. *Fluid Phase Equilib.* 340, 77–84. https://doi.org/10.1016/j.fluid.2012.12.001.

Cheenkachorn K, Douzou T, Roddecha S, Tantayotai P, Sriariyanun M. 2016. Enzymatic saccharification of rice straw under influence of recycled ionic liquid pretreatments. *Ener. Procedia.* 100, 160–165.

Chen L., Deng J., Hong S., Lian H. 2018a. Deep eutectic solvents-assisted cost- effective synthesis of nitrogen-doped hierarchical porous carbon xerogels from phenol- formaldehyde by two-stage polymerization. *J. Sol-Gel Sci. Technol.* 86. doi:10.1007/s10971-018-4660-8.

Chen Z., Reznicek W.D., Wan C. 2018b. Deep eutectic solvent pretreatment enabling full utilization of switchgrass. *Bioresour. Technol.* 263, 40–48. https://doi.org/10.1016/j.biortech.2018.04.058.

Cheng Y.S., Mutrakulcharoen P., Chuetor S., Cheenkachorn K., Tantayotai P., Panakkal E.J., Sriariyanun M. 2020. Recent situation and progress in biorefining process of lignocellulosic biomass: Toward green economy. *Appl. Sci. Eng. Prog.* 13(4), 299–311.

Dan L., Yan X., Zhuo S., Si M., Liu M., Wang S., Ren L., Chai L., Shi Y. 2018. Pandoraea Sp. B-6 assists the deep eutectic solvent pretreatment of rice straw via promoting lignin depolymerization. *Bioresour. Technol.* 257, 62–68. https://doi.org/10.1016/j.biortech.2018.02.029.

de Souza Moretti M.M., Perrone O.M., da Costa Carreira Nunes C., Taboga S., Boscolo M., da Silva R., Gomes E. 2016. Effect of pretreatment and enzymatic hydrolysis on the physical-chemical composition and morphologic structure of sugarcane bagasse and sugarcane straw. *Bioresour. Technol.* 219, 773–777. https://doi.org/10.1016/j.biortech.2016.08.075.

Dorez G., Ferry L., Sonnier R., Taguet A., Lopez-Cuesta J.M. 2014. Effect of cellulose, hemicellulose and lignin contents on pyrolysis and combustion of natural fibers. *J. Anal. Appl. Pyrol.* 107, 323–331.

Emami S. and Shayanfar A. 2020. Deep eutectic solvents for pharmaceutical formulation and drug delivery applications. *Pharm. Dev. Technol.* 25, 779–796. https://doi.org/10.1080/10837450.2020.1735414.

Eric L., Decaen P., Jacquet P., Coativy G., Pontoire B., Reguerre A.L., and Lourdin, D. 2012. Deep eutectic solvents as functional additives for starch based plastics. *Green Chem.* 14 (11), 3063–3066. https://doi.org/10.1039/C2GC36107H.

Esmaeili M., Anugwom I., Mänttäri M., Kallioinen M. 2018. Utilization of DES-lignin as a bio- based hydrophilicity promoter in the fabrication of antioxidant polyethersulfone membranes. *Membranes* 8, 80.

Ferreira V.R.A., Azenha M.A., Pinto A.C., Santos P.R.M., Pereira C.M., Silva A.F. 2019. Development of mesoporous polysaccharide/sol-gel composites with two different templating agents: Surfactants and choline chloride-based deep eutectic solvents. *Express Polym. Lett.* 13, 261–275.

Francisco M., Bruinhorst A.V.D., Kroon M.C. 2012. New natural and renewable low transition temperature mixtures (LTTMs): Screening as solvents for lignocellulosic biomass processing. *Green Chem.* 14, 2153–2157. https://doi.org/10.1039/C2GC35660K.

Fu N., Li L., Liu K., Kim C.K., Li J., Zhu T., Li J., Tang B. 2019. A choline chloride-acrylic acid deep eutectic solvent polymer based on Fe_3O_4 particles and MoS_2 sheets (poly(ChCl-AA DES)@Fe_3O_4@MoS_2) with specific recognition and good antibacterial properties for β-lactoglobulin in milk. *Talanta* 197, 567—577. https://doi.org/10.1016/j.talanta.2019.01.072.

Galbe M. and Wallberg O. 2019. Pretreatment for bio refineries: A review of common methods for efficient utilization of lignocellulosic materials. *Biotechnol. Biofuels* 12, 294. https://doi.org/10.1186/s13068-019-1634-1.

Galvis-Sánchez A.C., Sousa A.M.M., Gonçalves L.H.M.P., Souza H.K.S. 2016. Thermocompression molding of chitosan with a deep eutectic mixture for biofilms development. *Green Chem.* 18(6), 1571–1580. https://doi.org/10.1039/C5GC02231B.

Gan K., Tang W., Zhu T., Li W., Wang H., Liu X. 2016. Enhanced extraction of cleistanthol from Phyllanthus flexuosus by deep eutectic solvent-modified anion-exchange resin. *J. Liq. Chromatogr. Relat. Technol.* 39, 882–888. https://doi.org/10.1080/10826076.2017.1278609.

Genies E.M. and Tsintavis C. 1985. Redox mechanism and electrochemical behaviour or polyaniline deposits. *J. Electroanal. Chem. Interfacial Electrochem.* 195, 109–128. https://doi.org/10.1016/0022-0728(85)80009-7.

Gómez-Patiño M.B., Gutiérrez-Salgado D.Y., García-Hernández E., Mendez-Mendez J.V., Adame J.A.A., Campos-Terán J., Arrieta-Baez D. 2015. Polymerization of 10,16- dihydroxyhexadecanoic acid, main monomer of tomato cuticle, using the lewis acidic ionic liquid choline chloride·2ZnCl2. *Front. Mater.* 2. https://doi.org/10.3389/fmats.2015.00067.

Grylewicz A., Spychaj T., Zdanowicz M. 2019. Thermoplastic starch/wood biocomposites processed with deep eutectic solvents. *Compos. Part A Appl. Sci. Manuf.* 121, 517–524. https://doi.org/10.1016/j.compositesa.2019.04.001.

Guizhen L., Dai Y., Wang X., Row K. 2019. Molecularly imprinted polymers modified by deep eutectic solvents and ionic liquids with two templates for the simultaneous solid-phase extraction of fucoidan and laminarin from marine Kelp. *Anal. Lett.* 52(3), 511–525. https://doi.org/10.1080/00032719.2018.1471697.

Guizhen L., Tang W., Cao W., Wang Q., Zhu T. 2015. Molecularly imprinted polymers combination with deep eutectic solvents for solid-phase extraction of caffeic acid from hawthorn. *Se Pu Chin. J. Chromatogr.* 33(8), 792–798. https://doi.org/10.3724/sp.j.1123.2015.03025.

Gunny A.A.N., Arbain D., Nashef E.M., Jamal P. 2015. Applicability evaluation of deep eutectic solvents- cellulase system for lignocellulose hydrolysis. *Bioresource Technology* 181, 297–302. https://doi.org/10.1016/j.biortech.2015.01.057.

Guo H., Chang Y., Lee D.J. 2018. Enzymatic saccharification of lignocellulosic bio refinery: Research focuses. *Bioresour. Technol.* 252, 198–215. https://doi.org/10.1016/j.biortech.2017.12.062.

Hakimin S.M., Samsudin D., Yusof R., Gan C.Y. 2018. Characterization of bio-based plastic made from a mixture of momordica charantia bioactive polysaccharide and choline chloride/glycerol based deep eutectic solvent. *Int. J. Biol. Macromol.* 118 (Pt A), 1183–1192. https://doi.org/10.1016/j.ijbiomac.2018.06.103.

Hosu O., Barsan M., Cristea C., Sandulescu R., Brett C. 2017a. Nanocomposites based on carbon nanotubes and redox-active polymers synthesized in a deep eutectic solvent as a new electrochemical sensing platform. *Microchim. Acta* 184. https://doi.org/10.1007/s00604-017-2420-z.

Hosu O., Barsan M., Cristea C., Sandulescu R., Brett C. 2017b. Nanostructured electropolymerized poly(methylene blue) films from deep eutectic solvents. Optimization and characterization. *Electrochim. Acta* 232, 285–295. https://doi.org/10.1016/j.electacta.2017.02.142.

Huang W.C., Zhao D., Guo N., Xue C., Mao X. 2018. Green and facile production of chitin from crustacean shells using a natural deep eutectic Solvent. *J. Agric. Food Chem.* 66, 11897–11901. https://doi.org/10.1021/acs.jafc.8b03847.

Isikgor F. and Remzi B.C. 2015. Lignocellulosic biomass: A sustainable platform for the production of bio-based chemicals and polymers. *Polym. Chem.* 6, 4497–4559. https://doi.org/10.1039/C5PY00263J.

Jablonský M., Majová V., Šima J., Hrobonová K., Lomenova A. 2020. Involvement of deep eutectic solvents in extraction by molecularly imprinted polymers—A minireview. *Crystals* 10, 217. https://doi.org/10.3390/cryst10030217.

Jablonský M., Škulcová A., Kamenská L., Vŕska M., Šíma J. 2015. Deep eutectic solvents: Fractionation of wheat straw. *Bioresources* 10, 8039–8047.

Jablonský M., Škulcová A., Šima J. 2019. Use of deep eutectic solvents in polymer chemistry- A review. *Molecules* 24. https://doi.org/10.3390/molecules24213978.

Jakub A., Spychaj T., Zdanowicz M., Jędrzejewski R. 2018. Thermoplastic starch with deep eutectic solvents and montmorillonite as a base for composite materials. *Ind. Crops Prod.* 123, 278–284. https://doi.org/10.1016/j.indcrop.2018.06.069.

Ji Q., Yu X., Yagoub A.E., Chen L., Zhou C. 2020. Efficient removal of lignin from vegetable wastes by ultrasonic and microwave-assisted treatment with ternary deep eutectic solvent. *Ind. Crops Prod.* 149, 112357.

Kamarullah S.H., Mydin M.M., Noor B.H.M., Alias N.Z.A., Manap S., Mohamad R. 2015. Surface morphology and chemical composition of napier grass fibers. *Malays. J. Anal. Sci.* 19, 7.

Kohli R. 2019. Chapter 16- Applications of ionic liquids in removal of surface contaminants, in: Kohli, R., Mittal, K.L. (Eds.), *Developments in Surface Contamination and Cleaning: Applications of Cleaning Techniques*. Elsevier, Amsterdam, pp. 619–680. https://doi.org/10.1016/B978-0-12-815577-6.00016-5.

Kumar A.K., Parikh B.S., Pravakar M. 2016. Natural deep eutectic solvent mediated pretreatment of rice straw: Bio analytical characterization of lignin extract and enzymatic hydrolysis of pretreated biomass residue. *Environ. Sci. Pollut. Res. Int.* 23, 9265–9275. https://doi.org/10.1007/s11356-015-4780-4.

Longo Jr. L.S., Craveiro M.V., Longo Jr. L.S., Craveiro M.V. 2018. Deep eutectic solvents as unconventional media for multicomponent reactions. *J. Braz. Chem. Soc.* 29, 1999–2025. https://doi.org/10.21577/0103-5053.20180147.

Li A.L., Hou X.D., Lin K.P., Zhang X., Fu M.H. 2018. Rice straw pretreatment using deep eutectic solvents with different constituents molar ratios: Biomass fractionation, polysaccharides enzymatic digestion and solvent reuse. *J. Biosci. Bioeng.* 126, 346–354. https://doi.org/10.1016/j.jbiosc.2018.03.011.

Liang S, Yan H, Cao J, Han Y, Shen S, Bai L. 2017. Molecularly imprinted phloroglucinol-formaldehyde- melamine resin prepared in a deep eutectic solvent for selective recognition of clorprenaline and bambuterol in urine. *Anal. Chim. Acta* 951, 68–77. https://doi.org/10.1016/j.aca.2016.11.009.

Liang X., Fu Y., Chang J. 2019. Effective separation, recovery and recycling of deep eutectic solvent after biomass fractionation with membrane-based methodology. *Sep. Purif. Technol.* 210, 409–416. https://doi.org/10.1016/j.seppur.2018.08.021.

Liang Y., Duan W., An X., Qiao Y., Tian Y., Zhou H. 2020. Novel betaine-amino acid based natural deep eutectic solvents for enhancing the enzymatic hydrolysis of corncob. *Bioresour. Technol.* 310, 123389. https://doi.org/10.1016/j.biortech.2020.123389.

Li C., Li D., Zou S., Li Z., Yin J., Wang A., Cui Y., Yao Z., Zhao Q. 2013. Extraction desulfurization process of fuels with ammonium-based deep eutectic solvents. *Green Chem.* 15, 2793–2799. https://doi.org/10.1039/C3GC41067F.

Liew, S.Q., Ngoh G.C., Yusoff R., Teoh W.H. 2018. Acid and deep eutectic solvent (DES) extraction of pectin from pomelo (Citrus Grandis (L.) Osbeck) Peels. *Biocatal. Agr. Biotechnol.* 13, 1–11. https://doi.org/10.1016/j.bcab.2017.11.001.

Li G., Wang W., Wang Q., Zhu T. 2016. Deep eutectic solvents modified molecular imprinted polymers for optimized purification of chlorogenic acid from honeysuckle. *J. Chromatogr. Sci.* 54, 271–279. https://doi.org/10.1093/chromsci/bmv138.

Li G., Zhu T., Row K.H. 2017. Deep eutectic solvents for the purification of chloromycetin and thiamphenicol from milk. *J Sep Sci* 40, 625–634. https://doi.org/10.1002/jssc.201600771.

Li X. and Row K.H. 2017. Application of deep eutectic solvents in hybrid molecularly imprinted polymers and mesoporous siliceous material for solid-phase extraction of levofloxacin from green bean extract. *Anal. Sci. Int. J. Jpn. Soc. Anal. Chem.* 33, 611–617. https://doi.org/10.2116/analsci.33.611.

Lin W., Xing S., Jin Y., Lu X., Huang C., Yong Q. 2020. Insight into understanding the performance of deep eutectic solvent pretreatment on improving enzymatic digestibility of bamboo residues. *Bioresour. Technol.* 306, 123163. https://doi.org/10.1016/j.biortech.2020.123163.

Liu Y., Zheng J., Xiao J., He X., Zhang K., Yuan S., Peng Z., Chen Z., Lin X. 2019. Enhanced enzymatic hydrolysis and lignin extraction of wheat straw by triethylbenzyl ammonium chloride/lactic acid-based deep eutectic solvent pretreatment. *ACS Omega* 4, 19829–19839. https://doi.org/10.1021/acsomega.9b02709.

Lo S.L., Huang Y.F., Chiueh P.T., Kuan W.H. 2017. Microwave pyrolysis of lignocellulosic biomass. *Ener. Procedia* 105, 41–46.

Lynam J.G., Kumar N., Wong M.J. 2017. Deep eutectic solvents' ability to solubilize lignin, cellulose, and hemicellulose; thermal stability; and density. *Bioresour. Technol.* 238, 684–689. https://doi.org/10.1016/j.biortech.2017.04.079.

Magdalena Z. and Johansson C. 2016. Mechanical and barrier properties of starch- based films plasticized with two- or three component deep eutectic solvents. *Carbohydr. Polym.* 151, 103–112. https://doi.org/10.1016/j.carbpol.2016.05.061.

Magdalena Z., Staciwa P., Jędrzejewski R., Spychaj T. 2019. Sugar alcohol-based deep eutectic solvents as potato starch plasticizers. *Polymers* 11(9). https://doi.org/10.3390/polym11091385.

Magdalena Z., Spychaj T., Mąka H. 2016. Imidazole-based deep eutectic solvents for starch dissolution and plasticization. *Carbohydr. Polym.* 140, 416–423. https://doi.org/10.1016/j.carbpol.2015.12.036.

Miao C. and Hamad W.Y. 2013. Cellulose reinforced polymer composites and nanocomposites: A critical review. *Cellulose* 20, 2221–2262. https://doi.org/10.1007/s10570-013-0007-3.

Mamilla J.L.K., Novak U., Grilc M., Likozar B. 2019. Natural deep eutectic solvents (DES) for fractionation of waste lignocellulosic biomass and its cascade conversion to value-added bio-based chemicals. *Biomass Bioenergy* 120, 417–425.

Mano F., Martins M., Sá-Nogueira I., Barreiros S., Borges J.P., Reis R.L., Duarte A.R.C., Paiva A. 2017. Production of electrospun fast-dissolving drug delivery systems with therapeutic eutectic systems encapsulated in gelatin. *AAPS PharmSciTech* 18, 2579–2585. https://doi.org/10.1208/s12249-016-0703-z.

Maugeri Z., Leitner W., de María P.D. 2013. Chymotrypsin-catalyzed peptide synthesis in deep eutectic solvents. *Eur. J. Org. Chem.* 2013, 4223–4228. https://doi.org/10.1002/ejoc.201300448.

Mbous Y.P., Hayyan M., Hayyan A., Wong W.F., Hashim M.A., Looi C.Y. 2017. Applications of deep eutectic solvents in biotechnology and bioengineering—Promises and challenges. *Biotechnol. Adv.* 35, 105–134. https://doi.org/10.1016/j.biotechadv.2016.11.006.

Morais E.S., Lopes A.M.D.C., Freire M.G., Freire C.S.R., Coutinho J.A.P., Silvestre A.J.D. 2020. Use of ionic liquids and deep eutectic solvents in polysaccharides dissolution and extraction processes towards sustainable biomass valorization. *Molecules* 25. https://doi.org/10.3390/molecules25163652.

Mota-Morales J.D., Gutiérrez M.C., Sanchez I.C., Luna-Bárcenas G., del Monte F., 2011. Frontal polymerizations carried out in deep-eutectic mixtures providing both the monomers and the polymerization medium. *Chem. Commun.* 47, 5328–5330. https://doi.org/10.1039/C1CC10391A.

Mukesh C., Gupta R., Srivastava D.N., Nataraj S.K., Prasad K. 2016a. Preparation of a natural deep eutectic solvent mediated self polymerized highly flexible transparent gel having super capacitive behaviour. *RSC Adv.* 6, 28586–28592. https://doi.org/10.1039/C6RA03309A

Mukesh C., Upadhyay K.K., Devkar R.V., Chudasama N.A., Raol G.G., Prasad K. 2016b. Preparation of a noncytotoxic hemocompatible ion gel by self- polymerization of HEMA in a green deep eutectic solvent. *Macromol. Chem. Phys.* 217, 1899–1906. https://doi.org/10.1002/macp.201600122.

Ngammuangtueng P., Jakrawatan N., Gheewala S.H. 2020. Nexus resources efficiency assessment and management towards transition to sustainable bioeconomy in Thailand. *Resour. Conserv. Recycl.* 160, 104945.

Oh Y., Park S., Jung D., Oh K.K., Lee S.H. 2020. Effect of hydrogen bond donor on the choline chloride- based deep eutectic solvent-mediated extraction of lignin from pine wood. *Int. J. Biol. Macromol.* 165, 187–197. https://doi.org/10.1016/j.ijbiomac.2020.09.145.

Patil U., Shendage S., Nagarkar J. 2013. One-pot synthesis of nitriles from aldehydes catalyzed by deep eutectic solvent. *Synthesis* 45, 3295–3299. https://doi.org/10.1055/s-0033-1339904.

Park T.J. and Lee S.H. 2017. Deep eutectic solvent systems for $FeCl_3$-catalyzed oxidative polymerization of 3-octylthiophene. *Green Chem.* 19, 910–913. https://doi.org/10.1039/C6GC02789J.

Prathish K.P., Carvalho R.C., Brett C.M.A. 2016. Electrochemical characterisation of poly(-3,4- ethylenedioxythiophene) film modified glassy carbon electrodes prepared in deep eutectic solvents for simultaneous sensing of biomarkers. *Electrochimica Acta.* doi:10.1016/j.electacta.2015.11.092.

Procentese A., Johnson E., Orr V., Garruto Campanile A., Wood J.A., Marzocchella A., Rehmann L. 2015. Deep eutectic solvent pretreatment and subsequent saccharification of corncob. *Bioresour. Technol.* 192, 31–36. https://doi.org/10.1016/j.biortech.2015.05.053.

Procentese A., Raganati F., Olivieri G., Russo M.E., Rehmann L., Marzocchella A. 2017. Low- energy biomass pretreatment with deep eutectic solvents for bio-butanol production. *Bioresour. Technol.* 243, 464–473. https://doi.org/10.1016/j.biortech.2017.06.143.

Procentese A., Raganati F., Olivieri G., Russo M.E., Rehmann L., Marzocchella A. 2018. Deep eutectic solvents pretreatment of agro-industrial food waste. *Biotechnol. Biofuels* 11, 37. https://doi.org/10.1186/ s13068-018-1034-y.

Qin H. and Panzer M.J. 2017. Chemically cross-linked poly(2-hydroxyethyl methacrylate)-supported deep eutectic solvent gel electrolytes for eco-friendly supercapacitors. *Chem. Electro. Chem.* 4, 2556–2562. https://doi.org/10.1002/celc.201700586.

Rachmontree P., Douzou T., Cheenkachorn K., Sriariyanun M. 2020. Furfural: A sustainable platform chemical and fuel. *Appl. Sci. Eng. Prog.* 13(1), 3–10.

Rahman A.A.U. and Lynam J.G. 2020. Deep eutectic solvent extracted lignin from waste biomass: Effects as a plasticizer in cement paste. *Case Stud. Constr. Mater.* 13, e00460. https://doi.org/10.1016/j.cscm.2020.e00460.

Rastogi M. and Shrivastava S. 2017. Recent advances in second-generation bioethanol production: An insight to pretreatment, saccharification and fermentation processes. *Renew. Sustain. Ener. Rev.* 80, 330–340.

Rezania S., Alizadeh H., Park J., Din M.F., Darajeh N., Shafiei Ebrahimi S., Saha B., Kamyab H. 2019. Effect of various pretreatment methods on sugar and ethanol production from cellulosic water hyacinth. *Bioresources* 14, 592–606. https://doi.org/10.15376/biores.14.1.592-606.

Roda A., Matias A.A., Paiva A., Duarte A.R.C. 2019. Polymer science and engineering using deep eutectic solvents. *Polymers* 11, 912. https://doi.org/10.3390/polym11050912.

Rodiahwati W., Sriariyanun M. 2016. Lignocellulosic biomass to biofuel production: Integration of chemical and extrusion (Screw Press) pretreatment. *KMUTNB Int. J. Appl. Sci. Technol.* 9(4), 289–298.

Rorke D. and Kana E.B. 2016. Biohydrogen process development on waste sorghum (Sorghum bicolor) leaves: Optimization of saccharification, hydrogen production and preliminary scale up. *Int. J. Hydrogen Ener.* https://doi.org/10.1016/j.ijhydene.2016.06.112.

Salasinska K., Polka M., Gloc M., Ryszkowska J. 2016. Natural fiber composites: The effect of the kind and content of filler on the dimensional and fire stability of polyolefin- based composites. *Polimery Warsaw* 61(04), 255–265.

Sánchez-Condado A., Carriedo G.A., Presa Soto A., Rodríguez-Álvarez M.J., García-Álvarez J., Hevia E. 2019. Organolithium-initiated polymerization of olefins in deep eutectic solvents under aerobic conditions. *ChemSusChem* 12, 3134–3143. https://doi.org/10.1002/cssc.201900533.

Sánchez-Leija, R.J., Torres-Lubián J.R., Reséndiz-Rubio A., Luna-Bárcenas G., Mota-Morales J.D. 2016. Enzyme-mediated free radical polymerization of acrylamide in deep eutectic solvents. *RSC Adv.* 6(16), 13072–13079. https://doi.org/10.1039/C5RA27468K.

Sapir L., Stanley C.B., Harries D., 2016. Properties of polyvinylpyrrolidone in a deep eutectic solvent. *J. Phys. Chem. A* 120, 3253–3259. https://doi.org/10.1021/acs.jpca.5b11927.

Shamsuri A.A., Daik R., Zainudin E.S., Tahir P.M. 2014. Compatibilization of HDPE/agar biocomposites with eutectic-based ionic liquid containing surfactant. *J. Reinf. Plast. Comp.* https://doi.org/10.1177/0731684413516688.

Shawky B.T., Mahmoud M.G., Ghazy E.A., Asker M.M.S, Ibrahim G.S. 2011. Enzymatic hydrolysis of rice straw and corn stalks for monosugars production. *J. Genet. Eng. Biotechnol.* 9, 59–63. https://doi.org/10.1016/j.jgeb.2011.05.001.

Shen X.J., Wen J.L., Mei Q.Q., Chen X., Sun D., Yuan T.Q., Sun R.C. 2019. Facile fractionation of lignocelluloses by biomass-derived deep eutectic solvent (DES) pretreatment for cellulose enzymatic hydrolysis and lignin valorization. *Green Chem.* 21, 275–283. https://doi.org/10.1039/C8GC03064BSfs.

Shishov A., Pochivalov A., Nugbienyo L., Andruch V., Bulatov A. 2020. Deep eutectic solvents are not only effective extractants. *TrAC Trends Anal. Chem.* 129, 115956. https://doi.org/10.1016/j.trac.2020.115956.

Sirviö J.A., Visanko M., Ukkola J., Liimatainen H. 2018. Effect of plasticizers on the mechanical and thermomechanical properties of cellulose-based biocomposite films. *Ind. Crops Prod.* 122, 513–521.

Sivagnanam S.P., Cho Y.N., Woo H.C., Chun B.S. 2018. Green and efficient extraction of polysaccharides from brown seaweed by adding deep eutectic solvent in subcritical water hydrolysis. *J. Cleaner Prod.* 198, 1474–1484. https://doi.org/10.1016/j.jclepro.2018.07.151.

Siwina S. and Leesing R. 2020. Bioconversion of durian (Durio zibethinus Murr.) peel hydrolysate into biodiesel by newly isolated oleaginous yeast Rhodotorula mucilaginosa KKUSY14. *Renew. Ener.* 163, 237–245.

Smith E.L., Abbott A.P., Ryder K.S. 2014. Deep eutectic solvents (DESs) and their applications. *Chem. Rev.* 114, 11060–11082. https://doi.org/10.1021/cr300162p.

Sokolova M.P., Smirnov M.A., Samarov A.A., Bobrova N.V., Vorobiov V.K., Popova E.N., Filippova E., Geydt P., Lahderanta E., Toikka A.M. 2018. Plasticizing of chitosan films with deep eutectic mixture of malonic acid and choline chloride. *Carbohydr. Polym.* 197, 548–557. https://doi.org/10.1016/j.carbpol.2018.06.037.

Sousa A.M.M., Souza H.K.S., Latona N., Liu C.K., Gonçalves M.P., Liu L. 2014. Choline chloride based ionic liquid analogues as tool for the fabrication of agar films with improved mechanical properties. *Carbohydr Polym* 111, 206–214. https://doi.org/10.1016/j.carbpol.2014.04.019.

Sriariyanun M., Heitz J.H., Yasurin P., Asavasanti S., Tantayotai P. 2019. Itaconic acid: A promising and sustainable platform chemical? *Appl. Sci. Eng. Prog.* 12(2), 75–82.

Sriariyanun M. and Kitsubthawee K. 2020. Trends in lignocellulosic biorefinery for production of value-added biochemicals. *Appl. Sci. Eng. Prog.* 13(4), 283–284.

Tan Y.T., Chua A.S.M., Ngoh G.C. 2020. Deep eutectic solvent for lignocellulosic biomass fractionation and the subsequent conversion to bio-based products - A review. *Bioresour. Technol.* 297, 122522. https://doi.org/10.1016/j.biortech.2019.122522.

Taubert J. and Raghavan S. 2014. Effect of composition of post etch residues (PER) on their removal in choline chloride-malonic acid deep eutectic solvent (DES) system. *Microelectr. Eng.* 114, 141–147. https://doi.org/10.1016/j.mee.2012.12.009.

Thi S. and Lee K.M. 2019. Comparison of deep eutectic solvents (DES) on pretreatment of oil palm empty fruit bunch (OPEFB): Cellulose digestibility, structural and morphology changes. *Bioresour. Technol.* 282, 525–529. https://doi.org/10.1016/j.biortech.2019.03.065.

Tsegaye B., Balomajumder C., Roy P. 2017. Alkali pretreatment of wheat straw followed by microbial hydrolysis for bioethanol production. *Environ. Technol.* 40, 1–28. https://doi.org/10.1080/09593330.2017.1418911.

Vigier K.D.O., Chatel G., Jérôme F. 2015. Contribution of deep eutectic solvents for biomass processing: Opportunities, challenges, and limitations. *ChemCatChem* 7, 1250–1260. https://doi.org/10.1002/cctc.201500134.

Vilaplana F., Strömberg E., Karlsson S. 2010. Environmental and resource aspects of sustainable biocomposites. *Polymer Degradation and Stability, 2nd International Conference on Biodegradable Polymers and Sustainable Composites - Alicante 2009*, 95, pp. 2147–2161. https://doi.org/10.1016/j.polymdegradstab.2010.07.016

Vuoti S., Narasimha K., Reinikainen K. 2018. Green wastewater treatment flocculants and fixatives prepared from cellulose using high-consistency processing and deep eutectic solvents. https://doi.org/10.1016/J.JWPE.2018.09.003.

Wang D. and Tang R.C. 2018. Dissolution of wool in the choline chloride/oxalic acid deep eutectic solvent. *Mater. Lett.* 231, 217–220. https://doi.org/10.1016/j.matlet.2018.08.056.

Wang D., Yang X.H., Tang R.C., Yao F. 2018. Extraction of keratin from rabbit hair by a deep eutectic solvent and its characterization. *Polymers* 10. https://doi.org/10.3390/polym10090993.

Wang J, Han J, Khan MY, He D, Peng H, Chen D, Xie X, Xue Z. 2017. Deep eutectic solvents for green and efficient iron-mediated ligand-free atom transfer radical polymerization. *Polym. Chem.* 8(10), 1616–1627. https://doi.org/10.1039/C6PY02066F.

Wang S., Peng X., Zhong L., Jing S., Cao X., Lu F., Sun R. 2014. Choline chloride/urea as an effective plasticizer for production of cellulose films. *Carbohydr. Polym.*, 117, 133–139.

Wang Y., Wang C., Yi Y., Wang H., Zeng L., Li M., Yang Y., Tan Z. 2020. Comparison of deep eutectic solvents on pretreatment of raw ramie fibers for cellulose nanofibril production. *ACS Omega* 5(10), 5580–5588. https://doi.org/10.1021/acsomega.0c00506.

Wan Y.L. and Mun Y.J. 2018. Assessment of natural deep eutectic solvent pretreatment on sugar production from lignocellulosic biomass. *MATEC Web Conf.* 152, 01014. https://doi.org/10.1051/matecconf/201815201014.

Xiaoxia L., and Row K.H. 2017. Purification of antibiotics from the millet extract using hybrid molecularly imprinted polymers based on deep eutectic solvents. *RSC Adv.* 7 (28), 16997–17004. https://doi.org/10.1039/C7RA01059A.

Xu F., Sun J., Wehrs M., Kim K.H., Rau S.S., Chan A.M., Simmons B.A., Mukhopadhyay A., Singh S. 2018. Biocompatible choline-based deep eutectic solvents enable one-pot production of cellulosic ethanol. *ACS Sustain. Chem. Eng.* 6, 8914–8919. https://doi.org/10.1021/acssuschemeng.8b01271.

Xu G.C., Ding J.C., Han R.Z., Dong J.J., Ni Y. 2016a. Enhancing cellulose accessibility of corn stover by deep eutectic solvent pretreatment for butanol fermentation. *Bioresour. Technol.* 203, 364–369. https://doi.org/10.1016/j.biortech.2015.11.002.

Xu G., Li H., Xing W., Gong L., Dong J., Ni Y. 2020. Facilely reducing recalcitrance of lignocellulosic biomass by a newly developed ethylamine based deep eutectic solvent for biobutanol fermentation (preprint). In Review. https://doi.org/10.21203/rs.3.rs-46954/v1.

Xu K., Wang Y., Li Y., Lin Y., Zhang H., Zhou Y. 2016b. A novel poly(deep eutectic solvent)-based magnetic silica composite for solid-phase extraction of trypsin. *Anal. Chim. Acta* 946, 64–72. https://doi.org/10.1016/j.aca.2016.10.021.

Yang D., Hou M., Ning H., Zhang J., Ma J., Yang G., Han B. 2013. Efficient SO_2 absorption by renewable choline chloride–glycerol deep eutectic solvents. *Green Chem.* 15, 2261–2265. https://doi.org/10.1039/C3GC40815A.

Yan-Hui G., Shu H., Xu X.Y., Guo P.Q., Liu R.L., Luo Z.M., Chang C., Fu Q. 2019. Combined magnetic porous molecularly imprinted polymers and deep eutectic solvents for efficient and selective extraction of aristolochic acid I and II from rat urine. *Mater. Sci. Eng. C Mater. Biol. Appl.* 97, 650–657. https://doi.org/10.1016/j.msec.2018.12.057.

Yan X., Cheng J.R., Wang Y.T., Zhu M.J. 2020. Enhanced lignin removal and enzymolysis efficiency of grass waste by hydrogen peroxide synergized dilute alkali pretreatment. *Bioresour. Technol.* 301, 122756.

Yt T., Gc N., Asm C. 2018. Evaluation of fractionation and delignification efficiencies of deep eutectic solvents on oil palm empty fruit bunch. *Ind. Crops Prod.* 123, 271–277. https://doi.org/10.1016/j.indcrop.2018.06.091.

Yt T., Gc N., Asm C. 2019. Effect of functional groups in acid constituent of deep eutectic solvent for extraction of reactive lignin (WWW Document). *Bioresour. Technol.* https://doi.org/10.1016/j.biortech.2019.02.

Zdanowicz M., Wilpiszewska K., Spychaj T. 2018. Deep eutectic solvents for polysaccharides processing: A review. *Carbohydr. Polym.* 200, 361–380. https://doi.org/10.1016/j.carbpol.2018.07.078.

Zhang C.W., Xia S.Q., Ma P.S. 2016a. Facile pretreatment of lignocellulosic biomass using deep eutectic solvents. *Bioresour. Technol.* 219, 1–5. https://doi.org/10.1016/j.biortech.2016.07.026.

Zhang H., Lang J., Lan P., Yang H., Lu J., Wang Z. 2020a. Study on the dissolution mechanism of cellulose by ChCl-based deep eutectic solvents. *Materials* 13, 278. https://doi.org/10.3390/ma13020278.

Zhang L.S., Gao S.P., Huang Y.P., Liu Z.S. 2016b. Green synthesis of polymer monoliths incorporated with carbon nanotubes in room temperature ionic liquid and deep eutectic solvents. *Talanta* 154, 335–340. https://doi.org/10.1016/j.talanta.2016.03.088.

Zhang W., Cheng S., Zhai X., Sun J., Hu X., Pei H., Chen G. 2020b. Green and efficient extraction of polysaccharides from Poria Cocos F.A. Wolf by deep eutectic solvent. *Nat. Prod. Commun.* 15(2). https://doi.org/10.1177/1934578X19900708.

Zhang Y., Cao H., Huang Q., Liu X., Zhang H. 2018. Isolation of transferrin by imprinted nanoparticles with magnetic deep eutectic solvents as monomer. *Anal. Bioanal. Chem.* 410, 6237–6245. https://doi.org/10.1007/s00216-018-1232-2.

Zhao Z., Chen X., Ali M.F., Abdeltawab A.A., Yakout S.M., Yu G. 2018. Pretreatment of wheat straw using basic ethanolamine-based deep eutectic solvents for improving enzymatic hydrolysis. *Bioresour. Technol.* 263, 325–333. https://doi.org/10.1016/j.biortech.2018.05.016.

Zhou E. and Liu H. 2014. A novel deep eutectic solvents synthesized by solid organic compounds and its application on dissolution for cellulose. *Asian J. Chem.* 26, 3626–3630. https://doi.org/10.14233/ajchem.2014.16995.

Zoghlami A., Paës G. 2019. Lignocellulosic biomass: Understanding recalcitrance and predicting hydrolysis. *Front. Chem.* 7. https://doi.org/10.3389/fchem.2019.00874.

11 Chitosan-Based Biocomposites for Biomedical Application
Opportunity and Challenge

Chong-Su Cho, Soo-Kyung Hwang, and Hyun-Joong Kim
Seoul National University

CONTENTS

11.1	Introduction	289
11.2	Requirement of Wound Dressings	290
11.3	Chitosan-Based Biocomposites for Wound Dressing Application	291
	11.3.1 CS-Based Biocomposites Having Acceptable Mechanical Properties	291
	11.3.2 CS-Based Biocomposites Having Antibacterial Properties	294
	11.3.3 CS-Based Biocomposites Having Anti-inflammation and Antioxidant	296
	11.3.4 CS-Based Biocomposites Having Multifunctional Properties	296
11.4	Chitosan-Based Biocomposites for Bone Tissue Engineering	299
	11.4.1 Requirement of Bone Tissue Engineering	299
	11.4.2 CS-based Biocomposite for Bone Tissue Engineering	300
	11.4.2.1 CS-Based Biocomposites Having Appropriate Mechanical Properties	300
	11.4.2.2 CS-Based Biocomposites Having Appropriate Porosity	303
	11.4.3 CS-Based Biocomposites Having Appropriate Biological Functions	308
11.5	Conclusion, Opportunity, and Challenge	310
References		311

11.1 INTRODUCTION

Many polymers have been used for biomedical applications. The polymers include synthetic polymers such as poly(lactic acid) (PLA), poly(glycolic acid) (PGA), poly (lactic-co-glycolic acid) (PLGA), polycaprolactone (PCL), poly(ethylene glycol) (PEG),

DOI: 10.1201/9781003137535-11

polyurethane, poly(vinyl alcohol) (PVA), and natural polymers like alginate (AL), gelatin (GEL), starch (ST), collagen (COL), and chitosan (CS) [1]. Among them, natural polymers are especially interested in biomedical uses because their biological and chemical properties are similar to natural tissues due to the natural components of living structures [2]. Among naturally derived polymers, CS has been much attracted in biomedical applications because it has unique biological properties such as biocompatibility, biodegradability, nontoxicity, antibacterial, and anti-fungistatic properties [1].

The CS can be obtained from the source of chitin as one of the most abundant materials being second only to cellulose (CEL) produced annually by biosynthesis. The chitin is an important component of the exoskeleton in animals and the principal fibrillar polymer in the cell wall of fungi [3]. The CS as one of the linear polysaccharides is composed of glucosamine and N-acetyl glucosamine residues linked via $\beta(1,4)$ glycosidic bonds. The content of glucosamine should be over 60 mol.% to be called the CS after deacetylation from the chitin named as the degree of deacetylation (DD) although the molecular weight of the CS ranges from 300 to over 1,000 kD with a DD from 30% to 95% according to the source and preparation procedure [4]. The chemical structure of the CS provides many possibilities for ionic and covalent modifications to allow extensive adjustment of physicochemical and biological properties because the CS has three kinds of reactive functional groups such as amino group as well as both primary and secondary hydroxyl groups [1]. The CS has various biomedical applications in wound dressing [5,6], tissue engineering [7,8], and drug delivery carriers [9,10] because it provides several advantages of easily processed into sponge-like forms [11], scaffolds [12], microparticles [13], nanoparticles (NPs) [14], nanofibers [15], beads [16], and membranes [17].

Much attention has been focused on the use of CS and CS derivatives in biomedical applications over the last few decades. However, it is very difficult to make any available products using them due to several limitations such as insolubility in neutral pH, brittleness, weak mechanical property, and instability at environmental conditions [18]. In this review, we are aimed to overcome the limitations of CS alone for wound dressing and bone tissue engineering applications using CS-based biocomposites.

11.2 REQUIREMENT OF WOUND DRESSINGS

Wound healing is a complex and regulated physiological process that involves the activation of various cell types through several subsequent such as homeostasis, inflammation, proliferation, and tissue remodeling [19]. Therefore, it is important to use well-designed wound dressings to meet the wound healing cascade for ensuring optimal healing. There are several key parameters to design the optimal wound dressing. First, a moist wound environment for the wound dressing is very important because it facilitates the recruitment of immune cells to promote wound healing by the elaboration of several growth factors and it decreases pain during dressing changes [20]. Second, absorption of excess exudates and blood at the wound site is very critical for the wound dressing because the excess exudates contain degrading enzymes in the tissue which affect the activity and proliferation of cells with losing function of growth factors thus delay of the wound healing process [21]. Third,

TABLE 11.1
The Requirement of Wound Dressings

Requirement	Ref.
Moist wound environment for recruitment of immune cells and decrease of pain	[20]
Absorption of excess exudates and blood at the wound site for removing degraded enzymes	[21]
Prevention of infection and protection of bacterial invasion for favorable host repair	
Adequate water and oxygen exchange for cell metabolism	[22]
Low adherence for prevention of trauma and pain for easy removal	[20]
Acceptable mechanical properties for topical application and controllable degradation	[23]
Should not induce a toxic or inflammatory response	
Cost-effective and minimal frequency of wound dressing change	

prevention of infection and protection of bacterial invasion should be controlled because the infected wound gives an unpleasant odor, delays extracellular matrix synthesis, and prolongs the inflammatory phase which results in further microbial contamination to the tissue. Theoretically, wound dressings should be acted as a barrier between the wound area and the outside environment. Fourth, adequate water and oxygen exchange should be controlled because oxygen is one of the essential nutrients for cell metabolism, and exudates can be managed by the permeability of wound dressings to the water vapor [22]. Fifth, the wound dressings should be low adherent to the wound site for prevention of trauma, and pain because the strongly adherent wound dressings induce further tissue damage at the wound site and should be easy on removal. Sixth, mechanical properties should be acceptable to be compatible for topical application of wound dressings and degradation should be adjusted to match the timeline with the healing process. Seventh, the wound materials should not induce a toxic or inflammatory response. Finally, the wound dressings should be cost-effective and minimal frequency of change in terms of economic condition. The desirable requirement of wound dressings is summarized in Table 11.1.

11.3 CHITOSAN-BASED BIOCOMPOSITES FOR WOUND DRESSING APPLICATION

In this section, we discuss CS-based biocomposites to meet the requirement of wound dressings mentioned in the previous section. Among the several requirements, we focus on how to increase the mechanical property, how to prevent bacterial invasion, how to get anti-inflammation and antioxidants, how to keep water absorption, and how to have a multifunction property in the CS-based biocomposites for the application of wound dressing.

11.3.1 CS-Based Biocomposites Having Acceptable Mechanical Properties

Acceptable mechanical properties for the application of wound dressings should be compatible according to the physical forms such as film, lint, gauze, hydrocolloid, hydrogel, and skin scaffold [24].

CS films have poor mechanical properties which limit their applications for wound dressings. The CS-based biocomposites should be used to improve mechanical properties. Among CS-based biocomposites, CEL has been widely used to blend with the CS [25] because it is the most abundant and renewable polymer available in the world [26]. However, it is insoluble in water and general organic solvents due to the intermolecular hydrogen bonding among hydroxyl groups of the β-D-glucopyranose unit. Many types of research have been tried to achieve the solubility of CEL. Different CEL derivatives have been used to solve the solubility of the CEL by chemical substitution. Among them, hypromellose succinate (HPMCS) obtained by grafting of succinic and with hypromellose was prepared by Jiang et al. [27] because the HPMCS has a good film-forming ability due to the free carboxyl acid groups after dissolving in water. And then, the HPMCS-CS hydrogel films were prepared by amide bond formation among carboxyl acid groups in the HPMCS and amino groups in the CS using 1-ethyl-3-(3-dimethyl aminopropyl)carbodiimide (EDC) and N-hydroxysuccinimide (NHS) as a condensing agent and were assessed the applicability of the hydrogel films as a wound dressing. The results indicated that the mechanical properties of HPMCS-CS hydrogel films were significantly increased both in the dry and swollen state due to the crosslinking between HPMCS and CS compared with those of HPMCS/CS blend film, a suggestion of potential wound dressing although it is not cost-effective due to the several steps to prepare the hydrogel films.

Cho's group prepared the semi-interpenetrating polymer network (semi-IPNs) composed of CS and to enhance the mechanical properties of CS because the brittleness of the CS itself limits the wound dressing application [28]. The results indicated that the formation of Semi-IPNs between CS and poloxamer as shown in Figure 11.1 [28], the mechanical strength of semi-IPNs sponge remarkably increased due to the intermolecular hydrogen bonding between CS and poloxamer compared with CS/poloxamer blend. Also, they evaluated wound healing in a mouse skin defect model using CS/poloxamer semi-IPNs [29]. The results indicated that the wounds covered with CS/poloxamer semi-IPNs were filled with new epithelium without any adverse reactions, an indication of potential wound dressing biocomposites.

Akhavan-Kharazian et al. prepared and characterized CS/GEL)/nanocrystalline CEL (NCC)/calcium peroxide (CP) films to improve the mechanical properties of CS itself for potential wound dressing applications [30]. The results indicated that CS-based biocomposites with the combination of CP and NCC improved the hydrogen bonding between functional groups than CS itself with the antibacterial activity against E *coli* although the addition of CP and NCC particles reduced the amount of water vapor transmission rate and swelling, a suggestion of potential wound dressing materials.

Gao et al. prepared and characterized minocycline (MIN)-loaded carboxymethyl CS (CM-CS) gel/AL nonwoven biocomposites to overcome weak mechanical properties of CM-CS gel by coating MIN/CM-CS on the surface of plasma-treated calcium AL fiber needle-punched nonwovens and were crosslinked with EDC/NHS for wound dressing applications [31]. The wound dressing increased the mechanical properties of CM-CS itself due to the crosslinking and provided quickly absorbed wound exudates with the anti-bacterial property due to the porous biocomposite structure, a suggestion of a new functional wound dressing although it is very

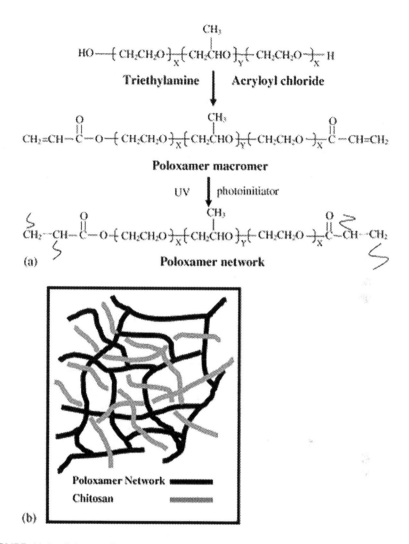

FIGURE 11.1 Scheme of poloxamer macromer and poloxamer networks from poloxamer macromer (a), and structure of CS/poloxamer semi-IPNs (b).

difficult to apply for the clinical healing of wounds due to the several steps of the preparation.

Patholamuthu et al. prepared CS-based electrospun biocomposites composed of CS, poly(ethylene oxide), and aloe vera to get mechanically strong enough to endure mechanical stimulus and to induce wound healing in *vivo* [32]. The results indicated that the mechanical properties of the biocomposite measured by a novel spirograph-based mechanical system (SBMS) were improved compared with those prepared by the static system due to the preparation of the uniformity electrospun mat by the SBMS, a suggestion of the importance of uniformity in the mechanical properties of the electrospun mat for wound dressing application.

Rahmani et al. prepared CS-based biocomposites composed of CS, PVA, and poly(vinyl pyrrolidone) (PVP) using 1,6-diamino carboxy sulfonate) (HMDACS) as a cross-linking agent for wound dressing agents [33]. The results indicated that the mechanical properties of the biocomposites depended on formulations of biocomposite films having the highest mechanical properties when formulated with CS (50 wt%), PVA (30 wt%), PVP (20 wt%), and HMDACS (2 wt%) although the best antibacterial activity against E *coli* was found in different formulation.

Khalili et al. prepared CS-based biocomposites containing CS, poly (phenyl sulfide) (PPS), and reduced graphene oxide (GO) for wound dressing application because the PPS has high mechanical properties due to the semi-crystalline polymer and the GO enhance the cellular activity [34]. The results indicated that the stress shown in Figure 11.6A [34] and compressive modulus shown in Figure 11.6B [34] were enhanced with the PPS/GO addition to the CS although the nonlinear behavior of the compression modulus in the biocomposites was found due to the water molecule present in the hydrogel structure [35].

11.3.2 CS-Based Biocomposites Having Antibacterial Properties

The wound dressings should have antibacterial properties because they can be used as the topical application in the outside microenvironment and gross microbial contamination delays the wound healing.

Diez-Pascual et al. prepared biocomposites containing castor oil (CO) polymeric films mixed with CS-modified ZnO nanoparticles (NPs) as shown in Figure 11.2 [36] to get antibacterial properties for the application of wound dressings [36] because the ZnO has antibacterial activity against both Gram-positive and Gram-negative bacteria even in the absence of light [37].

The results indicated that the antibacterial activity of the films against bacteria was increased with an increase of CS-ZnO content although the biocomposite films showed cytotoxicity *in vitro* when CS-ZnO NPs were mixed over 0.5wt%.

Amalraj et al. prepared PVA/gum Arabic (GA)/CS biocomposite films contained with black pepper essential oil (BPEO) and ginger essential (GEO) to have antibacterial activity for the wound dressing application [38] because both BPEO and GEO have strong antibacterial activity [39]. The results indicated that the BPEO/GEO-loaded PVA/GA/CS films significantly inhibited the growth of Gram-positive and Gram-negative bacteria, an indication of promising wound dressing and food packaging materials.

Haider et al. fabricated CEL-based biocomposite comprising adsorbed CS and silver (Ag)NPs to get the antibacterial property for application as wound dressings [40] because the AgNPs have an effective anti-microbial property [41]. The results indicated that AgNPs-loaded CS/CEL biocomposites exhibited good antibacterial activity against both Gram-positive and Gram-negative bacterial strains, an indication of the potential application of the wound dressing agent.

Sathiyaseelan et al. prepared CS-based biocomposites containing fungal CS (FCS), aloe vera extract (ALE), and Cuscuta reflexa-mediated AgNPs (CUS-AgNPs) to get the antibacterial property for wound dressing application because the CUS-AgNPs have less toxicity than general AgNPS [42]. The results indicated that CUS-AgNPs-loaded

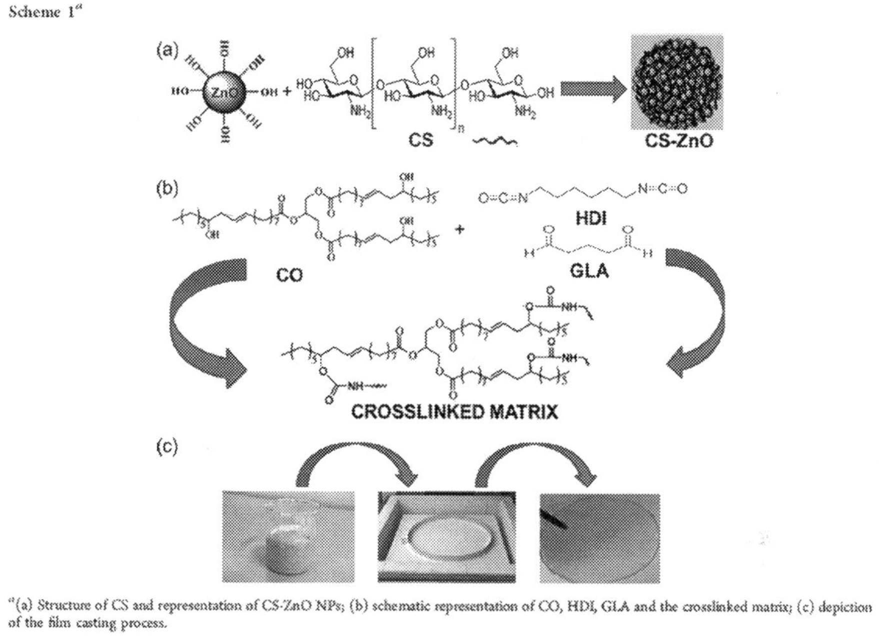

Chitosan-Based Biocomposites

Scheme 1*

FIGURE 11.2 (a) Structure of CS and representation of CS-ZnO NPs. (b) Schematic representation of CO, HDI, GLA, and the cross-linked matrix. (c) Depiction of the film casting process.

FCS/ALE sponges showed higher antibacterial activity against Gram-positive and Gram-negative bacteria compared with other groups without CUS-AgNPs due to the antibacterial property of the CUS-AgNPs, suggestion of potential wound dressing.

Cahu et al. prepared and evaluated CS-based biocomposite films containing GEL, chondroitin-4-sulfate (C4S), and ZnO NPs for wound dressing application [43] because the ZnO NPs have antibacterial and anti-inflammatory properties [44]. The results indicated that the CS/C4S/GEL/ZnO NPs films highly inhibited the growth of staphylococcus aureus compared with other groups and significantly enhanced wound contraction of rat skin with full-thickness when compared with other groups after 6 days, an indication of potential wound dressing application.

Sergi et al. prepared CS-based biocomposites composed of CS and several metal-containing bioactive glass gauges for wound dressing application [45] because the metal-containing bioactive glasses have antimicrobial activity [46]. The results indicated that CS-based composites containing bioactive glass gauge showed enhanced cell adhesion and proliferation and wound healing compared with CS itself due to the release of Sr, Mg, and Zn ions from the bioactive glass although they did not check the antibacterial property.

Xia et al. prepared CS-based biocomposites containing quaternary ammonium CS NPs, and CS to have asymmetric wettable surfaces [47] because asymmetric property enables the CS-based biocomposites to have a hydrophobic outer surface for showing waterproof and antiadhesion contamination properties and to have a hydrophilic inner surface for preserving water-absorbing ability. The results indicated that the QACS NPs/CS biocomposites promoted wound healing and angiogenesis with the effective prevention of wound infection, an indication of a promising dressing biomaterial for chronic wounds.

11.3.3 CS-BASED BIOCOMPOSITES HAVING ANTI-INFLAMMATION AND ANTIOXIDANT

In severe pathological conditions during wound healing, the general cascade of the wound healing process is lost and the wounds are locked in chronic inflammation with abundant neutrophil infiltration and release of reactive oxygen species (ROS) and reactive nitrogen species [48]. Therefore, anti-inflammatory and antioxidant should be used to mitigate the deregulated chronic inflammation during wound healing.

Negi et al. prepared thymoquinone (TQ)-loaded CS-lecithin NPs to incorporate them into Carbopol hydrogel for the wound dressing application [49] because the TQ has anti-inflammatory and antioxidant properties [50]. The results indicated that the Carbopol hydrogel incorporated with TQ-loaded CS-lecithin NPs exhibited superior wound healing efficacy and wound reduction in wound model mice compared with TQ or silver sulfadiazine although they did not check the anti-inflammatory and antioxidant properties.

Ehterami et al. prepared CS/AL hydrogels containing vitamin E (V-E) for wound dressing application in a rat model [51] because the V-E has been known as an antioxidant by protecting cell membranes from ROS attack [52]. The results showed that V-E-loaded CS/AL hydrogel-based biocomposites had a higher wound contraction than the gauge-treated wound as the control, a suggestion of potential wound dressing material.

Augustine et al. prepared electrospun PCL membranes containing CS ascorbate for wound dressing application [53] because the CS ascorbate promoted periodontal regeneration due to the reduced migration of inflammatory cells [54]. The results indicated that CS ascorbate-loaded electrospun membranes showed better cell adhesion and cell viability than PCL membranes although they did not perform wound healing properties.

Zhu et al. prepared CS-based electrospun biocomposites containing asiaticoside (AS), AL, PVA, and CS to evaluate the healing effect on deep partial-thickness rat burn injury [55] because the AS has anti-inflammatory and antioxidant activities [56]. The results indicated that wound healing on deep partial-thickness burn injury of a rat was significantly improved by the AS-loaded AL/PVA/CS electrospun nanofibers due to the downregulation of tumor necrosis factor and interleukin-6 by the loaded AS.

11.3.4 CS-BASED BIOCOMPOSITES HAVING MULTIFUNCTIONAL PROPERTIES

Wound healing is a complex and dynamic process by the defense mechanism of the body through the wound healing cascade for ensuring optical healing [21]. Therefore, multifunctional properties such as biocompatibility, antibacterial activity, anti-inflammatory, and antioxidant properties, ability to promote wound healing,

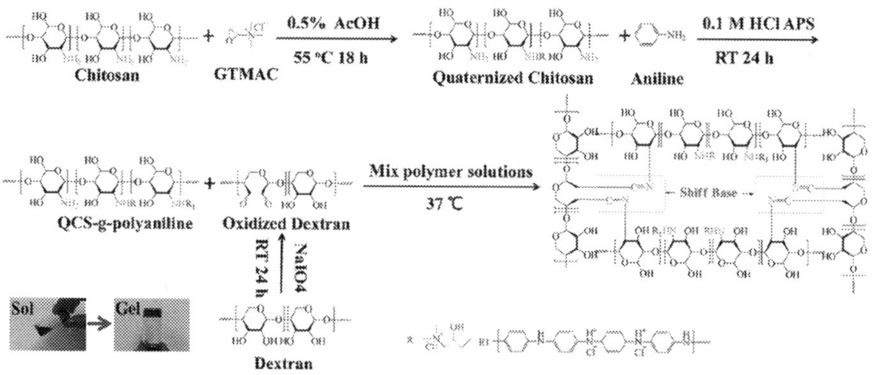

FIGURE 11.3 Schematic representation of the formation of QCS–polyaniline/oxidized dextran hydrogel. QCS represented quaternized chitosan and GTMAC is short for glycidyl trimethylammonium chloride.

mechanical properties, and a low frequency of wound dressing change should be designed. In this section, we discuss recent researches on CS-based biocomposites with multifunctional properties.

Zhao et al. prepared CS-based biocomposites composed of quaternized chitosan (QCS)-graft-polyaniline/oxidized dextran as shown in Figure 11.3 [57] to have antibacterial, conductive, and injectable hydrogels for joints skin wound healing because the electrical stimulation by polyaniline contributes to good cellular behaviors of electrical signal sensitive cells [58]. The results indicated that the hydrogels containing polyaniline showed higher antibacterial activity for *E. coli* and *S. aureus in vitro*, and showed enhanced antibacterial activity for *E. coli in vivo* with better cytocompatibility compared to the hydrogels without polyaniline due to the electroactive activity of the hydrogels, a suggestion of a new way to fabricate in situ forming antibacterial and electroactive hydrogels for skin tissue regeneration applications. They also prepared CS-based biocomposites by mixing QCS and benzaldehyde-terminated Pluronic F127 as shown in Figure 11.4 [59] to have antibacterial, injectable, rapid self-healing, extensibility, and compressibility hydrogels for joints skin wound dressing applications. The results indicated that the hydrogel dressings showed stretchable and compressive properties with good adhesive and fast self-healing ability to bear deformation. Also, the curcumin-loaded hydrogel showed antioxidant ability and accelerated wound healing with upregulation of vascular endothelial growth factor (VEGF) in a full-thickness mouse skin defect model, a suggestion of the possibility of wound dressing materials for joints skin wound healing. Furthermore, they prepared another CS-based biocomposite hydrogels based on QCS-graft-cyclodextrin (QCS-CD) and QCS-graft-adamantane (QCS-AD) as shown in Figure 11.5 [60] to have antibacterial, injectable self-healing, and photoconductive properties. The results indicated that the hydrogels had a conductivity value similar to that of the skin, rapid-healing property, and good antibacterial activity against *E. coli in vitro* and significantly accelerated the healing process of a full-thickness wound in *vivo*, indication of a promising wound dressing for full-thickness skin repair.

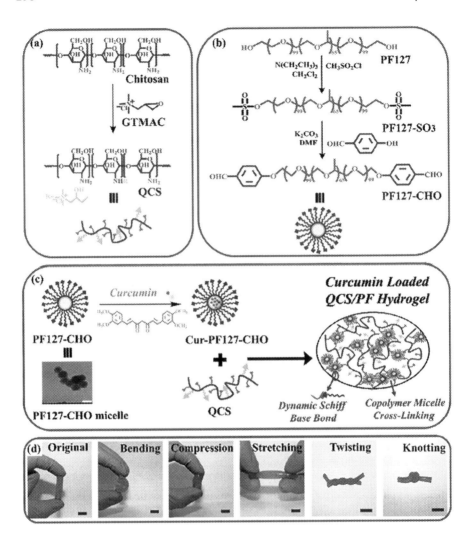

FIGURE 11.4 Schematic representation of Cur-QCS/PF hydrogel synthesis. (a) The synthesis scheme of QCS polymer. (b) PF127-CHO polymer. (c) Schematic illustration of Cur-QCS/PF hydrogel and TEM image of PF127-CHO micelles. Scale bar: 200 nm. (d) The original, bending, compression, stretching, twisting, and knotting shapes of rhodamine B dyed QCS/PF1.0 hydrogels. Scale bar: 1 cm.

Song et al. prepared CS-based biocomposite hydrogels containing cordycepin (CY) and CS after cross-linking by noncovalent bonds through a one-step freezing-thawing method for wound dressing application [61] because the CY as Chinese medicine has antibacterial, antioxidant, and suppression of inflammatory responsibility [61]. The results indicated that the hydrogels exhibited good biocompatibility, suitable water absorption, and remarkable antimicrobial effect with the desired mechanical strength *in vitro*. Also, the hydrogels showed a quicker re-epithelization of rat skin wounds, increased collagen deposition, and increased expression of epithelial regeneration

Chitosan-Based Biocomposites

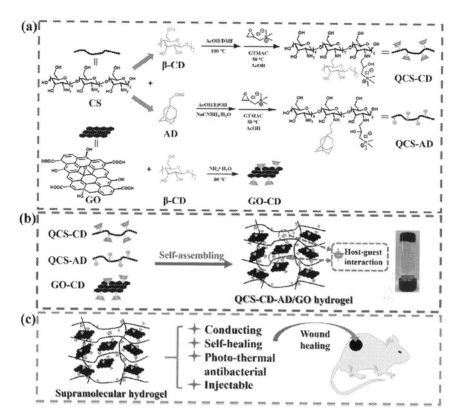

FIGURE 11.5 Schematic representation of QCS-CD-AD/GO supramolecular hydrogels preparation. (a) Preparation scheme of QCS-CD, QCS-AD, and GO-CD polymer. (b) QCS-CD-AD/GO supramolecular hydrogel. (c) Characteristic of QCS-CD-AD/GO hydrogel and the application in wound healing.

markers of laminin and involucrin in *vivo* compared with CS hydrogel due to the self-healing ability of the hydrogels.

Sundaram et al. prepared injectable CS-based biocomposites containing bioglass NPs and CS to control effective bleed when there is severe blood loss from major surgeries or skin wounds [62] because the bioglass can initiate the coagulation cascade [63]. The hydrogels showed injectable rapid blood clottable and cytocompatible properties in *vitro* and in *vivo*, an indication of potential hydrogel for getting effective bleeding control during critical situations.

11.4 CHITOSAN-BASED BIOCOMPOSITES FOR BONE TISSUE ENGINEERING

11.4.1 Requirement of Bone Tissue Engineering

Tissue engineering consisted of multidisciplinary science, including material engineering, molecular biology, and the clinical part to develop biological substitutes for defective tissues or organs has recently become an important therapeutic strategy for the present and future medicine [1]. Among defected tissues, bone defect treatments are the most urgent problem in orthopedic surgery although autogenous and allogenous bone grafts, bone substitute material transplantation, and metal implants are currently performed [64]. However, several problems such as limited bone sources, potential infection and immune responses, biocompatibility, and mechanical properties limit their use in clinical practice. Therefore, bone tissue engineering will provide a new therapeutic method to solve the above-mentioned problems as an alternative.

In this section, we discuss the requirement of an ideal biomaterial scaffold among the indispensable components such as cells, growth factors, and scaffolds in bone tissue engineering. The biomaterials should have mechanical and appropriate properties, mimic natural bone structure to mineralize in *vivo*, function transporting and exchange to grow blood vessels, and be conductive to cell adhesion and to get normal proliferation and differentiation ability of the cells [65].

11.4.2 CS-BASED BIOCOMPOSITE FOR BONE TISSUE ENGINEERING

In this section, we cover CS-based biocomposite to meet the requirement of bone tissue engineering mentioned in the previous section. Among the requirements, we focus on how to get the appropriate mechanical property, how to meet appropriate porosity, and how to function biological properties by the CS-based biocomposites for bone tissue engineering.

11.4.2.1 CS-Based Biocomposites Having Appropriate Mechanical Properties

Shokri et al. prepared CS-based biocomposites containing CS, bioactive glass (BG), and carbon nanotube (CNT) to overcome low mechanical strength and Young's modulus of CS itself for application of bone tissue engineering [66] because the presence of CNT can increase the compressive strength of the scaffolds. The results indicated that the compressive strength of CS was increased by adding CNT and was more increased with an increase of CNT in the scaffolds with an increase of attachment and proliferation of MG63 osteoblast cells on CS/BG/CNT scaffolds due to the surface formation of hydroxyapatite (HA) by the BG [67] although they did not perform *in vivo* study.

Nazeni et al. similarly prepared CS-based biocomposites containing CS, BG, and PLGA NPs to increase in mechanical strength of the scaffolds for bone tissue engineering application [68] because the PLGA as one of the synthetic polymers approved for clinical use due to the biocompatibility has relatively good processability [69]. The results indicated that the incorporation of the PLGA NPs increased the compression strength without affecting the morphologies of the scaffolds, a suggestion of a potential to be used as a controlled-release platform of related growth factor for bone tissue regeneration because the PLGA NPs have been used for sustained controlled drug release.

Pourhaghgouy et al. also prepared CS-based biocomposites containing CS and BG NPs by the freeze-casting method to increase the compressive strength and compressive modulus of the nanocomposite scaffolds for bone tissue engineering application [70].

Chitosan-Based Biocomposites

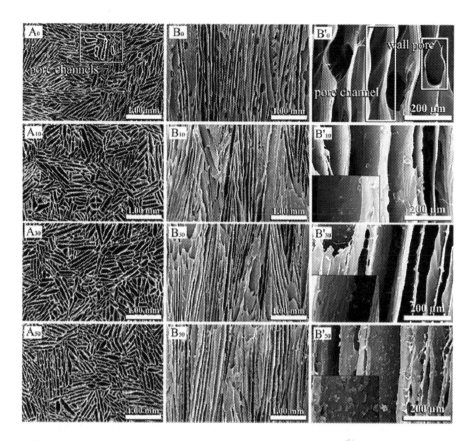

FIGURE 11.6 The SEM images taken from both (A) perpendicular and (B and B') parallel directions to the ice growth during the freeze-casting process. The images inserted in the B' series are the high magnification pictures of the B series which illustrate scaffolds' wall surfaces and the distribution of BGNPs on them. The subscripts indicate the BGNP contents of each scaffold (0, 10, 30, and 50 wt%).

The results indicated that the compressive strength and compressive modulus of the biocomposites increased 12 and 26 times, respectively, when 50 wt% of BG NPs were added into CS due to the unidirectional structure with a homogeneous distribution of BG NPs into CS scaffolds as shown in Figure 11.6 [70].

Zhang et al. prepared CS-based biocomposites containing CS, HA, and poly(3-hydroxybutyrate-co-3-hydroxy valerate) (PHBV) to enhance the mechanical property and biocompatibility of the biocomposite scaffolds [71] because the HA acts as a chelating agent for organizing the apatite-like mineralization [71] and the PHBV is a highly biocompatible polymer with high toughness [72]. The results indicated that the ultimate tensile strength of CS/PHBV/HA electrospun biocomposite nanofibers increased compared with that of CS/PHBV ones although the tensile strength of the CS/PHBV/HA depended on the content of HA and CS/PHBV/HA scaffold showed higher mineral deposition than that of PHBV one due to the synergistic effect of CS and HA, an indication of the potential to promote the regeneration of bone tissue.

Maji et al. also prepared CS-based biocomposites containing CS, GE, and HA to increase the mechanical strength of CS scaffold for bone scaffolds [73] because of having favorable mechanical properties by HA and having bioactivities by GE. The results indicated that the compressive strength of the biocomposites depended on the content of CS:GEL:HA in the composites and the highest compressive strength of the composites was obtained with a CS:GEL:HA weight ratio of 49-30-21 with having conductivity to mesenchymal stem cells adhesion due to the bioactive property of GEL, a suggestion of a successful contribution to the development of superior scaffolds for application of bone tissue engineering.

Kim et al. prepared CS-based biocomposites containing CS, AL, and HA NPs to increase the compressive strength and elastic modulus of CS/AL composite scaffolds for bone tissue engineering [74] because HA NPs can increase the mechanical properties due to homogenously dispersion of HA NPs in the biocomposites. The results indicated that the compressive strength and the elastic modulus of the biocomposites increased about 16 and about 20 times, respectively, compared with CS/AL composites with more differentiation and mineralization of the MC3T3-E1 cells when 70wt% of HA NPs were added in the composite scaffolds. Similarly, Acevedo et al. added HA and titania NPs to increase the mechanical properties of CS-based biocomposites containing CS and GEL for application in bone regeneration [75] because both NPs increase thermophysical and mechanical properties of the composites. The results indicated that both NPs were homogeneously distributed in the CS/GEL composite membranes and Young's modulus of NPs-contained CS/GEL composite membranes increased by UV-irradiated cross-linking with the increase of the differentiation of MEF cells due to the osteoconductive property of both NPs although they did not check the mechanical properties of the composites without both NPs. Furthermore, Teimouri et al. added zirconia (ZrO_2) NPs to increase enhanced mechanical properties of CS-based biocomposites containing CS and silk fibroin (SF) [76] because ZnO_2 NPs have remarkable mechanical properties [77]. The results indicated that the compressive strength of CS/SF scaffolds increased with an addition of ZnO_2 without cytotoxicity although they used human gingival fibroblast cells *in vitro* cytotoxicity and cell attachment for dental tissue engineering.

HA-contained biocomposites have been used for their application in bone tissue engineering because of the HA-induced osteoconductivity [78]. However, beta-tricalcium phosphate (β-TCP) as an alternative ceramic has been used because the β-TCP has a ten times higher degradation rate than HA [78] with the promotion of osteogenesis and improvement of bone regeneration [79].

Serra et al. prepared and characterized CS-based biocomposites containing CS, GEL, and β-TCP to get the osteogenesis of CS/GEL composite scaffolds for bone tissue engineering due to the fast dissolution and absorption of the β-TCP [80] although low mechanical properties of the biocomposites can be overcome by ionic crosslinking with sodium tripolyphosphate. The results indicated that the incorporation of GEL and/or β-TCP in the CS scaffolds increased their compressive strength by about 70% and enhanced mineral deposition on the biocomposite scaffolds immersed in standard simulated body fluid (SBF) solution with antimicrobial activity against *S. aureus*, a suggestion of production of biomimetic scaffolds to improve bone regeneration. Similarly, Puvaneswary et al. prepared CS-based biocomposites containing

CS, fucoidan (FU), and β-TCP to increase proliferation and mineralization in human bone marrow stromal cells for bone tissue engineering [81] because the FU showed mineralization in human adipose-derived stem cells (hADSs) [82]. The results indicated that CS/FU/β-TCP biocomposite scaffolds showed higher compressive strength and modulus than those of CS/β-TCP scaffolds due to the addition of FU and increased osteogenic differentiation of hMSCs due to the increase of released osteocalcin by the FU.

11.4.2.2 CS-Based Biocomposites Having Appropriate Porosity

The CS-based biocomposites for bone tissue engineering application should have the ability to develop a cell-based repairing biomaterial for the regeneration of bone defect [83] because the main function of the scaffolds is to facilitate the making of bone tissues of preferred size, shape, and function by serving as a structural template [84] with having adequate porosity to get cell adhesion, proliferation, and nutrient transfer. Several approaches such as gas foaming, freeze-drying, particle leaching, thermally induced phase separation, electrospinning, and three-dimensional (3D) printing have been tried to fabricate the scaffolds having the appropriate porosity. In this section, we want to discuss the characteristics of each method.

11.4.2.2.1 Gas Foaming

The gas foaming method was used to avoid organic solvents by using inert gas foaming agents such as carbon dioxide and nitrogen to pressurize molded polymers with water until they are saturated and full of gas bubbles [85]. This method generally makes sponge-like structures with pore sizes of 30~700 μm and porosity of up to 85% [86]. The disadvantages of this method have the use of excessive heat during compression molding, non-interconnected pore structures, and nonporous skin layers at the scaffold surface [85].

Gravel et al. prepared macroporous CS-based biocomposites containing CS and coral [87] by gas foaming because the coral mainly composed of calcium carbonate can produce carbon dioxide by the reaction between the coral and acidic CS solvent. The results indicated that the average pore sizes of the CS/coral scaffolds were from 80 to 400 μm according to the weight of coral from 0 to 75 wt% whereas the porosity decreased from 91% to 75 wt% with an increase of compressive modulus and fast MSCs adhesion due to the remained coral particles in the scaffolds. They also investigated responses of MSCs to the CS/coral biocomposites prepared by the gas-forming agent to check their scaffolding potential *in vitro* bone regeneration [88]. The results indicated that the CS/coral biocomposite scaffolds with a high content of coral showed higher cell number, alkaline phosphatase (ALP) activity, and osteocalcin (OC) protein expression compared to CS itself.

11.4.2.2.2 Freeze-Drying

The freeze-drying method known as lyophilization can be applied by several processes such as the dissolving of used scaffolds in a suitable solvent, cooling down of solved scaffolds below their freezing point for leading to the solidification of the solvent, and finally evaporation of the solvent via sublimation for making dry scaffolds [85]. The advantages of this method are to avoid high temperatures that can affect the activity of the incorporated growth factors and easily control pore size by tuning the freezing regime. On the other hand, the disadvantages of this method are lengthy time scales, the use of cytotoxic organic solvents, high energy consumption, and the production of small and irregular pore sizes [89].

Kalanthai et al. prepared CS-based biocomposite scaffolds containing graphene oxide (GO), CS, COL, and cross-linked AL by the freeze-drying method for bone tissue engineering [90] because the addition of GO in the crosslinked AL/CS/COL scaffolds increased the mechanical strength and improved osteogenic differentiation *in vivo* [91]. The results indicated that the GO/CS/COL/AL scaffolds exhibited interconnected pores of 0~250 µm range and a significant MC3T3 cell attachment compared to CS/COL/AL ones although they did not show any effect on the osteogenic ability of osteoblasts.

Demir et al. prepared CS-based biocomposite scaffolds containing CS, montmorillonite (MMT), and strontium (ST) by the freeze-drying method as bone tissue engineering scaffold [92] because the MMT has a good cation-exchange ability [93] and the ST stimulated pro-osteoblast proliferation and activity [94]. The results indicated that the CS/MMT/ST scaffolds showed highly porous morphologies with interconnected pores and displayed significantly higher DNA concentrations of the human osteoblast cells due to the incorporation of ST in the scaffolds.

Pineda-Castillo et al. prepared and characterized CS-based biocomposite scaffolds containing CS, PVA, and HA by the freeze-drying method for bone tissue regeneration [95] because the HA improved osteoconductivity [96]. The results indicated that the CS/PVA/HA scaffolds showed uniform pore sizes of 142~519 µm range that had been described as optimal bone defect regeneration without cytotoxicity due to pore formation with interconnected pores by the presence of CS in the scaffolds.

Peng et al. prepared CS-based biocomposite scaffolds containing CS, mesoporous calcium silicate (MCS), and lanthanum (LA) via the freeze-drying method for bone tissue engineering [97] because the MCS accelerated *in vivo* bone tissue regeneration [98] and the LA ions enhanced the proliferation and osteogenic differentiation of rat bone marrow-derived mesenchymal stem cells (BMSCs). The results indicated that the CS/MCS/LA scaffolds showed 3D macropores with a size of around 200 µm and significantly induced the osteogenic differentiation of BMSCs *in vitro* and promoted new bone deposition *in vivo* rat cranial bone defect models, a suggestion of application potential for the bone defect.

Shi et al. prepared CS-based CS biocomposites containing CS, dopamine-modified AL (DA-AL), and HA NPs by integrative layering method with further crosslinking by Ca^{2+} ions for bone tissue engineering [99] because the as-prepared CS/DA-AL/HA NPs scaffolds make gradient scaffolds for appropriate degradation rate to get fast bone regeneration. The results indicated that the CS/DA-AL/HA NPs scaffolds had integrated layer structures and high porosity at around 77.5% and showed good adhesion of chondrocytes and fibroblasts *in vitro*, and promoted the regeneration of the bone tissue with the acceleration of the repair of the bone defects in white New Zealand rabbits.

Recently, Sadeghinia et al. prepared CS-based biocomposites containing CS, GEL, clinoptilolite (CLN), and HA NPs by the freeze-drying method for bone tissue engineering [100] because the CLN showed an immunostimulatory effect with an increase of mechanical strength [101]. The results indicated that the CS/GEL/CLN/HA NPs scaffolds showed highly porous morphologies with pore sizes of 200 ± 100 μm and showed increased biomineralization and enhanced mechanical strength without cytotoxicity due to the presence of CLN and HA NPs in the scaffolds.

11.4.2.2.3 Particle Leaching

The particle leaching method is firstly to dissolve the scaffolds by the solvent with uniformly distributed salt particles, to evaporate the solvent with leaving the salt particles-loaded scaffolds, and finally to immerse in water for leaching out to make a porous structure [85]. The advantages of this method are relatively easy to make pores with the sustainable equipment cost and to make high scaffold porosity with the feasibility for tuning pore size [85]. On the other hand, the disadvantage of this method is only to form simple shape scaffolds and harmful to cells by the remained solvent [102].

Jamalpoor et al. prepared CS-based biocomposite scaffolds containing CS, GEL, and HA NPs by the particle leaching method using NaCl as porogen for bone tissue engineering [103] to make 3D scaffolds with optimum porosity and pore size with the bone matching mechanical strength. The results indicated that the CS/GEL/HA NPs scaffolds showed highly interconnected porous structures with a mean pore size of 140–190 μm and increased GEL content in the scaffolds improved attachment, infiltration, and proliferation of Saos 2 cells.

Ruixin et al. prepared CS-based biocomposite scaffolds containing CS and HA microparticles by the particle leaching method using spherical paraffin as porogen for bone regeneration because the HA can improve the bioactivity and bone-bonding ability. The results indicated that the pore of the CS/HA microparticle scaffolds showed interconnected spherical macropores with the increase of pore by the increase of porogen although there were some pores deformed due to the deformation of the paraffin by the stirring. Also, they prepared CS/HA NPs or CS/HA microparticle scaffolds using the same porogen to compare biocompatibility between both scaffolds [104]. The results indicated that both scaffolds showed interconnected spherical pores without differences in structural parameters and good biocompatibility in MC3T3-E1 cells without significant difference of cell viabilities between both scaffolds.

Wang et al. prepared hydroxyethyl CS (HCS)-based biocomposite scaffolds containing HCS and chemical crosslinked CEL by the particle leaching method using silicon dioxide particles as porogen for bone tissue engineering [105] because the crosslinked CEL can enhance the compression modulus and elasticity. The results indicated that HCS/crosslinked CEL scaffolds showed bubble-like macropore structure with a pore size of 100~250 μm by the removal of porogen SiO_2 particles and micropore structure with a pore size of several tens of microns by the sublimation of ice crystals formed during freeze-drying and facilitated the attachment, spreading, and osteoblastic MC3T3-E1 cells due to the addition of HCS in the scaffolds, suggestion of promising scaffolds for bone tissue engineering application.

11.4.2.2.4 Thermally Induced Phase Separation

The TIPS is one of the low-temperature processes because a scaffold solution is quenched and undergoes a liquid/liquid phase separation: one scaffold-rich and the other scaffold-poor [85]. The scaffold-rich phase solidifies whereas the scaffold poor one is removed, leaving a highly porous and nanoscale fibrous network [85]. The advantage of this method is a low-temperature process that favors the incorporation of growth factors in the nanoscale structure for serving as a template of drug-loaded NPs in the scaffolds. On the other hand, the disadvantage of this method is to combine with another method for making macroporous structures in the scaffolds.

Zhang et al. prepared biomimetic osteochondral scaffolds containing the oriented cartilage layer designed to mimic native cartilage tissue and fabricated with cartilage matrix-CS using TIPS, a compact layer designed to mimic the calcified-layer structure of natural cartilage and 3D-printed core-sheath structured-bone layer fabricated with PLGA/β-TCP-COL by low-temperature deposition method [106]. The results indicated that the three part-combined scaffolds exhibited good mechanical properties with hydrophilicity and BMSC-loaded scaffolds regenerated trabecular bone formed in the subchondral bone defect model of goat, a suggestion of possibility for future clinical application in bone-defect repair.

Rahman et al. prepared crosslinked-CS-based biocomposite scaffolds containing CS, COL 1, and HA by the TIPS method for restoration of defected maxillofacial mandible bone [107]. The results indicated that the scaffolds exhibited irregular porous structures with moderate interconnected structures with a pore diameter of 111.8~212.6 µm for CS/COL 1/HA although the pore diameters of the scaffolds were decreased after cross-linking and de-hydrothermal cross-linked CS/COL 1/HA scaffolds showed the restoration of defected bone in the rabbit.

Recently, Erickson et al. prepared CS-based biocomposite bilayer scaffolds consisted of CS-HYA (hyaluronic acid) cartilage layer and CS-AL HA bone layer by the TIPS method for osteochondral tissue regeneration [108]. The results indicated that the scaffolds showed an open pore network with interconnected structures although the pore diameter of the scaffolds depended on the content of used CS, AL, and HA, and gene expression related with osteogenesis and chondrogenesis increased after co-culture with chondrocyte-like SW-1353 and osteoblast-like MG 63 in the scaffolds.

11.4.2.2.5 3D Printing

The 3D printing method has been used to be a rational strategy to make 3D scaffolds for overcoming the limitations of traditional methods because direct or indirect 3D printing methods can provide precise control over pore interconnectivity, size, internal architecture, and external shape of the 3D scaffolds, and can create the 3D structure with the size of the defective part and the correct anatomical shape [109]. The advantages of this 3D printing method enable not only mass customization of goods on a large scale but also smaller production runs with a high degree of customization [110]. On the other hand, the disadvantages of this 3D printing method are how to get reproducibility of the various scales and complexities of engineered tissues.

Demirtas et al. prepared CS-based biocomposite hydrogel scaffolds containing CS and HA NPs by the extruder-based bioprinter for bone tissue engineering because 3D patterning of cells and growth factors as a bioprintable form can fabricate living tissue and organs for tissue engineering [111]. The results indicated that bioprinted-hydrogels MC3T3-E1 pre-osteoblast cell-laden CS/HA NPs showed peak expression levels for early and late stages osteogenic markers with high cell viability and the loaded cells in the CS/HA NPs hydrogels had higher cell proliferation and differentiation compared with AL itself, a suggestion of applicability and printability of CS/HA NPs hydrogel as a bioprinting solution.

Dong et al. prepared CS-based biocomposite hydrogel scaffolds containing CS and PCL by 3D printing to improve the cell seeding efficiency and osteoinductivity in the PCL scaffolds because an injectable thermo-sensitive CS hydrogel can be incorporated into 3D-printed PCL scaffold for bone tissue engineering application [112]. The results indicated that greater retention and proliferation in rabbit BMSCs and bone morphogenetic protein-2 (BMP-2)-laden CS/PCL hydrogel scaffolds were obtained than PCL itself and stronger osteogenesis with the bone-matrix formation was shown in the CS/PCL hybrid system than PCL one *in vitro* after 2-week, a suggestion of a promising platform for bone tissue engineering due to their ability to load cells and drugs, and excellent mechanical strength.

Yang et al. prepared CS-based biocomposite scaffolds containing quaternized CS (QCS)-grafted PLGA and HA by the 3D printing method to inhibit bacterial infection and to promote bone regeneration [111] because the infection is the pivotal cause of nonunion in the bone defect. The results indicated that QCS-grafted PLGA/HA scaffolds significantly exhibited improved antimicrobial and osteoconductive properties *in vitro*, and enhanced anti-infection and bone regeneration abilities in infected bone defect rats or rabbit models, a suggestion of a promising dual-functional scaffold for repairing bone defect under infection.

Tsai et al. prepared CS-based biocomposite scaffolds containing CS, titanium alloy (TA), and magnesium-calcium silicate (MCS) by the 3D printing method for orthopedic application [113] because the bioactivity of the TA/MCS can be improved using the simple immersion method by the CS. The results indicated that CS/TA/MCS scaffolds exhibited enhanced cell adhesion, proliferation, and differentiation *in vitro*, and enhanced bone regeneration and in growth at the critical size bone defects in the rabbit model, an indication of induction of micro-environment for bone regeneration by the simple immersion method.

Chen et al. prepared CS-based biocomposite scaffolds containing CS, GEL, and Mg (Mg)-substituted HA (Mg-HA) prepared by biomimetic mineralization of COL 1 and citric acid as the bi-template via 3D printing method for bone regeneration [114] because the substitution of Mg for cations reduce the crystallinity of HA without affecting the size and structure of HA. The results indicated that CS/GEL/Mg-HA scaffolds exhibited higher cell attachment, proliferation rate, increased expression of ALP activity, and osteogenic related genes such as osteocalcin, runt-related transcription factor 2, and COL 1, as an indication of a potential candidate of biocomposite scaffolds in bone tissue engineering.

Recently, Chen et al. prepared CS-based biocomposite scaffolds containing carboxymethyl chitosan (CMCS), HA, and polydopamine (PDA) by the 3D printing method [115] for repairing bone defects because the PDA enhanced cell adhesion [116] with high biocompatibility and improved stability of the bound materials in the surfaces. The results indicated that the CMCS/HA/PDA scaffolds exhibited a porous structure with the size of $415 \pm 87\,\mu m$ and $69.5\% \pm 4.6\%$ porosity and effectively stimulated new bone formation within the femoral lacuna defect site of rabbits after 12 weeks, a suggestion of a remarkable potential new scaffold for repair of bone defects.

11.4.2.2.6 Electrospinning

The electrospinning method is to use electrical charges for drawing fine fibers up to the nanometer scale and creating a nanofibrous architecture [85]. Generally, there are four major components such as a spinner with a metallic needle, a high-voltage power supply, a syringe pump, and a grounded collector.

Electrospinning has been used to fabricate scaffolds with both micro and nanostructures for tissue engineering application although the electrospun scaffolds have weak mechanical properties due to the high porosity and nonaligned microfibers.

Jing et al. at first prepared parallel-aligned poly(propylene carbonate) (PPC) microfibers by electrospinning, treated oxygen plasma, and introduced CS nanofibers to increase the mechanical properties of the composite scaffolds for the application of tissue engineering [117]. The results indicated that Young's modulus of the PPC increased by about 26% after the treatment of the CS.

Nanofibers under dry conditions with a superior cell response were obtained whereas the difference between PPC and PPC/CS scaffolds was not much obtained under wet conditions although they used 3T3 cells instead of bone cells.

11.4.3 CS-Based Biocomposites Having Appropriate Biological Functions

Growth factors have been widely used for bone tissue engineering because they are responsible for cellular behaviors such as proliferation, migration, and differentiation. Therefore, various growth factors have been loaded in CS-based biocomposite scaffolds after physically or chemically loaded of growth factors into the scaffolds [118]. In this section, we discuss the biological functions of bone-related cells after loading growth factors into the scaffolds. Among the growth factors, BMP-2 has been extensively used in bone tissue engineering because it is approved by the USA FDA for bone graft fusion due to the safety and efficient bone growth better than any BMPs [119].

Sobhani et al. prepared calcium phosphate/polyphosphazene scaffolds containing BMP-2-loaded CS microspheres in bone tissue engineering [120] because loaded BMP-2 into the CS microspheres can be sustainably released to induce an osteoblast proliferation. The results indicated that BMP-2-loaded scaffolds increased the osteogenic differentiation ability of BMSCs compared with the scaffolds alone. Similarly, Bastami et al. prepared GEL/β-TCP/COL scaffolds containing BMP-2-loaded CS NPs for bone tissue engineering [121] because loaded BMP-2 into the CS NPs can be sustainably released to get differentiation of human buccal fat pad-derived stem cells (hBGPSCs). The results indicated that the BMP-2-loaded scaffolds showed an enhanced osteoinductive graft compared with the scaffolds alone due to the sustained delivery of BMP-2 in a therapeutic window. Also, Deng et al. prepared PLGA/HA NPs scaffolds containing BMP-2-loaded CS NPs for bone tissue engineering [122] because the sustained release of BMP-2 from the CS NPs induced bone regeneration due to the osteogenic effect of the BMP-2. The results indicated that the BMP-2-loaded scaffolds showed faster new bone formation in a rabbit mandible bone defect model without any significant inflammatory response compared with scaffolds alone due to the osteogenesis effect by the released BMP-2.

Tong et al. prepared transforming growth factors-β1(TGF-β1)-loaded CS/SF 3D scaffolds for bone tissue engineering [123] because the TGF-β1 induces the differentiation and proliferation of osteoblasts and BMSCs [124]. The results indicated that the TGF-β1-loaded scaffolds significantly more enhanced the growth and proliferation of BMSC in a tissue-dependent manner *in vitro*, and exhibited extensive osteoconductivity with the host bone, and enhanced new bone formation after implant in rabbit mandibles model after 8 weeks compared with the scaffolds alone due to the effect of released TGF-β1 from the scaffolds, a suggestion of a promising potential to be applied in orthopedic surgery.

Oryan et al. prepared platelet gel (PG)-loaded CS/GEL biocomposite scaffolds to regenerate bone defect [125] because the PG contains angiogenic, mitogenic, and osteogenic growth factors in their α-granules [126]. The results indicated that the PG-loaded CS/GEL scaffolds showed significantly higher new bone formation, bone volume, the density of osseous and cartilaginous tissue, and numbers of osteons in critical-sized radial bone defect of a rat after 8 weeks compared with the CS/GEL scaffolds alone due to the regenerative effect by the incorporated PG in the scaffolds. Similarly, Liao et al. prepared PRP-loaded thermo-gelling hydrogel, HYA-g-CS-g-poly (*N*-isopropyl acrylamide) (HYA-CS-PNIP) after embedding of biphasic calcium phosphate (BCP) and rabbit adipose-derived stem cells (rASCs) to get osteoinductive properties [127]. The results indicated that the PRP- and rASCs-loaded injectable thermo-sensitive HYA-CS-PNIP/BCP scaffolds showed increased cell proliferation, alkaline phosphatase activity, increased calcium deposition, and upregulated expression of osteogenesis *in vitro*, and induced more new bone formation at the rabbit critical size calvarial bone defect model, a suggestion of a promising biocomposite hydrogel scaffolds for bone tissue engineering.

In some cases, combined growth factors can be used to get synergistic or additive effects of growth factors for promotion of the bone regeneration. Wang et al. prepared stromal cell-derived factor (SDF-1)- and BMP-2-loaded CS/agarose (AG)/GEL scaffolds synthesized via gelation method using cross-linked CS, AG, and GEL, after modified by CS/HEP NPs [128] because the SDF-1 plays a critical role in the mobilization of MSCs and the BMP-2 plays a critical role in osteogenesis of MSCs. The results indicated that both growth factors-loaded and CS/HEP NPs-modified scaffolds retained migration activity of MSCs and strongly induced differentiation towards osteoblasts *in vitro*, and showed a continuous chemotactic response of MSCs in nude mice after subcutaneous implantation of two growth factors-loaded and CS/HEP NPs-modified scaffolds into the back of the mouse, a suggestion of attractive scaffolds to promote bone repair and regeneration. Similarly, Dou et al. prepared COL/HA scaffolds containing VEGF- and BMP-2-loaded CMCS microspheres for bone tissue engineering [129] because the VEGF promotes vascular regeneration and improves the activity of osteoblasts, and the BMP-2 promotes bone regeneration [130]. The results indicated that both growth factors-loaded scaffolds showed more conductivity to the differentiation of pre-osteoblasts *in vitro* and promoted the formation of blood vessels and the formation of COL *in vivo* due to the sequential release of the double growth factors. Furthermore, Sadeghinia et al. prepared CS/GEL/HA NPs biocomposite scaffolds combined with PRP and fibrin glue (FG) to enhance proliferation and differentiation of seeded human dental pulp stem cells

(HDPSCs) for dental bone tissue engineering [100] because the PRP enhanced osteogenesis and bone formation [131], and the FG showed increase osteoconductivity and biocompatibility [132]. The results indicated that both growth factors-loaded scaffolds improved adhesion formation of bone minerals, and BMP-2 gene expression of seeded HDPSCs compared with the scaffolds alone.

11.5 CONCLUSION, OPPORTUNITY, AND CHALLENGE

The CS and CS derivatives have been used in various biomedical applications such as wound dressing, tissue engineering, and drug delivery carriers due to the unique biological properties, extensive adjustment of physicochemical properties, and easy processability although they have relatively poor mechanical, thermal, not-enough biological and barrier properties. In this regard, CS-based biocomposites after the addition of two or more biomaterials should be desirable.

Considering the acceptable mechanical property, prevention of bacterial invasion, absorption of excess exudates, adequate water and oxygen exchange, and the existence of anti-inflammation and antioxidants for wound dressing application, no single wound dressing can meet all requirements, thus the challenge is to develop novel CS-based wound dressings that can positively affect all or most wound types. The current challenge is how to develop multi-functional wound dressings for simultaneously providing therapeutic properties such as adhesion, absorption, mechanical strength, antibacterial property, and a moist wound environment using CS-based biocomposites because the variation in the rate of production of wound exudates and the variation in the appearance of the wound surface should be considered, although the advancement of tissue engineering technologies can help limitation of the single wound dressing. Also, the emergence of novel CS-based biocomposites to mimic the skin environment, conditions and structure is very important in the management of various wound types. Furthermore, drug- and growth factors-loaded CS-based biocomposites can stimulate wound healing responses to promote optimal treatment.

Tissue engineering technique to mimic the ECM for the regulation of cellular behaviors is generally satisfied with 3D constructs as well as more closely to mimic the *in vivo* micro/nanoarchitecture for improving the function of tissue-engineered constructs because highly porous and fortified 3D molds are very critical for the bone regeneration. Biodegradability of the 3D scaffolds is very important in harmony with bone regeneration although the choice of appropriate biomaterials depends on the particular site of application. Multifunctional injectable scaffolds including micro-/nano-hydrogels to deliver cells and growth factors for minimal surgical intervention are very promising because traditional tissue-engineered scaffolds provide painful procedures to create lesions and longer healing times. In this regard, material scientists to design scaffold architecture, polymer chemists to get optimum physicochemical properties, cell biologists for the regulation of cellular behaviors, and clinicians for the successful implant in clinical trials should be harmonized.

This review covers an overview of the current status of CS-based biocomposites in wound dressing and bone tissue engineering applications. It also discussed their current challenges and opportunities for future researches. We expect that this review will be truly helpful for the researchers working in the field of CS-based biocomposite-related approaches.

REFERENCES

1. Kim I.Y., Seo S.J., Moon H.S., Yoo M.K., Park I.Y., Kim B.C., Cho C.S. Chitosan and its derivatives for tissue engineering applications, *Biotechnology Advances* 2008;26:1–21.
2. Krajewska B. Membrane-based processes performed with use of chitin/chitosan materials. *Separation and Purification Technology* 2005;41:305–312.
3. Eugene K., Lee Y.L. Implantable applications of chitin and chitosan, *Biomaterials* 2003;24:2339–2349.
4. Domish M., Kaplan D., Skaugrud O. Standards and guidelines for biopolymers in tissue-engineered medical products: ASTM alginate and chitosan standard guides. *Annals of the New York Academy of Sciences* 2001;944:388.
5. Muzzarelli R.A.A. Chitins and chitosans for the repair of woundedskin, nerve, cartilage and bone. *Carbohydrate Polymers* 2009;76:167–182.
6. Sudheesh Kumar P.T., Abhilash S., Manzoor K., Nair S.V., Tamura H., Jayakumar R. Preparation and characterization of novel-chitin/nano silver composite scaffolds for wound dressing applications. *Carbohydrate Polymers* 2010;80:761–777.
7. Vacanti C.A. The history of tissue engineering. *Journal of Cellular and Molecular Medicine* 2006;10:569–576.
8. Krajewska B. Membrane-based processes performed with use of chitin/chitosan materials. *Separation and Purification Technology* 2005;41:305–312.
9. Dev A, Binulal NS, Anitha A, Nair SV, Furuike T, Tamura H, Jayakumar R. Preparation of novel poly(lactic acid)/chitosan nanoparti-cles for anti HIV drug delivery applications. *Carbohydrate Polymers* 2010;80:833–838.
10. Yi H., Wu L.Q., Bentley W.E., Ghodssi R., Rubloff G.W., Culver J.N., Payne G.F. Biofabrication with chitosan. *Biomacromolecules* 2005;6:2881–2894.
11. Muramatsu K., Masuda S., Yoshihara S., Fujisawa A. In vitro degradation behavior of freeze-dried carboxymethyl-chitin sponges processed by vacuum-heating and gamma irradiation. *Polymer Degradation and Stability* 2003;81:327–332.
12. Drury J.L., Mooney D.J. Hydrogels for tissue engineering: Scaffold design variables and applications. *Biomaterials* 2003;24:4337–4351.
13. Prabaharan M., Mano J.F. Chitosan-based particles as controlled drugdelivery systems. *Drug Delivery* 2005;12:41–57.
14. Anitha A., Divya R.V.V., Krishna R., Sreeja V., Selvamurugan N., Nair S.V., Tamura H., Jayakumar R. Synthesis, characterization, cyto-toxicity and antibacterial studies of chitosan, O-carboxymethyl, N, O-carboxymethyl chitosan nanoparticles. *Carbohydrate Polymers* 2009;78:672–677.
15. Jayakumar R., Prabaharan M., Nair S.V., Tamura H. Novel chitin and chitosan nanofibers in biomedical applications. *Biotechnology Advances* 2010;28:142–150.
16. Jayakumar R., Reis R.L., Mano J.F. Synthesis and characterization of pH-sensitive thiol-containing chitosan beads for controlled drug delivery applications. *Drug Delivery* 2007;14:9–17.
17. Ehrlich H., Krajewska B., Hanke T., Born R., Heinemann S., Knieb C., Worch H. Chitosan membrane as a template for hydroxyap-atite crystal growth in a model dual membrane diffusion system. *Journal of Membrane Science* 2006;273:124–128.
18. Khalil H.P.S.A., Saurabh C.K., Adnan A.S., Fazita M.R.N., Syakir M.I., Davoudpour Y., Rafatullah M., Abdullah C.K., Haafiz M.K.M., Dungani R. A review on chitosan-cellulose blends and nanocellulose reinforced chitosan biocomposites: Properties and their applications. *Carbohydrate Polymers* 2016;150:216–226.

19. Suarato G., Rosalia Bertorelli R., Athanassiou A. Borrowing From Nature: Biopolymers and Biocomposites as Smart Wound Care Materials. *Frontiers in Bioengineering and Biotechnology* 2018;6:137. doi:10.3389/fbioe.2018.00137W-6.
20. Parsons D., Bowler P.G., Myles V., Jones S. Silver antimicrobial dressings in wound management, a comparison of anti-bacterial physical and chemical characteristics. *Wounds* 2005;17:222–232.
21. Mayet N., Choonara Y.E., Kumar P., Tomar L.K., Tyagi C., Du Toit L.C., Pillay V. A comprehensive review of advanced biopolymeric wound healing systems. *Journal of Pharmaceutical Sciences* 2014;103:2211–2230. doi:10.1002/jps.24068.
22. Kallaiinen L.K., Gordillo G.M., Schlanger R.K., Sen C. Topical oxygen as adjunct to wound healing, a clinical case series. *Pathophysiology* 2003;9:208–216.
23. Lloyd A.W. Interfacial bioengineering to enhance surface biocompatibility. *Medical Device Technologies* 2002;13:18–21.
24. Boateng J.S., Matthews K.H., Stevens N.E., Eccleston G.M. Wound healing dressings and drug delivery systems: A review. *Journal of Pharmaceutical Sciences* 2008;97:2892–2923.
25. Harkins A.L., Duri S., Kloth L.C., Tran, C.D. Chitosan-cellulose composite for wound dressing material. Part 2. Antimicrobial activity, blood absorption ability, and biocompatibility. *Journal of Biomedical Materials Research Part B* 2014;102(6):1199–1206.
26. Leuner C., Dressman J. Improving drug solubility for oral delivery using solid dispersions. *European Journal of Pharmaceutics and Biopharmaceutics* 2000;50(1):47–60.
27. Jiang Q., Zhou W., Wang J., Tang R., Zhang D., Wang X. Hypromellose succinate-crosslinked chitosan hydrogel films for potential wound dressing. *International Journal of Biological Macromolecules* 2016;91:85–91.
28. Kim I.Y., Yoo M.K., Kim B.C., Kim S.K., Lee H.C., Cho C.S. Preparation of semi-interpenetrating polymer networks composed of chitosan and poloxamer. *International Journal of Biological Macromolecules* 2006;38:51–58.
29. Kim I.Y., Yoo M.K., Seo J.H., Park S.S., Na H.S., Lee H.C., Kim S.K., Cho C.S. Evaluation of semi-interpenetrating polymer networks composed of chitosan and poloxamer for wound dressing application. *International Journal of Pharmaceutics* 2007;341:35–43.
30. Akhavan-Kharazian N., Izadi-Vasafi H. Preparation and characterization of chitosan/gelatin/nanocrystalline cellulose/calcium peroxide films for potential wound dressing applications. *International Journal of Biological Macromolecules* 2019;133:881–891.
31. Gao Y., Zhang X., Jin X. Preparation and properties of minocycline-loaded carboxymethyl chitosan gel/alginate nonwovens composite wound dressings. *Marine Drugs* 2019;17(10):575. doi:10.3390/md17100575.
32. Pathalamuthu P., Siddharthan A., Giridev V.R., Victoria V., Thangam R., Sivasubramanian S., Savariar V., Hemamalini T. Enhanced performance of Aloe vera incorporated chitosan-polyethylene oxide electrospun wound scaffold produced using novel Spirograph based collector assembly. *International Journal of Biological Macromolecules* 2019;140:808–824.
33. Rahmani H., Najafi S.H.M., Ashori A., Fashapoyeh M.A., Mohseni F.A., Torkaman S. Preparation of chitosan-based composites with urethane cross linkage and evaluation of their properties for using as wound healing dressing. *Carbohydrate Polymers* 2020;230:115606. doi:10.1016/j.carbpol.2019.115606.
34. Khalili R., Zarrintaj P., Jafari S.H., Vahabi H., Saeb M.R. Electroactive poly (p-phenylene sulfide)/r-graphene oxide/chitosan as a novel potential candidate for tissue engineering. *International Journal of Biological Macromolecules* 2020;154:18–24.
35. Atoufi Z., Zarrintaj P., Motlagh G.H., Amiri A., Bagher Z., Kamrava S.K. A novel bio electro active alginate-aniline tetramer/agarose scaffold for tissue engineering:

Synthesis, characterization, drug release and cell culture study. *Journal of Biomaterials Science, Polymer* 2017;15:1617–1638.
36. Diez-Pascual A.M., Diez-Vicente A.L. Wound healing bionanocomposites based on castor oil polymeric films reinforced with chitosan-modified ZnO nanoparticles. *Biomacromolecules* 2015;16:2631–2644.
37. Díez-Pascual A.M., Xu C.P., Luque R. Development and characterization of novel poly(ether ether ketone)/ZnO bionanocomposites. *Journal of Materials Chemistry B* 2014;2:3065–3078.
38. Amalraj A., Haponiuk J.T., Thomas S., Gopi S. Preparation, characterization and antimicrobial activity of polyvinyl alcohol/gum arabic/chitosan composite films incorporated with black pepper essential oil and ginger essential oil. *International Journal of Biological Macromolecules* 2020;151:366–375.
39. Rakmai J., Cheirsilp B., Mejuto J.C., Torrado-Agrasar A., Simal-Gandara J. Physicochemical characterization and evaluation of bio-efficacies of black pepper essential oil encapsulated in hydroxypropyl-betacyclodextrin. *Food Hydrocolloids* 2017;65: 157–164.
40. Haider A., Haider S., Kang I.K., Kumar A., Kummara M.R., Kamal T., Han S.S., A novel use of cellulose based filter paper containing silvernanoparticles for its potential application as wound dressing agent. *International Journal of Biological Macromolecules* 2018;108:455–461.
41. Alavi M., Rai M. Recent advances in antibacterial applications of metal nanoparticles (MNPs) and metal nanocomposites (MNCs) against multidrug-resistant (MDR) bacteria. *Expert Review of Anti-infective Therapy* 2019;17(6):419–428.
42. Sathiyaseelan A., Shajahan A., Kalaichelvan P.T., Kaviyarasan V. Fungal chitosan based nanocomposites sponges—An alternative medicine for wound dressing. *International Journal of Biological Macromolecules* 2017;104:1905–1915.
43. Cahú T.B., Silva R.A., Silva R.P.F., Silva M.M., Arruda I.R.S., Silva J.F., Costa R.M.P.B., Santos S.D., Nader H.B., Bezerra R.S. Evaluation of chitosan-based films containing gelatin, Chondroitin 4-sulfate and ZnO for wound healing. *Applied Biochemistry and Biotechnology* 2017;183:765–777. doi:10.1007/s12010-017-2462-z.
44. Kim, M.S., Park, S.J., Gu, B.K., Kim C.H. Fabrication of chitosan nanofibers scaffolds with small amount gelatin for enhanced cell viability. *Applied Mechanics and Materials* 2015;749:220–224.
45. Sergi R., Bellucci D., Salvatori R., Cannillo V. Chitosan-based bioactive glass gauze: Microstructural properties, in vitro bioactivity, and biological tests. *Materials* 2020;13(-12):2819. doi:10.3390/ma13122819.
46. Luz, G.M., Mano, J.F. Chitosan/bioactive glass nanoparticles composites for biomedical applications. *Biomedical Materials* 2012;7(5):054104.
47. Xia G., Zhai D., Sun Y., Hou L., Guo X., Wang L., Li Z., Wang F. Preparation of a novel asymmetric wettable chitosan-based sponge and its role in promoting chronic wound healing. *Carbohydrate Polymers* 2020;227:115296. doi:10.1016/j.carbpol.2019.115296.
48. Mohanty C., Das M., Sahoo S.K. Sustained wound healing activity of curcumin loaded oleic acid based polymeric bandage in a rat model. *Molecular Pharmaceutics* 2012;9(10):2801–2811.
49. Negi P., Sharma G., Verma C., Garg P., Rathore C., Kulshrestha S., Lal U.R., Gupta B., Pathania D. Novel thymoquinone loaded chitosan-lecithin micelles for effective wound healing: Development, characterization, and preclinical evaluation. *Carbohydrate Polymers* 2020;230:115659.
50. Mariod A.A., Ibrahim R.M., Ismail M., Ismail N. Antioxidant activity and phenolic content of phenolic rich fractions obtained from black cumin (Nigella sativa) seedcake. *Food Chemistry* 2009;116:306–312.

51. Ehterami A., Salehi M., Farzamfar S., Samadian H., Vaez A., Ghorbani S., Ai J., Sahrapeyma H. Chitosan/alginate hydrogels containing Alpha-tocopherol for wound healing in rat model. *Journal of Drug Delivery Science and Technology* 2019;51:204–213.
52. Biesalski H.K. Polyphenols and inflammation: basic interactions. *Current Opinion in Clinical Nutrition & Metabolic Care* 2007;10(6):724–728.
53. Augustine R., Dan P., Schlachet I., Rouxel D., Menu P., Sosnik A. Chitosan ascorbate hydrogel improves water uptake capacity and cell adhesion of electrospun poly(epsilon-caprolactone) membranes. *International Journal of Pharmaceutics* 2019;559:420–426.
54. Wang X., Jia H.C., Feng Y.M., Hong L.H. Chitosan-ascorbate for periodontal tissue healing and regeneration in a rat periodontitis model: An effectiveness validation. *Journal of Clinical Rehabilitative Tissue Engineering Research* 2010;14(12):2268–2272.
55. Zhu L., Liu X., Du L., Jin Y. Preparation of asiaticoside-loaded coaxially electrospinning nanofibers and their effect on deep partial-thickness burn injury. *Biomedicine & Pharmacotherapy* 2016;83:33–40.
56. Qiu J., Yu L., Zhang X., Wu Q., Wang D., Wang X., Xia C., Feng H. Asiaticoside attenuates lipopolysaccharide-induced acute lung injury via down-regulation of NF-kB signaling pathway. *International Immunopharmacology* 2015;26:181–187.
57. Zhao X., Li P., Guo B., Ma P.X. Antibacterial and conductive injectable hydrogels based on quaternized chitosan-graft-polyaniline/oxidized dextran for tissue engineering. *Acta Biomaterialia* 2015;26:236–248.
58. Jun I., Jeong S., Shin H. The stimulation of myoblast differentiation by electrically conductive sub-micron fibers. *Biomaterials* 2009;30:2038–2047.
59. Qu J., Zhao X., Liang Y., Zhang T., Ma P.X., Guo B. Antibacterial adhesive injectable hydrogels with rapid self-healing, extensibility and compressibility as wound dressing for joints skin wound healing. *Biomaterials* 2018;183:185–199.
60. Zhang B., He J., Shi M., Liang Y., Guo B. Injectable self-healing supramolecular hydrogels with conductivity and photo-thermal antibacterial activity to enhance complete skin regeneration. *Chemical Engineering Journal* 2020;400:125994.
61. Song R., Zheng J., Liu Y., Tan Y., Yang Z., Song X., Yang S., Fan R., Zhang Y., Wang Y. A natural cordycepin/chitosan complex hydrogel with outstanding self-healable and wound healing properties. *International Journal of Biological Macromolecules* 2019;134:91–99.
62. Sundaram M.N., Amirthalingam S., Mony U., Varma P.K., Jayakumar R. Injectable chitosan-nano bioglass composite hemostatic hydrogel for effective bleeding control. *International Journal of Biological Macromolecules* 2019;129:936–943.
63. Ostomel T.A., Shi Q., Stucky G.D. Oxide hemostatic activity. *Journal of the American Chemical Society* 2006;128:8384–8385.
64. Tao F., Cheng Y., Shi X., Zheng H., Du Y., Xiang W., Deng H. Applications of chitin and chitosan nanofibers in bone regenerative engineering. *Carbohydrate Polymers* 2020;230:115658. doi:10.1016/j.carbpol.2019.115658.
65. Zhang M., Matinlinna J.P., Tsoi J.K.H., Liu W., Cui X., Lu W.W., Pan H. Recent developments in biomaterials for long-bone segmental defect reconstruction: A narrative overview. *Journal of Orthopaedic Translation* 2020;22: 26–33.
66. Shokri S., Movahedi B., Rafieinia M., Salehi H. A new approach to fabrication of Cs/BG/CNT nanocomposite scaffold towards bone tissue engineering and evaluation of its properties. *Applied Surface Science* 2015;357:1758–1764.
67. Jones J.R. Acta biomaterialia review of bioactive glass: From Hench to hybrids. *Acta Biomaterialia* 2013;9:4457–4486.
68. Nazemi K., Azadpour P., Moztarzadeh F., Urbanska A.M., Mozafari M. Tissue-engineered chitosan/bioactive glass bone scaffolds integrated with PLGA nanoparticles: A therapeutic design for on-demand drug delivery. *Materials Letters* 2015;138:16–20.

69. Jalali N., Moztarzadeh F., Mozafari M., Asgari S., Motevalian M., Alhosseini S.N. Surface modification of poly(lactide-co-glycolide) nanoparticles by D-α-tocopheryl polyethylene glycol 1000 succinate as potential carrier for the delivery of drugs to the brain. *Colloids and Surfaces A: Physicochemical and Engineering Aspects* 2011;392(1):335–342.
70. Pourhaghgouy M., Zamanian A., Shahrezaee M., Masouleh M.P. Physicochemical properties and bioactivity of freeze-cast chitosan nanocomposite scaffolds reinforced with bioactive glass. *Materials Science and Engineering C* 2016;58:180–186.
71. Zhang S., Prabhakaran M.P., Qin X., Ramakrishna S. Biocomposite scaffolds for bone regeneration: Role of chitosan and hydroxyapatite within poly-3-hydroxybutyrate-co-3-hydroxyvalerate on mechanical properties and in vitro evaluation. *Journal of the Mechanical Behavior of Biomedical Materials* 2015;51:88–98.
72. Zhao D., Cai L., Wu J., Li M., Liu H., Han J., Zhou J., Xiang H. Improving polyhydroxyalkanoate production by knocking out the genes involved in exopolysaccharide biosynthesis in Haloferax mediterranei. *Applied Microbiology and Biotechnology* 2013;97:3027–3036.
73. Maji K., Dasgupta S., Kundu B., Bissoyi A. Development of gelatin-chitosan-hydroxyapatite based bioactive bone scaffold with controlled pore size and mechanical strength. *Journal of Biomaterials Science, Polymer Edition* 2015;26(16):1190–1209.
74. Kim H.L., Jung G.Y., Yoon J.H., Han J.S., Park Y.J., Kim D.G., Zhang M., Kim D.J. Preparation and characterization of nano-sized hydroxyapatite/alginate/chitosan composite scaffolds for bone tissue engineering. *Materials Science and Engineering C* 2015;54:20–25.
75. Acevedo C.A., Olguín Y., Briceño M., Forero J.C., Osses N., Díaz-Calderón P., Jaques A., Ortiza R. Design of a biodegradable UV-irradiated gelatin-chitosan/nanocomposed membrane with osteogenic ability for application in bone regeneration. *Materials Science and Engineering C* 2019;99:875–886.
76. Teimouri A., Ebrahimi R., Emadi R., Beni B.H., Chermahini A.N. Nano-composite of silk fibroin–chitosan/Nano ZrO_2 for tissue engineering applications: Fabrication and morphology. *International Journal of Biological Macromolecules* 2015;76:292–302.
77. Pattnaik S., Nethala S., Tripathi A., Saravanan S., Moorthi A., Selvamurugan N. Chitosan scaffolds containing silicon dioxide and zirconia nano particles for bone tissue engineering. *International Journal of Biological Macromolecules* 2011;49:1167–1172.
78. Zhou Y., Xu L., Zhang X., Zhao Y., Wei S., Zhai M. Radiation synthesis of gelatin/CM-chitosan/β-tricalcium phosphate composite scaffold for bone tissue engineering. *Materials Science and Engineering C* 2012;32(4):994–1000.
79. Yaszemski M.J., Payne R.G., Hayes W.C., Langer R., Mikos A.G. Evolution of bone transplantation: Molecular, cellular and tissue strategies to engineer human bone. *Biomaterials* 1996;17:175–185.
80. Serra I.R., Fradique R., Vallejo M.C.S., Correia T.R., Miguel S.P., Correia I.J. Production and characterization of chitosan/gelatin/β-TCP scaffolds for improved bone tissue regeneration. *Materials Science and Engineering C* 2015;55:592–604.
81. Puvaneswary S., Talebian S., Raghavendran H.B., Muralia M.R., Mehrali M., Afifi A.M., Kasim N.H.B.A., Kamarul T. Fabrication and in vitro biological activity of βTCP-Chitosan-Fucoidan composite for bone tissue engineering. *Carbohydrate Polymers* 2015;134:799–807.
82. Park S.J., Lee K.W., Lim D.S., Lee S. The sulfated polysaccharide fucoidan stimulates osteogenic differentiation of human adipose-derived stem cells. *Stem Cells Development*, 2012;21(12):2204–2211.
83. Vacanti J.P., Vacanti C.A., Lanza R.P., Langer R., Vacanti J. *Principles of Tissue Engineering*, Second ed., Academic Press, Cambridge, MA, 2000, pp. 3–9.
84. Langer R., Vacanti J.P., Tissue engineering. *Science* 1993;260:920–926.

85. Roseti L., Parisi V., Petretta M., Cavallo C., Desando G., Bartolotti I., Grigolo B. Scaffolds for bone tissue engineering: State of the art and new perspectives. *Materials Science and Engineering C* 2017;78:1246–1262.
86. Thavornyutikarn B., Chantarapanich N., Sitthiseripratip K., Thouas G.A., Chen Q., Bone tissue engineering scaffolding: Computer-aided scaffolding techniques. *Progress in Biomaterials* 2014;3:61–102.
87. Gravel M., Vago R., Tabrizian M. Use of natural coralline biomaterials as reinforcing and gas-forming agent for developing novel hybrid biomatrices: Microarchitectural and mechanical studies. *Tissue Engineering* 2006;12(3):589–600.
88. Gravel M., Gross T., Vago R., Tabrizian M. Responses of mesenchymal stem cell to chitosan–coralline composites microstructured using coralline as gas forming agent. *Biomaterials* 2006;27:1899–1906.
89. Matassi F., Nistri L., Paez D.C., Innocenti M. New biomaterials for bone regeneration. *Clinical Cases in Mineral and Bone Metabolism* 2011;8(1):21–24.
90. Kolanthai E., Sindu P.A., Khajuria D.K., Veerla S.C., Kuppuswamy D., Catalani L.H., Mahapatra D.R. Graphene oxide-a tool for the preparation of chemically crosslinking free alginate–chitosan–collagen scaffolds for bone tissue engineering. *ACS Applied Materials Interfaces* 2018;10:12441–12452.
91. Zhou T., Li G., Lin S., Tian T., Ma Q., Zhang Q., Shi S., Xue C., Ma W., Cai X. Electrospun poly (3-Hydroxybutyrate-Co-4-Hydroxybutyrate)/Graphene oxide scaffold: Enhanced properties and promoted in vivo bone repair in rats. *ACS Applied Materials Interfaces* 2017;9:42589–42600.
92. Demir A.K., Elçin A.E., Elçin Y.M. Strontium-modified chitosan/montmorillonite composites as bone tissue engineering scaffold. *Materials Science and Engineering C* 2018;89:8–14.
93. Aguzzi C., Cerezo P., Viseras C., Caramella C. Use of clays as drug delivery systems: Possibilities and limitations. *Applied Clay Science* 2007;36:22–36.
94. Bonnelye E., Chabadel A., Saltel F., Jurdic P. Dual effect of strontium ranelate: Stimulation of osteoblast differentiation and inhibition of osteoclast formation and resorption in-vitro. *Bone* 2008;42:129–138.
95. Pineda-Castillo S., Bernal-Ballén A., Bernal-López C., Segura-Puello H., Nieto-Mosquera D., Villamil-Ballesteros A., Muñoz-Forero D., Munster L. Synthesis and characterization of poly(Vinyl Alcohol)-chitosan-hydroxyapatite scaffolds: A promising alternative for bone regeneration. *Molecules* 2018;23:2414. doi:10.3390/molecules 23102414.
96. Mi Zo S., Singh D., Kumar A., Cho Y.W., Oh T.H., Han S.S. Chitosan-hydroxyapatite macroporous matrix for bone tissue engineering. *Current Science* 2012;102:1438–1446.
97. Peng X., Hu M., Liao F., Yang F., Ke Q., Guo Y., Zhu Z., La-Doped mesoporous calcium silicate/chitosan scaffolds for bone tissue engineering. *Biomaterials Science* 2019;7:1565–1573.
98. Shahsavari R., Hwang S.H. Size- and shape-controlled synthesis of calcium silicate particles enables self-assembly and enhanced mechanical and durability properties. *Langmuir* 2018;34:12154–12166.
99. Shi D., Shen J., Zhang Z., Sci C., Chen M., Gu Y., Liu Y. Preparation and properties of dopamine-modified alginate/chitosan–hydroxyapatite scaffolds with gradient structure for bone tissue engineering. *Journal of Biomedical Materials Research Part A* 2019;107(8):1615–1627. doi:10.1002/jbm.a.36678.
100. Sadeghinia A., Davaran S., Salehi R., Jamalpoor Z. Nano-hydroxy apatite/chitosan/gelatin scaffolds enriched by a combination of platelet-rich plasma and fibrin glue enhance proliferation and differentiation of seeded human dental pulp stem cells. *Biomedicine & Pharmacotherapy* 2019;109:1924–1931.

101. Kokubo T., Kim H.M., Kawashita M. Novel bioactive materials with different mechanical properties. *Biomaterials* 2003;24(13):2161–2175.
102. Cao H., Kuboyama N. A biodegradable porous composite scaffold of PGA/beta-TCP for bone tissue engineering. *Bone* 2010;46:386–395.
103. Jamalpoor Z., Mirzadeh H., Joghataei M.T., Zeini D., Bagheri-Khoulenjani S., Nourani M.R. Fabrication of cancellous biomimetic chitosan-based nanocomposite scaffolds applying a combinational method for bone tissue engineering. *Journal of Biomedical Materials Research Part A* 2015:103A:1882–1892.
104. Ruixin L., Cheng X., Yingjie L., Hao L., Caihong S., Weihua S., Weining A., Yinghai Y., Xiaoli Q., Yunqiang X., Xizheng Z., Hui L. Degradation behavior and compatibility of micro, nanoHA/chitosanscaffolds with interconnected spherical macropores. *International Journal of Biological Macromolecules* 2017;103:385–394.
105. Wang Y., Qian J., Zhao Z., Liu T., Xu W., Suo A. Novel hydroxyethyl chitosan/cellulose scaffolds with bubble-likeporous structure for bone tissue engineering. *Carbohydrate Polymers* 2017;167:44–51.
106. Zhang T., Zhang H., Zhang L., Jia, S. Liu J., Xiong Z., Sun W. Biomimetic design and fabrication of multilayered osteochondral scaffolds by low-temperature deposition manufacturing and thermal-induced phase-separation techniques. *Biofabrication* 2017;9:025021.
107. Rahman S., Rana M., Spitzhorn L.S., Akhtar N., Hasan Z., Choudhury N., Fehm T., Czernuszka J.T., Adjaye J., Asaduzzaman S.M. Fabrication of biocompatible porous scaffolds based on hydroxyapatite/collagen/chitosan composite for restoration of defected maxillofacial mandible bone. *Progress in Biomaterials* 2019;8:137–154.
108. Erickson A.E., Sun J., Levengood S.K.L., Swanson S., Chang F.C., Tsao C.T., Zhang M. Chitosan-based composite bilayer scaffold as an in vitro osteochondral defect regeneration model. *Biomedical Microdevices* 2019;21:34.
109. Cima M.J., Sachs E., Cima L.G., Yoo J., Khanuja S., Borland S., Wu B., Giordano R. Computer derived microstructures by 3D printing: bio-and structural materials. *International Solid Freeform Fabrication Symposium Proc: DTIC Doc* 1994;181–190.
110. Chia H.N., Wu B.M. Recent advances in 3D printing of biomaterials. *Journal of Biological Engineering* 2015;9:4. doi:10.1186/s13036-015-0001-4.
111. Yang Y., Chu L., Yang S., Zhang H., Qin L., Guillaume O., Eglin D., Richards R.G., Tang T. Dual-functional 3D-printed composite scaffold for inhibiting bacterial infection and promoting bone regeneration in infected bone defect models. *Acta Biomaterialia* 2018;79:265–275.
112. Dong L., Wang S.J., Zhao X.R., Zhu Y.F., Yu J.K. 3D- printed poly(ε-caprolactone) scaffold integrated with cell-laden chitosan hydrogels for bone tissue engineering. *Scientific Reports* 2017;7:13412. doi:10.1038/s41598-017-13838-7.
113. Tsai C.H., Hung C.H., Kuo C.N., Chen C.Y., Peng Y.N., Shie M.Y. Improved bioactivity of 3D printed porous titanium alloy scaffold with chitosan/magnesium-calcium silicate composite for orthopaedic applications. *Materials* 2019;12:203. doi:10.3390/ma12020203.
114. Chen S., Shi Y., Zhang X., Ma J. Biomimetic synthesis of Mg-substituted hydroxyapatite nanocomposites and three-dimensional printing of composite scaffolds for bone regeneration. *Journal of Biomedical Materials Research* 2019;107:2512–2521.
115. Chen T., Zou Q., Du C., Wang C., Li Y., Fu B. Biodegradable 3D printed HA/CMCS/PDA scaffold for repairing lacunar bone defect. *Materials Science & Engineering C* 2020;116:111148.
116. Liu Y., Ai K., Lu L. Polydopamine and its derivative materials: Synthesis and promising applications in energy, environmental, and biomedical fields. *Chemical Reviews* 2014;114(9):5057–5115.

117. Jing X., Mi H.Y., Peng J., Peng X.F., Turng L.S. Electrospun aligned poly(propylene carbonate) microfibers with chitosan nanofibers as tissue engineering scaffolds. *Carbohydrate Polymers* 2015;117:941–949.
118. Venkatesan J., Anil S., Kim S.K., Shim M.S. Chitosan as a vehicle for growth factor delivery: Various preparations and their applications in bone tissue regeneration. *International Journal of Biological Macromolecules* 2017;104:1383–1397.
119. Even J., Eskander M., Kang J. Bone morphogenetic protein in spine surgery: Current and future uses. *Journal of the American Academy of Orthopaedic Surgeons* 2012;20:547–552.
120. Sobhani A., Rafienia M., Ahmadian M., Naimi-Jamal M.R. Fabrication and characterization of polyphosphazene/calcium phosphate scaffolds containing chitosan microspheres for sustained release of bone morphogenetic protein 2 in bone tissue engineering. *Tissue Engineering and Regenerative Medicine* 2017;14(5):525–538.
121. Bastami F., Paknejad Z., Jafari M., Salehi M., Rad M.R., Khojasteh A. Fabrication of a three-dimensional β-tricalcium-phosphate/gelatin containing chitosan-based nanoparticles for sustained release of bone morphogenetic protein-2: Implication for bone tissue engineering. *Materials Science and Engineering C* 2017;72:481–491.
122. Deng N., Sun J., Li Y., Chen L., Chen C., Wu Y., Wang Z., Li L. Experimental study of rhBMP-2 chitosan nano-sustained release carrierloaded PLGA/nHA scaffolds to construct mandibular tissue-engineered bone. *Archives of Oral Biology* 2019;102:16–25.
123. Tong S., Xu D.P., Liu Z.M., Du Y., Wang X.K. Synthesis of and *in vitro* and *in vivo* evaluation of a novel TGF-β1-SF-CS three-dimensional scaffold for bone tissue engineering. *International Journal of Molecular Medicine* 2016;38:367–380.
124. Baylink D.J., Finkelman R.D., Mohan S. Growth factors to stimulate bone formation. *Journal of Bone and Mineral Research* 1993;8(suppl 2):S565–S572.
125. Oryan A., Alidadi S., Bigham-Sadegh A., Moshiri A., Kamali A. Effectiveness of tissue engineered chitosan-gelatin composite scaffold loaded with human platelet gel in regeneration of critical sized radial bone defect in rat. *Journal of Controlled Release* 2017;254:65–74.
126. Intini G. The use of platelet-rich plasma in bone reconstruction therapy. *Biomaterials* 2009;30:4956–4966.
127. Liao H.T., Tsai M.J., Brahmayya M., Chen J.P. Bone regeneration using adipose-derived stem cells in injectable thermo-gelling hydrogel scaffold containing platelet-rich plasma and biphasic calcium phosphate. *International Journal of Molecular Sciences* 2018;19:2537. doi:10.3390/ijms19092537.
128. Wang B., Guo Y., Chen X., Zeng C., Hu Q., Yin W., Li W., Xie H., Zhang B., Huang X., Yu F. Nanoparticle-modified chitosan-agarose-gelatin scaffold for sustained release of SDF-1 and BMP-2. *International Journal of Nanomedicine* 2018;13:7395–7408.
129. Dou D.D., Zhou G., Liu H.W., Zhang J., Liu M.L., Xiao X.F., Fei J.J., Guan X.L., Fan Y.B. Sequential releasing of VEGF and BMP-2 in hydroxyapatite collagen scaffolds for bone tissue engineering: Design and characterization. *International Journal of Biological Macromolecules* 2019;123:622–628.
130. Krishnan L., Priddy L.B., Esancy C., Klosterhoff B.S., Stevens H.Y., Tran L., Guldberg R.E. Delivery vehicle effects on bone regeneration and heterotopic ossification induced by high dose BMP-2. *Acta Biomaterialia* 2017;49:101–112.
131. Choi B.H., Zhu S.J., Kim B.Y., Huh J.Y., Lee S.H., Jung J.H. Effect of plateletrich plasma (PRP) concentration on the viability and proliferation of alveolar bone cells: an in vitro study. *International Journal of Oral and Maxillofacial Surgery* 2005;34(4):420–424.
132. Nihouannen D.L., Saffarzadeh A., Gauthier O., Moreau F., Pilet P., Spaethe R., Layrolle P., Daculsi G. Bone tissue formation in sheep muscles induced by a biphasic calcium phosphate ceramic and fibrin glue composite. *Journal of Materials Science: Materials in Medicine* 2008;19(2):667–675.

12 Starch-Based Biocomposites
Opportunity and Challenge

Ankit Manral
Netaji Subhas Institute of Technology

Ranjana Mishra and Rahul Joshi
Netaji Subhas University of Technology

CONTENTS

12.1	Introduction	320
12.2	Starch	321
	12.2.1 Physical and Chemical Properties of Starch	322
	12.2.2 Starch Modification	322
	12.2.2.1 Esterification	324
	12.2.2.2 Cross-Linking	324
	12.2.2.3 Stabilization	325
	12.2.3.4 Pregelatinization	325
12.3	Processing of Starch-Based Materials	325
	12.3.1 Rheological Properties of Starch-Based Polymers	325
	12.3.2 Effects of Plasticizers and Additives	326
	12.3.3 Techniques Used in Processing of Starch-Based Materials	326
	12.3.3.1 Sheet/Film Extrusion	326
	12.3.3.2 Foaming Extrusion	328
	12.3.3.3 Injection Molding	328
	12.3.3.4 Compression Molding	328
	12.3.3.5 Film Casting	328
	12.3.3.6 Reactive Extrusion	328
12.4	Starch-Based Biocomposites and Nano-Biocomposites	329
	12.4.1 Starch-Based Nano-Biocomposites Reinforced by Phyllosilicates	329
	12.4.2 Starch-Based Nano-Biocomposites Reinforced by Polysaccharide Nanofillers	329
	12.4.3 Starch-Based Nano-Biocomposites Reinforced by Carbonaceous Nanofillers	331
12.5	Starch Bio-Composites: Present Trends and Challenges	331
12.6	Conclusion	334
References		334

DOI: 10.1201/9781003137535-12

12.1 INTRODUCTION

Polymers along with composites have evolved into indispensable components of mankind over the last few decades. The high aspect ratios of polymeric materials have caught the attention of various industries around the world [1]. Compared to metals and alloys, they can render better mechanical, thermal, insulating, conducting, flame retardant, and corrosion-resistant properties with substantial weight reduction. They are used in several fields such as construction, automobile, aerospace, agriculture, sports, etc. Increasing population along with evolution of industrial products has escalated the production of synthetic polymers and allied products [1]. In the ecosystem, these synthetic polymers accumulate, posing a growing ecological threat to wild terrestrial and marine life. According to survey polyethylene is the most used synthetic polymer with a current global supply of approximately 140 million tons annually [2]. To overcome this problem and looking into the recent environmental requirements, researchers have intensified the search for newer green materials and technologies.

Over several decades, biopolymers are gaining popularity in the search for biodegradable and sustainable materials. Broadly defined, biopolymers or biocomposites are composite materials made from natural fiber and petroleum-derived nonbiodegradable polymers or biodegradable polymers. Since both components are biodegradable, the composite is also considered to be biodegradable as an integral element [3,4]. Figure 12.1 shows the classification of biopolymers based on their synthesis. According to classification most of the polymers are obtained from biomass except fourth category of polymer, which is of fossil origin [5,6]. By minimizing reliance on fossil fuels and through the associated positive environmental impacts, such as decreased carbon dioxide emissions, bio-based polymers make significant contributions. The Japan Bioplastic Association predicted that by the year 2020, the demand for biocomposite products in the market will reach 20% of the overall use of plastic [1]. Biopolymers have numerous special properties although it possesses some deficiencies like poor process ability, high water sensitivity, weak mechanical properties, etc. It is possible to substitute and reinforce such defects with a wide variety of fillers, like nanoparticles, layered silicates in specific polymers, etc., that have resulted in significant enhancement in its structural and thermal properties [7]. Starch is the second most abundant biopolymer after cellulose and, because of its unique properties; it has gained abundant recognition by the researchers for the development of eco-friendly polymers and composites. With the properties of starch like biodegradability, availability, renewability, and inexpensiveness, it has almost replaced conventional plastic. In 2003, the demand for starch-based bioplastics accounted for about 25,000 tons [1]. Around 70% of the global market for bioplastics accounted for the share of these products. In 2007, global consumption of starch-based biodegradable polymers raised up to 114,000 tons. However, according to the latest data published by the University of Utrecht, the production capacity shows a forecast increase to 810,000 tons by 2020 [8].

The total use of starch and starch derivatives in Europe in 2002 was around 7.9 million tons, 54% of which was utilized in food industry and 46% in industries other than food [9]. In the European Union, the main consumers of starch are the paper, cardboard, and corrugating industries approx. 30% of total production. Textiles,

Starch-Based Biocomposites

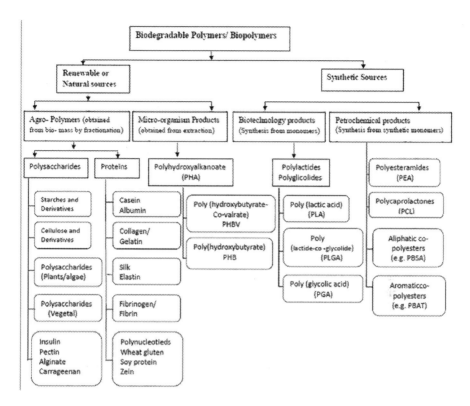

FIGURE 12.1 Classification of biodegradable polymers.

cosmetics, pharmaceuticals, manufacturing, and paint are other important areas of starch application [9]. Starch will play an increasing role in the field of renewable and biodegradable composites in the nearby future. A brief description of starch biopolymer nanocomposites and their properties is included in this chapter. The different processing techniques for the production of these nanocomposites are also highlighted. Different characterization techniques were also addressed in this chapter to study the effect of starch on properties of polymer and to analyze the morphology of fabricated parts.

12.2 STARCH

Starch is a polymer of renewable carbohydrates that can be produced from a wide variety of crops at a low cost [10]. The primary sources are waxy maize, corn, wheat, rice, cassava, potatoes, peas, amyloma, etc. [11]. The photosynthesis process initiates starch production in plants [1]. Starch granules exist predominantly in seeds, roots, and tubers of various origins [12]. The word "starch" has originated from German root that means strong or stiff. Pure starch is a white powder having no taste and odor andis generally consumed as carbohydrate [7]. Key source of starch is maize or maize starch, which makes up 80% of the global market [9]. Rice is the most widely consumed food products in the world and 500 million tons of it is produced annually [1,13]. Rice starches have specific qualities dependent on paddy varieties [14], results in biodegradable composites with different properties. Due its economical and ample supply along with comparable mechanical strength, it is being utilized for the manufacture decomposable films in place of synthetic polymers [15–17]. However, these rice starch composites have some shortcomings which restrict their uses [18]. These barrier properties can be improved by blending the rice starch with various proteins [1,19]. After cellulose, starch is the most produced organic material. The primary structure of starch is almost similar to cellulose, although the secondary and tertiary structures are the deciding factors that differentiate the characteristics of both the materials [20]. Starch granules are usually between 1 and 100 μm in size and are also available in various shapes [16]. It consists of a 6:10:5 [$C_6H_{10}O_5$] proportion of carbon, hydrogen, and oxygen [7,20]. There is a significant use of starch as raw material in the production of biodegradable plastics due to its inexpensiveness and ease of access for products such as detergent and insecticide water-soluble pouches, bags, and devices for medical delivery [1,9,10,21,22].

12.2.1 Physical and Chemical Properties of Starch

Starch is a polysaccharide produced as a means of storing energy by mostly higher-order plants. Physically, most native starches are semi-crystalline, having a crystallinity of about 20%–45% [23]. Generally, starch contains two microstructures—linear microstructure amylose and branched microstructure amylopectin. The ratio of amylose/amylopectin depends on the extraction source, method, and age of the starch. Amylose and the amylopectin branching points form amorphous regions. The primary crystalline component in granular starch is the short-branched chains associated with amylopectin. Crystalline regions are found in the form of double helices and the lengths of helices are approximately ~5 nm. Amylopectin molecular weight is around 100 times greater than that of amylose [24]. Starch granules also contain small amounts of lipids and proteins. Chemically, starch is a polymeric carbohydrate containing anhydro glucose units which are primarily linked together through α–d–(1→4) glucosidic bonds. Although the extensive microstructures of various starches are still being interpreted, it has been deduced that starch is a heterogeneous substance comprising of two microstructures—linear (amylose) and branched (amylopectin).

Amylose is primarily a linear microstructure of α-1,4 linked D-glucose units having chain length between 102 and 104 glucose unites [24,25]. Amylopectin is a network structure of short α-1,4 chains linked by α-1,6 bonds and the length of chain lies between 104 and 105 glucose units. In starch, relative amount of amylose and amylopectin ranges from 18% to 28% amylose and 72%–82% amylopectin [24,26]. Table 12.1 shows the chemical composition of different types of starch [5].

12.2.2 Starch Modification

Native starch has some limitations and barrier characteristics such as retrogradation, uncontrolled/highly viscous, brittleness, and insolubility in the cold water, etc., for its

TABLE 12.1
Chemical Composition of Different Types of Starch [5–7]

Starch	Amylopectin Content (%)	Amylose Content (%)	Protein Content (%)	Lipid Content (%)	Granule diameter (mm)	Crystallinity (%)	Phosphorus Content (%)	Moisture Content (%)
Waxy Starch	99	<1	0.10	0.23	15	39	0.01	N.d
Potato	79–74	20–25	0.05	0.03	40–100	25	0.08	18–19
Wheat	72–73	26–27	0.30	0.63	25	36	0.06	13
Maize	71–73	26–28	0.30	0.63	15	39	0.02	12–13
Amylomaizes	20–50	50–80	0.50	1.11	15	19	0.03	N.d

Note: N.d-not determined.

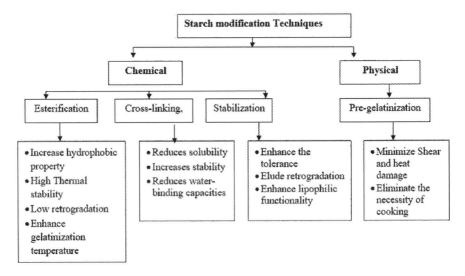

FIGURE 12.2 Modification methods of starch and their impact on starch granules.

end products. Owing to its unfavorable properties, native starch has a restricted range of applications. Starches are chemically and physically modified in order to mend their barrier properties. Researchers have studied various modification strategies, such as pregelatinization, esterification, cross-linking, and stabilization [27,28]. Figure 12.2 depicts generally used modification methods of and their impact on starch granules.

12.2.2.1 Esterification

Esterification is a chemical technique to modify and improve the anatomy of starch granules. In the modification, hydroxyl groups are replaced by ester groups [29]. A lot of starch esters, like starch phosphates, starch sulfates, starch nitrates, starch alkenyl succinate, and starches of fatty acids, are available. The aim of modifying is to improve the hydrophobic property and the resistivity of water so that it can be beneficially utilized in the pharmaceutical and beverage industries. Esterification also improves the gelatinization temperature, mechanical properties, and high temperature performance of composites [29,30]. It also results in slowing the degradation rate of esterified starch/PLA than unmodified starch/PLA [31].

12.2.2.2 Cross-Linking

It is one of the most frequently utilized chemical starch modifications techniques. Starch has been chemically treated in this process to link the starch molecules with the cross-bridges together. Most glucose units in starch consist of one main and two secondary groups of hydroxyls that interact effectively with various compounds, such as acid anhydrides, ethylene compounds, organic chloro compounds, and epoxy [7,27]. Cross-linking increases the heat resistance or resistant to gelatinization and it depends on number of cross-links [32]. It also decreases solubility of starch derivatives in water that provides more stability [33].

Starch-Based Biocomposites

12.2.2.3 Stabilization

Stabilization, another starch modification technique, is substituted on the starch to avoid the linear, dispersed fragments to retrograde, substantial groups include octenyl succinate, etc. [34]. The effect of stabilization is determined by the number of groups substituted. Due to the presence of the polyhydroxyl groups in the chain, the native starch is hydrophilic in character. It does not contain the required affinity to be an efficient emulsifier for hydrophobic compounds (e.g., oil) but the lipophilic functionality of starch can be improved by chemical modifications [35]. Stabilization enhances the various shortcomings of native starch such as retrogradation and the resistance of starch to temperature variations which helps in increase the shelf life of starch products [36].

12.2.3.4 Pregelatinization

Some research shows that many starches need cooking in order to produce the necessary properties [27]. Pregelatinization is the process of starch modification which eliminates the cooking requirement. In pregelatinization, mainly two methods, drum drying and spray cooking processes are generally used to achieve a wide range of cold thickening starch. This method overcomes the need for over-cooking, and minimizes shear and heat damage [7].

12.3 PROCESSING OF STARCH-BASED MATERIALS

Thermal processing properties (TPPs) are the special features of starch-based polymers and are much more intricate than conventional polymers. Generally, it occurs due to the different chemical and physical reactions in polymers. The presence of the intra- and intermolecular hydrogen bonds between the hydroxyl groups in starch molecules are responsible for the nonplastic behavior of native starch, which shows its crystallinity [24]. Thermal processing spontaneously disrupts and transforms the semi-crystalline structure of starch granules for the fabrication of an amorphous and homogeneous material. Gelatinization agents or plasticizers [37–46] are used in a very small amount to accomplish this transformation. The material obtained from this transformation is called thermoplastic starch (TPS). According to numerous research studies, extrusion [47–52], injection/compression molding [53–57], intensive mixing [58–61], and heat pressing [62] are the techniques used for manufacturing TPS materials.

12.3.1 RHEOLOGICAL PROPERTIES OF STARCH-BASED POLYMERS

As discussed above, TPPs are far more multifaceted than traditional polymers, due to occurrence of several physical and chemical reactions during processing, such as gelatinization, melting, crystallization, granular expansion, water diffusion, and decomposition. The material properties that govern the specific way in which deformation or flow behaviors occur are called rheological properties. Capillary rheometry is the most widely used approach to investigate the rheological properties of TPS. The rheological behavior of TPS is examined by the power-law model [41,44,45,49,55,57] as

$$\eta = K\gamma^{n-1}$$

where η = molten viscosity, K = consistency, γ = shear rate, and n = power-law (pseudo-plasticity) index. The apparent viscosity η of TPS usually decreases at constant temperature with a rise in plasticizer content [41,44,45,49,55]; the value of η also decreases with an increase in temperature at the same weight ratio of plasticizer [44,45,57]. Moreover, starches exhibit higher melt viscosities with higher amylose contents [41]. The value of K decreases with an increase of plasticizer at the same weight content because plasticizers can form a strong bonding interaction with starch which weakens the interaction of starch molecules and enables the movement among starch molecules [63–68]. As per reports the influence of plasticizer on n has been observed and it is found that n decreased with an increase in citric acid and glycerol as a plasticizer weight content [44,49,63]. On the other hand, n increases at higher temperature and thus make a starch more Newtonian, melt less, and pseudo-plastic [63,66].

12.3.2 Effects of Plasticizers and Additives

Without a plasticizer or gelatinizing agent, a starch-based polymer cannot be thermally processed because its decomposition temperature is lower than its pregelatinization melting temperature [24,69]. During thermal processing, multiple plasticizers and additives have been evaluated and developed to gelatinize starch. In the thermal processing of starch-based polymers water plasticizer is the most popular plasticizer. TPS, which contains only water as a plasticizer, limits the value of TPS in practical application because of poor mechanical properties, especially the brittleness occurring from fast retrogradation [70–72]. Various other plasticizers and polyols have been evaluated and used such as glycerol, sorbitol, glycol, sugars, ethanolamine, urea, and citric acid [41–44,46,49,56,58,59,61,62] to enhance the product performance and processing properties of TPS.

12.3.3 Techniques Used in Processing of Starch-Based Materials

Injection molding, extrusion, and film casting are the starch processing techniques and are similar to conventional petroleum-based plastics processing techniques. The processing of starch is much more complex and rigorous for conventional polymers, due to the undesirable properties such as unique phase transitions, fast retrogradation, water evaporation, high viscosity, etc. Extrusion is the most widely used technique for processing starch-based polymers. This technique has the ability to handle high-viscosity polymers without using any solvent, large operational flexibility, control of both residence time (distribution) and the degree of mixing and the feasibility of multiple-injection [27,73]. Injection and other processing techniques are often combined with extrusion. Figure 12.3 shows the various techniques used in processing of starch-based materials and their area of application.

12.3.3.1 Sheet/Film Extrusion

Using a twin-screw extruder with a slit or flat film die, accompanied by a takeoff mechanism for orientation and collection, is a basic and firmly established technique for creating sheets or films by extrusion [24]. In this process, viscoelastic starch

Starch-Based Biocomposites

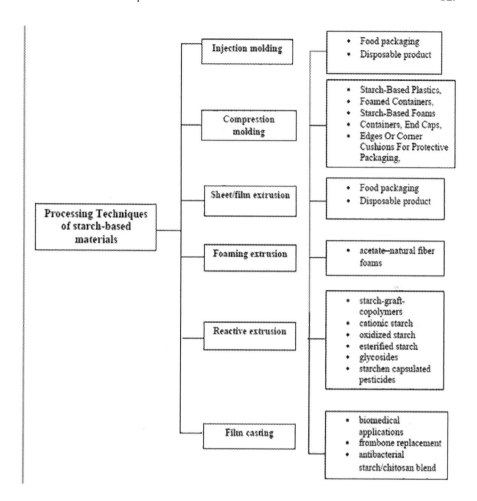

FIGURE 12.3 Processing techniques of starch-based materials and applications impact on starch granules.

material is enforced through the die to produce a film or sheet. Some researchers have already practiced a two-stage sheet/film extrusion processing technique [74–77].

In this method, first of all, starch blends are extruded in a twin-screw extruder to make ribbons and followed by the drying and grinding process. Then with the help of a single-screw extruder accompanied by a slit die is used to extrude flat sheets or films. Although this two-stage extrusion technique may be more time-consuming but various shortcomings such as poor processibility and high viscosity of starch-based materials can be removed by the single-screw extruder by the virtue of its high-pressure capacity, this makes the extrusion of starch sheets more stable and easier. The outlet of slit die is adjustable which helps in controlling the thickness of the extruded material. Two-stage technique can produce much thinner blown films as compare to slit/flat film die [74–82].

12.3.3.2 Foaming Extrusion

Foaming extrusion mainly consists of a single-screw extrusion [24]. However, twin-screw extrusion [24] offers more ease of feeding, more extensive shear, more flexible temperature control, and longer residence time. Cylindrical extrudates are produced by using a die nozzle. A spinning cutter mounted on the front of the extrusion die to cut those extrudates into finite lengths. This method is used to produced loose fill packaging materials like extruded expanded snack foods [24].

12.3.3.3 Injection Molding

Injection molding techniques are often combined with extrusion. This method is used to produce food packaging and disposable products. In injection molding lack of reliable parameters, poor flow properties, and high viscosity of starch based materials makes it difficult to design the optimum processing conditions.

12.3.3.4 Compression Molding

Compression molding involves mainly three processes, starch gelatinization, expanding of starch, and drying process. This technique has been intensively examined to process starch based plastics. The main product formed by this process is foamed containers. Several agents are used in formulations to prevent the starch sticking to compression mold such as steric and stearate acid along with gelatinization agents.

12.3.3.5 Film Casting

In film casting technique of processing starch based materials solution processing, gelatinization, casting, and drying processes are usually included [24,83]. Products formed by this process are mainly antibacterial starch/chitosan blend, including biomedical applications such as bone replacement materials. Starch and plasticizer are mixed together in water, maintaining the solid concentration range between 3% and 10% approximately. Later, this film/sheet forming solution is poured into a container like Brabender visco graph cup, and then the solution is heated up to 95°C temperature from room temperature, where it is kept for at least 10 min while being continually blended or shaken. While adding another plasticizer like glycerol or any to the formulation, it is mandatory to use a higher temperature. Later these gelatinized suspensions are poured onto any flat plates like acrylic or Teflon plates. Then keep this suspension to dry for about 42 h in the oven, maintaining the temperature of 40°C–75°C at a constant weight ratio. Cast film thickness generally ranges between 0.02 and 0.10 mm, which depends and is controlled by the quantity calculation of suspension poured onto the plate [24].

12.3.3.6 Reactive Extrusion

Unlike other processing techniques, reactive extrusion (REX) involves mainly two processes—concurrent reaction and extrusion. The technology behind this technique is already applied in numerous areas such as grafting, polymerization, and cross linking. Due to several advantages such as low-cost product manufacturing, high-efficiency, and increased demand in the field of biocomposites, the use of REX in starch modification received focus in this area also. A typical REX system includes twin-screw or single-screw REX systems for starch modification.

12.4 STARCH-BASED BIOCOMPOSITES AND NANO-BIOCOMPOSITES

During the last decade, starch-based nano-biocomposite materials have gathered a lot of attention. Polysaccharide nanofillers, phyllosilicates, carbonaceous nanofillers, and many other nanofillers and natural fillers have been considered to be used in the formation of starch-based biocomposites. They all are different in surface chemistry and shape-size geometry; therefore they reflect the different reinforcing capability and a lot of variation in the starch matrix and nanofiller interactions [12,83]. Due to attractive advantages such as wide availability, low toxicity, and low cost, phyllosilicates (layered silicates) are the most frequently used nanofillers. Montmorillonite (MMT) is the most commonly used phyllosilicates. Polysaccharide nanofillers consist of similar chemical structure to starch which makes it the second popular group of nanofiller for starch [83]. Similarly, other nanofillers such as carbonaceous nanofillers, polysaccharide nanofillers, etc., are different and highly considered in the fabrication of starch based composites. Several nanofillers and their advantages and enhanced properties description in starch biocomposites are discussed below.

12.4.1 STARCH-BASED NANO-BIOCOMPOSITES REINFORCED BY PHYLLOSILICATES

As discussed above, phyllosilicates possess some significant advantages such as abundant, eco-friendliness, low cost, versatility, and low toxicity. Phyllosilicates (layered silicates) are a prime group of minerals that include serpentine, clay minerals, micas, talc, serpentine, etc. They display different textures, structures, and/or morphology. Particularly, some phyllosilicates do not display a normal layered structure. Some displays fibrous structure such as sepiolite, and some forms spheroidal aggregates (halloysite). Some chemical modification (cationic exchange) of the phyllosilicate surface is carried out to enhance the interlayer spacing. Various organo-modified MMTs (OMMTs) are formed by using this chemical modification technique, which are used for starch-based nano-biocomposites [83]. Normally, in situ intercalative polymerization, solution intercalation, or melt intercalation are three main techniques of composite preparation which incorporates phyllosilicate nano-layers into a polymer matrix [12, 84,85]. Phyllosilicate-reinforced starch biocomposites exhibit many improvements in the end product like increased young modulus, storage modulus, and tensile strength. It also shows improvement in Tg, biodegradation rate [83,84], moisture uptake, water vapor permeability, oxygen barrier property [12], and generally means improved performance and reduction in elongation at break (eb). Starch-based nano-biocomposites reinforced by phyllosilicates exhibit changes in different properties are shown in Table 12.2.

12.4.2 STARCH-BASED NANO-BIOCOMPOSITES REINFORCED BY POLYSACCHARIDE NANOFILLERS

Polysaccharide nanofillers are the second important group of nanofillers which are used in fabrication of starch-based nano-biocomposites. Several polysaccharides such as cellulose, starch, chitin, and chitosan have similar chemical structure and cellulose nano-whiskers (CNWs) and starch nanoparticles (SNPs) are the different

TABLE 12.2
Matrix Used, Preparation Techniques, Composite Structures, and Various Properties of Phyllosilicates Reinforced by Starch-Based Nanocomposites

Matrix (Type of Starch)	Nanofiller Type	Preparation Technique	Composite Structure	Thermal Properties T_g (°C); T_d (°C)	Mechanical Properties σ_u (MPa); σ_b (MPa); σ_y (MPa); E (MPa); εb (%)	Moisture Sensitivity M (%); M∞ (%); D (mm²s⁻¹); P (g m⁻¹s⁻¹ Pa⁻¹); θc (°)	References
Maize	MMT-Na⁺ (pretreated with glycerol by a HSEM)	ME, and CM	Intercalated	—	σ_u: 4.0→9.2; εb:108→53	P: $5.0 \times 10^{-10} \to 1.9 \times 10^{-10}$	[87]
Sweet Potato	OMMT (modified by 12-OREC)	ME, and CM	—	—	σ_u: 4.2→6.8; E: 42→102; εb: 90→50	—	[88]
Potato	MMT-Na⁺	SC	Intercalated/exfoliated,	T_d: 318→315	—	M∞: 10.3→7.4	[89]
Cassava	Natural sodium bentonite,	ME, and CM	Intercalated	—	σ_u: 0.96→1.45; E: 16→42; εb: 63.3→72.93	P: $2.1 \times 10^{-7} \to 1.7 \times 10^{-10}$	[90]
Amylomaize	MMT-Na⁺	ME, and CM	Intercalated by glycerol	T_d:344→349	—	—	[91]
Wheat	MMT-Na⁺	MM	Intercalated by glycerol	T_g:11.7→23.9	σ_b: 2.3→1.8; E*: 28→39; εb: 32→21	P:$4.68 \times 10^{-10} \to 3.96 \times 10^{-10}$	[92,93]

σ_u, ultimate tensile strength; σ_b, breaking tensile strength; σ_y, yield tensile strength; εb, elongation at break M (M∞), moisture uptake at specific RH after a specific time (at equilibrium); D, water diffusion coefficient; P, water vapour permeability; θc, contact angle; ME/MM-melt extrusion/mixing; CM/IM, compression/injection moulding; SC, solution casting; HSEM, high-speed emulsifying machine; CC, conventional composite.

forms of nanofillers produced by them. Generally, solution casting method (except few, used a melt mixing process) is used in the formation of polysaccharide reinforced starch-based nano-biocomposites [86].

These three main types of starch nano-composites produced by polysaccharide nanofillers are discussed in this chapter [12]:

a. Nano-biocomposites reinforced by CNWs
b. Nanocomposites reinforced by SNPs
c. Nanocomposites reinforced by chitin/chitosan nanoparticles

Cellulose nano-fibres (CNWs) are normally extracted from biomass like cassava bagasse, hemp, flax, peal hull, ramie, and tunicate. It can also be isolated through other sources such as microcrystalline cellulose, acid hydrolysis, etc. Several promising characteristics of CNWs (high crystallinity, higher E) encourage the use of this nanofiller, which improves the mechanical property of reinforced biocomposites.

Starch nanocrystals can be procured by acid hydrolysis or ethanol precipitation of native starch granules. Normally sulfuric acids are used for the hydrolysis of starch granules and SNPs can be prepared by constant stirring of ethanol precipitation into a gelatinized starch solution.

CNWs and SNPs reflect improvement in various properties such as mechanical properties, thermal property (Tg), and moisture resistance. Table 12.3 contains the summarized details of changed properties after inclusion of polysaccharide nanofillers.

12.4.3 STARCH-BASED NANO-BIOCOMPOSITES REINFORCED BY CARBONACEOUS NANOFILLERS

It is extremely interesting to introduce this community of nanofillers into preparation of starch based materials, but they have not been thoroughly studied so far. Carbon nanotubes (CNTs), carbon black (CB), and graphite oxide (GO) are the promising group of nanofillers. These nanofillers provide new functionalities especially electroactivity and electrical conductivity along with performance improvement [86].

CNTs-reinforced biocomposites are already widely used in several fields such as biomedical applications stimulators of bone cells, etc. [98–101]. Starch-based nano-biocomposites reinforced by GO show tremendous quality improvement in biocomposites. GO content enhances the mechanical, thermal properties, and decreases the water uptake. Generally, CB-reinforced starch-based nano-biocomposites are prepared by the solution casting method. Reenforcing of CB into starch biocomposites reflects better consequences in conductivity and WVP [83]. Table 12.4 shows the effect of carbonaceous nanofillers on starch.

12.5 STARCH BIO-COMPOSITES: PRESENT TRENDS AND CHALLENGES

Starch exhibits many promising and unique characteristics apart from biodegradability and low cost. This makes the development of starch-based biocomposites and its industrial use more than any other polymer composites. Uses of biodegradable

TABLE 12.3
Matrix Used, Preparation Techniques, Composite Structures, and Various Properties of Polysaccharide Reinforced by Starch-Based Nanocomposites

Matrix (Type of Starch)	Nanofiller Type	Plasticizer Type	Preparation Technique	Thermal Properties T_g (°C); T_d (°C) T_m (°C)	Mechanical Properties σ_u (MPa); σ_b (MPa); σ_y (MPa); E (MPa); εb (%)	Moisture Sensitivity M (%); M∞ (%); D (mm²s⁻¹); P (g m⁻¹s⁻¹ Pa⁻¹); θc (°)	References
Wheat	Cellulose nano-whiskers Ramie,	Glycerol	SC	T_g: 26.8→55.7; Tm: not Observed	σ_u: 2.8→6.9 E: 56→480; εb:94.2→13.6	M∞: 63→45 (98% RH)	[94]
Pea	Flax	Glycerol	SC	T_g: 43.3→48.8	σ_u: 3.9→11.9; E: 32→498; εb: 98.2→7.2	M: 70→57 (98% RH, 72 h)	[95]
Potato	MC	Sorbitol	SC	T_g: 55→70	σ_y: 370→460; σ_y: 11.3→13.7; εb: 25→32	—	[96]
Waxy maize	Tunicin	Glycerol	SC	T_g: 0.9→57.9; T_m: 132.4→134.4	σ_u: 1.2→2.0; E*: 23→105; εb: 19→10	M∞: 62→40; D:1.76×10⁻⁷→1.59×10⁻⁷ (98% RH)	[83,97]

MC, microcrystalline cellulose; T_m, melting temperature.

TABLE 12.4
Matrix Used, Preparation Techniques, Composite Structures, and Various Properties of Phyllosilicates Reinforced by Starch-Based Nanocomposites

	Nanofiller Type	Plasticiser Type	Preparation Technique	Thermal Properties T_g (°C); T_d (°C) T_m (°C)	Mechanical Properties σ_u (MPa); σ_y (MPa); E (MPa); εb (%)	Moisture Sensitivity M (%); M∞ (%); D (mm²s⁻¹); P (g m⁻¹s⁻¹ Pa⁻¹); θc (°)	References
Tapioca	MWCNTs	Glycerol	SC	T_g: 0→40	σ_u: 1.1→1.5; E: 2.5→4.2; εb: 80→90	P: increased	[102,103]
Pea	GO	Glycerol	SC	T_d: 311.5→318.4	σ_b: 4.6→13.8; E: 110→1050; εb: 36.1→12.1	M∞: 63.0→41.5 (98% RH)	[104]
Maize	CB	Glycerol	SC	–	σ_y*: 3.8→10.6; εb: 34→8	P*: 5.7×10⁻¹⁰→ 2.6×10⁻¹⁰	[105,106]

MWCNTs, multiwalled carbon nanotubes; GO, graphene oxide; CB, carbon black.

polymers are increased due to societal concern and environmental regulations. There is lot of opportunities for green composite to be used and produced. Starch is one of the most promising polymers among all. Several researches have been already done in the field of biocomposites but many more are still left to be explored.

Apart from all goodness there are few shortcomings in such biocomposites based on starch. Few properties such as heterogeneous dispersion of the nanofiller in starch, resulting in dissociation of two phases, compromise the efficacy of nanofillers. This shortcoming can be overcome by choosing the right filler which possess chemical affinity to starch or are relatively hydrophilic in nature. Modification and processing of starch enhance the properties of starch but still could not greatly influence the mechanical and other required properties. It is worth to carry out studies regarding preparation method of starch based biocomposites. Crystallinity, Tg, and phase distribution are the promising factors that greatly influence the mechanical properties, reflects that there is a great need to explore mechanism and fabrication techniques. However, there is no consent in the literature which reflects how nanofiller affects these properties [83]. Further research into how multiple variables can synergistically lead to improved mechanical properties is probably worth carrying out [83]. Further research into how multiple variables can synergistically lead to improved mechanical properties is probably worth carrying out. Though there is great opportunity to develop new bio-based products, but the real challenge is to develop a design sustainable bio-based products. Other major limitations of present biodegradable polymers are the high cost and it can be reduced with their large scale uses. Further research should overcome the performance limitations of starch biocomposites. Recent developments in genetic engineering, the production of natural fibers, and composite science provide substantial prospects for sustainable energy to boost value-added products with increased support for global sustainability. In the 21st century field of renewable materials, their unusual combination of properties will open up new business growth possibilities for bio-composites.

12.6 CONCLUSION

In this chapter, starch-based biocomposites and effect of various fillers or nanofillers on starch polymer has been discussed. In order to enhance the properties of starch biocomposites, various processing techniques are used which are also explained in details. New trends and opportunities in the field of biocomposites are in wave these days which increases the opportunities to develop new products. Challenges which limit the uses of starch-based biocomposites are also discussed.

REFERENCES

1. Prabhu T.N, Prashantha K.P. (2018). A review on present status and future challenges of starch based polymer films and their composites in food packaging applications. *Polym. Compos.* 39(7):2499–2522.
2. Sivan A. (2011). New perspectives in plastic biodegradation. *Curr. Opin. Biotechnol.* 22(3):422–426.

3. John M.J., Thomas S. (2008). Biofibres and biocomposites. *Carbohydr. Polym.* 71(3): 343–364.
4. Fowler P., Hughes M., Elias R.M. (2007). Review biocomposites: Technology, environmental credentials and market forces. *J. Sci. Food Agric.* 86(12):1781–1789.
5. Averous L., Boquillon N. (2004). Biocomposites based on plasticized starch: Thermal and mechanical behaviours. *Carbohydr. Polym.* 56(2):111–122.
6. Gurunathan T., Mohanty S., Nayak S.K. (2015). A review of the recent developments in biocomposites based on natural fibres and their application perspectives. *Compos.* 77:1–25.
7. Mishra R., Manral A. (2020). Graphene functionalized starch biopolymer nanocomposites: Fabrication, characterization, and applications. In: B. Sharma and P. Jain (eds.) *Graphene Based Biopolymer Nanocomposites.* Springer, Singapore, pp. 173–189.
8. Shen L., Worrell E., Patel M. (2010). Present and future development in plastics from biomass. *Biofuels, Bioprod. Biorefining.* 4(1): 25–40.
9. Babu R.P., Connor K.O., Seeram R. (2013). Current progress on bio-based polymers and their future trends. *Prog Biocomp.* 2:8.
10. Zheng P., Ma T., Ma X. (2013). Fabrication and properties of starch-grafted graphene nanosheet/plasticized-starch composites. *Indus. Engi. Chem. Res.* 52(39):14201–14207.
11. Averous L. (2004). Biodegradable multiphase systems based on plasticized starch: A review. *J. Macromo. Sci. Part C: Poly. Rev.* 44(3):231–274.
12. Xie F, Pollet E, Halley PJ, Averous L. (2013). Starch-based nano-biocomposites. *Prog. Poly. Sci.* 38(10–11):1590–1628.
13. Feder G., Umali D.L. (1993). The adoption of agricultural innovations: A review. *Technol. For. Soc. Change* 43(3–4):215–239.
14. Wani A.A., Singh P., Shah M.A., et al. (2013). Physico-chemical, thermal and rheological properties of starches isolated from newly released rice cultivars grown in Indian temperate climates. *LWT-Food Sci. Technol.* 1:176–183.
15. Xu Y.X., Kim K.M., Hanna M.A., Nag D. (2005) Chitosan–starch composite film: preparation and characterization. *Indus. Crop. Prod.* 21(2):185–192.
16. Stading M., Hermansson A.M., Gatenholm P. (1998) Structure, mechanical and barrier properties of amylose and amylopectin films. *Carbo. Poly.* 36(2–3):217–224.
17. Pagella C., Spigno G., De Faveri D.M. (2002) Characterization of starch based edible coatings. *Food Bioprod. Proc.* 80(3):193–198.
18. Kester J.J., Fennema O.R. (1986). Edible films and coatings: A review. *Food technol. (USA).* 213–255.
19. Jagannath J.H., Nanjappa C., Das Gupta D.K., Bawa A.S. (2003). Mechanical and barrier properties of edible starch–protein-based films. *J. Appl. Poly. Sci.* 88(1):64–71.
20. Ashori A. (2014). Effects of graphene on the behavior of chitosan and starch nanocomposite films. *Poly. Eng. Sci.* 54(10):2258–2263.
21. Ma X.F., Jian R.J., Chang P.R., Yu J.G. (2008). Fabrication and characterization of citric acid-modified starch nanoparticles/plasticized-starch composites. *Biomacromol.* 9:3314–3320.
22. Kalambur S., Rizvi S.S.H. (2006). An overview of starch-based plastic blends from reactive extrusion *J. Plas. Film Shtg.* 22(1):39–58.
23. Whistler R.L., BeMiller J.N., Paschall E.F. (2012). *Starch: Chemistry and Technology.* Academic Press, Cambridge, MA.
24. Liu H., Xie F., Yu L., Chen L., Li L. (2009). Thermal processing of starch-based polymers. *Prog. Poly. Sci.* 34(12):1348–1368.
25. Ramesh M., Mitchell J.R., Jumel K., Harding S.E. (1999) Amylose content of rice starch. *Starch-Starke* 51(8–9):311–313.
26. Tomasik P., Schilling C.H. (2004). Chemical modification of starch. *Adv. Carbo. Chem. Biochem.* 59:175–403.

27. Khan B., Niazi M., Samin G., Jahan Z. (2017). Thermoplastic starch: A possible biodegradable food packaging material—A review. *J. Food Pro. Eng.* 40(3): doi:10.1111/jfpe.12447.
28. Zhu F. (2015). Composition, structure, physicochemical properties, and modifications of cassava starch. *Carbo. Poly.* 122:456–480.
29. Grommers H.E., Van der Krogt D.A. (2009) Potato starch: production, modifications and uses. In: J. BeMiller and R. Whistler (eds.) *Starch: Chemistry and Technology.* Academic Press, Cambridge, MA, pp. 511–539.
30. Wang Z.F., Peng Z., Li S.D., et al. (2009). The impact of esterification on the properties of starch/natural rubber composite. *Compo. Sci. and Techno.* 69(11–12):1797–1803.
31. FengZuo Y., Gu J., Qiao Z., Tan H., Cao J., Zhang Y. (2015). Effects of dry method esterification of starch on the degradation characteristics of starch/polylactic acid composites. *Inter. J. Biol. Macro.* 72:391–402.
32. Ratnayake W.S., Jackson D.S. (2008). Starch gelatinization. *Advan. Food Nutri. Research.* 55:221–268.
33. Zhong K., Lin Z.T., Zheng, X.L. et al. (2013). Starch derivative-based super absorbent with integration of water-retaining andcontrolled-release fertilizers. *Carbohydr. Polym.* 92:1367–1376.
34. Phillips G.O., Williams P.A., (2000). *Handbook of Hydrocolloids.* CRC Press, Boca Raton, FL.
35. Fisk I.D., Linforth R., Trophardy G., Gray D. (2013). Entrapment of a volatile lipophilic aroma compound (d-limonene) in spray dried water-washed oil bodies naturally derived from sunflower seeds (Helianthus annus). *Food Res. Int.* 54(1):861–866.
36. Beinecke C.R., McFarland V.L. et al. (2012). Meat-Like Product and its Method of Production. United States patent application US 13/357,168.
37. Yang J.H., Yu J.G., Ma X.F. (2006). Preparation and properties of ethylenebisformamide plasticized potato starch (EPTPS). *Carbohydr. Polym.* 63:218–223.
38. Yang J.H., Yu J.G., Ma X.F. (2006). Study on the properties of ethylenebisformamide and sorbitol plasticized corn starch (ESPTPS). *Carbohydr. Polym.* 66:110–116.
39. Ma X.F., Yu J.G, Wan J.J. (2006). Urea and ethanolamine as a mixed plasticizer for thermoplastic starch. *Carbohydr Polym.* 64:267–273.
40. Yang J.H., Yu J.G., Ma X. (2006). Preparation of a novel thermoplasticstarch (TPS) material using ethylenebisformamide as the plasticizer. *Starch/Starke.* 58:330–337.
41. Thuwall M., Boldizar A., Rigdahl M. (2006). Extrusion processing of high amylose potato starch materials. *Carbohydr. Polym.* 65:441–6.
42. Keszei S., Szabo A., Marosi G. et al. (2006). Use of thermoplasticstarch in continuous pharmaceutical process. *Macromol Symp.* 239:101–104.
43. Follain N., Joly C., Dole P., Roge B., Mathlouthi M. (2006). Quaternary starchbased blends: influence of a fourth component addition to thestarch/water/glycerol system. *Carbohydr. Polym.* 63:400–407.
44. Yu J., Wang N., Ma X. (2005). The effects of citric acid on the propertiesof thermoplastic starch plasticized by glycerol. *Starch/Starke* 57:494–504.
45. Ma X.F., Yu J.G., Ma Y.B. (2005). Urea and formamide as a mixed plasticizer for thermoplastic wheat flour. *Carbohydr. Polym.* 60:111–116.
46. Teixeira E.M., Da Roz A.L., Carvalho A.J.F., Curvelo A.A.S. (2005). Preparation and characterisation of thermoplastic starches from cassava starch, cassavaroot and cassava bagasse. *Macromol. Symp.* 229:266–275.
47. Ma X., Yu J. (2004). Formamide as the plasticizer for thermoplastic starch. *J. Appl. Polym. Sci.* 93:1769–1773.
48. Ma X., Yu J. (2004). The effects of plasticizers containing amide groups on theproperties of thermoplastic starch. *Starch/Starke* 56:545–551.

49. Rodriguez-Gonzalez F.J., Ramsay B.A., Favis B.D. (2004). Rheological and thermalproperties of thermoplastic starch with high glycerol content. *Carbohydr. Polym.* 58:139–147.
50. Ma X., Yu J., Feng J. (2004). Urea and formamide as a mixed plasticizer forthermoplastic starch. *Polym. Int.* 53:1780–1785.
51. Ma X.F., Yu J.G., (2004). The plasticizers containing amide groups for thermoplasticstarch. *Carbohydr. Polym.* 57:197–203.
52. Ma X.F., Yu J.G. (2004). Studies on the properties of formamide plasticized thermoplastic starch. *Acta. Polym. Sin.* 2:240–245.
53. Stepto R.F.T. (2006). Understanding the processing of thermoplastic starch. *Macromol. Symp.* 245:571–577.
54. Stepto R.F.T. (2003). The processing of starch as a thermoplastic. *Macromol. Symp.* 201: 203–212.
55. Onteniente J.P., Abbes B., Safa L.H. (2000) Fully biodegradable lubricated thermoplastic wheat starch: mechanical and rheological properties of an injection grade. *Starch/Starke* 52:112–117.
56. Van Soest J.J.G., Borger B.D. (1997). Structure and properties of compression molded thermoplastic starch materials from normal and high amylosemaize starches. *J. Appl. Polym. Sci.* 64:631–644.
57. Stepto R.F.T. (1997). Thermoplastic starch and drug delivery capsules. *Polym. Int.* 43:155–8.
58. Shi R., Zhang Z., Liu Q., Han Y., et al. (2007). Characterizationof citric acid/glycerol co-plasticized thermoplastic starch preparedby melt blending. *Carbohydr. Polym.* 69: 748–755.
59. Teixeira E.M., Da Roz A.L., Carvalho A.J.F., Curvelo A.A.S. (2007). The effect ofglycerol/sugar/water and sugar/water mixtures on the plasticizationof thermoplastic cassava starch. *Carbohydr. Polym.* 69:619–624.
60. Da Roz A.L., Carvalho A.J.F., Gandini A., Curvelo A.A.S. (2006). The effect of plasticizers on thermoplastic starch compositions obtained by melt processing. *Carbohydr. Polym.* 63:417–24.
61. Forssell P.M., Mikkila J.M., Moates G.K., Parker R. (1997). Phase and glass transition behaviour of concentrated barley starch–glycerol–watermixtures: A model for thermoplastic starch. *Carbohydr. Polym.* 34:275–282.
62. Zhang S.D., Zhang Y.R., Zhu J. et al. (2007). Modified corn starches with improved comprehensive properties forpreparing thermoplastics. *Starch/Starke* 59:258–268.
63. Lai L.S., Kokini J.L. (1990). The effect of extrusion operating conditions onthe on-line apparent viscosity of 98% amylopectin (amioca) and70% amylose (Hylon 7) corn starches during extrusion. *J. Rheol.* 8:1245–1266.
64. Della Valle G., Vergnes B., Tayeb J. (1992) Measurements of the viscosity of low hydrated molten starches with a new in-line rheometer. *Entropie.* 169:59–63.
65. Willet J.L., Jasberg B.K., Swanson C.L. (1995). Rheology of thermoplastic starch: effects of temperature, moisture content, and additives on melt viscosity. *Polym. Eng. Sci.* 35:202–210.
66. Della Valle G., Colonna P., Patria A., Vergnes B. (1996). Influence of amylosecontent on the viscous behavior of low hydrated molten starches. *J. Rheol.* 40:347–362.
67. Aichholzer W., Fritz H.G. (1998). Rheological characterization of thermoplasticstarch materials. *Starch/Starke* 50:77–83.
68. Della Valle G., Buleon A., Carreau P.J. et al. (1998). Relationship between structure and viscoelastic behavior of plasticizedstarch. *J. Rheol.* 42:507–525.
69. Wiedmann W., Strobel E. (1991). Compounding of thermoplastic starch with twin-screw extruders. *Starch-Stärke* 43(4):138–145.

70. Bulkin B.J., Kwak Y., Dea I.C.M. (1987). Retrogradation kinetics of waxy-cornand potato starches: A rapid, Raman-spectroscopic study. *Carbohydr. Res.* 160:95–112.
71. Gudmundsson M. (1994). Retrogradation of starch and the role of its components. *Thermo. Chim. Acta.* 246:329–341.
72. Liu Q., Thompson D.B. (1998). Effects of moisture content and different gelatinization-heating temperatures on retrogradation of waxy-typemaize starches. *Carbohydr. Res.* 314:221–235.
73. Van Duin M., Machado A.V., Covas J. (2001). A look inside the extruder: Evolution of chemistry, morphology and rheology along the extruderaxis during reactive processing and blending. *Macromol. Symp.* 170:29–39.
74. Myllymaki O., Myllarinen P., Forssell P. et al. (1998). Mechanical and permeability properties of biodegradable extruded starch/polycaprolactone films. *Pack. Technol. Sci.* 11:265–274.
75. Matzinos P., Tserki V., Gianikouris C. et al. (2002). Processing and characterization of starch/polycaprolactone products. *Polym. Degrad. Stabil.* 77:17–24.
76. Ratto J.A., Stenhouse P.J., Auerbach M. et al. (1999) Processing, performance and biodegradability of a thermoplastic aliphaticpolyester/starch system. *Polymer* 40:6777–6788.
77. Fishman M.L., Coffin D.R., Onwulata C.I. et al. (2006). Two stage extrusionof plasticized pectin/poly(vinyl alcohol) blends. *Carbohydr. Polym.* 65:421–429.
78. Arevalo K., Sandoval C.F., Galan L.J. et al. (1996). Starch-based extruded plastic films and evaluation of theirbiodegradable properties. *Biodegradation* 7:231–237.
79. Fanta G.F., Swanson C.L., Shogren R.L. (1992). Starch–poly(ethylene-*co*-acrylicacid) composite films: effect of processing conditions on morphologyand properties. *J. Appl. Polym. Sci.* 44:2037–2042.
80. Otey F.H, Westhoff R.P., Doane W.M. (1987). Starch-based blown films. *Ind. Eng. Chem. Res.* 26:1659–1663.
81. Stenhouse P.J., Ratto J.A., Schneider N.S. (1997). Structure and properties ofstarch/poly(ethylene-*co*-vinyl alcohol) blown films. *J. Appl. Polym. Sci.* 64:2613–2622.
82. Glenn G.M., Orts W.J. (2001). Properties of starch-based foam formedby compression/explosion processing. *Ind. Crop. Prod.* 13:135–143.
83. Xie F., Averous L., Halley P.J., Liu P. (2015). Mechanical performance of starch-based biocomposites. In: *Biocomposites*, pp. 53–92. doi:10.1016/B978-1-78242-373-7.00011-1
84. Giannelis E.P. (1996). Polymer layered silicate nanocomposites. *Adv. Mater.* 8:29–35.
85. Mallapragada S.K., Narasimhan B. (2005). *Handbook of Biodegradable Polymeric Materials and Applications: Materials.* American Scientific Publishers, Valencia, CA, pp. 154–197.
86. Teixeira E.D.M, Pasquini D., Curvelo A.A.S. et al. (2009). Cassava bagasse cellulose nanofibrils reinforced thermoplastic cassava starch. *Carbohydr. Polym.* 78:422–431.
87. Wang N., Zhang X., Han N., Bai S. (2009). Effect of citric acid and processing on the performance of thermoplastic starch/montmorillonite nanocomposites. *Carbohydr. Polym.* 76:68–73.
88. Ren P., Shen T., Wang F. et al. (2009) Study on biodegradable starch/OMMT nanocomposites for packaging applications. *J. Poly. Envi.* 17:203–207.
89. Zeppa C., Gouanve F., Espuche E. (2009) Effect of a plasticizer on the structure of biodegradable starch/clay nanocomposites: Thermal, water-sorption, and oxygen-barrier properties. *J. Appl. Poly. Sci.* 112:2044–2056.
90. Muller C.M.O., Laurindo J.B., Yamashita F. (2011). Effect of nanoclay incorporation method on mechanical and water vapor barrier properties of starch-based films. *Ind. Cro. Prod.* 33:605–610.
91. Zhang Q.X., Yu Z.Z., Xie X.L. et al. (2007) Preparation and crystalline morphology of biodegradable starch/clay nanocomposites. *Poly.* 48:7193–7200.

92. Chivrac F., Pollet E., Schmutz M., Averous L. (2008). New approach to elaborate exfoliated starch-based nano biocomposites. *Biomacromol.* 9:896–900.
93. Chivrac F, Pollet E, Schmutz M, Averous L. (2010) Starch nano biocomposites based on needle-like sepiolite clays. *Carbohydr. Polym.* 80:145–153.
94. Lu Y., Weng L., Cao X. (2006). Morphological, thermal and mechanical properties of ramie crystallites—reinforced plasticized starch biocomposites. *Carbohydr. Polym.* 63:198–204.
95. Cao X., Chen Y., Chang P.R., Muir A.D., Falk G. (2008). Starch-based nanocomposites reinforced with flax cellulose nanocrystals. *Exp. Poly. Lett.* 2(7):502–510.
96. Kvien I., Sugiyama J., Votrubec M., Oksman K. (2007). Characterization of starch based nanocomposites. *J. Mater. Sci.* 42:8163–8171.
97. Angles M.N., Dufresne A. (2001). Plasticized starch/tunicin whiskers nanocomposite materials. 2. Mechanical behavior. *Macromol.* 34:2921–2931.
98. Harrison B.S., Atala A. (2007). Carbon nanotube applications for tissue engineering. *Biomaterials* 28:344–353.
99. Wang J. (2005). Carbon-nanotube based electrochemical biosensors: A review. *Electroanalysis* 17:7–14.
100. Lahiff E., Lynam C., Gilmartin N., et al. (2010). The increasing importance of carbon nanotubes and nanostructuredconducting polymers in biosensors. *Ana. Bio. Chem.* 398:1575–1589.
101. Xiao Y., Li C.M., (2008) Nano composites from fabrications to electrochemical bio applications. *Electroanalysis* 20:648–662.
102. Fama L.M., Pettarin V., Goyanes S.N., Bernal C.R. (2011). Starch/multi-walled carbon nanotubes composites with improved mechanical properties. *Carbohydr. Polym.* 83:1226–1231
103. Fama L., Rojo P.G, Bernal C., Goyanes S. (2012). Biodegradable starch based nanocomposites with low water vapor permeability and high storage modulus. *Carbohydr. Polym.* 87:1989–1993.
104. Li R., Liu C., Ma J. (2011). Studies on the properties of graphene oxide-reinforced starch biocomposites. *Carbohydr. Polym.* 84:631–637.
105. Ma X., Chang P.R., Yu J., Lu P. (2008). Characterizations of glycerol plasticized-starch (GPS)/carbon black (CB) membranes prepared by melt extrusion and microwave radiation. *Carbohydr. Polym.* 74:895–900.
106. Ma X., Chang P.R., Yu J., Lu P. (2008). Electrically conductive carbon Black (CB)/glycerol plasticized-starch (GPS) composites prepared by microwave radiation. *Starch/Starke* 60:373–375.

13 Current Status of Utilization of Agricultural Waste and Prospects in Biocomposites

Pervinder Kaur and Harshdeep Kaur
Punjab Agricultural University

CONTENTS

13.1	Biocomposites	342
13.2	Natural Fiber	342
13.3	Structure and Chemical Composition of Natural Fibers	343
13.4	Physical and Mechanical Properties	348
13.5	Drawbacks of Natural Fibers	349
	13.5.1 *Hydrophilicity*	349
	13.5.2 Inconsistency in Fiber Property	352
	13.5.3 Poor Compatibility and Wettability of Matrix	352
13.6	Matrix	352
	13.6.1 Petroleum-Based Polymer Matrix	352
	13.6.2 Bio-Based Polymer Matrix/Resins	357
	13.6.3 Petroleum-Based Polymer Matrix	357
	13.6.3.1 Poly(Alkylene Dicarboxylate)	357
	13.6.3.2 Poly(ϵ-Caprolactone)	357
	13.6.4 *Natural-Based Polymer Matrix*	357
	13.6.4.1 Polyhydroxy Lactic Acid (PLA)	357
	13.6.4.2 Poly(β-Hydroxyalkanoate)	359
	13.6.5 *Polysaccharide-Based Biocomposites*	360
	13.6.5.1 Starch and Its Biocomposites	360
	13.6.5.2 Cellulose and Its Composites	363
	13.6.5.3 Polypeptide Based	363
	13.6.6 *Soya Protein*	363
	13.6.7 *Wheat Gluten*	364
13.7	Approaches to Improve the Performance of Biocomposites by Modifications	365
	13.7.1 Chemical Treatment Methods	365
	13.7.1.1 Mercerization	366
	13.7.1.2 Acetylation (Esterification)	366

DOI: 10.1201/9781003137535-13

	13.7.1.3	Silane Treatment	366
	13.7.1.4	Maleated Coupling	372
	13.7.1.5	Acrylation and Acrylonitrile Grafting	372
	13.7.1.6	Graft Copolymerization	372
13.7.2	Grafting via Living Polymerizaiton Technique		372
	13.7.2.1	Ring Opening Polymerization for Biocomposites	372
	13.7.2.2	Free Radical Grafting	372
13.7.3	Biological Methods		373
13.7.4	Physical Treatment		374
	13.7.4.1	Addition of Other Materials	374
	13.7.4.2	Addition of Compatibilizers and Cross-Linkers	376

13.8 Current Trends ... 378
    13.8.1 *Hybrid Composites* ... 378
    13.8.2 *HyperBranched Polymers* ... 381
        13.8.2.1 Grafting from Method ... 381
        13.8.2.2 Grafting to Method ... 381
    13.8.3 *Nanocellulose Biocomposites* ... 382
13.9 Applications ... 385
References ... 387

13.1 BIOCOMPOSITES

Composites are compounds composed of two or more constituents' materials as matrices and fibers. These materials when blended together are stronger compared to individual materials by themselves. Biocomposites are the composites in which either the fiber or matrix or both matrix and fiber are derived from biological resources (Figure 13.1).

Due to their environmentally friendly nature they are getting higher attention in recent years and have great potential to address the needs of sustainability (Al-Qqla and Salit 2017). The properties of biocomposites are customized according to specific application. Proper selection of natural fiber, matrix, filler/coupler and manufacturing method are important factors affecting their properties, biodegradability and application of biocomposites.

13.2 NATURAL FIBER

Natural fibers are produced from plants and animals and are environmentally friendly materials used as reinforcement in making biocomposites. The plants from which natural fibers are produced are characterized as primary and secondary. The primary plants (hemp, jute) are grown only for their fiber while secondary plants (pineapple, banana) are cultivated for their fruits and the fibers are byproducts of these plants. The plants which produce natural fibers can be classified into bast fiber (flax, ramie, jute, kenaf, hemp); leaf fibers (banana, abaca, pineapple and sisal); seed fibers (cotton, kapok and coir); core fibers (jute, kenaf and hemp); grass and seed fibers (rice, wheat and corn), stalk fibers (straw, wood and bamboo) and other types (roots and wood) (Figure 13.1).

Agricultural Waste & Prospects in Biocomposites 343

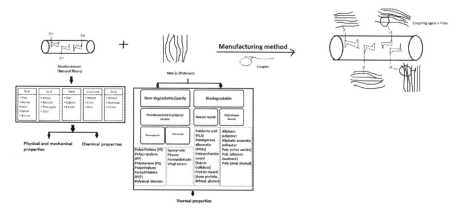

FIGURE 13.1 Important constituents and factors for manufacturing bio-composites.

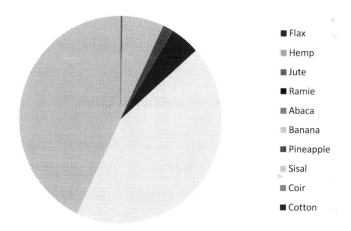

FIGURE 13.2 Natural plant fibers and their world production (tons). (*Source*: http://faostat3.fao.org/.)

Bast fiber are found in inner bark of certain plant stems and are made up of overlapping cell of bundle in which fibers are bounded together by pectin. Fibers of leaf origin are part of fibro vascular system of leaves. Bast and leaf fibers are mostly used in composite application due to their high mechanical and thermal properties. The fibers found in fruits and seeds (coir, oil palm, cotton and kapok) cannot be assembled in bundles. The most common and commercially available natural plant fibers and their global production are presented in Figure 13.2.

13.3 STRUCTURE AND CHEMICAL COMPOSITION OF NATURAL FIBERS

Natural fiber structure and its micro-structural organization are presented in Figure 13.3.

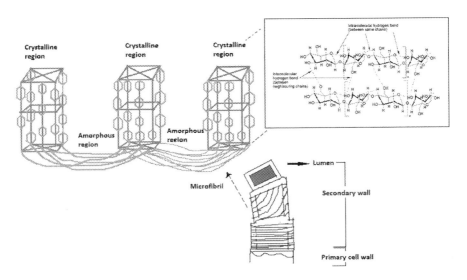

FIGURE 13.3 Natural fiber structure and microstructural organization (Kabir et al., 2012).

The natural fiber has primary cell at the peripheral, three secondary walls developed onto the inner surface of primary wall and lumen in the center. The secondary wall has number of cylindrical and anisotropic cellulose microfibrils which are surrounded and joined by a loose and complicated molecular network of lignin-hemicellulose matrix.

The microfibrils present in the inner secondary wall have spiral arrangement about the fiber at an angle called the microfibrillar angle (MFA) (angle between the axis of the fiber and the interior of the fibers) and it varies from fiber to fiber (Al-Qqla and Salit 2017). The MFA plays an important role in controlling the smoothness and dexterity (quality) of the fiber. The thickness of secondary layer also plays an important role in the mechanical behavior of the fiber. The amorphous matrix in cell wall is complex consisting of hemicelluloses, pectin and lignin. The hemicellulose molecule is hydrogen bonded to cellulose microfibrils and forms the main structural elements of fiber cell called as cellulose-hemicellulose network (Yu 2014; Al-Qqla and Salit 2017). The chemical composition of various natural fibers along with microfibrillar angle is shown in Table 13.1.

Cellulose, hemicelluloses and lignin are three major chemical constituents of natural fibers.

All plant-based fibers contain high proportion of cellulose (40%–90%). Cellulose is a hydrophilic polysaccharide in which hundreds of β-(1,4)-glycopyranosyl units are connected by β-(1,4) bond (Figure 13.3). Due to numerous hydroxyl groups extensive inter- and intra-molecular hydrogen bonding occurs thereby forming crystalline cellulose microfibrils and the hydrogen bonding is responsible for stiffness, high strength, crystallinity, biocompatibility and durability. The degree of polymerization for cellulose in primary and secondary wall is about 3,000 and >15,000, respectively. Cellulose microfibrils have both crystalline and amorphous regions (Figure 13.3) (Roy et al. 2009).

TABLE 13.1
Chemical Composition of Natural Fibers

Origin	Natural Fiber	Cellulose (%)	Hemicellulose (%)	Lignin (wt%)	Pectin (wt%)	Wax (%)	Ash	Microfibrillar Angle (°)	References
Bast	Jute	61.00–72.00	13.6–22.00	9.00–13.00	0.2	0.5	0.5–2.0	8–13.7	Mwaikambo et al. (2006) and Akil et al. (2011)
	Kenaf	45–57	8.00–21.5	8.00–22.00	0.6–5	0.8	2–5	2–6.2	Ramamoorthy et al. (2015), Komuraiah et al. (2014), Mwaikambo et al. (2006), and Akil et al. (2011)
	Hemp	70–92	0.9–22.4	3–6	0.8–0.9	0.8–6.2	0.8	2–12	Mwaikambo (2006) and Bismarck et al. (2005)
	Flax	70.50	16.50–20.6	2.20–2.50	2.3	1.7	–	5–10	Komuraiah et al. (2014), Mwaikambo et al. (2006), and Akil et al. (2011)
	Ramie	68.6–76.2	13.1–16.7	0.6–0.7	1.9	2	–	10–22	Bismarck et al. (2005) and Akil et al. (2011)
Grass	Bagasse	54.3–55.2	16.8–29.7	24.3–25.3	10	–	1.9	–	El Tayeb et al. (2012) and Siqueira et al. (2010)
Seed	Cotton	85–90	5.7	21.5	0–1	0.6	–	20–30	Pecas et al. (2018) and Bimarck et al. (2005), and Akil et al. (2011)
	Wheat	32.9	24.0	8.9	–	–	6.7	–	Nigam et al. (2009), Bharath and Basavarajappa (2016), and Petroudy (2017)
	Kapok	70.50	16.50–20.6	2.20–2.50	2.3	1.7	–	5–10	Kumar and Singh (2020), Sharma et al. (2020), Bharath and Basavarajappa (2016), and Petroudy (2017)
	Coir	32–43	0.15–0.25	40–45	3–4	–	–	30–49	Bismarck et al. (2005) and Akil et al. (2011)
	Rice straw	36.2–45	23.5	15.6	0–1	8–38	12.4–20	–	Kumar and Singh (2020) and Petroudy (2017)

(Continued)

TABLE 13.1 (Continued)
Chemical Composition of Natural Fibers

Origin	Natural Fiber	Cellulose (%)	Hemicellulose (%)	Lignin (wt%)	Pectin (wt%)	Wax (%)	Ash	Microfibrillar Angle (°)	References
Leaf	Abaca	56–63	21–25	12–131	1	5–10	–	–	Pecas et al. (2018) and Akil et al. (2011)
	Sisal	66–78	10–14	8–14	10	2	1	10–22	Bismarck et al. (2005) and Akil et al. (2011)
	Banana	63–83	6–19	5	–	11	–	11–12	Kumar and Singh (2020), Sharma et al. (2020), and Petroudy (2017)
	Pineapple	70–92	0.9–22.4	3–6	0.8–0.9	0.8–6.2	0.8	2–12	Kumar and Singh (2020), Sharma et al. (2020), Bharath and Basavarajappa (2016), and Petroudy (2017)
Grass	Corn	61.2	19.3	6.9	1.9	–	10.8	–	Sharma et al. (2020) Bharath and Basavarajappa (2016), and Petroudy (2017)
	Bamboo	26–43	30	1–31	3–4	22		–	Kumar and Singh (2020), Sharma et al. (2020), and Petroudy (2017)

In addition to cellulose, other two major constituents of natural fibers are hemicelluloses and lignin. Hemicellulose is heterogeneous biopolymer and is made of monomers. It is an amorphous polysaccharide, highly cross linked molecular complex, partially soluble in alkaline solution and water. It is very hydrophilic in nature and hardness and strength of plant fibers are intrinsically linked with extent and percentage of monomer in the polymer. Some monomers include mannose, galactose, arabinose, xylose an glucose besides monomeric acid sugars like glucuronic acids. Lignins are combination of three-dimensional heterogenous polymer based on alcohol monomers: p-coumoryl, sinapyl and coniferyl alcohol. It has amorphous structure, a high molecular weight, less polar than cellulose and act as chemical adhesive within and between fibers. It absorbs less water, is insoluble in most of the solvents and cannot be broken down into monomeric units. High contents of hemicelluloses and lignin in fiber have several associated disadvantages which cause high moisture adsorption thereby reducing interfacial adhesion between the fiber and the matrix. Fibers like bamboo, coir, wheat, rice straw and cotton have highest composition of hemicellulose and lignin content and this phenomenon shows significant drawback in the composites derived from these fibers. On exposure to water/moisture, the water/moisture penetrates into fiber and gets attached to hemicellulose and lignin by intermolecular hydrogen bonding in fiber thereby reducing interfacial adhesion between matrix and fiber (Figure 13.4a).

Degradation occurs due to development of stress in swelled fiber at the interface and creates micro-cracking conditions in matrix around the fiber which results in capillary and transport through the micro-cracks (Figure 13.4b).

Thereafter due to excessive water absorption bound water increases and water soluble substance in natural fiber begins to leach from fiber resulting in debonding phenomenon between fiber and matrix (Figure 13.4c).

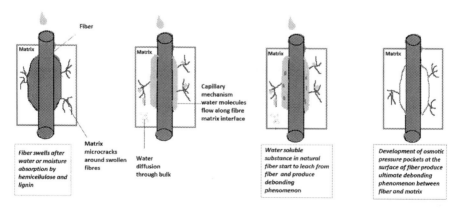

FIGURE 13.4 Effect of water or moisture absorption by natural fiber. (a) Fiber swells after water or moisture absorption by hemicellulose and lignin; (b) water diffusion through bulk; (c) water soluble substance in natural fiber starts to leach from fiber and produce debonding phenomenon; (d) development of osmotic pressure pockets at the surface of fiber procedure forms ultimate debonding phenomenon between fiber and matrix.

Osmotic pressure pockets are developed at the surface of fiber which forms the ultimate debonding situation between fiber and matrix (Figure 13.4d).

Considering this, fibers with high cellulose and low hemicelluloses and lignin content preferred are preferred for use as reinforcement in synthesis of biocomposites (Jamiluddin et al. 2018).

13.4 PHYSICAL AND MECHANICAL PROPERTIES

The physical and mechanical properties of fibers vary depending upon their geometric properties, type and amount of chemical constituents, crystallinity and degree of polymerization of cellulose and microfibrillar orientation (Table 13.2). Natural fiber geometric properties include length, width and good aspect ratio. Amongst these, aspect ratio is the most important parameter. For good reinforcement, the fiber aspect ratio should be above the critical value to transfer maximum stress to the fiber. The value below the critical value limits the reinforcing ability due to sparse stress transfer to fiber. If the aspect ratio is above the critical value, fiber may entangle during processing resulting in poor mechanical properties due to poor dispersion and fiber may act as filler rather than reinforcement. Thus, some minimum fiber length called as critical fiber length is required for efficient utilization of fiber strength. Critical fiber length is defined as the shortest length of the fiber in a given composite that allows maximal fiber loading. Fiber aspect ratio in the range of 10–50 and moduli ratio of 10^2–10^3 are sufficient for good adhesion between fiber and matrix. Increased fiber volume fraction and aspect ratio improve flexural and tensile strength and impact performance by providing space for interaction of fiber with matrix (Mohanty et al. 2005; Yilmaz et al. 2015).

MFA affects the inherent fiber strength properties. Low MFA and high cellulose content result in fiber with superior thermal and mechanical properties. Fiber with higher cellulosic content and low MFA when pulled in tension possess higher tensile strength and initial modulus (Petroudy 2017) while low cellulose content and higher MFA impart ductility to the fiber and it shows high failure strain as fibers twist when stretched. The small value of MFA is indicative of anisotrophic properties of fibers. The Young's modulus depends on the average MFA and volume fraction of cellulose with respect to the loading direction. Generally, the MFA of plant fibers is inversely proportional to fiber stiffness and varies from 6° to 10°.

The orientation and direction of microfibrils relative to each other help to predict ductility of fiber. Failure occurs if the orientation and direction of fiber are not in line of the applied stress, failure is bound to occur easily. Depending on the orientation of fibers in the matrix, reinforcement can be (a) transversely aligned fiber-filled composites, (b) longitudinally aligned fiber-filled composites and (c) randomly oriented short fiber composites (Kabir et al. 2012; Joseph et al. 2002). The longitudinally filled composites have high tensile strength and low compression strength due to buckling of fibers while transversely filled composites have low tensile strength. Due to complexities of the distribution of load along the interface of the fiber and the matrix, in the randomly oriented composites prediction of mechanical properties is difficult. Hence, careful control over dispersion, orientation and aspect ratio of the fibers helps to considerably improve mechanical properties of the composites.

The mechanical properties and thermal degradation temperature of natural fibers are also affected by cellulose content and crystallinity of the fiber. Crystalline cellulose has notably better stiffness than other constituents. High cellulose and high crystallinity are desirable factors for selecting natural fibers to be used as reinforcement in biocomposites. The increase in crystallinity increases thermal decomposition temperature of natural fibers. Different natural fibers have varying density and most of the natural fibers have maximum density (1,600 kg/m^3) of higher than that of water. However, the maximum density of these fibers is far less than that of glass fiber (inorganic fibers). Due to low density of reinforcement and flexibility in structural designs, it has found to be used in applications where weight is in consideration, viz., aerospace, marine, automotive and constructive industries.

The high moisture content of natural fibers is a major obstacle restricting its use in different applications. This is due to the presence of free hydroxyl and other polar groups on the surface of natural fiber which leads to decrease in mechanical properties and dimensional stability. Water absorption is dependent on factors such as surface orientation, fiber orientation, exposed surfaces, loading, temperature, diffusivity, permeability, void content and hydrophilicity.

Cellulose fibrils in natural fibers have diameter in the range of 1–30 nm and are made up of 30–100 cellulose molecules in extended chain formation which provide mechanical strength to fiber. Generally increase in cellulose content increases Young's modulus and tensile strength of natural fiber. The reinforcing elements of natural fibers are cellulose microfibrils that are surrounded by hemicelluloses and lignin. One application of load microfibril gets aligned with fiber axis. The failure of fiber takes place due to loss of bonding between hemicellulose and lignin and reinforcing fibrils. Hence, lower is the cellulose and hemicellulose content of the fiber, lower will be its tensile strength. The orientation of fibrils with respect to axis determines stiffness. Spiral orientation of fiber axis gives ductile plant fibers while parallel orientation gives inflexible and rigid fiber with high tensile strength. Young's modulus and tensile strength of natural fibers are lower than that of e-glass fiber. The specific strength and specific moduli of some of the natural fibers compared with the main type of glass fiber (e-glass) is given in Table 13.2.

To summarize, the critical requirement for selection of natural fibers in biocomposites fabrication includes high degree of polymerization, high cellulose content and low microfibril angle. These attributes are expected to yield biocomposites with higher tensile strength and Young's modulus.

13.5 DRAWBACKS OF NATURAL FIBERS

13.5.1 *Hydrophilicity*

Natural fibers are proficient, environment friendly attractive materials which can be utilized to replace synthetic fiber but the use of natural fibers for the synthesis of biocomposites is not free from problems and they have notable deficits in properties. Some of the disadvantages with natural fibers are that they are not durable as synthetic fibers as it degrades on exposure to light. They are all easily attacked by a variety of microorganisms in a conducive environment like temperature and high

TABLE 13.2
Physical and Mechanical Properties of Natural Fibers and Synthetic Fibers

Origin	Fiber Type	Diameter (μm)	Moisture Content (%)	Density (g/cm³)	Elongation Break (%)	Tensile Strength (MPa)	Young's Modulus (GPa)	Cellulose Crystallinity (%)	References
Bast	Jute	25–350	10–13.7	1.31–1.5	1.16–1.8	393–800	13–26.5	73.4	Akil et al. (2011) and Mohanty et al. (2005)
	Kenaf	70–250	12	1.3	1.6	284–930	21–60	–	Rouison et al. (2004), Petroudy (2017), Mohanty et al. (2005), and Gholampour and Ozbakkaloglu (2019)
	Hemp	25–500	6.2–12	1.47–1.48	1.6–4.0	550–900	70	93.8	Rouison et al. (2004), Mohanty et al. (2005), and Thygesen et al. (2011)
	Flax	40–600	8–12	1.4–1.5	1.2–3.2	345–1,500	27.6–80	96.9	Akil et al. (2011), Petroudy (2017), Mohanty et al. (2005), and Gholampour and Ozbakkaloglu (2019)
	Ramie	50	10	1.5	2	500	44	64	Kumar and Singh (2020), Petroudy (2017), and Mohanty et al. (2005)
Seed	Cotton	12–38	8.5	1.51–1.6	3–10	287–800	5.5–12.6	73	Akil et al. (2011), Kumar and Singh (2020), Mohanty et al. (2005), and Gholampour and Ozbakkaloglu (2019)
	Coir	100–460	10–12	1.15–1.46	15–40	131–220	4–6	39.06	Petroudy (2017), Mohanty et al. (2005), and Gholampour and Ozbakkaloglu (2029)
	Kapok	25–500	9.86	1.47–1.48	1.6–4.0	550–900	70	–	Ramamoorthy et al. (2015), Mohanty et al. (2005), and Gholampour and Ozbakkaloglu (2019)

(*Continued*)

TABLE 13.2 (Continued)
Physical and Mechanical Properties of Natural Fibers and Synthetic Fibers

Origin	Fiber Type	Diameter (μm)	Moisture Content (%)	Density (g/cm³)	Elongation Break (%)	Tensile Strength (MPa)	Young's Modulus (GPa)	Cellulose Crystallinity (%)	References
	Wheat	240–330	18–20	0.6–1.25	–	290	11–17	66.60	Petroudy (2017), Gholampour and Ozbakkaloglu (2019) and Qasim et al. (2019)
	Corn	12–38	–	1.51–1.6	3–10	287–800	5.5–12.6	66.18	Mohanty et al. (2005) and Gholampour and Ozbakkaloglu (2019)
	Rice husk	40–600	14	1.4–1.5	1.2–3.2	345–1,500	27.6–80	58.73	Petroudy (2017) and Mohanty et al. (2005)
Grass	Bagasse	320–400	20–28	0.89	5.5	350	22	–	Ramamoorthy et al. (2015) and Gholampour and Ozbakkaloglu (2019)
	Bamboo	240–330	11.7	0.6–1.25	2.5–3.7	290	11–17	–	Ku et al. (2011), Petroudy (2017), Mohanty et al. (2005), and Gholampour and Ozbakkaloglu (2019)
Leaf	Abaca	150–180	5–10	1.5	2.9–10	430–813	31.1–33.6	65	Mohanty et al. (2005), Gholampour and Ozbakkaloglu (2019) and Saragih et al. (2018)
	Sisal	50–300	10–22	1.3–1.45	2.3–7	390–700	12–41	83.12	Akil et al. (2011), Rouison et al. (2004), and Petroudy (2017)
	Banana	50–250	16	1.35	2.6–5.9	529–914	27–32	–	Mohanty et al. (2005) and Gholampour and Ozbakkaloglu (2019)
	Pineapple	105–300	11.8	1.44	14.5	413–1,627	60–82	–	Petroudy (2017) and Gholampour and Ozbakkaloglu (2019)
	e-glass	–	25–30	2.5	2.5	2,000–3,000	70	–	Pecas et al. (2018), Rouison et al. (2004), and Petroudy (2017)

humidity. They permit water absorption from surroundings due to their structure. In the crystalline region of cellulose, hydroxyl groups are strongly linked which provide hydrophilicity to the fiber. Most of the hydrophobic matrices are incomparable with hydrophilic fibers and results in weak bonding at the interfaces, poor interfacial interactions between fiber and polymer matrix causing detrimental consequences to the mechanical properties. Moreover, due to the hydrophilicity of the fiber they swell thereby resulting in cracks and hence huge decrease in mechanical strength. Natural fibers have low thermal stability, impact strength and poor compatibility.

13.5.2 INCONSISTENCY IN FIBER PROPERTY

Natural fibers do not have consistent properties; they vary depending upon the variety, harvesting region, harvesting time, soil condition, climatic conditions and extraction methods. Figure 13.5 summarizes some of the factors responsible for fiber inconsistency at different stages of biocomposites fabrication. These conditions may also vary with the variety and maturity of the plant which causes inconsistency of their mechanical properties as compared to synthetic fibers. Fiber obtained from various parts of same plant shows high variability in the properties. Pickering et al. (2016) has reported that strength of natural fibers reduced by 15% over 5 days after optimum harvest time and 20% higher strength is found in manually extracted flax fibers as compared to those extracted mechanically. Varieties with high cellulose content and cellulose microfibrils aligned in the fiber direction show high performance for fabrication of biocomposites.

13.5.3 POOR COMPATIBILITY AND WETTABILITY OF MATRIX

Due to poor fiber-matrix interface, stress distribution capacity is reduced which decrease the compatibility of fiber with the matrix. To obtain a suitable performance of the composite, weak boundary layers from the fibers are required to be eliminated.

13.6 MATRIX

Matrix is an important component of fiber-reinforced composite. It provides a barrier against adverse environments, protects the surface of the fiber from mechanical abrasion and transfers the load to fiber (Maslowski et al. 2019). The most commonly used matrix in natural fiber composites can broadly be classified into party degradable petroleum-based polymer matrix and fully degradable bio-based matrix (Figure 13.1).

13.6.1 PETROLEUM-BASED POLYMER MATRIX

Petroleum-based matrices (thermoplastic and thermoset) have been extensively used for production of biocomposites. These conventional thermoplastic matrix include polystyrene, polyvinyl chloride, polypropylene, polyethylene and poly(ethylene terephthlate) while vinyl esters and epoxy resin phenol formaldehyde are thermoset

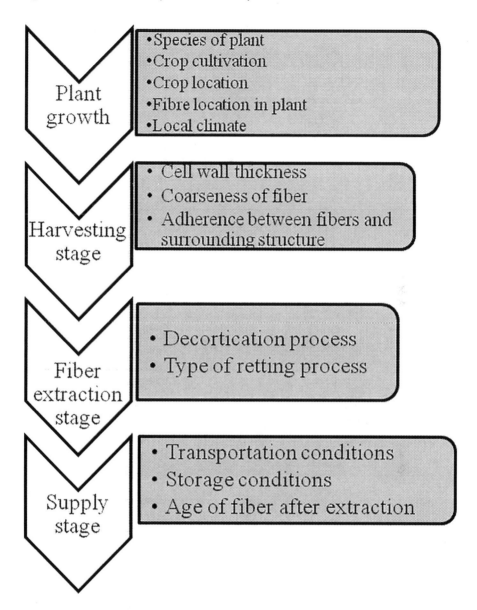

FIGURE 13.5 Factors contributing on quality of fibers at each stage to be the reinforcement agent in biocomposites formation.

matrices (Table 13.3). The selection of matrix is based on the degradation temperature of natural fiber. Most of the natural fibers used for reinforcement in biocomposites are thermally unstable above 200°C while quite few of them can withstand higher temperature for short time period. Thus, thermoplastics that soften below this temperature like polyolefin, polyethylene, polystyrene and polyvinyl chloride and thermoset are used mostly as matrix for formation of biocomposites.

TABLE 13.3
Mechanical Properties of Some Biocomposites

Matrix	Reinforcement	Fabrication Method	Fiber Content (%)	Tensile Strength (MPa)	Young's Modulus (GPa)	References
			Thermoplastics			
PP (Propylene)	Casurana	Blending	5.0–20.0	29.5–31.2	1.4–3.5	Lima et al. (2015)
	Flax	Trimming	30.0–40.0	39.0–109.0	4.9–10.1	Ngo et al. (2016) and Bledzki et al. (2008)
	Hemp	Injection molding	30.0–40.0	44.0–52.0	3.9–4.0	Burgstaller and Stadlbauer (2010), Burgstaller (2013), and Sain et al. (2005)
	Kenaf	Compression molding	30.0–50.0	46.0–53.0	5.0–7.5	Zampaloni et al. (2007)
	Rice husk	Injection molding	30.0	32.1	27.5	Burgstaller and Stadlbauer (2010) and Burgstaller (2013)
	Sisal	Injection molding	30.0	45.5	3.9	Burgstaller and Stadlbauer (2010) and Burgstaller (2013)
Epoxy	Bamboo	Crushing	57.0	392.0	29.0	Biswas et al. (2015)
	Cotton coir-sisal		40.0	56.0	–	Girisha et al. (2012)
	Flax yarn	Retting and spinning	45.0	133.0	28.0	Van de et al. (2003)
	Hemp	Compression molding	65.0	165.0	17.0	Islam et al. (2010)
	Jute	Retting and spinning	52.0	216.0	31.0	Van de et al. (2003)
	Rice husk	Washing and drying	25.0	117.1	6.76	Jaffer and Jawad (2011)
	Sisal	Dew retting	48.0–73.0	210.0–410.0	6.0–20.0	Rong et al. (2001)
			Thermosets			
PF (Phenol formaldehyde)	Banana	Chopping	45.0	7.0–23.0	175.0–398.0	Joseph et al. (2002)
	Grewia optiva	Dewaxing and washing	30.0	37.9	0.8	Pai and Jagtap (2015)
UPE	Flax	Rovings or pre-yarns	34.0	143.0	14.0	Goutianos et al. (2006)
	Jute	Dewaxing and molding	35.0	50.0	8.0	Rodriguez et al. (2007)
	Rice husk	Washing and drying	25.0	73.4	6.0	Jaffer and Jawad (2011)
VE (vinyl esters)	Flax	Rovings or pre-yarns	24.0	248.0	24.0	Goutianos et al. (2006)

(Continued)

TABLE 13.3 (Continued)
Mechanical Properties of Some Biocomposites

Matrix	Reinforcement	Fabrication Method	Fiber Content (%)	Tensile Strength (MPa)	Young's Modulus (GPa)	References
			Biobased			
PHB	Kenaf	Chopping, film stacking method	40.0	70.1	–	Graupner and Mussig (2011)
	Bast	Melt mixing	High	–	1,067–2,788	Avella et al. (2007)
	Straw/husk	Blending	–	7.5–45	1.2–3.8	Avella et al. (2000) and Wu (2014)
	Bamboo fibers	Compression molding	–	High	–	Wang and Darzal (2012)
	Flax	Injection molding	>15	27.2	2.5	Wei et al. (2015)
PLA	Abaca	Injection molding	30.0	74.0	8.0	Sawpan et al. (2011)
	Bamboo	Rinsing and mercurization	20.0	90.0	1.8	Kalia et al. (2009)
	Flax	Cutting	30.0	53.0, 100.0	8.0	Oksman et al. (2003) and Bodros et al. (2007)
	Hemp	Blending	45.0	65.0	–	Ibrahim et al. (2011)
	Jute	Injection molding	35.0	50.0	–	Yang et al. (2012)
	Kenaf	Melting	40.0	52.9	7.1	Choi and Lee (2012)
	Cotton	Injection molding	30.0	58.5	4.1	Burgstaller and Stadlbauer (2010) and Burgstaller (2013)
Starch	Coir-alkaline, silane	Washing and drying	10.0–30.0	42.1–47.8	2.0	Zaman and Beg (2014)
	Ramie	Biosizing	70.0	45–75	–	Marsyahyo et al. (2011)
	Jute	Mixing	–	2.43–10.50	3.02–7.43	Toprakci et al. (2019)
		Film casting	60	21.3	2.2	Soykeabkaew et al. (2012)
	Bamboo	Plasticizing and degassing	–	Low	–	Chen et al. (2019)
	Sisal	Plasticizing and degassing	–	Low	–	Chen et al. (2019)
	Cotton	Plasticizing and degassing	–	Low	–	Chen et al. (2019)

(Continued)

TABLE 13.3 (Continued)
Mechanical Properties of Some Biocomposites

Matrix	Reinforcement	Fabrication Method	Fiber Content (%)	Tensile Strength (MPa)	Young's Modulus (GPa)	References
	Cellulose	Film casting	50	80.9	5.2	Soykeabkaew et al. (2012)
		Cutting and heating	High	–	–	Hafizulhaq et al. (2018)
	Hemp	Blending	–	2.1	–	Cordoba et al. (2019)
Starch PCL blend	Sisal	Blending	High	–	–	Franco et al. (2004)
		Injection molding	–	123	–	Cyras et al. (2003)
		Alkaline Treatment	20	2.3–3.8	–	Campos et al. (2012)
		Belching and Extrusion	20	–	–	Campos et al. (2013)
Starch PLA blend	Jute	Extrusion	1–25	404	High	Yoksan et al. (2015)
	Sisal	Injection molding	–	High	–	Encalada et al. (2018)
	Bamboo	Blending and drying	–	33.1	–	Guan et al. (2019)
	Cellulose	Cutting and extrusion	–	High	High	Masmoudi et al. (2016)
Protein	Hemp	Blending	30	550–900	70	Daud et al. (2013)
		Injection molding				Mohanty et al. (2005)
	Jute	Injection molding	–	393–800	10–30	Daud et al. (2013)
	Flax	Blending	–	800–1,500	27.6–80	Daud et al. (2013)
	Kenaf	Melting	–	930	53	Daud et al. (2013)
		Compression molding	50	High	–	Liu et al. (2007)
		Molding and hot pressing	High	11.2	–	Liang et al. (2012)
		Dewaxing	–	17–23	–	Won et al. (2015)

13.6.2 Bio-Based Polymer Matrix/Resins

Bio-based resins, produced from renewable sources, are classified on the basis of their physical properties. These include natural-based polymers such as polylactic acids (PLA), polyhydroxyalkanoate (PHA), polysaccharide-based polymers (starch and cellulose) and protein-based polymers (soya protein, wheat, gluten) and petroleum-based polymers include poly(ε-caprolactone) (PCL), poly(alkylene dicarboxylate), aliphatic polyester, aliphatic aromatic polyester, poly(ester amide), poly(alkyene succinate) and poly(vinyl alcohol) (Figure 13.1).

13.6.3 Petroleum-Based Polymer Matrix

13.6.3.1 Poly(Alkylene Dicarboxylate)

It is a biodegradable aliphatic polyester produced by polycondensation reaction of glycols with aliphatic dicarboxylic acids. For preparation of high molecular weight, coupling reaction is carried out in the presence of small amount of coupling agents. The most commonly used aliphatic polyester includes polybutylene succinate (PBS); poly(ethylene succinate); poly(butylene succinate-co-butylene adipate). Methods like polymer blending, copolymerization and integration with fillers have been used to improve its properties.

13.6.3.2 Poly(ε-Caprolactone)

Poly(ε-caprolactone) (PCL) is a partially crystalline linear polyester with a low T_g of 60°C and a low melting temperature of 60°C (Table 13.4). It is prepared by a ring opening polymerization (ROP) of cyclic seven membered lactone monomer in the presence of catalyst such as stannous octanoate. In the initiation step, stannous octanoate is converted into tin alkoxide by reaction with alcohol and other protic compound by coordination insertion mechanism. PCL is grafted on fiber cellulose by surface initiated ROP using organic acid with amino acid as catalyst Alternatively, surface-initiated ROP of ε-PCL has also been performed in the presence of benzyl alcohol as free-radical initiator (Scheme 13.1).

PCL has modulus in between those of high-density polyethylene and low-density polyethylene. It is degraded by enzymes and lipases secreted from microorganisms. PCL is compatible with many other polymers and organic materials and has been used in many polymer formulations. Blending of starch with PCL improves water resistance of starch-based formulations and this can be used in number of future applications. Modification including blending or cross-linking with other polymers has been used to overcome low transition temperature, strong hydrophobicity and tendency to break of PCL-based polymers.

13.6.4 Natural-Based Polymer Matrix

13.6.4.1 Polyhydroxy Lactic Acid (PLA)

It is the most commonly used biopolymer and is synthesized from renewable resources such as starch, potato, beets and sugarcane. It is hydrophobic polymer with lactic acid as basic polymer unit. This monomer can be polymerized to PLA by direct

TABLE 13.4
Common Characteristics of Polymers

Biopolymer	Density (g/cm³)	Melting Temperature (°C)	Crystallization Rate	Glass Transition Temperature (°C)	Tensile Strength (MPa)	Elongation at Break (%)	Young Modulus (MPa)	Degradation Temperature (°C)
Poly(ε-caprolactone) PCL	1.06–1.13	59–64	Low	−60	4–35	80–500	343.9–363.4	380
Poly(lactic acid) PLA, P(L)LA and P(D)LA	1.21–1.29	150–170	Low	53/63	48–66	3–30	30.1–46.5	320
Poly(3-hydroxybutyrate) PHB	1.18–1.26	170–180	Low	−5/5	15–40	4–10	48	220
Poly(butylene succinate) PBS	1.22–1.28	114–116	High	−32/−30	25–34	330–660		405
Starch	1.5	160	–	25	4.48–8.14	60	116.42–294.98	300
Starch PLA blend	0.96–1.19	120–160	Low	4–6	2.1–2.8	42–63	28–74	
Starch PCL blend	–	150–190		−1 to −66	7.6–13.2	11–49	240	
Cellulose	1.5	260–270	–	200–250	–	–	78,000	315
Soy protein	0.13–0.44	–	–	−113	1.6–3.5	–	21.8–26.5	92
Gluten	0.625	–	–	123.85–144.85	3.09–5.4	23	5,000	130

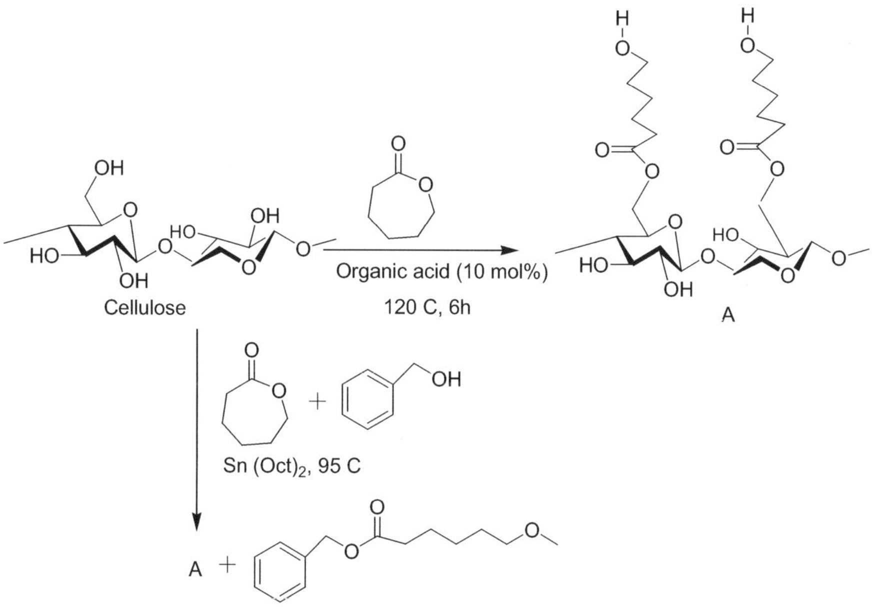

SCHEME 13.1 Organocatalytic ROP process of ε-caprolactone with cellulose.

polycondensation, ROP and azeotropic condensation polymerization. The physical properties and bio-degradability of PLA can be regulated by stereochemistry of lactic acid and lactide monomer (Scheme 13.2). It can be amorphous, crystalline or semicrystalline depending on composition of stereoisomers. The biodegradability, biocompatibility, renewability, high mechanical strength and modulus make it a good alternative for typical petroleum-based polymers.

Various biofibers such as coir, flax, hemp, agricultural waste and starch are used as reinforcement to PLA matrices (Guan et al. 2019; Oksman 2000; Masirek et al. 2007). Wong et al. (2004) resulted in improvement in modulus, tensile and flexural strength, water resistance (Table 13.3).

13.6.4.2 Poly(β-Hydroxyalkanoate)

Poly(β-hydroxyalkanoate) (PHAs) are synthesized biochemically by fermentation of glucose, sugar or lipids by microorganisms. Poly(hydroxybutyrate) (PHB) is a biotechnologically produced polyester and is a homopolymer of 3-hydroxybutyrate and are readily get degraded. However, brittleness and a narrow processability window are the important disadvantages that need to be tackled. To improve these properties, various copolymers poly(3-hydroxy- butyrate-co-3-hydroxyvalerate) (PHBV), poly(hydroxybutyrate-co-3-hydroxy-hexanoate) (PHBHA), poly(hydroxyl butyrate-co-hydroxy octanate) and poly(hydroxyl butyrate co-hydroxy octadecanoate) have been biosynthesized. Cellulose and PHB/PHBV composites are prepared by in situ reactive extrusion in presence of peroxide like dicumyl peroxide (DCP) for free radical initiation. Grafting reaction occurs between the hydrophilic fiber surface and hydrophobic PHB/PHBV matrix (Scheme 13.3).

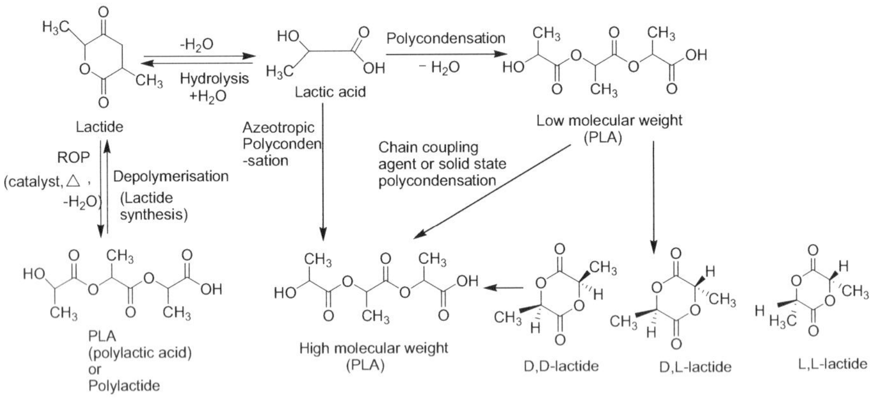

SCHEME 13.2 Chemistry of interconversions between lactic acid, lactide and poly(lactic acid) (PLA).

PHBV polymers are ideal candidates for the substitution of nonbiodegradable polymeric materials in number of commercial applications. Biofibers incorporated with PHA polymer matrix showed improved thermal and mechanical properties (Mohanty et al. 2005; Yu et al. 2006; Barud et al. 2011). Fiber length, aspect ratio, fiber source, fiber treatment and interfacial bonding were detrimental factor for improvement in toughness and tensile strength (Table 13.3). However, the high cost, low impact resistance and small difference between thermal degradation and melting temperature and relatively high glass transition (T_g) are major hindrances in its commercial application.

Godbole et al. (2003) blended films of PHB with starch had a single glass transition temperature and all prepared combinations were crystalline in nature. This blending helps to reduce cost of PHB as starch is abundantly available at a very low cost. Use of 30% of starch in PHB reduced the price of bioplastics, without sacrificing the physical properties.

13.6.5 POLYSACCHARIDE-BASED BIOCOMPOSITES

Polysaccharides are natural polymeric carbohydrates found in plants and consist of long chain monosaccharide units bound by glycosidic linkages. Polysaccharides are categorized as structural (cellulose) and storage (starch).

13.6.5.1 Starch and Its Biocomposites

Starch is least expensive biodegradable material and is a mixture of branched amylopectin (poly-α-1,4-D-glucopyranoside and 1,6-D-glucopyranoside) and linear amylose (poly-α-1,4-D-glucopyranoside). The amount of amylose and amylopectin in plant varies with the source. Typically the content of amylase and amylopectin in starch ranged from 17% to 35% and 65% to 85%, respectively. These components give crystalline properties to starches (Table 13.4).

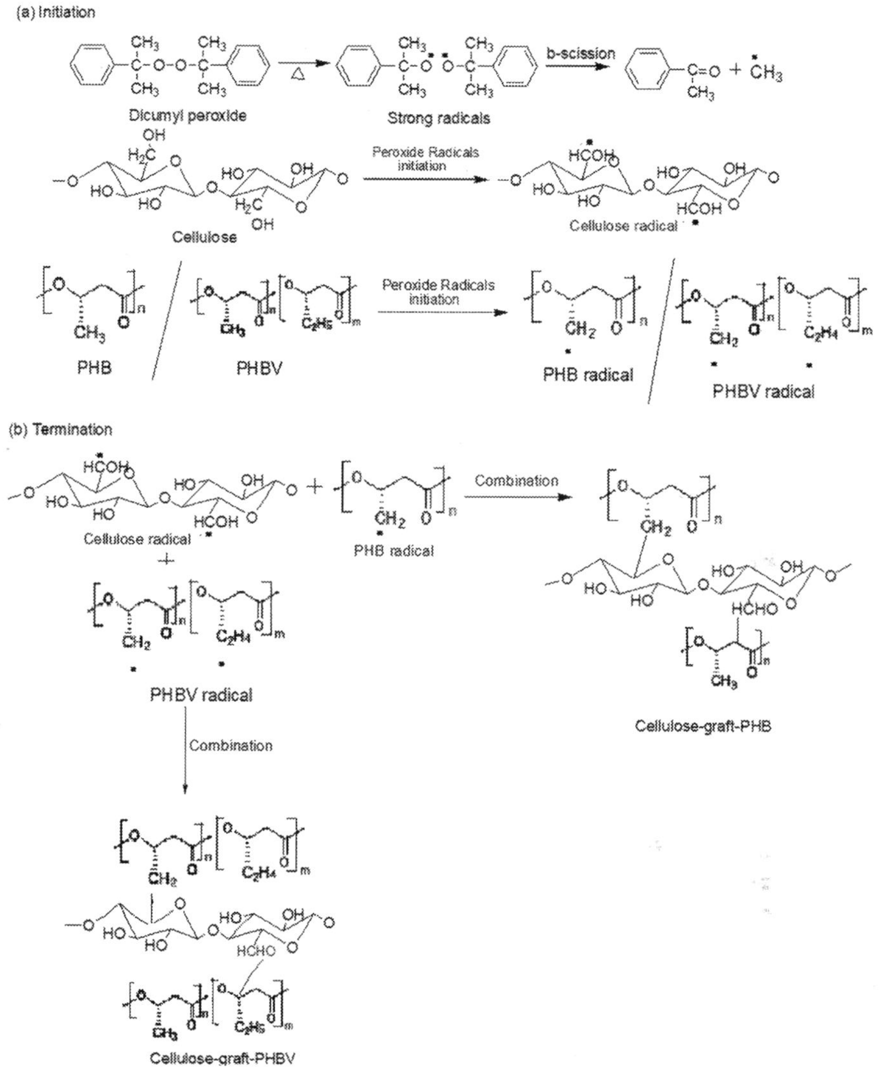

Agricultural Waste & Prospects in Biocomposites 361

SCHEME 13.3 Peroxide radical-initiated grafting of PHBV onto cellulose.

Corn, potato, wheat and rice are primary source of starch.

The internal interaction and morphology of starch vary readily by water molecules, dimension and mechanical properties, and glass transition temperature (T_g) depends on the water content. Native starch has low T_g (60°C–80°C) at water content of 0.12–0.14. This allows the use of starch by injection mould to obtain thermoplastic starch polymers. Destructurization in the presence of plasticizers under specific extrusion conditions convert starch into thermoplastic. Starch is hydrophilic due to the presence of large number of hydroxyl groups and it can be derivatized and reduced to overcome these problems and thus decreasing matrix polarity and increasing adhesion to fiber surface.

The hydrophilicity of starch can be used to improve the degradation rate of some degradable hydrophobic polymers. Starch is totally biodegradable as it is hydrolyzed to glucose by microorganism or enzymes, and then metabolized into carbon dioxide and water. Starch has low processability, poor mechanical properties and dimensional stability to be used directly for commercial applications. Chemical derivatization has been an important tool to solve these problems. Natural fibers from sisal, jute, flax and palm have been reinforced with starch after chemical alkali treatment to produce stronger interactions with starch matrix. Because of the treatment fibril fibrillation occurs which is expected to increase the mechanical anchoring of fiber with the matrix. Reinforcing natural fibers like casura, cellulosic husk, date palm and flax with starch-based polymer increased modulus and impact toughness and reduce water uptake. Corn starch reinforced with flax by molding method after soaking fiber in NaOH displayed strong adhesion between fiber and matrix. Such composites are completely biodegradable and hence environmentally safe. Tensile properties of starch-based biocomposites are dependent on nature and type of fiber, their content and orientation and type of blending. Processing methods with high stress and more efficient mechanical mixing promotes the opening of the fiber bundles increasing the aspect ratio of fiber. Table 13.4 enlists some of the starch-based composites.

Starch has been blended with PLA and PCL to overcome associated hurdles. The PLA starch blend was obtained by free radical reaction between maleic anhydride and PLA starch in the presence of suitable free radical initiator (Scheme 13.4). The formation of hydrogen bonds between the hydroxyl group of starch and carboxylic group of hydrolyzed anhydride and carboxyl group of PLA improves compatibility of PLA starch blend. Kalambur and Rizvi blended starch with PCL by cross-linking oxidized starch (replacement of OH group by carboxyl) with ester group in PCL. Fenton agent was used as oxidizing/cross-linking agent (Scheme 13.4). The blending significantly improved thermal and mechanical properties, enhanced interfacial adhesion and performance of biocomposites.

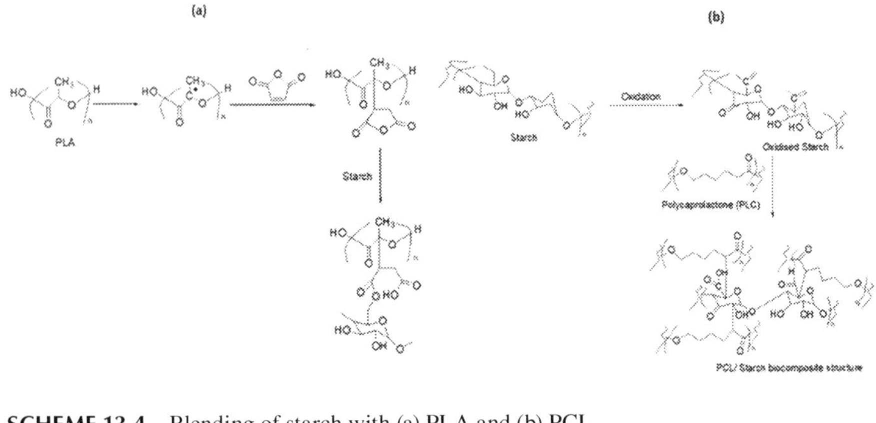

SCHEME 13.4 Blending of starch with (a) PLA and (b) PCL.

13.6.5.2 Cellulose and Its Composites

It is a linear chain of ringed glucose molecules linked by β(1 → 4) glycosidic linkage. Various plant-based waste materials such as peel, husk, and bagasse are used as source of cellulose. Various properties of cellulose are presented in Table 13.4. However, its insolubility in water and other organic solvents is major hindrance limiting its use. These drawbacks have been overcome by derivatization. Ether derivatives (methyl cellulose, carboxy methyl cellulose, hydroxyl propyl methyl/propyl cellulose) and esters (cellulose acetate) have been exploited for different applications (Table 13.3).

13.6.5.3 Polypeptide Based

There are naturally occurring linear unbranched polymer chains in which 50–2,000 amino acids linked together by peptide bonds. α-Helix and β-sheets are two major possible conformation of peptides. The hydrogen bonding between amino and carboxy groups of amino acids stabilize both conformation. Linking of β-strand by hydrogen bonding creates β-sheets. Polypeptide chains fold so that the hydrophobic side chains are buried within structure while polar (hydrophilic) side chains are at the surface. The tertiary structure of protein is considered the global configuration formed as an arrangement of secondary structure in space. Proteins from plants, viz., soya protein and wheat gluten being of renewable nature and their biodegrability, are attractive materials for the application in fabrication of biocomposites for industrial applications.

13.6.6 SOYA PROTEIN

Soya is plant-based protein obtained from soyabean plant. It has mainly global storage protein that contain nonpolar amino acids, viz., valine, leucine and alanine; basic amino acids, viz., lysine, arginine and noncharged polar amino acid cysteine and glycine. These groups are cross-linked with different additives. In dried soyabean soya protein content is 36% and same can be used as biopolymer. Plasticizer is necessary in soya protein as matrix because they are brittle and sensitive to water due to hydrophilic nature. With plasticizer shear modulus of soya protein is 1.769, tensile strength range from 10 to 40 mPa with elongation break at about 1.3%–4.8%. With increase in plasticizer content elongation increases and Young modulus decreases.

Environment friendly biocomposites using jute, kenaf, sisal and flax grass fiber reinforced with soya protein resin with or without any chemical plasticizer have been fabricated to act as replacement for nonbiodegradable materials (Table 13.3). These biocomposites have excellent tensile strength, flexural strength and modulus and showed high mechanical and thermal properties. The heat of deflection temperature of biocomposites increased with increase in fiber length and content. Hydrogen bonding interactions are often insufficient resulting in inadequate mixing of lignin with soya protein.

Huang and Netravali (2007) reinforced jute fiber with soya protein modified with glutaraldehyde and nanoclay by applying tension to minimize shrinkage and misalignment. Blending of kraft lignin with protein using methylene diphenyl isocyanate

(MDI) as compatibilizer showed that the addition of MDI in 2% enhanced tensile strength, modulus and elongation at break of polymer due to cross-linking and graft copolymerization.

13.6.7 WHEAT GLUTEN

Wheat gluten contains 75% proteins with 45%–50% gliadins and 35%–45% glutenins which are hydrophobic. Gliadins are low molecular weight, single-chain compact polypeptide and are globar shaped and give gluten its strength whereas glutenins are high molecular multiple chain polymeric proteins interlinked by intermolecular disulfide and hydrogen and give elasticity to gluten. Glutenin contains free SH groups while gliadin with only intramolecular S–S bond and no free SH moieties. As temperature exceeds 90°C the free SH group of glutenin induces covalent linkage with gliadin through heat-induced SH-SS exchange mechanism resulting in gliadin-glutenin cross-linking (gluetin network) (Figure 13.6). The free SH group carries out nucleophilic attack on sulfur of disulfide. Addition of reducing agent and high temperature increases the level of free SH group thus increases gliadin-glutenin cross-linking.

Plasticizers are prerequisite for fabrication of protein-based composites as it reduce intermolecular forces, decreased glass transition temperature and increased polymeric chain mobility. Common plasticizer used include glycerol, xylan, 1,4-butanediol, water and octanoic acid. Protein-protein based cross-links play an important role in design of biodegradable biocomposites with suitable rheological and mechanical properties. L-cysteine, glutaraldehyde and formaldehyde have been used predominantly to form cross-linked polymer network because it reacts with

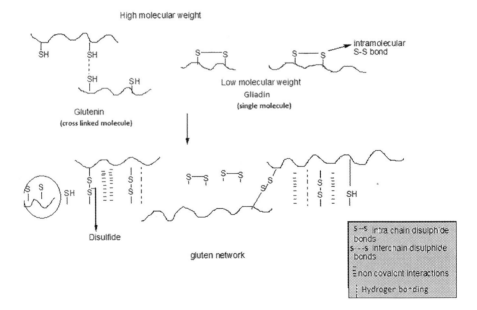

FIGURE 13.6 Cross-linking of gliadin-glutenin.

Agricultural Waste & Prospects in Biocomposites

functional groups in both proteins and carbohydrates and can provide material with substantial improvement in tensile strength. Various studies found that with the use of chemicals as cross-linking agents improved the tensile strength and oxygen barrier properties while elongation at break and water vapor properties decreased.

13.7 APPROACHES TO IMPROVE THE PERFORMANCE OF BIOCOMPOSITES BY MODIFICATIONS

The obstacles for the practical application of natural fibers for production of biocomposites can be overcome by using surface modification methods which may be either physical treatments (i.e., ultraviolet, plasma and ultrasound), biological methods or chemical treatments (mercerization, acetylation, esterification and coupling) (Figure 13.7).

Physical treatment has no effect on chemical composition of the fibers; however, it modifies the structural properties thus affecting the mechanical bonding to polymers. Chemical modification of the fibers changes the chemical composition and surface properties of fibers to increase the wetting with polymer matrix. It also enhances the adhesion of fiber with the matrix due to elimination of unwanted materials like lignin, hemicelluloses and pectin from the surface of fiber.

13.7.1 Chemical Treatment Methods

The chemical treatments decrease the hydrophilic nature (Ahmad et al. 2019) and improve physicochemical and mechanical properties of natural fibers by modifying their properties and microstructure. These modifications improve the wettability,

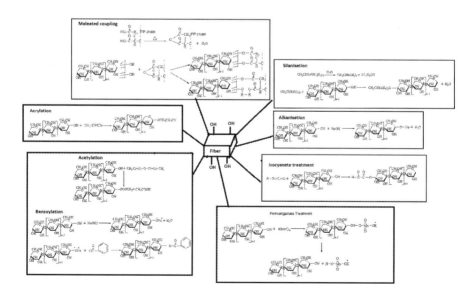

FIGURE 13.7 Different surface treatment methods for natural fibers.

surface morphology, enhance thermal stability, moisture resistance properties and tensile strength of the fiber.

13.7.1.1 Mercerization

Mercerization is the extensively used treatment in which the natural fiber is treated with alkali (Table 13.5).

Natural fiber absorbs moisture due to the presence of hydroxyl group, i.e., cellulose, hemicelluloses and lignin on its surface. Alkaline treatments weaken the hydrophilic nature of the fibers by reducing the number of hydroxyl group. This treatment also removes some fraction of hemicellulose, lignin, pectin, wax and oil from the surface of fibers. Due to alkali treatment the surface of natural fibers becomes rough and fibers are reduced to fibrils. It improves aspect ratio and increases interfacial bonding and finally improves mechanical strength and thermal behavior of natural fibers.

The another advantage of alkali treatment is that chemical modifications with other functional groups are feasible on the exposed OH groups in cellulose chains. The concentration of alkali, time and temperature of treatment plays decisive role in attainment of the optimal effectiveness of the fiber. High alkali concentration results in elimination of covering materials from surface thereby delignifying the fiber and negatively affecting its strength (Sahu and Gupta 2019).

13.7.1.2 Acetylation (Esterification)

Acetylation is the process of introduction of acetyl group onto the surface of the fibers. Acetyl group reacts with hydrophilic hydroxyl groups of the fiber to become hydrophobic thereby providing stability to composite due to reduction in hydrophilic nature of fiber and increased fiber-matrix adhesion. The strong bond results in good properties of natural fiber-based composites (Thomas 2003; Rong et al. 2001). Normally, alkali treatment is provided to fiber before treatment with glacial acetic acid. Acetic acid treatment improves interfacial bonding, tensile and flexural strength, stiffness, dimensional and thermal stability and resistance to fungal attack (Table 13.5). Overtreatment results in deleterious effect on mechanical properties due to degradation of cellulose and cracking of fiber. Other esterification methods include benzoylation (benzoyl group interact with hydroxyl group of natural fibers) and esterification with fatty acids like stearic acid and oleoyl chloride and peroxide (Table 13.5).

13.7.1.3 Silane Treatment

Silane is a multifunctional molecule that is used as coupling agent to modify the fiber surface. The silane group form chemical link between the fiber surface and matrix and improves the fiber-matrix adhesion and mechanical properties of natural fibers. Silane groups increase hydrophobicity and strength of natural fibers and large increase occurs when silane and matrix are covalently bonded (Table 13.5). Most commonly reported silane includes amino, methane, glycidoxy and alkyl silanes. The vinyltrimethoxysilane and aminopropyltriethoxy silane are commonly used silanes to obtain reliable modification of natural fibers. Bodur et al. (2016) observed significant improvement in strength and Young's modulus of silane-treated fibers.

TABLE 13.5
Different Treatment Methods Its Effects on the Natural Fibers

Treatment Method	Treatment	Fiber Name	Specific Conditions	Tensile Strength (MPa)	Young's Modulus (GPa)	Elongation (%)	Thermal Stability	References
Chemical treatment	Alkali treatment	Cotton	Soaked in 5% (w/v) NaOH solution for 15, 30, 45, 60, 75 and 90 min	357–1,809	10.45–87.57	1.91–5.88	High	Rajkumar et al. (2016)
		Jute	5% NaOH for 2 h up to 72 h at room temperature	High	2.915–12.850	—	—	Ray et al. (2001)
		Sisal	5% NaOH for 2 h up to 72 h at room temperature	460.10–612.75	106.81–213.17	3.55–4.42	—	Mishra et al. (2001)
		Hemp	Pretreatment (12% w/w NaOH, 2 h), acid hydrolysis (1 M HCl, 80°C, 1.5 h). Alkaline treatment (2% w/w NaOH, 2 h, 80°C)	High	—	—	—	Wang et al. (2007)
		Kapok, sisal, jute and hemp	Fibers placed in water bath at 20°C ± 2°C for 48 h and removed using 1% acetic acid	High	Low	—	Low	Mwaikambo and Ansell (2002)
		Hemp	5 wt% sodium hydroxide solution maintaining a solution: fiber ratio of 20:1 (by weight)	600	34	—	—	Sawpan et al. (2011)
		Bamboo	Two modes of loading: loading on the side with higher vascular bundle density and loading on the side with lower vascular bundle density	1,430–1,690	High	—	—	Chen et al. (2020)

(*Continued*)

TABLE 13.5 (Continued)
Different Treatment Methods Its Effects on the Natural Fibers

Treatment Method	Treatment	Fiber Name	Specific Conditions	Tensile Strength (MPa)	Young's Modulus (GPa)	Elongation (%)	Thermal Stability	References
		Okra	Pretreatment: 10% NaClO$_2$ with 1:80 fiber liquor ratio for 90 min, with sodium meta bisulfate with 1:20 fiber liquor ratio up to 15 min	200.6–620.4	1.9–16.9	–	Low	Rosa et al. (2011)
Acetic acid treatment		Jute, flax, hemp, ramie	Fiber treated with acetic acid soaked in acetic anhydride	345–4570	9.4–85	1.2–8.0	High	Kabir et al. (2012)
		Ramie, sisal, cotton	5, 10, and 15% (w/v) acetic acid solution for 2 h at room temperature, maintaining a liquor ratio of 30:1	88.40–121.16	13.15–16.21	0.99–1.36	High	Kommula et al. (2016)
		Sisal	NaOH (18%), glacial acetic acid and acetic anhydride	450–700	7–13	4–9	High	Naira et al. (2001)
Silane treatment		Sisal	Soaked fiber in aminosilane (2%) for 5 min	375.8–535.1	4.5–15.5	2.5–8.3	–	Rong et al. (2001)
		Oil palm	Vinyl silane solution (1%) in water-ethanol	111	1120	13.5	High	Sreekala et al. (1997)
Benzoyl peroxide treatment		Sisal	soaked with 11 of a 6% solution of DCP in acetone for 30 min	400–700	–	5–14	–	Joseph et al. (1996)
		Oil palm	Fibers were coated with benzoyl peroxide from acetone solution	10	3.75	2	–	Sreekala et al. (1997)

(Continued)

TABLE 13.5 (Continued)
Different Treatment Methods Its Effects on the Natural Fibers

Treatment Method	Treatment	Fiber Name	Specific Conditions	Tensile Strength (MPa)	Young's Modulus (GPa)	Elongation (%)	Thermal Stability	References
	Potassium permanganate treatment	Banana	KMnO$_4$ treatment: soaked in 0.5% in acetone for half an hour)	600–750 (banana)	–	26/2–4	–	Paul et al. (2008)
		Oil palm	Washed in a 2% surf detergent solution at 708°C	130	3.58	9.7	–	Rahman et al. (2007)
		Sisal	Dipped in permanganate solution at conc. (0.033%, 0.0625% and 0.125%) in acetone for 1 min	8.82	0.137	200	–	Paul et al. (1997)
	Stearic acid treatment	Sisal	Dissolved in ethyl alcohol (1%)	8.82, 400–700	0.137	200, 5–14	–	Paul et al. (1997)
		Jute	Soaked in 0.5% in acetone for 30 min	–	–	–	High	Saravanakumar et al. (2014b)
	Cellulose powder treatment	Kapok, hemp, jute	5 wt% NaOH for 30 min, washed with diluted hydrochloric acid	34.0	1.61	1.40	–	Indran et al. (2015)
	Polymer Coating	Hemp and sisal	Fibers immersed in stirring solution for 2 h at 90°C	Low	High	–	High	Hajiha et al. (2014)
	Bleaching	Sisal	Dipped in textone (sodium chlorite) solution with a liquor ratio 25:1 for 2 h	High	–	–	–	Mishra et al. (2001)
	Isocyanate treatment	Jute, flax, hemp, coir, sisal, ramie	Electric discharge, Mercerization and wettability	140–4570	5–240	–	–	Bledzki et al. (1996)

(*Continued*)

TABLE 13.5 (Continued)
Different Treatment Methods Its Effects on the Natural Fibers

Treatment Method	Treatment	Fiber Name	Specific Conditions	Tensile Strength (MPa)	Young's Modulus (GPa)	Elongation (%)	Thermal Stability	References
		Hemp, flax, ramie, sisal, cotton, kapok, coir, banana	Stretching, calendering, thermotreatment and wettability	1.17–1627	4–128	1.16–9.0	–	Malkapuram et al. (2008)
		Sisal	6% solution of peroxide in acetone for 30 min)	8.82	137.2	200	–	Paul et al. (1997)
	Maleated coupling	Jute	–	20.60–24.20	27.61–34.31	–	Low	Mohanty et al. (2004)
		Banana, hemp and sisal	–	2.79–9.46	0.33–1.30	1.56–7.70	–	Mishra and Naik (2005)
		Jute, flax, hemp, coir, sisal, ramie	Electric discharge, mercerization and wettability	140–4570	5–240	–	–	Bledzki et al. (1996)
	Graft copolymerization	Jute, hemp, kenaf, flax, ramie, sisal, cotton, kapok, coir	Stretching, calendering, thermotreatment and wettability	1.17–1627	4–128	1.16–9.0	–	Malkapuram et al. (2008)
		Sisal	Washed with 5% NaOH and 3% orthophosphoric acid followed by washing with double distilled water	460.10–612.75	106.81–213.17	3.55–4.22	–	Mishra et al. (2001)
		Hemp, kenaf	Graft copolymerization and removal of homopolymer and determination of graft level	–	–	–	High	Thakur et al. (2013)

(*Continued*)

TABLE 13.5 (Continued)
Different Treatment Methods Its Effects on the Natural Fibers

				Properties				
Treatment Method	Treatment	Fiber Name	Specific Conditions	Tensile Strength (MPa)	Young's Modulus (GPa)	Elongation (%)	Thermal Stability	References
Biological methods	Fungal mix	Abaca	Fibers were dried at 80°C in an air circulating oven for 24 h	—	High	Low	Low	Bledzki et al. (2010)
	Enzyme	Hemp	Washed using 2% (w/v) industrial detergent for 1 h at 70°C	High	—	—	Low	George et al. (2014)
	Enzyme	Hemp	Fabric was washed at 60°C in 217°C detergent solution for 30 min and air dried	Low	—	—	—	Buschle-Diller et al. (1999)
Physical treatment	Plasma	Flax	Fibers were put inside the reactor, up to a final pressure of 10^{-4} Pa and gas was introduced in a controlled flow	70	3.7–73	3	—	Marais et al. (2005)
			Initial moisture content of the seeds was determined by oven drying at 105°C ± 1°C for 24 h	—	—	—	—	Wang et al. (2007)
	Ultraviolet irradiation	Jute	—	High	—	—	—	Gassan et al. (2000)
			—	—	—	—	—	Gassan and Gutowski (2000)
	X-ray	Oil seed	Fibers immersed in 5% sodium hydroxide solution for 20, 30, and 40 min and immediately washed with distilled water	High	Low	—	High	Foruzanmehr et al. (2015)

This was due to the formation of strong bonds between silanol (Si–OH) and –OH groups of the fibers. The remaining Si–OH undergoes condensation with adjacent Si–OH groups. The hydrophobic polymerized silane get linked to matrix by van der Waals force thus providing good interfacial interaction between fiber and matrix (Bodur et al. 2016).

13.7.1.4 Maleated Coupling

Maleated coupling has been extensively used to improve interfacial shear strength between fiber and matrix. It involves use of maleic anhydride to modify the surface of fiber and polymeric matrix to achieve improved fabric matrix interfacial interactions and enhance mechanical properties. Maleic anhydride reacts with hydroxyl group of natural fiber by condensation reaction (covalent bond) and physical interactions (hydrogen bonding). These interactions improve fiber-matrix compatibility by reducing hydrophilic nature of fiber. Combination of maleic anhydride polypropylene (MAH-PP) treatment with mercerization showed synergistic effect. Hemp fibers modified by MAH-PP coupling agent are used for the formation of epoxy and PP composites with high tensile modulus and improved mechanical properties.

13.7.1.5 Acrylation and Acrylonitrile Grafting

Cellulose in the presence of high energy radiation induced free radical and initiate acrylation. Acrylic acid reacts with cellulosic hydroxyl group of natural fiber and promotes the formation of polymerization medium. Another alternative of modification is by acrylonitrile grafting. This involves dehydration and oxidation followed by interaction with activated free radical sites.

13.7.1.6 Graft Copolymerization

Another technique to modify the surface of natural fibers is grafting by copolymerization which involve attachment of polymer chain directly to natural fiber. Graft copolymerization is done either by grafting of fiber with a single or mixture of two or more monomers or with the polymer. Living polymerization (ROP and free radical polymerization), photochemical, plasma radiation induced grafting, plasma induced initiation and enzyme grafting have been used to graft copolymerization monomer onto backbone of biofiber.

13.7.2 GRAFTING VIA LIVING POLYMERIZAITON TECHNIQUE

13.7.2.1 Ring Opening Polymerization for Biocomposites

ROP catalyzed by organic acid produced PCL-graft-cellulose directly. Caprolactone and lactide were grafted onto cellulose surface via covalent bonding via ROP process catalyzed by Sn(Oct)$_2$ (Scheme 13.5). The reaction was carried out using benzyl alcohol as an initiator. Ratio of initiator to monomers controlled the grafting density.

13.7.2.2 Free Radical Grafting

The grafting by free radical initiators has been used to graft various monomers onto cellulose fibers and their derivatives. Free radicals are generated by thermal (organic

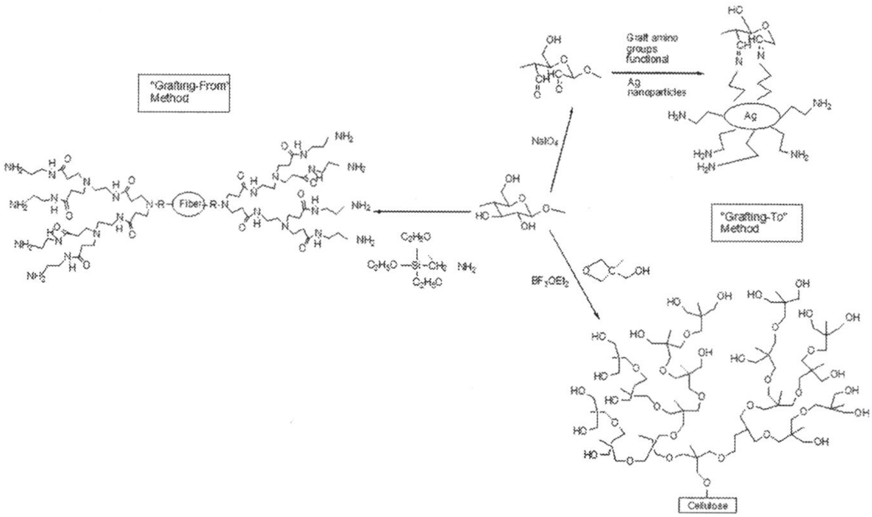

SCHEME 13.5 "Grafting To" and "Grafting From" method of compatibilization.

peroxide and azobisisobutyronitrile), chemical (various persulfates, ceric ammonium nitrate and Fenton reagent (Fe (II)–H_2O_2)) or irradiation (gamma, UV and electron/plasma ion beam) initiation. Poly methylacrylate (PMA)-grafted cellulose copolymer was formed by grafting cellulosic optiva fibers with ferrous ammonium sulfate-potassium per sulfate as redox free radical initiator.

Atom transfer radical polymerization was used to graft Gum rosin to lignin by initial formation of 2-bromoisobutyryl ester-modified lignin. The lignin was esterified with three different monomers of dehydroabietic acid (DA). Hydrophobicity of lignin increases on grafting with DA and rosin polymers. Cellulose radical through chain scission was generated by high energy radiation (e.g., rays from electron beams or radioactive isotopes). Sisal fibers were grafted with co-monomer of styrene and ethyl acrylate initiated by gamma irradiation (El-Naggar et al. 1992).

Grafted cellulose PHB/PHBV biocomposites using DCP as free radical initiator (Scheme 13.3) showed better mechanical properties due to interactions between hydrophobic PHB or PHBV matrix and hydrophilic fibers. The reduction of crystallinity was indicative of the fact that grafting reaction occurred not only at the amorphous region but also in the crystalline domains of the cellulose fiber. Grafting decreased brittleness and improved flexibility. The grafted copolymer showed improved structural, chemical, thermal and morphological properties and has been potentially used for biocomposite applications.

13.7.3 BIOLOGICAL METHODS

These methods involve selective degradation of all the components of the fiber, i.e., hydrophilic pectin and hemicelluloses except cellulose using biological organism such as enzyme and fungi. The selective degradation was probably due to high crystallinity of cellulose. Biological treatment of natural fibers provides fiber with better thermal and mechanical properties and rough interface for better interlocking with the matrix.

The main role of the enzymes in the treatment process is to degum the natural fibers while improving the homogeneity, cleanliness, fineness and softness (Konczewicz and Kozłowski 2012). The enzyme treatment is performed at the atmospheric pressure below 100°C and pH was maintained at 4–8. The efficiency of the treatment increased by addition of chelators like ethylene diamine tetraacetic acid (Konczewicz and Kozłowski 2012). However, this treatment has restricted use due to high cost associated and appropriate choice of enzymatic agents (Konczewicz and Kozłowski 2012).

13.7.4 Physical Treatment

Surface modification by physical methods involves the use of corona, plasma, ultraviolet irradiation, vacuum, ozone, X-ray and laser treatment to change the fiber surface physically. These create rough surfaces that are beneficial for mechanical interlocking. Plasma treatment modifies the surface of fiber and introduces chemical functionality on it. This leads to improved adhesion mechanism between polymer matrix and surface of the fiber. This introduces rigidity and stiffness to the fibers thereby increasing its applicability for suitable structural applications. Steam explosion enhances dispersibility and adhesion of lignocellulosic material with polymer matrix. Table 13.5 enlists some of the physical treatment methods and their effect on natural fibers.

13.7.4.1 Addition of Other Materials

13.7.4.1.1 Incorporation of Nanoparticles

The addition of nanoparticles to biocomposites has gained more popularity in recent time. Nanoparticles incorporation into polymer composites improves Young's modulus and their strength (Dorigato et al. 2012). The modifications of the particles further showed improved chemical interactions when added to the composites thereby improving the crystallinity resulting in better mechanical interlocking of the fiber-polymer interface. Fillers or additives are materials added in low percentage (0.1–5 wt%) to improve performance and impact properties. Low-cost fillers are preferred as they provide cost-effective enhancement of performance properties. Commonly used nanofillers can be broadly classify as inorganic fillers which include layer silicates, silica, carbon nanotubes, alumina, ZnO, TiO_2, Ag and organic nanofillers (starch nanocrystals, chitin/chitosan, cellulose and zein) (Table 13.6).

Mohanty and Srivastava (2015) observed better interfacial adhesion between fiber/matrix after the addition of clay and aluminum powder filler at low wt%. Use of high wt% of filler deteriorated the properties of biocomposites due to agglomeration of filler and weak interfacial adhesion between fiber and matrix. Natural fibers incorporated with nanofillers like TiO_2, SiO_2, carbon nanotube, ZnO, graphene oxide at low concentration 0.1%–1% have significant influence on mechanical

TABLE 13.6
Effect of Nanofiller Addition on Biocomposites

Nanofiller	Polymer	\multicolumn{7}{c}{Effect of Nanofiller Addition}	References
		TS	EAB	Thermal stability	WVP	YM	MC	Others	
AgNPs nanocellulose	PVA	Increase (51%)	Increase (28%)	Increase (42%)	Reduction (21%)	–	–	–	Sarwar et al. (2018)
TiO$_2$ NPs	Starch	Increase (45%)	Reduction (28%)	–	Reduction (51%)	–	Reduction (31%)	–	Oleyaei et al. (2016)
Fe$_2$O$_3$ NPs	Cellulose	Increase (10%)	–	Increase (38%)	–	Increase (15%)	–	–	Yadav et al. (2018)
α-Fe$_2$O$_3$ NPs FeNPs	PVA	Increase	Reduction (35%)	–	–	–	–	Improved magnetic properties	Hoque et al. (2018)
ZnO nanorods	Starch	Increase (30%)	Reduction (54%)	–	–	–	–	–	Marvizadeh et al. (2017)
ZnO nanorods	Polysaccharide	Reduction (18%)	Increase (41%)	–	Reduction (32%)	–	–	Increment of heat seal strength	Akbariazam et al. (2016)
Bacterial cellulose nanocrystal	PVA	Increase	–	–	–	Increase (28%)	–	Improved toughness	Rouhi et al. (2017)
Cellulose nanofibers	Soy protein	Increase (400%)	Reduction (56%)	–	–	Increase (767%)	–	–	Gonzalez et al. (2019)
Cellulose nanocrystals	PVA	Increase (83%)	–	–	Reduction (82%)	–	–	–	Achaby et al. (2017)
Cellulose nanocrystals	Starch	Increase (5.6%)	–	–	Reduction (43%)	–	Reduction (42%)	–	Ma et al. (2017)
Multiwalled carbon nanotube	PVA	–	–	Increase (47%)	–	–	–	–	Mallakpour et al. (2017)
Graphene oxide	Cellulose	Increase (100%)	Reduction (64%)	Increase	–	–	–	–	Gan et al. (2017)
Graphene oxide	Amylase	Increase (16.5%)	–	Increase	–	–	Reduction	–	He et al. (2013)
ZnS NPs	PVA	Increase (28%)	–	Increase (56%)	–	–	–	Reduction in SR	Yun et al. (2017)

properties of the biocomposites. The tensile modulus increased and impact strength decreased due to polymer matrix improved cross-link density. Modification of clay has significant effect on impact strength of the composite. These composites showed less voids and high density, improved stiffness and result in better interfacial adhesion between matrix and fiber (Wu et al. 2016; Halder et al. 2017). The addition of filler to sisal fiber increased moisture absorption by reinforced biocomposites (Halder et al. 2017).

13.7.4.2 Addition of Compatibilizers and Cross-Linkers

Biopolymers often have inferior properties. As a consequence, they are often modified to meet the expectations of the market. Techniques like blending and copolymeric grafting allows considerable improvement in their properties. Physical blending is simple mixing of polymeric material in melt state with no chemical reaction. Blending improves T_g, fracture resistance and flexibility. However, very few polymer pairs are miscible or compatible with each other. Moreover, physically blended materials lose integrity after some time. As a consequence, chemical routes are often used to achieve properties required for specific applications. PLA and starch are most often commonly used blend. The melt blending of PCL and PLA results in marginal improvement in toughness but stiffness and strength decrease due to weak interfacial adhesion.

Poor interfacial interaction between hydrophilic starch and hydrophobic PLA is a major problem of this blend. To improve the compatibility, suitable compatibilizers are added to the blend. Different categories of compatiblization have been distinguished. One approach generally involves application of reactive homopolymer which has groups to fit to create bond with other blend (copolymer) compounds. It is also called as reactive compatibilization (Table 13.7; Figure 13.8).

It can also react permitting grafting into polymer backbone or terminal group as in PLA/PP blend with epoxy or blending of PE/iPP (polyethylene and isolactic polypropylene).

The use of block copolymers, such as PCL, PCA, diblock, triblock, random block, a PCL-PEG copolymer, has helped to achieve some success. This type of compatiblization is also referred to as ex-situ compatibilization. Alternatively, as biopolymers also have number of reactive groups, there is excellent possibility of reactive compatibilization. This involves introduction of double functionalized low molecular weight compound that has the capability to link to two polymers. It is called reactive compatibilization in situ (Table 13.7; Figure 13.8). A multifunction compatiblizer is generally required as it acts physically and chemically at interface linking the different modified matrix filler and fiber phase.

Methylene diphenyl disiocyanate has been added to improve the interface of PLA-wheat resin blend. It enhanced the mechanical properties of blend at temperature above T_g. Other compatiblizers used include dioctyl maleate, polyvinyl alcohol, maleic anhydride grafted PLA chains enhance interfacial adhesion with granular starch and the reaction proceeds through radical reaction route using peroxide initiators. Basically, this improves occurrence of chemical reaction/modification of one of

TABLE 13.7
Blending of Polymers

Component			Properties			
Base Polymer	Second Polymer	Method	Mechanical Property	Toughness	Water sensitivity	References
Physical Blending						
PLA	PHAs	Melt blending	High	High	–	Zhang and Thomas (2011), Qiang et al. (2011), and Takagi et al. (2004)
	PCL	Physical Blending	–	High	–	Na et al. (2002)
	NR	Melt blending	–	High	–	Jaratrotkamjorn et al. (2011)
	Starch	Melt blending	–	–	–	Martin and Averous (2001)
	PBS	Blending	–	High	–	Wu et al. (2012)
	Starch	Melt blending	High	High	Low	Imre and Pukanszky (2013)
Starch	PHB	Mixing and blending	High	–	–	Lai et al. (2005)
	PLA	Mixing and extrusion	High	–	–	Yokesahachart and Yoksan (2011)
PHB	Starch	Mixing and blending	High	–	–	Zhang and Thomas (2009)
In situ compatibilization						
PLA	PA 11	Blending, Extrusion and injection	High	–	Low	Landreau et al. (2009)
	PU	Heating, stirring and blending	–	1.8–7.1 MPa	Low	Cao et al. (2003); Lu et al. (2005)
	EVA	Melt compounded	High	High	–	Zhang and Lu (2016)
	HDA	Hot pressing melt compounded	High	High	–	Liu et al. (2019)
	PUEP	Dynamic vulcanization	High	High	–	Lu et al. (2014)
Ex situ compatibilization						
PLA	PBS	Blown film	High	High	–	Supthanyakul et al. (2016)
PLLA	PBS	Compression molding	High	High	–	Zhang et al. (2017)
	PCL	Compression molding	High	High	–	Chavalitpanya et al. (2013)
PCL	PHB	Compression molding	High	4.68–26.99	–	Shuai et al. (2001)

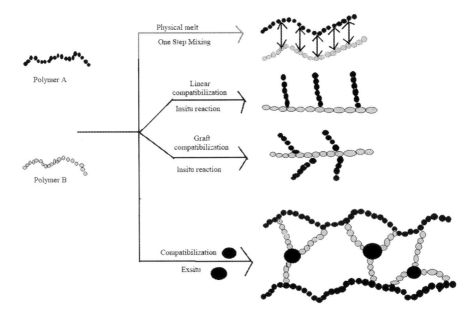

FIGURE 13.8 Different types of compatibilization of polymers.

the blend component in first step followed by reaction with second component during blending.

Karmakar and Youngquist (1996) reported drastic improved mechanical properties of Jute/PP composites after addition of maleic anhydride grafted polypropylene as compatibilizers at 3 wt%. Epoxidized natural rubber has been used to make natural rubber more compatible. Good filler dispersion and good bonding quality between filler and matrix have been achieved. Incorporation of maleated natural rubber in biocomposites reduces interfacial tension and increased interfacial adhesion to yield strong matrix filler interaction. Soy protein resin absorbs large amount of moisture and use of glutaraldehyde vinyl alcohol; benzilic acid decreased the moisture absorption of the resin.

13.8 CURRENT TRENDS

13.8.1 *Hybrid Composites*

Hybridization of composite with fibers of different matrix is one of the strategies used to fabricate composite with good mechanical properties comparable to glass and carbon fiber composites. Hybrid composites have two or more reinforcing phases in the single matrix. Natural fibers have been reinforced with natural fiber, synthetic fibers and filler for tailoring the properties of the composites. A hybrid composite with fiber-like carbon, basalt and Kevlar has been extensively reported but biodegradable

TABLE 13.8
Hybrid Composites

Fiber-1	Fiber-2	Matrix	Tensile Strength (MPa)	Tensile Modulus (GPa)	Flexural Strength (MPa)	Flexural Modulus (GPa)	Impact Property	Composition/ Ratio	Density (g/cm^{-3})	Total Fiber Load	References
Flax	Glass	Soyabean oil based	123.3	3.5	130	6.9	2.71	1:0, 4:1, 3:2, 2:3, 0:1	1.5–2.6	31%–40%	Morye and Wool (2005)
Coir	Sisal	Natural rubber	15.46	0.0012–00.0038	120	5.9	1.5	1:1	0.96–1.02	10–50 parts per 100 parts of rubber by weight	Haseena et al. (2005)
Kapok	Cotton	Polyester	52087	1.64	55.34	0.7	119.2 kJ/m^2	2:3	1.14–1.23	58%–65%	Mwaikambo and Bisanda (1999)
Jute	Hemp	Polyester	35% inc.	26% inc.	–	–	86% inc.	1:5	1.3–1.42	42%–64%	Mehta et al. (2005)
Oil palm	Kenaf	Polyhydroxy butyrate	53.30	5.4	77.9	7.3	42.2 J/m	1:3	–	62%–68%	Khoshnava et al. (2017)
Jute	GF	Woven polyester	26.3	0.931	49	1.92	752 J/mm	1:5	1.2–1.5	–	Rajesh and Pitchaimani (2017)
Glass	Banana	Woven polyester	31	1.290	30	1.275	326 J/mm	1:2	1.3–1.8	–	Rajesh and Pitchaimani (2017)
Banana	Red mud	Polyester	54 to 77	–	175 to 294		90–360 kJ/m^2	1:2, 1:4, 1:6, 1:8, 1:10		0.3%	Prabu et al. (2014)
Coir	Red mud	Polyester	18.7	200–1,200 MPa	31.8	1.815	–	1:1	1.014	0.20%	Rachchh et al. (2015)
Jute	Glass	Polyester	36	–	–	1.844	–	1:2	1.1	72.0%	Ahmed et al. (2007)
Jute	Glass	Polyester (isothalic)	5400	4.1	825	1.714	–	–	–	78.0%	Ahmed et al. (2007)
Banana	Kenaf	Polyester	7300	5.3	15250	6.27	13–16 kJ/m^2	1:5	1:6	62%–72%	Thiruchitrambalam et al. (2009)

(*Continued*)

TABLE 13.8 (Continued)
Hybrid Composites

Fiber–1	Fiber–2	Matrix	Tensile Strength (MPa)	Tensile Modulus (GPa)	Flexural Strength (MPa)	Flexural Modulus (GPa)	Impact Property	Composition/ Ratio	Density (g/cm^{-3})	Total Fiber Load	References
Sisal	Kapok	Unsaturated polyester	6200	4.5	720	2.64	12–14 kJ/m^2	1:3	1:2	32%–41%	Reddy et al. (2008)
Kenaf	Glass	Natural rubber	5400	3.6	654	3.70	13–15 kJ/m^2	–	–	–	Busu et al. (2010)
Jute	Biomass	Bisphenol–C–formaldehyde	6200	6.9	420	4.25	10 kJ/m^2	1:4	1:4	6.35%	Mehta and Parsania (2005)
Bagasse	Henequen	Vinyl ester/ unsaturated polyester	150	–	159	–	335 J/m	–	–	–	Prasanna et al. (2019)
PALF	Coir fiber (CF)	PLA	19.15	4.75	–	4.86	4.24 kJ/m^2	–	–	30%	Siakeng et al. (2019)
Kenaf	Bamboo	PLA	187	–	199–206	15	–	–	–	–	Yusoff et al. (2016)
Alkali treated sisal	Coir	PLA	–	–	–	–	–	7:3	–	–	Duan et al. (2018)
Treated kenaf fiber	Montmorillonite clay	PLA	–	–	–	–	–	–	–	–	Ramesh et al. (2019)
Kenaf	Oil palm	PHB	53.3	5.4	77.90	7.3	40.6 J/m	–	–	–	Khoshnava et al. (2017)
Bamboo	PLF	Polyester resin	136	–	93	–	–	1:1	–	–	Rihayat et al. (2018)

hybrid biocomposites have been less explored. An overview of hybrid composites of different type is presented in Table 13.8. Hybrid composites show deviation from expected mechanical properties due to synergistic and negative hybrid effect. Fiber length, fiber loading, layering pattern and orientation of fibers, degree of mixing of fiber and dispersion of fiber in matrix are some of the parameters that influence the mechanical properties of biocomposites.

13.8.2 HYPERBRANCHED POLYMERS

These are highly branched polymers having low viscosity and high solubility and reactivity and have number of advantages like less molecular enlargement, high solubility, low viscosity and reactivity. The use of hyperbranched polymers has significant impact on the composites. These composites are well dispersed with reinforcement and there is improved compatibility as the interface between the fiber and the matrix resin is improved. The natural fibers are modified with hyperbranched polymer by "grafting to" and "grafting from" method (Sun 2019). The hyperbranched polymer is first synthesized and then grafted onto the surface of natural fiber via chemical reaction in grafting to method while hyperbranched polymer is achieved by step wise growth on the surface of plant fiber in grafting from method. Both methods have associated merits and demerits. In grafting to method, the steric hindrance of hyperbranched polymer reduces the efficiency of modification by grafting onto the fiber surface. The grafting from method helps to achieve high graft density due to easy access of the reactive groups to the chain ends of growing polymer.

13.8.2.1 Grafting from Method

Amino terminated hyperbranched polymers were grafted on cellulose fibers using grafting from method. Cellulose fibers were treated with cyano ethyl cellulose under alkaline conditions. Then cyano group was hydrolyzed into amino group and thereafter amino group again reacts with acrylonitrile to form cyano derivative which was hydrolyzed. The steps are repeated to graft hyperbranched polymer. Amino group was incorporated onto the fiber surface by γ-amino propyl-triethoxysilane modifications and then grafted the hyperbranched polyamide onto sisal fiber surface by repeated reaction with methyl acrylate and ethylene diamine (Sun 2019).

13.8.2.2 Grafting to Method

The hyperbranched polymer with terminal amino group was synthesized by polycondensation reaction of diethylenetriamine with methyl acrylate. This hyperbranched polyamide was treated with aqueous solution of $AgNO_3$ to prepare silver nanoparticles with amino group. This hyperbranched polymer with silver nanoparticles as core was grafted onto oxidized cotton fiber having aldehyde group. Oxidation time, concentration of oxidant and concentration of hyperbranched polymer with active nanoparticles as core and treatment time have detrimental role on the mechanical properties and grafting efficiency (Sun 2019).

Poly-3-methyl-3-oxelane methanol and boron trifluoride diethyl ether were used to synthesize poly-3-methyl-3-oxelane methanol. Hyperbranched polymer which was grafted onto cellulose fiber surface by surface induced ROP. The results showed that increase in number of hydroxyl group on the surface of cellulose laid a solid technical basis for subsequent functionalization of cellulose surface. Usually the hyperbranched polymer reduces the viscosity of composite system due to less intermolecular enlargement. In addition, there are large number of reactive terminal groups on their surface which is very conducive for improvement in compatibility of the interface between fiber and matrix.

13.8.3 NANOCELLULOSE BIOCOMPOSITES

After the removal of noncellulosic part of natural fiber, cellulosic content is exposed to different chemical treatment (ionic liquid, bleaching, mercerization, pulping and enzyme), mechanical treatments including sonication, grinding and homogenization or by using combination of treatments. All these treatments have been already discussed expect ionic liquids. Ionic liquids provide novel route for dissolution and regeneration of lignocellulosic material. Commonly used ionic liquid is 1-butyl-3-methylimidazolium chloride. Ionic liquid dissolved cellulose easily passed through homogenizer without clogging. After cellulose is precipitated by addition of water and nanocellulose is generated by freeze drying, the crystallinity of cellulose obtained by this method depends on dissolution time, regeneration kinetics and choice of nonsolvent.

Nanocellulose is cellulose in the form of fibers or crystal having length in few micrometers and diameter <100 nm. As cellulose sources are variable, degree of crystallinity influences the dimensions of liberated nanocellulose. Modifications in nanocellulose are relatively easier due to the presence of large number of reactive OH groups at surface making them suitable for surface functionalization for use in a variety of applications. Nanocellulose materials are mainly of three types: nanofibrillated cellulose, cellulose nanocrystal (CNC) and bacterial nanocellulose (Table 13.9). Among these CNCs have received great attention to fabricate high-performance nanocomposites due to exceptional properties conferred by high rigidity, stiffness, crystallinity, mechanical strength, optical properties as well as surface chemistry. The nanocellulose is biodegradable, light weight (density 1.6 g m/L), high tensile strength (10 GPa), high rigidity resulting in high thermal strength and Young's modulus. However, CNC has several commercial implications as the cost of producing CNC is higher than other nanocellulose materials. Thus to achieve sustainable and economic production there is in-between class referred as nanowhiskers. They are similar to CNC but without the specification of crystallinity. Nanocellulose biocomposites have number of industrial, structural and packing applications and exhibit better performance as compared to traditional biocomposites.

Extensive studies have been conducted to study the incorporation of nanofillers to nanocellulosic biocomposites to provide environmental sustainability. TiO_2, polyhedral oligomeric silsesquioxane, SiO_2, ZnO, carbon nanotypes, fullerene and

TABLE 13.9
Summary of Nanocellulose-Based Nanocomposites

Types of Nanocelluloses	Inorganic Phase	Method of Production	Average Dimensions Diameter (nm)	Average Dimensions Length	Young's Modulus	References
Cellulose nanocrystals (CNC)	—	Acid hydrolysis	3–50	30–300 nm	8.3 ± 0.9	Rusli and Eichhorn (2008) and Hoeger et al. (2011)
Nanofibrillated Cellulose (NFC)	—	High-pressure homogenization, microfluidization	5–60	Several microns regardless of the cellulose source	17.2 ± 1.2 (NFC paper)	Cranston et al. (2011)
Microfibrillated cellulose (MFCs)	AgNPs	UV radiation	5	Several hundreds of Nanometers up to a micron	—	Dong et al. (2013)
	Titania, chitosan and AgNPs	UV radiation	4–20		—	Xiao et al. (2013)
	AgNPs	Adsorption	—	—	—	Diez et al. (2011)
	TiO$_2$ NPs	Mixing in dispersion	—	—	—	Schutz et al. (2012)
	TiO$_2$ thin layer	CVD	3–15	—	80–130	Kettunen et al. (2011)
	AuNPs	Reduction by NaBH$_4$	—	—	80–130	Koga et al. (2010)
	Au-Pd bimetallic NPs	Reduction by NaBH$_4$	—	—	—	Azetsu et al. (2011)
Bacterial cellulose (BC)	AgNPs	Hydrolytic decomposition/ reduction by TEA	2–4	30.50 mm	—	Barud et al. (2011)
	AgNPs	Reduction by NaBH$_4$	<10	—	80–130	Yang et al. (2012)
	AgNPs	Reduction by hydroxyls	15	—	—	Yang et al. (2012)

(Continued)

TABLE 13.9 (Continued)
Summary of Nanocellulose-Based Nanocomposites

Types of Nanocelluloses	Inorganic Phase	Method of Production	Average Dimensions Diameter (nm)	Average Dimensions Length	Young's Modulus	References
	AgNPs	Reduction by NH_2NH_2, NH_2OH or $C_6H_8O_6$	30	–	–	Maria et al. (2010)
	AgCl NPs	Precipitation	–	–	–	Hu et al. (2009)
	AuNPs	Reduction by PEI	60, 40	–	–	Wang et al. (2011), Zhang et al. (2010), and Wang et al. (2010)
	AuNPs	Reduction by hydroxyls	<15 nm	–	–	Park et al. (2013)
	AgNPs	Reduction by sodium citrate	–	0.9 nm	90–120	Marques et al. (2008)
	Au-SiO$_2$ NPS	Reduction by sodium citrate	–	–	–	Pinto et al. (2007)
	AuNPs	Biosynthesis of glucose and alcohol	10–100	Mostly several tens of micrometers up to a mm	80–130 (individual BC nanofibril)	–
Nanocrystalline cellulose (NCC)	PdNPs	Reduction by H_2	–	100–300 nm	–	Cirtiu et al. (2011)
	AuNPs	Reduction by hydroxyls	23.6 ± 11.6	100–250 nm	–	Wu et al. (2014)
	CuO NPs	Reduction by $NaBH_4$	–	–	–	Zhou et al. (2013)
	Pt, Cu, Ag, AuNPs	CTAB stabilization/reduction by $NaBH_4$	–	–	50–110	Padalkar et al. (2010)

graphene oxide have been exploited for improving the applicability of nanofibers as biocomposites.

13.9 APPLICATIONS

Based on fiber type, global market has dominant share of flax, kenaf, cotton, hemp and others for synthesis of biocomposites in addition to wood. About 54.8% of the market in 2019 was dominated by the use of wood due to its high strength and solidity. Flax and kenaf are the most widely used material after wood. Seed fiber cotton is most commonly used in textile industry. Growing demand for renewable and biodegradable products has increased the use of biopolymers like PLA, PHB, starch and soya protein as matrices and it is anticipated to grow faster over the coming years. North America, Latin America, Europe, Asia-Pacific and Middle East and Africa are main regions of global biocomposites market. Asia Pacific has the largest market (40%) of biocomposite due to numerous use in industries such as automotive, building, construction consumer goods and others and expansion project in China, India, Japan, South Korea and South East Asian countries. Asia Pacific is followed by North America (33.0%), Europe (27%), Latin America and Middle East and Africa have moderate growth in biocomposites market.

Biocomposites have been used extensively in automobile, aerospace, packaging, military, naval, sports, medical, buildings and constructions and almost all major sectors (Ciccarelli et al. 2019). Among these, the application of use of biocomposites is most prevalent in construction and automobile industry has a global share of 56.0% and 28.2% (Mohammed et al. 2015; Huda et al. 2008). Bio-based composites provide lighter weight body parts for cars and airplanes, along with protection against heat and any external impacts have good mechanical properties, reduce emission of CO_2 and improve fuel efficiency. The tremendous interest and research studies on this sector are also gradually reducing the processing costs. Henry Ford fabricated the first bio-based composite using hemp fibers in the year of 1940 (Witayakran et al. 2017). Later, several car manufacturers reported the fabrication of car bodies and other associated parts from biocomposites in the 1950s and in between 1990 and 1996, which gradually increased and gained popularity over the years (Huda et al. 2008). The Toyota motor company proposed an eco-friendly bio-based car concept through designing polyester reinforced with hemp for lighter weight seats, body panels, carpets, and different interior parts.

Biocomposites also have very good potentiality for sustainable manufacturing through using green materials instead of traditional petroleum-based composites. Electronics and mobile handsets have also drawn attention for implementing the green concept through reducing the harmful ingredients by replacing them with natural fiber-based products (Kalia et al. 2011; Dicker et al. 2014). Natural fiber hybrid nanocomposites are also used in various industries, viz., biomedical, filtration, drug delivery, tissue delivery, tissue template, wound dressing and cosmetics. Nanocellulose is gaining popularity in the biomedical industry for its use in scaffolds in tissue engineering, bone reconstruction, systems for drug release, the replacement of skin due to burning and wound dressings (Kalia et al. 2011). Some of the potential applications are listed in Table 13.10.

TABLE 13.10
Applications of Natural Fibers in Different Sectors

Reinforcing Fiber	Polymeric matrix	Method of Manufacturing	Application	References
Cotton	PLA, silane	Extrusion and injection molding	Food packaging automotive, building	Mahdi and Dean (2020) and Battegazzore et al. (2019)
Jute	Polyester	Compression molding and hand lay-up	Door panels, chairs, helmets, chest guards, ropes	Ashraf et al. (2019) and Dinesh et al. (2019)
Flax	PLA and polyester	Hand lay-up and vacuum infusion	Automotive, textile and structural	Aliotta et al. (2019) and Zhang et al. (2020)
Coir	Polyester	Injection molding and extrusion	Roofing sheets, insulation boards	Munde et al. (2018) and Mishra et al. (2020)
Sisal	Polyester	Compression molding	Automobile parts and roofing sheets	Chaitanya et al. (2019) and Saxena et al. (2011)
Kenaf	PLA	Compression molding and pultrusion	Automotive parts and bearings	Asumani and Paskaramoorthy (2020)
Sugarcane bagasse	PLA	Injection molding and extrusion	Automotive parts	Wirawan and Sapuan (2018)
Bamboo	PLA	Compression molding	Electronics parts and furniture	Lokesh et al. (2020)
Ramie	PLA	Injection molding	Bulletproof vests	Djafar et al. (2020)
Flax, cotton	Cellulose	Thermopressing and casting	Interface melting	Ilyas and Sapuan (2020)
Cellulose, kenaf	Rubber	Vulcanization and mixing	Rubbery materials	Sapuan et al. (2018)
Flax, cotton	PLA	Twin-screw extruder	Film blowing and packaging	Ilyas and Sapuan (2020)
Flax, cotton	Starch	Blending	Materials for food packaging	Ilyas and Sapuan (2020)
Kenaf, cellulose	PVA	Casting	Diagnosis of wounds and biosensor scaffolds	Sapuan et al. (2018)
Flax	PLA, polyester	Hand laying and vacuum infusion	Floor trays, insulation mats, paper, textile industries	Dhaliwal (2020) and Dunne et al. (2016)
Coir	Polyester	Injection molding and extrusion	Insulation boards, brush making, sacks and horticulture products	Mishra et al. (2020) and Dunne et al. (2016)
Cotton	PLA	Injection molding and extrusion	Insulation building, packaging of food, furniture	Dhaliwal (2020)

(*Continued*)

TABLE 13.10 (Continued)
Applications of Natural Fibers in Different Sectors

Reinforcing Fiber	Polymeric matrix	Method of Manufacturing	Application	References
Hemp	Polyester	Compression molding	Seat back panels, load floors and furniture	Witayakran et al. (2017)
Sisal	PLA	Hand laying and compression molding	Interior door linings and automobile body parts	Huda et al. (2008)
Kenaf	PLA	Compression molding	Bearings, mats, package trays and	Dunne et al. (2016)
Wood	PLA, PHA, starch	Compression molding	Spare tires, seatback cushions	Huda et al. (2008)
Ramie	PLA	Injection molding	Bulletproof vests and textile fabric	Chen et al. (2019)
Jute	PLA, polyester	Injection molding and compression	Sacking, roofing and carpet backing	Das (2016)
Kapok	PLA	Extrusion	Matresses and pillow	Dhaliwal (2020)
Sugarcane bagasse	Polyester	Injection molding and extrusion	Automobile parts and plastic composites	Marichelvam et al. (2020)
Bamboo	Rubber	Compression molding	Pillar materials, furniture and building	Siti et al. (2013)

REFERENCES

Ahmad, R., Hamid, R. and Osman, S. A. "Physical and chemical modifications of plant fibres for reinforcement in cementitious composites". *Advances in Civil Engineering* 2019 (2019): 1–18.

Ahmed, K. S., Vijayarangan, S. and Kumar, A. "Low velocity impact damage characterization of Woven Jute-glass fabric reinforced isothalic polyester hybrid composites". *Journal of Reinforced Plastics and Composites* 26, no. 10 (2007): 959–76.

Akbariazam, M., Ahmadi, M., Javadian, N. and Mohammadi, N. A. "Fabrication and characterization of soluble soybean polysaccharide and nanorod-rich ZnO bionanocomposite". *International Journal of Biology Macromolecules* 89 (2016): 369–75.

Akil, H. M., Omar, M. F., Mazuki, A. A. M., Safiee, A., Ishak, Z. A. M. and Bakar, A. A "Kenaf fiber reinforced composites: A review". *Materials and Designs* 32 (2011): 4107–21.

Aliotta, L., Gigante, V., Coltelli, M. B., Cinelli, P., Lazzeri, A. and Seggiani, M. "Thermomechanical properties of PLA/short flax fiber biocomposites". *Applied Science* 9 (2019): 3797.

Al-Qqla, F. M. and Salit, M. S. "Natural fiber composites." In F.M. Al-Qqla and M.S. Salit (Eds), Materials Selection for Natural Fiber Composites. Woodhead Publisher: Sawston (2017), pp. 23–48.

Ashraf, M. A., Zwawi, M., Mehran, M. T., Kanthasamy, R. and Bahadar, A. "Jute based bio and hybrid composites and their applications". *Fibers* 7 (2019): 77.

Asumani, O. and Paskaramoorthy, R. "Fatigue and impact strengths of kenaf fibre reinforced polypropylene composites: Effects of fibre treatments". *Advances in Composite Materials* 30 (2020): 1–13.

Avella, M., Martuscelli, E. and Raimo, M. "Review melt crystallisation of polymer materials: The role of the thermal conductivity and its influence on the microstructure". *Journal of Material Science* 35 (2000): 523.

Avella, M., Pace, E. D., Immirzi, B. Impallomni, G. "Addition of glycerol plasticizer to seaweeds derived alginates: Influence of microstructure on chemical-physical properties". *Carbohydrate Polymers* 69, no. 3 (2007): 503–11.

Azetsu, A., Koga, H., Isogai, A. and Kitaoka T. "Synthesis and catalytic features of hybrid metal nanoparticles supported on cellulose nanofibers" *Catalysts* 1, no. 1 (2011): 83–96.

Barud, H. S., Regiani, T., Marques, R. F., Lustri, W. R., Messaddeq, Y. and Ribeiro, S. J. "Antimicrobial bacterial cellulose-silver nanoparticles composite membranes". *Journal of Nanomaterials* 10 (2011): 1–8.

Battegazzore, D. Abt, T., Maspoch, M. L. and Frache, A. "Multilayer cotton fabric bio-composites based on PLA and PHB copolymer for industrial load carrying applications." *Composites Part B Engineering* 163 (2019): 761–68.

Bharath, K. N. and Basavarajappa, S. "Applications of biocomposite materials based on natural fibers from renewable resources: A review". *Science and Engineering of Composite Materials* 23, no. 2 (2016): 123–33.

Bismarck, A., Mishra, S. and Lampke, T. "Plant fibers as reinforcement for green composites". *Natural Fibers, Biopolymers and Biocomposites* (2005). doi: 10.1201/9780203508206.CH2.

Biswas, R., Uellelendahl, H. and Ahring, B. K. "Wet explosion: A universal and efficient pretreatment process for lignocellulosic biorefineries". *Bioenergy Research* 8, no. 3 (2015): 1101–16.

Bledzki, A. K., Reihmane, S. and Gassan, J. "Properties and modification methods for vegetable fibers for natural fiber composites". *Journal of Applied Polymer Science* 59 (1996): 1329–36.

Bledzki, A. K., Mamun, A. A., Gabor, M. L., Gutowski, V. S. "The effects of acetylation on properties of flax fibre and its polypropylene composites". *Polymer Letters* 2, no. 6 (2008): 413–22.

Bledzki, A. K., Mamun, A. A, Volk, J. "Physical, chemical and surface properties of wheat husk, rye husk and soft wood and their polypropylene composites". *Composites: Part A* 41 (2010): 480–88.

Bodros, E., Pillin, I., Montrelay, N. and Baley, C. "Could biopolymers reinforced by randomly scattered flax fibre be used in structural applications". *Composites Science and Technology* 67 (2007): 462–70.

Bodur, M. S., Bakkal, M. and Sonmez, H. E. "The effects of different chemical treatment methods on the mechanical and thermal properties of textile fiber reinforced polymer composites". *Journal of Composite Materials* 50 (2016): 3817–30.

Buschle-Diller, G., Fanter, C. and Loth, F. "Structural changes in hemp fibers as a result of enzymatics hydrolysis with mixed enzyme systems". *Textile Research Journal* 69 (1999): 244–51.

Busu, W. N. W., Anuar, H., Ahmad, S. H., Rasid, R. and Jamal, N. A. "The mechanical and physical properties of thermoplastic natural rubber hybrid composites reinforced with hibiscus cannabinus, L and short glass fiber". *Polymer-Plastics Technology and Engineering* 49 (2010): 1315–22.

Campos, A., Marconcini, J. M., Franchetti, S. M. M., Mattoso, L. H. C. "The influence of UV-C irradiation on the properties of thermoplastic starch and polycaprolactone biocomposite with sisal bleached fibers". *Polymer Degradation and Stability* 97 (2012): 1948–55.

Campos, A. D., Tonoli, G. H. D., Marconcini, J. M., Mattoso, L. H. C., Klamczynski, A., Gregorski, K. S., Wood, D., Williams, T., Chiou, B. S., Imam, S. H. "TPS/PCL composite reinforced with treated sisal fibers: Property, biodegradation and water-absorption". *Journal of Polymers and the Environment* 21 (2013): 1–7.

Cao, X., Zhang, L., Huang, J., Yang, G. and Wang, Y. "Structure–properties relationship of starch/waterborne polyurethane composites". *Journal of Applied Polymer Science* 90, no. 12 (2003): 3325–32.

Chaitanya, S., Singh, I. and Song, J. I. "Recyclability analysis of PLA/sisal fiber biocomposites". *Composites Part B Engineering* 173 (2019): 106895.

Chen, M., Ye, L., Li, H., Wang, G., Chen, Q., Fang, C., Dai, C. and Fei, B. "Flexural strength and ductility of moso bamboo". *Construction and Building Materials* 246 (2020): 118418.

Chen, Q., Liu, Y., Chen, G. "A comparative study on the starch-based biocomposite films reinforced by nanocellulose prepared from different non-wood fibers". *Cellulose* 26 (2019): 2425–35.

Choi, H. Y. and Lee, J. S. "Effects of surface treatment of ramie fibers in a ramie/poly(lactic acid) composite". *Fibers and Polymers* 13, no. 2 (2012): 217–23.

Ciccarelli, N., Milanini, B., Baldonero, E. and Fabbiani, M. "Cognitive impairment and cardiovascular disease related to alexithymia in a well-controlled HIV-infected population". *International Journal of Medical Informatics* 3 (2019): 266–82.

Cirtiu, C. M., Dunlop-Brière, A. F. and Moores, A. "Cellulose nanocrystallites as an efficient support for nanoparticles of palladium: Application for catalytic hydrogenation and heck coupling under mild conditions". *Green Chemistry* 13, no. 2 (2011): 288–91.

Compos, A., Teodoro, K. B. R., Teixeira, E. M., Correa, A. C., Marconcini, J. M., Wood, D. F., Mattoso, L. H. C. "Properties of thermoplastics starch and TPS/polycaprolactone blend reinforced with sisal whiskers using extrusion processing". *Polymer Engineering and Science* 53, no. 4 (2012): 800–8.

Cordoba, A. L., Areco, S. E. and Goyanes, S. "Potato starch-based biocomposites with enhanced thermal, mechanical and barrier properties comprising water-resistant electrospun poly(vinyl alcohol) fibers and yerba mate extract". *Carbohydrate Polymers* 215 (2019): 377–87.

Cranston, E. D., Eita, M., Johansson, E., Netrval, J., Salajková, M., Arwin, H. and Wagberg, L. "Determination of Young's modulus for nanofibrillated cellulose multilayer thin films using buckling mechanics". *Biomacromolecules* 12, no. 4 (2011): 961–69.

Cyras, V. P., Vallo, C., Kenny, J. M. and Zqezi, V. A. "Effect of chemical treatment on the mechanical properties of starch-based blends reinforced with sisal fibre". *Journal of Composite Materials* 38, no. 16 (2003): 1387.

Das, M. "Bamboo fiber-based polymer composites." In Z. Lin, A. Zhang and Y. Yang (Eds), *Polymer-Engineered Nanostructures for Advanced Energy Applications.* Springer Science and Business Media LLC: Cham, Switzerland (2016), pp. 627–45.

Dicker, M. P. M., Duckworth, P. F., Baker, A. B., Francois, G., Hazzard, M. K. and Weaver, P. M. "Green composites: A review of material attributes and complementary applications". *Composites Part A: Applied Science and Manufacturing* 56 (2014): 280–89.

Diez, I., Eronen, P., Osterberg M., Linder, M. B., Ikkala, O. and Ras R. H. "Functionalization of nanofibrillated cellulose with silver nanoclusters: Fluorescence and antibacterial activity". *Macromolecular Bioscience* 11, no. 9 (2011): 1185–91.

Dinesh, S., Kumaran, P., Mohanamurugan, S., Vijay, R., Singaravelu, D. L., Vinod, A., Sanjay, M., Siengchin, S., Bhat, K. S. "Influence of wood dust fillers on the mechanical, thermal, water absorption and biodegradation characteristics of jute fiber epoxy composites". *Journal of Polymer Research* 27 (2019): 9.

Djafar, Z., Renreng, I. and Jannah, M. "Tensile and bending strength analysis of ramie fiber and woven ramie reinforced epoxy composite". *Journal of Natural Fibers* (2020): 1–12. doi: 10.1080/15440478.2020.1726242.

Dong, H., Snyder, J. F., Tran, D. T. and Leadore, J. L. "Hydrogel, aerogel and film of cellulose nanofibrils functionalized with silver nanoparticles". *Carbohydrate Polymer* 95, no. 2 (2013): 760–67.

Dorigato, A., Sebastiani, M., Pegoretti, A. and Fambri, L. "Effect of silica nanoparticles on the mechanical performances of poly(lactic acid)". *Journal of Polymers and the Environment* 20, no. 3 (2012): 713–25.

Duan, J., Wu, H., Fu, W. and Hao, M. "Mechanical properties of hybrid sisal/coir fibers reinforced polylactide biocomposites". *Polymer Composites* 39 (2018): E188–199.

El Achaby, M., El Miri, N., Aboulkas, A., Zahouily, M., Bilal, E., Barakat, A. and Solhy, A. "Processing and properties of eco-friendly bio-nanocomposite films filled with cellulose nanocrystals from sugarcane bagasse". *International Journal of Biology Macromolecules* 96 (2017): 340–52.

El-Naggar, A. M., El-Hosamy, M. B., Zahran, A. H. and Zohdy, M. H. "Surface morphology/mechanical/dyeability properties of radiation-grafted sisal fibers". *American Dyestuff Reporter* 81 (1992): 40.

El-Tayeb, T. S., Abdelhafez, A. A., Ali, S. H. and Ramadan, E. M. "Effect of acid hydrolysis and fungal biotreatment on agro-industrial wastes for obtainment of free sugars for bioethanol production". *Brazilain Journal of Microbiology* 43, no. 4 (2012): 1523–35.

Encalada, K., Aldás, M. B., Proaño, E. and Valle, V. "An overview of starch-based biopolymers and their biodegradability". *Ciencia e Ingeniería* 39 (2018): 245–258.

Foruzanmehr, M., Vuillaume, P. Y., Robert, M., Elkoun, S. "The effect of grafting a nano-TiO$_2$ thin film on physical and mechanical properties of cellulosic natural fibers". *Materials and Designs* 85 (2015): 671–78.

Franco, C. R. D., Cyras, V. P., Busalmen, J. P., Ruseckaite, R. A., Zquez, A. V. "Degradation of polycaprolactone/starch blends and composites with sisal fibre". *Polymer Degradation and Stab* 86 (2004): 95–103.

Gan, S., Zakaria, S. and Syed Jaafar, S. N. "Enhanced mechanical properties of hydrothermal carbamated cellulose nanocomposite film reinforced with graphene oxide". *Carbohydrate Polymer* 172 (2017): 284–93.

Gassan, J. and Gutowski, V. S. (2000) "Effects of corona discharge and UV treatment on the properties of jute-fibre epoxy composites". *Composites Science and Technology* 60, no. 15 (2000): 2857–63.

Gassan, J., Gutowski, V. S. and Bledzki, A. K. "About the surface characteristics of natural fibres". *Macromolecular Materials and Engineering* 283 (2000): 132–9.

George, M., Mussone, P. G. and Bressler, D. C. "Surface and thermal characterization of natural fibres treated with enzymes". *Industrial Crops and Products* 53 (2014): 365–73.

Gholampour, A. and Ozbakkaloglu, T. "A review of natural fiber composites: Properties, modification and processing techniques, characterization, applications". *Journal of Material Science* 55 (2019):829–92.

Girisha, K., Shrikiran, A., Sakamoto, O., Gopinath, P., Satyamoorthy, K. and Bidchol, A. "Novel mutation in an Indian patient with Methylmalonic Acidemia. *Indian Journal of Human Genetics* 18, no. 3 (2015): 346.

Godbole, S., Gote, S., Latkar, M. and Chakrabarti, T. "Preparation and characterization of biodegradable poly-3-hydroxybutyrate-starch blend films". *Bioresource Technology* 86, no. 1 (2003): 33–37.

Gonzalez, A., Gastelu, G., Barrera, G. N., Ribotta, P. D. and Alvarez Igarzabal, C. I. "Preparation and characterization of soy protein films reinforced with cellulose nanofibers obtained from soybean by-products". *Food Hydrocolloids* 89 (2019): 758–64.

Goutianos, S., Peijs, T., Nystrom, B. and Skrifvars, M. "Development of flax fibre based textile reinforcements for composite applications". *Applied Composite Materials* 13 (2006): 199–215.

Graupner, N. and Mussig, J. "A comparison of the mechanical characteristics of kenaf and lyocell fibre reinforced poly(lactic acid) (PLA) and poly(3-hydroxybutyrate) (PHB) composites". *Composites: Part A* 42 (2011): 2010–19.

Guan, M., Zhang, Z., Yong, C. and Du, K. "Interface compatibility and mechanisms of improved mechanical performance of starch/poly(lactic acid) blend reinforced by bamboo shoot shell fibers". *Journal of Applied Polymer Science* 136, no. 35 (2019): 47899.

Hafizulhaq, F., Abral, H., Kasim, A., Arief, S. and Affi, J. "Moisture absorption and opacity of starch-based biocomposites reinforced with cellulose fiber from bengkoang". *Fibers* 6, no. 3 (2018): 62.

Hajiha, H., Sain, M. and Meic, L. H. "Modification and characterization of hemp and sisal fibers". *Journal of Natural Fibers* 11 (2014): 144–68.

Halder, D., Mandal, M., Chatterjee, S. S., Pal, N. K. and Mandal, S. "Indigenous probiotic lactobacillus isolates presenting antibiotic like activity against human pathogenic bacteria". *Biomedicines* 5, no. 4 (2017): 31.

Haseena, A. P., Dasan, K. P., Unnikrishnan, G. and Thomas, S. "Mechanical properties of sisal/coir hybrid fibre reinforced natural rubber". *Progress in Rubber, Plastics Recycling Technology* 21, no. 3 (2005): 155–181.

He, Y., Wang, X., Wu, D., Gong, Q., Qiu, H., Liu, Y., Wu, T., Ma, J. and Gao, J. "Biodegradable amylose films reinforced by graphene oxide and polyvinyl alcohol". *Materials Chemistry and Physics* 142 (2013): 1–11.

Hoeger, I., Rojas O. J., Efimenko, K., Velev, O. D. and Kelley, S. S. "Ultrathin film coatings of aligned cellulose nanocrystals from a convective-shear assembly system and their surface mechanical properties". *Soft Matter* 7, no. 5 (2011): 1957–67.

Hoque, M. A., Ahmed, M. R., Rahman, G. T., Rahman, M. T., Islam, M. A., Khan, M. A. and Hossain, M. K. "Fabrication and comparative study of magnetic Fe and α-Fe$_2$O$_3$ nanoparticles dispersed hybrid polymer (PVA+chitosan) novel nanocomposite film". *Results Physics* 10 (2018): 434–43.

Hu, W., Chen, S., Li, X., Shi, S., Shen, W., Zhang, X. and Wang, H. "In situ synthesis of silverchloride nanoparticles into bacterial cellulose membranes". *Materials Science and Engineering: C* 29, no. 4 (2009): 1216–19.

Huang, X. and Netravali, A. "Characterisation of flax fiber reinforced soyprotein resin based green composites modified with nano-clay particles". *Composites Science and Technology* 67 (2007): 2005–14.

Huda, M. S., Drzal, L. T., Ray, D., Mohanty, A. K. and Mishra, M. "Natural-fiber composites in the automotive sector". In K. Pickering (Ed.), *Properties and Performance of Natural-Fibre Composites.* Woodhead Publishing: Oxford, UK (2008), pp. 221–268.

Ibrahim, N. A., Yunus, W. M. Z. W., Othman, M. and Abdan, K. "Effect of chemical surface treatment on the mechanical properties of reinforced plasticized poly(lactic acid) biodegradable composites". *Journal of Reinforced Plastics and Composites* 30, no. 5 (2011): 381–88.

Ilyas, R. A. and Sapuan, S. M. "Biopolymers and biocomposites: Chemistry and technology". *Current Analytical Chemistry* 16, no. 5 (2020): 500–3.

Imre, B. and Pukánszky, B. "Compatibilization in bio-based and biodegradable polymer blends". *European Polymer Journal* 49 (2013): 1215–33.

Indran, S., Raj, R. E., Daniel, B. and Saravanakumar, S. S. "Cellulose powder treatment on Cissus quadrangularis stem fiber-reinforcement in unsaturated polyester matrix composites". *Journal of Reinforced Plastics and Composites* 35, no. 3 (2015): 212–227.

Islam, M. S., Pickering, K. L. and Foreman, N. J. "Influence of alkali treatment on the interfacial and physico-mechanical properties of industrial hemp fibre reinforced polylactic acid composites". *Composites: Part A* 41 (2010): 596–603.

Jaffer, H. I. and Jawad, M. K. (2011). "Tensile strength investigation of UPE and EP composites filled with rice husk fibers, AL-Mustansiriyah". *Journal of Science* 22, no. 5 (2011): 270–76.

Jamiluddin, J., Siregar, J., Tezara, C., Hamdan, M. and Sapuan, S. "Characterisation of cassava biopolymers and the determination of their optimum processing temperatures". *Plastics, Rubber and Composites*, 47, no. 10 (2018): 447–57.

Jaratrotkamjorn, R., Khaokong, C., Tanrattanakul, V. "Toughness enhancement of poly(lactic acid) by melt blending with natural rubber". *Journal of Applied Polymer Science* 124 (2011): 5027–36.

Joseph, K., Thomas, S. and Pavithran, C. "Effect of chemical treatment on the tensile properties of short sisal fibre-reinforced polyethylene composites". *Polymer* 33, no. 23 (1996): 5139–49.

Joseph, S., Sreekalab, M. S., Oommena, Z., Koshy, P. and Thomas, S. "A comparison of the mechanical properties of phenol formaldehyde composites reinforced with banana fibres and glass fibres". *Composites Science and Technology* 62 (2002): 1857–68.

Kabir, M. M., Wang, H., Lau, K. T. and Cardona, F. "Chemical treatments on plant-based natural fibre reinforced polymer composites: An overview". *Composites: Part B* 43 (2012): 2883–92.

Kalia, S., Kaith, B. S., Kaur, I. "Pretreatments of natural fibers and their application as reinforcing material in polymer composites-a review". *Polymer Engineering and Science* 49, no. 7 (2009): 1253–72.

Kalia, S., Dufresne, A., Cherian, B. M., Kaith, B. S., Averous, L., Njuguna, J. and Nassiopoulos, E. "Cellulose-based bio- and nanocomposites: A review". *International Journal of Polymer Science* 2011 (2011): 749.

Karmakar, A.C., and J. A. Youngquist. "Injection Molding of Polypropylene Reinforced with Short Jute Fibers". *Journal of Applied Polymer Science* 62 (1996): 1147–1151.

Kettunen, M., Silvennoinen, R. J., Houbenov, N., Nykanen, A., Ruokolainen, J., Sainio, J., Pore, V. and Kemell, M. M. "An kerfors and T. Lindström, Photoswitchable superabsorbency based on nanocellulose aerogels". *Advanced Functional Materials* 21, no. 3 (2011): 510–17.

Khoshnava, S. M., Rostami, R., Ismail, M., Rahmat, A. R. and Ogunbode, B. E. Woven hybrid Biocomposite: Mechanical properties of woven kenaf bast fibre/oil palm empty fruit bunches hybrid reinforced polyhydroxybutyrate biocomposite as non-structural building materials. *Construction and Building Materials* 154 (2017): 155.

Koga, H., Tokunaga, E., Hidaka, M., Umemura, Y., Saito, T., Isogai, A. and Kitaoka, T. "Topochemical synthesis and catalysis of metal nanoparticles exposed on crystalline cellulose nanofibers". *Chemical Communications* 46, no. 45 (2010): 8567–69.

Kommula, V. P., Reddy, K. O., Shukla, M., Marwala, T., Reddy, E. V. S and Rajulu, A. V. "Extraction, modification, and characterization of natural ligno-cellulosic fiber strands from napier grass". *International Journal of Polymer Analysis and Characterization* 21, no. 1 (2016): 18–28.

Komuraiah, A., Kumar, N. S. and Prasad, B. D. "Chemical composition of natural fibers and its influence on their mechanical properties". *Mechanics of Composite Materials* 50 (2014): 359–76.

Konczewicz, W. and Kozlowski, R. M. "Enzymatic treatment of natural fibres". R. Kozlowski (Ed.), *Handbook of Natural Fibres*. Woodhead Publishing: Oxford, UK (2012), pp. 168–84.

Ku, H., Wang, H., Pattarachaiyakoop, N. and Trada, M. "A review on the tensile properties of natural fiber reinforced polymer composites". *Composites: Part B* 42 (2011): 856–73.

Lai, S. M., Don, T. M. and Huang, Y. C. "Preparation and properties of biodegradable thermoplastic starch/poly(hydroxy butyrate) blends". *Journal of Applied Polymer Science* 100 (2005): 2371–79.

Landreau, E., Tighzert, L., Bliard, C., Berzin, F. and Lacoste, C. "Morphologies and properties of plasticized starch/polyamide compatibilized blends". *European Polymer Journal* 45 (2009): 2609–18.

Liang, S., Xu, Z., Xu X. X., Zhao, X., Huang, C. and Wei, Y. Quantitative proteomics for cancer biomarker discovery. *Combinatorial Chemistry & High Throughput Screening* 15 (2012): 221–31.

Lima, P., Silva, S. P. M. D., Oliveira, J. and Costa, V. "Rheological properties of ground tyre rubber based thermoplastic elastomeric blends". *Polymer Testing* 45 (2015): 58–67.

Liu, W., Drzal, L. T., Mohanty, A. K. and Misra, M. "Influence of processing methods and fiber length on physical properties of kenaf fiber reinforced soy based biocomposites". *Composites: Part B* 38 (2007): 352–59.

Liu, Z., Fu, M., Ling, F., Sui, G., Bai, H., Zhang, Q. "Stereocomplex-type polylactide with bimodal melting temperature distribution: Toward desirable melt-processability and thermomechanical performance". *Polymer* 169 (2019): 21–28.

Lokesh, P., Kumari, T. S., Gopi, R. and Loganathan, G. B. "A study on mechanical properties of bamboo fiber reinforced polymer composite". *Materials Today: Proceeding*. 22 (2020): 897–903.

Lu, Y., Tighzert, L., Berzina, F. and Rondot, S. "Innovative plasticized starch films modified with waterborne polyurethane from renewable resources". *Carbohydrate Polymer* 61 (2005): 174–82.

Lu, X., Wei, X., Huang, J., Yang, L., Zhang, G., He, G., Wang, M. and Qu, J. "Supertoughened poly(lactic acid)/polyurethane blend material by in situ reactive interfacial compatibilization via dynamic vulcanization". *Industrial Engineering and Chemistry Research* 53, no. 44 (2014): 17386–93.

Ma, X., Cheng, Y., Qin, X., Guo, T., Deng, J. and Liu, X. "Hydrophilic modification of cellulose nanocrystals improves the physicochemical properties of cassava starch-based nanocomposite films". *LWT* 86 (2017): 318–26.

Mahdi, E. and Dean, A. "The effect of filler content on the tensile behavior of polypropylene/cotton fiber and poly(vinyl chloride)/cotton fiber composites". *Materials* 13 (2020): 753.

Malkapuram, R., Kumar, V. and Negi, Y. S. "Recent development in natural fiber reinforced polypropylene composites". *Journal of Reinforced Plastics and Composites* 28 (2008): 1169–1189.

Mallakpour, S. and Nezamzadeh, E. A. "Preparation and characterization of chitosan-poly(vinyl alcohol) nanocomposite films embedded with functionalized multi-walled carbon nanotube". *Carbohydrate Polymer* 166 (2017): 377–86.

Marais, S., Gouanve, F., Bonnesoeura, A., Greneta, J., Epaillard, F. P., Morvan, C., Metayer, M. "Unsaturated polyester composites reinforced with flax fibers: Effect of cold plasma and autoclave treatments on mechanical and permeation properties". *Composites: Part A* 36 (2005): 975–86.

Maria, L., Santos, A. L., Oliveira, P. C., Valle, A. S., Barud, H. S., Messaddeq, Y. and Ribeiro, S. J. "Preparation and antibacterial activity of silver nanoparticles impregnated in bacterial cellulose". *Polymer Science Technology* 20, no. 1 (2010): 72–77.

Marichelvam, M. K., Manimaran, P., Verma, A., Sanjay, M. R., Siengchin, S., Kandakodeeswaran, K. and Geetha, M. "A novel palm sheath and sugarcane bagasse fiber based hybrid composites for automotive applications: An experimental approach". *Polymer Composites* 42 (2020): 1–10.

Marques, P. A., Nogueira, H. I., Pinto, R. J., Neto, C. P. and Trindade, T. "Silver-bacterial cellulosic sponges as active SERS substrates". *Journal of Raman Spectroscopy* 39, no. 4 (2008): 439–43.

Marsyahyo, E., Astuti, S. and Ruwana, I. "Mechanical improvement of ramie woven reinforced-starch based biocomposite using biosizing method". In P. Tesinova (Ed.) *Advances in Composite Materials - Analysis of Natural and Man-Made Materials*. InTech: Croatia (2011), pp. 297–306.

Martin, O. and Averous, L. "Poly(lactic acid): Plasticization and properties of biodegradable multiphase system". *Polymer* 42 (2001): 6209–19.

Marvizadeh, M. M., Oladzadabbasabadi, N., Mohammadi N. A. and Jokar, M. "Preparation and characterization of bionanocomposite film based on tapioca starch/bovine gelatin/nanorod zinc oxide". *International Journal of Biology Macromolecules* 99 (2017): 1–7.

Masirek, R., Kulinski, Z., Chionna, D., Piorkowska, E. and Pracella, M. "Composites of poly(L-lactide) with hemp fibers: Morphology and thermal and mechanical properties". *Journal of Applied Polymer Science* 105 (2007): 255–68.

Maslowski, A. K., LaCaille, L. J., Reich, C. M. and Klinger, J. "Effectiveness of mental health first aid: A meta-analysis". *Mental Health Review Journal* 24, no. 4 (2019): 245–61.

Masmoudi, F., Atef, B. A., Dammak, M., Jaziri, M. and Ammar, E. "Biodegradable packaging materials conception based on starch and polylactic acid (PLA) reinforced with cellulose". *Environmental Science and Pollution Research* 23 (2016): 20904–14.

Mehta, N. M. and Parsania, P. H. "Fabrication and evaluation of some mechanical and electrical properties of jute-biomass based hybrid composites". *Journal of Applied Polymer Science* 100 (2005): 1754–58.

Mehta, G., Mohanty, A. K., Thayer, K., Misra, M. and Drzal, L. T. (2005) "Novel biocomposites sheet molding compounds for low cost housing panel applications". *Journal of Polymers and the Environment* 13 (2): 169–75.

Mishra, S. and Naik, J. B. "Effect of treatment of maleic anhydride on mechanical properties of natural fiber: polystyrene composites". *Polymer-Plastics Technology and Engineering* 44, no. 4 (2005): 663–75.

Mishra, S., Misra, M., Tripathy, S. S., Nayak, S. K. and Mohanty, A. K. "Graft copolymerization of acrylonitrile on chemically modified sisal fibers". *Macromolecular Materials and Engineering* 286 (2001): 107–113.

Mishra, S., Nayak, C., Sharma, M. K. and Dwivedi, U. K. "Influence of coir fiber geometry on mechanical properties of SiC filled epoxy composites". *Silicon* 13 (2020): 1–7.

Mohammed, L., Ansari, M. N. M., Pua, G., Jawaid, M. and Islam, M. S. "A review on natural fiber reinforced polymer composite and its applications". *International Journal of Polymer Science* 2015 (2015): 1–15.

Mohanty, S., Nayak, S. K., Verma, S. K. and Tripathy, S. S. "Effect of MAPP as a coupling agent on the performance of jute-PP composites". *Journal of Reinforced Plastics and Composites* 23, no. 6 (2004): 625–737.

Mohanty, A. K., Misra, M., Drzal, L. T., Selke, S. E., Harte, B. R. and Hinrichsen, G. "Natural fibers, biopolymers, and biocomposites: An introduction". In: A. K. Mohanty, M. Misra and L. T. Drzal (Eds), *Natural Fibers, Biopolymers, and Biocomposites*. CRC Press: Boca Raton, FL (2005), pp. 1–36.

Mohanty, A., & Srivastava, V. K. (2015). "Effect of alumina nanoparticles on the enhancement of impact and flexural properties of the short glass/carbon fiber reinforced epoxy based composites". *Fibers and Polymers* 16, no. 1 (2015): 188–195.

Morye, S. S. and Wool, R. P. "Mechanical properties of glass/flax hybrid composites based on a novel modified soybean oil matrix material". *Polymer Composites* 26, no. 4 (2005): 407–416.

Munde, Y. S., Ingle, R. B. and Siva, I. "Investigation to appraise the vibration and damping characteristics of coir fibre reinforced polypropylene composites". *Advances in Materials Processing Technology* 4 (2018): 1–12.

Mwaikambo, L. Y. and Bisanda, E. T. N. Kapok/cotton fabric–polypropylene composites. *Polymer Testing* 18 (1999): 181.

Mwaikambo, L. Y. and Ansell, M. P. "Chemical modification of hemp, sisal, jute, and kapok fibers by alkalization". *Journal of Applied Polymer Science* 84 (2002): 2222–34.

Na, Y. H., He, Y., Shuai, X., Kikkawa, Y., Doi, Y. and Inoue, Y. "Compatibilization effect of poly(ecaprolactone)-b-poly(ethyleneglycol) block copolymers and phase morphology analysis in immiscible poly(lactide)/poly(E-caprolactone) blends". *Biomacromolecules* 3 (2002): 1179–86.

Naira, K. C. M., Thomas, S. and Groeninc, G. "Thermal and dynamic mechanical analysis of polystyrene composites reinforced with short sisal fibres". *Composites Science and Technology* 61 (2001): 2519–29.

Ngo, T. D., Nguyen, Q. T. and Tran, P. "Heat release and flame propagation in prefabricated modular unit with GFRP composite facades". *Building Simulation* 9 (2016): 607–616.

Nigam, S. N., Gupta, N. and Anthwal, A. "Pre-treatment of agro-industrial residues". In P. S. N. Nigam and A. Pandey (Eds.), Biotechnology Agro-Industrial Reseource Utilisation. Springer Science and Business Media: (2009), pp. 13–33.

Oksman, K. "Mechanical properties of natural fibre mat reinforced thermoplastic". *Applied Composite Materails* 7 (2000): 403–14.

Oksman, K., Skrifvars, M. and Selin, J. F. "Natural fibres as reinforcement in polylactic acid (PLA) composites". *Composites Science and Technology* 63, no. 9 (2003): 1317–24.

Oleyaei, S. A., Zahedi, Y., Ghanbarzadeh, B. and Moayedi, A. A. "Modification of physicochemical and thermal properties of starch films by incorporation of TiO_2 nanoparticles". *International Journal of Biology Macromolecules* 89 (2016): 256–64.

Padalkar S., Capadona, J., Rowan, S. J., Weder, C., Won, Y. H., Stanciu, L. A. and Moon, R. J. "Natural biopolymers: Novel templates for the synthesis of nanostructures." *Langmuir* 26, no. 11 (2010): 8497–502.

Pai, A. R. and Jagtap, R. N. "Surface morphology & mechanical properties of some unique natural fiber reinforced polymer composites: A review". *Journal of Materials and Environmentl Science* 6, no. 4 (2015): 902–17.

Park, M., Chang, H., Jeong, D. H. and Hyun, J. "Spatial deformation of nanocellulose hydrogel enhances SERS". *BioChip Journal* 7, no. 3 (2013): 234–41.

Paul, A., Joseph, K. and Thorna, S. "Effect of surface treatments on the electrical properties of low-density polyethylene composites reinforced with short sisal fibers". *Composites Science and Technology* 57 (1997): 67–79.

Paul, S. A., Boudenne, A., Ibos, L., Candau, Y., Joseph, K. and Thomas, S. "Effect of fiber loading and chemical treatments on thermophysical properties of banana fiber/polypropylene commingled composite materials". *Composites: Part A* 39 (2008): 1582–88.

Pecas, P., Carvalho, H., Salman, H. and Leite, M. "Natural fibre composites and their applications: A review". *Journal of Composites Science* 2, no. 4 (2018): 66.

Petroudy, S. R. D. "Physical and mechanical properties of natural fibers". *Advanced High Strength Natural Fibre Composites in Construction* 3 (2017): 59–83.

Pickering, K. L., Efendy, M. G. A. and Le, T. M. "A review of recent developments in natural fibre composites and their mechanical performance". *Composites Part A: Applied Science and Manufacturing* 83 (2016): 98–112.

Pinto, R. J., Marques, P. A., Martins, M. A., Neto, C. P. and Trindade, T. "Electrostatic assembly and growth of gold nanoparticles in cellulosic fibres". *Journal of Colloid Interface Science* 312, no. 2 (2007): 506–12.

Prabu, V. A., Uthayakumar, M., Manikandan, V., Rajini, N. and Jeyaraj, P. "Influence of redmud on the mechanical, damping and chemical resistance properties of banana/polyester hybrid composites". *Materials and Designs* 64 (2014): 270–79.

Prasanna, V. G., Jha, N. K. and Kumar, A. K. "Optimisation & mechanical testing of hybrid BioComposites". *Materials Today: Proceedings* 18 (2019): 3849–55.

Qiang, T., Yu, D. and Gao, H. "Wood flour/polylactide biocomposites toughened with polyhydroxyalkanoates". *Journal of Applied Polymer Science* 124 (2011): 1831–39.

Quasim, U., Osman, A., Al-Muhtaseb, A. H, Farrell, C., Al-Abri, M., Ali, M., Dai-Viet, N. V., Jamil, F. and Rooney, D. W. Renewable cellulosic nanocomposites for food packaging to avoid fossil fuel plastic pollution: A review. *Environmental Chemistry Letters* (2019). doi: 10.1007/s10311-020-01090-x.

Rachchh, N. V., Misra, R. K. and Roychowdhary, D. G. "Effect of red mud filler on mechanical and buckling characteristics of coir fibre reinforced polymer composite". *Iranian Polymer Journal* 24 (2015): 253–65.

Rahman, M. M., Mallik, A. K. and Khan, M. A. "Influences of various surface pretreatments on the mechanical and degradable properties of photografted oil palm fibers". *Journal of Applied Polymer Science* 105, no. 5 (2007): 3077–86.

Rajesh, M. and Pitchaimani, J. "Mechanical and dynamic mechanical behaviour of novel glass–natural fibre intra-ply woven polyester composites". *Sadhana* 42 (2017): 1215–23.

Rajkumar, R., Manikandan, A. and Saravanakumar, S. S. "Physicochemical properties of alkali treated new cellulosic fiber from cotton shell". *International Journal of Polymer Analysis and Characterization* 21, no. 4 (2016): 359–64.

Ramamoorthy, S. K., Skrifvas, M. and Persson, A. "A review of natural fibers used in biocomposites: Plant, animal and regenerated cellulose fibers". *Polymer Reviews* 55, no. 1 (2015): 107–162.

Ramesh, P., Prasad, B. D. and Narayana, K. L. "Morphological and mechanical properties of treated kenaf fiber/MMT clay reinforced PLA hybrid biocomposites". *Advances in Polymer Composites: Mechanics, Characterization and Applications* 2057 (2019): 020035.

Ray, D., Sarkar, B. K., Rana, A. K. and Bose, N. R. "Effect of alkali treated jute fibres on composite properties". *Bulletin of Materials Science* 24, no. 2 (2001): 129–35.

Reddy, G. V., Naidu, S. V. and Rani T. S. "Impact properties of kapok based unsaturated polyester hybrid composites". *Journal of Reinforced Plastics and Composites* 27, no. 16 (2008): 17.

Rihayat, T., Suryani, S., Fauzi, T., Agusnar, H., Wirjosentono, B., Syafruddin, Helmi, Zulkifli, Alam, P. N. and Sami, M. "Mechanical properties evaluation of single and hybrid composites polyester reinforced bamboo, PALF and coir fiber". *Materials Science and Engineering* 334 (2018): 012081.

Rodriguez, E. S., Stefani, P. M. and Vazquez, A. "Effects of fibers alkali treatment on the resin transfer molding processing and mechanical properties of Jute-Vinylester composites". *Journal of Composites Material* 41, no. 14 (2007): 1729–41.

Rong, M. Z., Zhang, M. Q., Liu, Y., Yang, G. C. and Zeng, H. M. "The effect of fiber treatment on the mechanical properties of unidirectional sisal-reinforced epoxy composites". *Composites Science and Technology* 61 (2001): 1437–47.

Rosa, I. M. D., Kenny, J. M., Maniruzzaman, M., Moniruzzaman, M., Monti, M., Puglia, D., Santulli, C. and Sarasini, F. "Effect of chemical treatments on the mechanical and thermal behaviour of okra (Abelmoschus esculentus) fibres". *Composites Science and Technology* 1 (2011): 246–54.

Rouhi, M., Razavi, S. H. and Mousavi, S. M. "Optimization of crosslinked poly(vinyl alcohol) nanocomposite films for mechanical properties". *Materials Science and Engineering: C* 71 (2017): 1052–63.

Rouison, D., Sain, M. and Couturier, M. "Resin transfer molding of natural fiber reinforced composites: Cure simulation". *Composites Science and Technology* 64, no. 5 (2004): 629–44.

Roy, P., Nei, D., Orikasa, T., Xu, Q., Okadome, H., Nakamura, N. and Shiina, T. "A review of life cycle assessment (LCA) on some food products". *Journal of Food Engineering* 90, no. 1 (2009): 1–10.

Rusli, R. and Eichhorn, S. J. "Determination of the stiffness of cellulose nanowhiskers and the fiber-matrix interface in a nanocomposite using Raman spectroscopy". *Applied Physics Letters* 93, no. 3 (2008): 033111.

Sahu, P. and Gupta, M. "A review on the properties of natural fibres and its bio-composites: Effect of alkali treatment. Proceedings of the Institution of Mechanical Engineers Part L". *Journal of Materials: Design and Applications* 234 (2019): 1–20.

Sain, M., Suhara, P., Law, S. and Bouilloux, A. "Interface modification and mechanical properties of natural fiber–polyolefin composite products". *Journal of Reinforced Plastics and Composites* 24, no. 2 (2005): 121–30.

Sapuan, S. M., Nazrin, A., Ilyas, R. A., Sherwani, S. F. K. and Syafiq, R. "Nanocellulose reinforced Thermoplastic Starch (TPS), Poly(lactic) Acid (PLA), and Poly(Butylene Succinate) (PBS) for food packaging applications". *Frontier Chemistry* 8 (2018): 2296–646.

Saragih, L. R., Dachyar, M., Zagloel, Y. M. and Satar, M. "The industrial IoT for Nusantara". *IEEE International Conference on Internet of Things and Intelligence System (IOTAIS)*, Bali, Indonesia (2018).

Saravanakumar, S. S., Kumaravel, A., Nagarajan, T. and Moorthy, I. G. "Effect of chemical treatments on physicochemical properties of Prosopis juliflora fibers". *International Journal of Polymer Analysis and Characterization* 19 (2014): 383–90.

Sarwar, M. S., Niazi, M. B. K., Jahan, Z., Ahmad, T. and Hussain, A. "Preparation and characterization of PVA/nanocellulose/Ag nanocomposite films for antimicrobial food packaging". *Carbohydrate Polymer* 184 (2018): 453–64.

Sawpan, M. A., Pickering, K. L. and Fernyhough, A. "Effect of various chemical treatments on the fibre structure and tensile properties of industrial hemp fibres". *Composites: Part A* 42 (2011): 888–95.

Saxena, M., Pappu, A., Haque, R. and Sharma, A. Sisal fiber based polymer composites and their applications. In S. Kalia and I. Kaur (Eds), *Cellulose Fibers: Bio- and Nano-Polymer Composites*. Springer Science and Business Media LLC: Berlin, Germany (2011), pp. 589–659.

Schutz, C., Sort, J., Bacsik, Z., Oliynyk, V., Pellicer, E., Fall, A., Wagberg, L., Berglund, L., Bergstrom, L. and Salazar-Alvarez, G. "Hard and transparent films formed by nanocellulose–TiO$_2$ nanoparticle hybrids". *PLoS One* 7, no. 10 (2012): 45828.

Sharma, V. K., Ghosh, S. K., Sakai, V. G. and Mukopadhyay, R. "Enhanced microscopic dynamics of a liver lipid membrane in the presence of an ionic liquid". *Frontier Chemistry* 8 (2020): 577508.

Shuai, X., He, Y., Na, Y. H. and Inoue, Y. "Miscibility of block copolymers of poly(ε caprolactone) and poly(ethylene glycol) with poly(3-hydroxybutyrate) as well as the compatibilizing effect of these copolymers in blends of poly(ε-caprolactone) and poly(3 hydroxybutyrate)". *Journal of Applied Polymer Science* 80 (2001): 2600.

Siakeng, R., Jawaid, M., Ariffin, H. and Sapuan, S. M. "Mechanical, dynamic, and thermomechanical properties of coir/pineapple leaf fiber reinforced polylactic acid hybrid biocomposites". *Polymer Composites* 40 (2018): 446–463.

Siqueira, G., Bras, J. and Durfresne, A. "Lufa cylindrica as a lignocellulosic source of fiber, microfibrillated cellulose and cellulose nanocrystals". *BioResources* 5, no. 2 (2010) 727–40.

Siti, S., Abdul, H., Wan, W. and Jawai, M. "Bamboo based biocomposites material, design and applications". In Y. Mastai (Ed.) *Materials Science: Advanced Topics*. IntechOpen: London, UK (2013), pp. 489–517.

Soykeabkaew, N., Laosat, N., Ngaokla, A., Yodsuwan, N., Tunkasiri, T. "Reinforcing potential of micro- and nano-sized fibers in the starch-based biocomposites". *Composites Science and Technology* 72 (2012): 845–52.

Sreekala, M. S., Kumaran, M. G. and Thomas, S. "Oil palm fibers: Morphology, chemical composition, surface modification, and mechanical properties". *Journal of Applied Polymer Science* 66 (1997): 821–35.

Sun, Z. "Hyperbranched polymers in modifying natural plant fibres and their applications in polymer matrix composites-a review". *Journal of Agricultural and Food Chemistry* 67 (2019):8715–24.

Supthanyakul, R., Kaabbuathong, N. and Chirachanchai, S. "Random poly(butylene succinate-co-lactic acid) as a multi-functional additive for miscibility, toughness, and clarity of PLA/PBS blends". *Polymer* 105 (2016): 1.

Takagi, Y., Yasuda, R., Yamaoka, M. and Yamane, T. "Morphologies and mechanical properties of polylactide blends with medium chain length poly(3-hydroxyalkanoate) and chemically modified poly(3-hydroxyalkanoate)". *Journal of Applied Polymer Science* 93 (2004): 2363–69.

Thakur, V. K., Thakur, M. K. and Guptac, R. K. "Development of functionalized cellulosic biopolymers by graft copolymerization". *International Journal of Biology Macromolecules* 62 (2013): 44–51.

Thiruchitrambalam, M., Alavudeen, A., Athijayamani, A., Venkateshwaran, N. and Perumal, A. E. "Improving mechanical properties of banana/kenaf polyester hybrid composites using sodium laulryl sulfate treatment". *Materials Physics and Mechanics* 8 (2009): 165–73.

Thomas, D. R. "A general inductive approach for qualitative data analysis". *American Journal of Evaluation* 27, no. 2 (2003): 237–46.

Thygesen, L. G, Hidayat, B. J., Johansen, K. S. and Felby, C. "Role of supramolecular cellulose structures in enzymatic hydrolysis of plant cell walls". *Journal of Industrial Microbiology and Biotechnology* 38 (2011): 975–83.

Toprakci, H. A. K., Turgut, A. and Toprakci, O. "A novel approach for fabrication of thermoplastic starch based biocomposites". *Journal of Textile Engineering* 26 (2019): 115.

Wang, T. and Drzal, L. T. "Cellulose-nanofiber-reinforced poly(lactic acid) composites prepared by a water-based approach". *Applied Materials and Interfaces* 4 (2012): 5079–85.

Wang, B., Sain, M. and Oksman, K. "Study of structural morphology of hemp fiber from the micro to the nanoscale". *Applied Composite Materials* 14 (2007a): 89–103.

Wang, S., Li, L. and Zhu, Z. H. "Solid-state conversion of fly ash to effective adsorbents for Cu removal from wastewater". *Journal of Hazardous Materials* 139 (2007b): 254–59.

Wang, W., Zhang, T. J., Zhang, D. W., Li, H. Y., Ma, Y. R., Qi, L. M., Zhou, Y. L. and Zhang, X. X. "Amperometric hydrogen peroxide biosensor based on the immobilization of heme proteins on gold nanoparticles–bacteria cellulose nanofibers nanocomposite" *Talanta* 84, no. 1 (2011): 71–77.

Wei, L., Liang, S. and McDonalda, A. G. "Thermophysical properties and biodegradation behavior of green composites made from polyhydroxybutyrate and potato peel waste fermentation residue". *Industrial Crops and Products* 69 (2015): 91–103.

Wirawan, R. and Sapuan, S. M. Sugarcane bagasse-filled poly(vinyl chloride) composites. In S. M. Sapuan, E. S. Zainudin and H. Ismail (Eds), *Natural Fibre Reinforced Vinyl Ester and Vinyl Polymer Composites*. Elsevier BV and Woodhead Publishing: Duxford, UK (2018), pp. 157–68.

Witayakran, S., Smitthipong, W., Wangpradid, R., Chollakup, R. and Clouston, P. L. "Natural fiber composites: Review of recent automotive trends". In *Reference Module in Materials Science and Materials Engineering*, Elsevier Publishing: Amherst, MA (2017), pp. 1–9.

Won, J. S., Lee, J. E., Jin, D. Y. and Lee, S. G., "Mechanical properties and biodegradability of the kenaf/soy protein isolate-PVA biocomposites". *International Journal of Polymer Science* 860617 (2015): 1–10.

Wong, D. W., Yuan, L. and Perlin, S. A. "Comparison of spatial interpolation methods for the estimation of air quality data". *Journal of Exposure Science and Environmental Epidemiology* 14 (2004): 404–15.

Wu, C. S. "Preparation and characterization of polyhydroxyalkanoate bioplastic-based green renewable composites from rice husk". *Journal of Polymers and the Environment* 22 (2014): 384–92.

Wu, D., Yuan, L., Laredo, E., Zhang, M. and Zhou, W. Interfacial properties, viscoelasticity, and thermal behaviors of poly(butylene succinate)/polylactide blend. *Industrial and Engineering Chemistry Research* 51, no. 5 (2012): 2290–98.

Wu, X., Lu, C., Zhou, Z., Yuan, G., Xiong, R. and Zhang, X. "Green synthesis and formation mechanism of cellulose nanocrystal-supported gold nanoparticles with enhanced catalytic performance". *Environmental Science Nanomaterials* 1, no. 1 (2014) 71–79.
Wu, L. X., Ding, X. X., Li, P. W., Du, X. H., Zhou, H. Y., Bai, Y. Z. H. and Zhang, L. X. "Aflatoxin contamination of peanuts at harvest in China from 2010 to 2013 and its relationship with climatic conditions". *Food Control* 60 (2016): 117–23.
Xiao, W., Xu, J., Liu, X., Hu, Q. and Huang, J. "Antibacterial hybrid materials fabricated by nanocoating of microfibril bundles of cellulose substance with titania/chitosan/silver nanoparticle composite films." *Journal of Materials Chemistry B* 1, no. 28 (2013): 3477–85.
Yadav, M. "Study on thermal and mechanical properties of cellulose/iron oxide bionanocomposites film". *Composites Communications* 10 (2018): 1–5.
Yang, G., Xie, J., Deng, Y., Bian, Y. and Hong, F. "Hydrothermal synthesis of bacterial cellulose/AgNPs composite: A "green" route for antibacterial application". *Carbohydrate Polymer* 87, no. 4 (2012): 2482–87.
Yang, Y., Murakami, M. and Hamada, H. "Molding method, thermal and mechanical properties of jute/PLA injection molding". *Journal of Polymers and the Environment* 20 (2012): 1124–33.
Yilmaz, N. D., Sulak, M., Yilmaz, K. and Khan, G. M. A. "Effect of chemical treatments on physico-chemical properties of fibres from banana fruit and bunch stems". *Industrial Journal of Fibre Textile Research* 42 (2015): 111–17.
Yokesahachart, C. and Yoksan, R. "Effect of amphiphilic molecules on characteristics and tensile properties of thermoplastic starch and its blends with poly(lactic acid)". *Carbohydrate Polymer* 83 (2011): 22–31.
Yoksan, R., Sane, A., Khanoonkon, N., Yokesahachart, C., Noivoil, N. and Dang, K. M. "Effect of starch and plasticizer types and fiber content on properties of polylactic acid/thermoplastic starch blend". *International Journasl of Materials, Metallurgy and Engineering* 9, no. 9 (2015): 1–5.
Yu, C. "Natural textile fibres: Vegetable fibres: Chapter 2". In R. Sinclair (Ed.), *Textiles and Fashion: Materials, Design and Technology*. Woodhead Publishing: Oxford, UK (2014), pp. 29–56.
Yu, L., Dean, K. and Li, L. "Polymer blends and composites from renewable resources". *Progress in Polymer Science* 31 (2006): 576–602.
Yun, Y. H., Youn, H. G., Shin, J. Y. and Yoon, S. D. "Preparation of functional chitosan-based nanocomposite films containing ZnS nanoparticles". *International Journal of Biology Macromolecules* 104 (2017): 1150–57.
Yusoff, R. B., Takagi, H. and Nakagaito, A. N. "Tensile and flexural properties of polylactic acid-based hybrid green composites reinforced by kenaf, bamboo and coir fibers". *Industrial Crops and Products* 94 (2016): 562–73.
Zaman, H. U. and Beg, M. D. H. "Preparation, structure, and properties of the coir fiber/polypropylene composites". *Journal of Composite Materials* 48, no. 26 (2014): 3293–301.
Zampaloni, M., Pourboghrat, F., Yankovich, S. A., Rodgers, B. N., Moore, J., Drzal, L. T., Mohanty, A. K. and Misra, M. "Kenaf natural fiber reinforced polypropylene composites: A discussion on manufacturing problems and solutions". *Composite: Part A* 38 (2007): 1569–80.
Zhang, M. and Thomas, N. L. "Preparation and properties of polyhydroxybutyrate blended with different types of starch". *Journal of Applied Polymer Science* 116 (2009): 688–94.
Zhang, M. and Thomas, N. L. "Blending polylactic acid with polyhydroxybutyrate: The effect on thermal, mechanical, and biodegradation properties". *Advances in Polymer Technology* 30, no. 2 (2011): 67–79.
Zhang, N. and Lu, X. "Morphology and properties of super-toughened bio-based poly(lactic acid)/poly(ethylene-co-vinyl acetate) blends by peroxide-induced dynamic vulcanization and interfacial compatibilization". *Polymer Testing* 56 (2016): 354–363.

Zhang, T., Wang, W., Zhang, D., Zhang, X., Ma, Y., Zhou, Y. and Qi, L. "Biotemplated synthesis of gold nanoparticle–bacteria cellulose nanofiber nanocomposites and their application in biosensing". *Advanced Functional Materials* 20, no. 7 (2010): 1152–60.

Zhang, B., Sun, B., Bian, X., Li, G. and Chen, X. "High melt strength and high toughness PLLA/PBS blends by copolymerization and insitu reactive compatibilization". *Industrial and Engineering Chemistry Research* 56 (2017): 52.

Zhang, X., Xiong, R., Kang, S., Yang, Y. and Tsukruk, V. V. "Alternating stacking of nanocrystals and nanofibers into ultra-strong chiral biocomposite laminates". *ACS Applied Nanomaterials* 14, no. 11(2020): 14675–85.

Zhou, Z., Lu, C., Wu, X. and Zhang, X. "Cellulose nanocrystals as a novel support for CuO nanoparticles catalysts: Facile synthesis and their application to 4-nitrophenol reduction". *RSC Advances*, 3, no. 48 (2013): 26066–73.

14 Applications of Biocomposites in Reduction of Environmental Problems

Chukwuma Chris Okonkwo
Nnamdi Azikiwe University

Francis Odikpo Edoziuno
Nnamdi Azikiwe University
Delta State Polytechnic

Adeolu Adesoji Adediran
Landmark University

Kenneth Kanayo Alaneme
Federal University of Technology Aapkure
University of Johannesburg

CONTENTS

14.1	Introduction	402
14.2	Concept of Biocomposites	403
14.3	Environmental Pollution and Challenges	405
14.4	Classification of Environmental Pollutants	407
14.5	Techniques for Environmental Remediation	409
14.6	Application of Biocomposite in Treating Pollutants	410
14.7	Organic Pollutants	410
	14.7.1 Dyes	410
	14.7.2 Phenols	412
	14.7.3 Hydrocarbons	413
14.8	Other Organic Pollutants	415
14.9	Inorganic Pollutants	416
	14.9.1 Metals	416
	14.9.2 Nonmetal	418
14.10	Conclusion	418
References		419

DOI: 10.1201/9781003137535-14

14.1 INTRODUCTION

The problem of environmental pollution can be largely attributed to past industrial activities when there was little or no awareness on the health and environmental effects of producing, using, and disposing hazardous substances (Luka, Highina, and Zubairu 2018). The growing rate of environmental pollution is an evident problem in the world today that needs to be prioritized and addressed adequately to ensure the earth's sustainability for habitation for years to come (Kanmani et al. 2017). For example, in Vietnam, the pollution of the air, water and soil environments are very much intense with contaminants such as total suspended particles, organic substances, heavy metals, nutrients including ammonium and phosphate exceeding the allowable values of national standard in most areas (Chu 2018). A similar scenario is obtainable in Malaysia where environmental health problems are mainly caused by atmospheric pollution, water pollution, climate change, ozone depletion, and toxic, chemical, and hazardous waste management (Qureshi et al. 2015). Environmental pollution is one of the greatest challenges facing the human race in the 21st century (Hagos et al. 2017), with the pollution of water, land and air due to human activities disrupting the natural habitat in many developing and developed nations (Ubaid et al. 2018). Air pollution constitutes a serious health challenge due to its negative impacts on the quality of life as its long term exposure is responsible for respiratory and cardiovascular problems contributing to the overall increase in hospital admissions and healthcare spending (Jung, Mehta, and Tong 2018). Water pollution is a major problem in the world today as many regions lack adequate clean water to meet human drinking and sanitation needs which is an obstacle to the quest to maintain a clean environment and healthy ecosystems. This impacts negatively on human health, productivity and economic development (Cosgrove and Loucks 2015). Land pollution is not left out, as the remediation and management of persistent organic pollutants contaminated soil is becoming a global priority. The toxic groups of chemical pollutants are potentially harmful to living organism, due to their higher degree of halogenations, inclination to bioaccumulate in the lipid component, and their resistance to natural degradation (Abhilash et al. 2013).

Several technologies have been explored in the past for environmental remediation and they include coagulation-flocculation, adsorption, ion-exchange, membrane separation, evaporation, floatation and electroprecipitation among others (Kanmani et al. 2017). These conventional methods are not environmentally sustainable as the remediating material utilized for their application tends to produce toxic byproducts which are hazardous to the environment (Aftab et al. 2020). To address this issue, a lot of research effort have been made to develop sustainable and eco-friendly materials for environmental remediation (Campagnolo et al. 2018). Biocomposites have emerged as valuable materials for environmental remediation owing to their unique adsorption potentials, chelating ability, nutrient recycling and antibacterial effects, and these characteristics make them very efficient in environmental remediation through coagulation and biosorption techniques (Sajid, Rafiq, and Nadeem 2018). Furthermore, the employment of agricultural and industrial byproducts such as chitosan, cellulose and hemicellulose in biocomposite materials can provide a sustainable and eco-friendly method for environmental remediation (Zia, Hartland, and Mucalo 2020).

14.2 CONCEPT OF BIOCOMPOSITES

The combination of two or more constituents insoluble in each other at a macroscopic level can be regarded generally as a composite material, with such combination improving the system with characteristics superior to that of the individual components (Ariadurai 2012; D'Souza et al. 2011). Basically, all composites are made up of two phases, the matrix or host phase and the reinforcement or fiber phase (Ariadurai 2012). The matrix is responsible for the shape of the composite and also acts as a load transfer medium to the filler, while the fiber is responsible for the optimization of selected mechanical or other properties of the composite which include electrical or thermal conductivity, electromagnetic interference shielding, etc. (D'Souza et al. 2011). In recent years, a lot of significant attention has been given by stakeholders in the field of composites regarding the use of renewable bio-based and biodegradable materials as a means to checkmate climate changes and address the growing regulatory demands for a clean and safe global environment (AL-Oqla, Almagableh, and Omari 2017). There are various definitions for the term "Biocomposites." One definition describes it as the combination of natural fibers such as wood, rice straw, hemp, banana, pine apple, sugar cane, oil palm, jute, sisal, flax, etc., with polymer matrices from both renewable and nonrenewable resources (Mitra 2014). Another definition extends the term to any composite materials with at least one bio-based or natural constituent (Mitra 2014). Mitra (2014) distinguished between three categories of biocomposites and they include bio-fiber-reinforced petroleum-derived polymers which are nonbiodegradable such as polyolefins polyester, epoxy, vinyl ester, phenolics; bio-polymers reinforced by synthetic fibers, e.g., glass or carbon; and bio-polymers such as polylactic acid reinforced by bio-fibers such as jute. As shown in Figure 14.1, Peças et al. (2018) provided a broader classification of

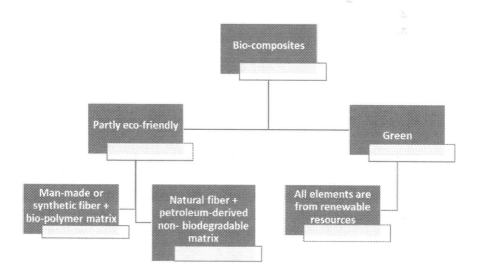

FIGURE 14.1 Classification of biocomposites based on nature of constituents. (Adapted from Peças et al. (2018).)

biocomposites based on the nature of the constituents. It was classified as green and partly eco-friendly biocomposites, with green biocomposites referring to those with all constituents made up of renewable resources and partly eco-friendly referring to those with one constituent made from nonrenewable resources such as synthetic fiber or petroleum-derived polymer.

Biocomposites like all polymer matrix composite materials are made up of two phases, namely, the matrix or continuous phase and reinforcement or discontinuous phase (Banga, Singh, and Choudhary 2015; Johnson et al. 2017). A biocomposite's effectiveness is hugely dependent on its ability to transfer load or stress from matrix phase to the reinforcement phase (Essabir et al. 2013). The matrix separates the reinforcement from one another to ensure there is no mechanical abrasion and that no new surface flaws are formed; it also holds the reinforcements in place by acting as a bridge (Verma et al. 2012). The role of the matrix is very significant in terms of the capacity of the composite structure to carry tensile load (Kispotta 2011). The quality of a matrix is measured by its deformation ability under applied load, its ability to transfer the load onto the reinforcements, and to distribute the stress concentration evenly (Verma et al. 2012). Different types of polymers have been used for the matrix phase of biocomposites and they include both biodegradable and nonbiodegradable ones. Polymers can generally be described as developed macromolecules from singular repeating units termed monomers which exhibit excellent properties that include corrosion and water resistant, flexibility and ease of manufacturing; this makes suitable alternatives to traditional materials (Lala, Deoghare, and Chatterjee 2018). Nonbiodegradable polymers such as thermoplastics and thermosets are used commonly as matrix for biocomposites, with the later attracting a lot of interest due to its mechanical properties, chemical resistance, thermal stability, durability and low processing temperature which helps prevent degradation of natural fibers (Testoni 2015). The waste generated by this type of polymers is a major problem with studies showing that <10% of these synthetic plastics generated are recycled (Vinod et al. 2020). To address this issue, biodegradable polymers are beginning to gain prominence and they include thermoplastic starch, polyhydoxyalkanoates, polylactides, lignin-based epoxy, soy-based resins and epoxidised linseed oil (Španić et al. 2019). However, in comparison to nonbiodegradable polymers, these biodegradable polymers exhibit poor thermomechanical properties (Vinod et al. 2020).

The role of the reinforcement phase in a composite material is to increase the mechanical properties of the matrix system, with different reinforcements having different properties and thus influencing the composite properties in different ways (Verma et al. 2012). The need to improve and stimulate rural economies as well as reduce the world's dependency on petroleum based materials has resulted in much interest and focus in the use of various varieties of natural fibers as reinforcing agents for composite materials (Fuqua, Huo, and Ulven 2012). The high strength, availability, low cost, sustainability and eco-friendly characteristics of natural materials such as agricultural waste makes them quite beneficial and efficient as reinforcement for composite materials (Dungani et al. 2016). Other advantages of natural fibers over synthetic fibers such as glass and carbon include low density, acceptable specific properties, ease of separation and enhanced energy recovery (Aigbodion 2019). These advantages of natural fibers over synthetic fibers has made Lignocellulosic fibre (flax,

Applications of Biocomposites

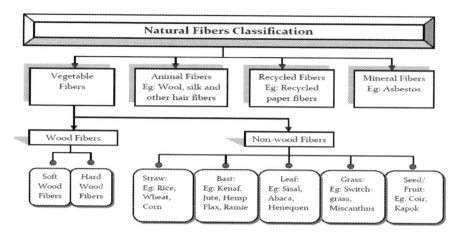

FIGURE 14.2 Types of natural fibers (Nagarajan 2012).

hemp, kenaf, henequen, banana, oil palm, and jute) substitute of huge potential for synthetic fibers (Gurunathan, Mohanty, and Nayak 2015). In contrast to synthetic fiber-based polymer composites, natural fiber based composites can be disposed of easily or composted at the end of their lifespan without polluting the environment (Gurunathan et al. 2015). As shown in Figure 14.2, Nagarajan (2012), classified natural fibers into vegetable, animal, recycled and mineral fibers. Vegetable fibers constitute mainly of polysaccharides and that of animal constitute mainly of proteins (Zini and Scandola 2011).

14.3 ENVIRONMENTAL POLLUTION AND CHALLENGES

As most part of the world continues to experience rapid economic growth, there is bound to be an increase in environmental pollution due to the role rapid urbanization and industrial growth plays in polluting the environment (Chu 2018). One factor that has been crucial to the progress of industrialization and urbanization is the use of fossil fuels. As shown in Figure 14.3, Bose (2010) used a chart to illustrate the world's energy generation scenario with oil and coal (both fossil fuels) dominating with 38% and 28%, respectively.

While fossil fuels have made significant contributions to energy generation in the world, the environmental issues it poses remains a major challenge (Bian et al. 2010). The negative environmental impacts associated with coal mining and utilization include coal mine accidents, land subsidence, pollution of water environments, disposal of mine waste and air pollution (Bian et al. 2010). In addition to spillage which is the main environmental issue associated with oil exploration, its production usually involves the discharge of wastewater which contains sodium, calcium, ammonia, boron, trace metals and high total dissolved solids (Obiechina and Joel 2018). In the Niger Delta region of Nigeria, pollution arising from the two major source of oil spillage and gas flaring has degraded the environment hugely and left the region clamoring for survival in more ways than one (Nnaemeka 2020).

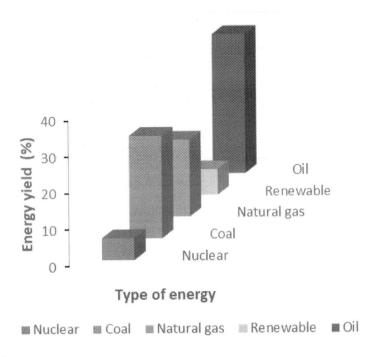

FIGURE 14.3 Global energy generation scenario (Bose 2010).

Urbanization and industrialization have intensified the rate of water pollution globally and this is most severe in developing countries wherein several toxic solid wastes, effluents and emission are being discharged into various waterbodies (Farid, Baloch, and Ahmad 2012). Water and land pollution are closely related with land-based activities responsible for 80% of pollutants in oceans, rivers, bays, streams, lakes and other bodies of water (Dhankhar and Hooda 2011). A major contributor to water and land pollution is the lack of proper waste management and this problem has been exacerbated by urbanization as population increase brings about an increase in waste generation. In China, surface water and groundwater are mostly polluted by unregulated sewage and municipal wastewater discharge, industrial wastewater discharge, agricultural fertilizers and pesticides (Han, Currell, and Cao 2016). In India, one of the major environmental challenges faced is waste management as the current systems in place cannot handle the volumes of waste generated by the increasing urban population, which has negatively affected the environment and public health (Kumar et al. 2017). In Lagos Nigeria, pollution in urban settlements as resorted to poor waste disposal facilities and the discharge of cattle waste in nearby rivers which greatly degrade the quality of water (Lawanson et al. 2012). In most developing countries, the lack of appropriate treatment plants, as resulted to a large percentage of the waste generation of discharged untreated (Ariffin and Sulaiman 2015). This untreated waste contains heavy metals and biological contaminants which constitutes an environmental challenge for man and aquatic lives (Khan and Ghouri 2011). Furthermore, waste-activated sludge generated from conventional sewage treatment processes is a major source of secondary environmental contamination due to the presence of various pollutants such as polycyclic aromatic hydrocarbons, dioxins, furans, heavy metals, among others, hence there is the need for innovative and cost-effective treatment processes to ensure the safe and environment-friendly disposal of sludge (Raheem et al. 2018). In Malaysia, Ariffin and Sulaiman (2015) observed that the use of outdated technology in treating sewage was one of the causes of sewage pollution and hence there is the need to explore other alternatives.

Among the several pollution problems facing the world today, air pollution has attracted a lot of attention with major concerns regarding its widespread nature, damage to our environment and potential health risk to humans (Leung 2015). Transportation and other industrial activities that involve the burning of fossil fuels remain the major reason for the deterioration of air quality in major cities (Mokthsim and Salleh 2014). The most established severe environmental issues associated with the burning of fossil fuel are the greenhouse effect due to CO_2 and other greenhouse gases emission (Ashraf et al. 2020). In Germany, air pollution is mainly observed in urban areas as traffic circulation and demographic density degrades the quality of environment, with high proportion of diesel trucks contributing about 13% of total air pollutants in the country (Bashir et al. 2020). In China, air quality in urban areas has been contaminated with high concentrations of SO_2 and total suspended particle for many years as a result of its coal-dominated energy structure (Wang and Hao 2012). Continuous efforts being made in recent years to decouple air pollution and carbon intensity and achieve a society based on sustainable development and ecological civilization (Lu et al. 2020).

An emerging environmental issue is plastic pollution. Plastic pollution poses a major threat to natural ecosystems and human health, with initial prediction of a two-fold increase in the number of plastic debris by 2030 subject to aggravation due the excessive use and consumption of single-use plastics (including personal protective equipment such as masks and gloves) during this pandemic era of COVID-19 (Silva et al. 2021). Ferraro and Failler (2020) pointed out that there are two major ways which plastics enter the marine environment, with the first being human activity at seas such as the plastic waste generated as a result of fishing activities. The second being land-based activities such as the transportation of plastic generated on land into the oceans via wind, tides, rivers and wastewater outflows. The pollution of the marine environment with waste debris had long been associated with land based waste discharge and various human activities, with a study showing that about 80% of the total debris in the marine environment come from land-based sources (Alimba and Faggio 2019). When these plastics find their way into waterbodies after several transformation processes, they induce toxicity and destroy the biota due to their ability to adsorb toxic contaminants and transmit to hostile species (Bilal and Iqbal 2020).

14.4 CLASSIFICATION OF ENVIRONMENTAL POLLUTANTS

An environmental pollutant can be described as any alien substance that is not usually present within an environment or that which is found in the environment in concentrations higher than the permissible limits (Mathew et al. 2017). Urbanization and industrialization have intensified the release of various pollutants into the environment with the contamination of air with CO_2 and other greenhouse gases. The presence of NO_x, SO_2, particulate matter; water with a variety of chemicals, nutrients, leachates, oil spills, among others, accounts for the lack of proper facilities for waste (Gavrilescu et al. 2015). In addition, the disposal of hazardous wastes in the soil, spreading of pesticides, sludge, as well as the use of disposable goods or non-biodegradable materials is an additional output caused by the lack of proper facilities for waste (Gavrilescu et al. 2015). Irrespective of whatever form an environmental pollutant is in, it has negative impacts on the health and well-being of human beings and other living organisms (Katole, Kumar, and Patil 2013).

A wide range of environmental pollutants that contain very toxic substances which threaten the health of humans and other living organisms even at very low concentrations are present in the environment (Song et al. 2019). As shown in Figure 14.4, environmental pollutants can broadly be classified into organic and inorganic pollutants (Bharagava, Saxena, and Mulla 2020).

Inorganic pollutants are persistent and nonbiodegradable in nature and may accumulate in the cells, tissues and specific body parts of the living organisms, hence constituting a public health concern (Verma and Ratan 2020). Inorganic pollutants can further be divided in metals and nonmetal. The metals include a wide range of toxic heavy metals such as cadmium, chromium, arsenic, lead and mercury (Bharagava et al. 2020). The nonmetals include minerals and their acids, inorganic salts, trace elements, cyanides and sulfates (Verma and Ratan 2020).

The organic pollutants include phenols, chlorinated phenols, endocrine-disrupting chemicals, azo dyes, polyaromatic hydrocarbons, polychlorinated biphenyls and pesticides. The major source of organic pollutants in the environment is wastewater

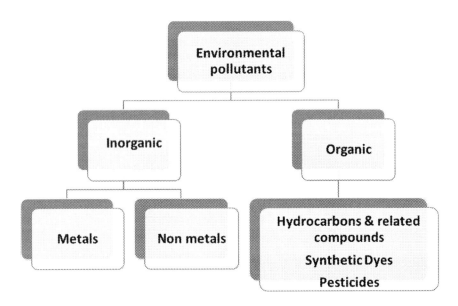

FIGURE 14.4 Classification of environmental pollutants. (Adapted from Kanmani et al. (2017).)

treatment plants as they are not designed to completely eradicate these organic compounds and as such many of these compounds persist without any alteration, even after undergoing tertiary treatment (Urbano et al. 2020). The most important organic pollutants of environmental concern are called persistent organic pollutants, a heterogeneous group of man-made compounds that spread throughout the environment and reach hazardous concentrations even in places where they have never been produced or used (Campos et al. 2008). This type of pollutant is resistant to environmental degradation (chemical, biological and photolytic reactions) and they consist of pesticides, industrial chemicals such as polychlorinated biphenyls, and byproducts of industrial processes such as dioxins and furans (Alharbi et al. 2018). When these pollutants find their way into the food chain, they accumulate in the fatty tissue of human body which constitutes a risk and causes adverse effects to the environment and human health (Gaur, Narasimhulu, and Pydisetty 2018).

14.5 TECHNIQUES FOR ENVIRONMENTAL REMEDIATION

The pollution of the environment with various toxic pollutants poses a significant threat to public health; hence there is an urgent need for suitable, effective and sustainable solutions (Falciglia et al. 2018). There are various techniques being used around the world for the remediation of environmental pollution (Hamadan et al. 2020). These remediation techniques for environmental pollution can be generally divided into Physical, Chemical and Biological techniques (Kuppusamy et al. 2016). It is also important to point out that some environmental remediation methods involve a combination of two or more of the aforementioned three for a more efficient and economical remediation (Khalid et al. 2016). To choose an appropriate remediation method, one needs to properly understand the state of the affected environment; the contaminant's nature, composition and properties; the contaminant's fate, transport and distribution; the degradation mechanism (Ossai et al. 2020). Also, the interactions and relationships with microorganisms, intrinsic and extrinsic factors influencing remediation and the potential effect of the possible remedial measure are among other essential factors (Ossai et al. 2020).

Physical techniques of environmental remediation include methods such as thermal desorption, cement kiln, air stripping soil leaching, steam extraction, off-site landfill and incineration for soil remediation (Karami and Shamsuddin 2010; Yang et al. 2019); and others such as membrane filtration, coagulation, adsorption, ion-exchange and irradiation for water remediation (Gusain et al. 2019). This technique is quite effective for highly contaminated environment but it is also laborious and very costly (Khalid et al. 2016). Furthermore, the effectiveness of this technique is very much limited when used to treat dense nonaqueous phase liquids and such treatment usually bring about byproducts (aqueous effluent) which needs further treatment or disposal (Okonkwo, Edoziuno, and Orakwe 2020).

Chemical techniques of environmental remediation can be described as those which employ the use of chemical substances to treat pollutants in contaminated environment (Okonkwo et al. 2020). This type of technique includes methods such as encapsulation, solvent extraction, neutralization, oxidation-reduction and precipitation for soil remediation (Karami and Shamsuddin 2010); and others such as electrochemical routes, photocatalysis, ozonation and Fenton's reagent for water remediation (Gusain et al. 2019). This type of technique produces very rapid and effective result but it is largely dependent on the type of environmental media, chemical used and pollutant involved (Khalid et al. 2016). The major drawbacks associated with this technique are its byproducts which are hazardous to the ecosystem, high cost and its inability to treat huge amounts of pollutants (Okonkwo et al. 2020).

Biological techniques of environmental remediation are those which employ the use of biological agents to alter the toxicity of environmental pollutants and make them less toxic or more easily biodegradable (Okonkwo et al. 2020). This type of techniques includes methods such as bioremediation and phytoremediation for soil remediation (Karami and Shamsuddin 2010); and anaerobic degradation and aerobic degradation for water remediation (Gusain et al. 2019). This technique is safe, economical, easy to use and eco-friendly but it is not suitable for highly contaminated environment due to the toxicity of high concentrations of pollutants to the bioremediating material (Khalid et al. 2016; Okonkwo et al. 2020).

14.6 APPLICATION OF BIOCOMPOSITE IN TREATING POLLUTANTS

The efficiency of various types of biocomposite in treating pollutants of both organic and inorganic nature has been researched extensively. This section reviews and discusses these works with emphasis on the biocomposite materials, mechanism of pollutant treatment and the results obtained.

14.7 ORGANIC POLLUTANTS

14.7.1 DYES

Faruk et al. (2017) studied the application of functional biocomposite materials in remediating methylene blue dye. The biocomposite materials were developed by the encapsulation of *Pseudomonas aeruginosa* bacteria within electrospun polyvinyl alcohol and polyethylene oxide nanofibrous webs. Fluorescence microscopy and viable cell counting assay were used to assess viabilities of the bacterial cell. The resulting biocomposite materials were studied for the removal of methylene blue in water and results revealed that the removal capabilities of the Biocomposites materials were dependent on the presence of bacteria. Results also showed that biological removal was responsible for pollutant removal rather than adsorption and through increased initial cell viability numbers or use of a more capable bacterial strain; the removal performances can be optimized. The biocomposites also showed good storage properties and hence they can be effective in methylene blue remediation in contaminated water with improvable properties. Jerold et al. (2017) used borohydride reduction method to develop a biocomposite material of nanoscale zero-valent iron and *Sargassum swartzii*. Scanning electron microscopy was used to characterize the material and FTIR spectroscopy was used to examine the role of its functional groups in adsorption. The ability of the biocomposite to adsorb crystal violet in aqueous solutions was studied, and at pH of 8, maximum biosorption capacity was achieved with a dye uptake of 200 mg/g. The study showed that the biocomposite is an excellent biosorbent for the adsorption of crystal violet in aqueous media.

Chacón-Patiño et al. (2013) investigated the efficiency of nanostructured MnO_2 and Fique fibers bionanocomposite in degrading indigo carmine dye. The bionanocomposite was prepared in situ by exposing the fiber surface to alkaline conditions and the use of coulombic interactions to implant permanganate anions (MnO_4^-) onto the resultant alkali cellulose. The permanganate anions were further reduced to MnO_2 nanoparticles via ultrasound-assisted procedure and precursor concentration; loading and reduction time's influence on the synthesis of the nanoparticles were examined using UV-Vis diffuse reflectance. The bionanocomposite was used remove indigo carmine dye in water samples with an efficiency of 98% recorded in <5 min; mass spectrometry showed that the degradation route was via oxidative and ring cleavage pathways. The study established that the biodegradable bionanocomposite was efficient in degrading pollutants from industrial effluents with its ease of synthesis and reusability as added advantages. Perullini et al. (2010) used a silica-alginate-fungi biocomposite to investigate the remediation of water contaminated with malachite green dye. To produce the biocomposites, alginate beads loaded with filamentous fungus *Stereum hirsutum* were immobilized inside nanoporous silica hydrogels using a two-step encapsulation method. The resulting biocomposite proved to be effective in degrading high concentrations of the dye via adsorption, regulated transport of the dye and dye degradation enzymes retention inside the hydrogel. Results of the study showed that bioremediation is possible without the discharge of harmful organisms into the environment.

Noreen et al. (2020) prepared polymeric biocomposites of polyaniline/sugarcane bagasse (Pan/SB), polypyrrole/sugarcane bagasse (PPy/SB), polyaniline/chitosan (PAn/Ch), polypyrrole/chitosan (PPy/Ch), polyaniline/starch (PAn/St) and polypyrrole/starch (PPy/St), and examined their application for acid black dye adsorption in batch and column modes. At 60 min reaction time and adsorbent dose of 0.05 g, efficient adsorption was achieved at an optimum pH of 2, 3, 4, 4, 5, 3 and 3 for SB, PPy/SB, PAn/SB, PPy/Ch, PAn/Ch, PPy/St and PAn/St, respectively. The biosorption processes of SB, PAn/SB, PAn/Ch, PPy/St and PAn/St fitted pseudo-first-order kinetic model, while that of PPy/SB and PPy/Ch fitted pseudo-second-order kinetic model better. Maximum adsorption capacities of 52.6, 100, 90.91, 90.0, 71.4, 66.6 and 62.5 were achieved, respectively, for SB, PPy/SB, PAn/SB, PPy/Ch, PAn/Ch, PPy/St and PAn/St with the process of biosorption showing satisfactory fitness for Langmuir isotherm. The study demonstrated the efficient use of polymeric biocomposites as adsorbents for the removal of dyes in textile wastewater. To modify natural luffa sponge with reduced graphene oxide, Li et al. (2016) used a one-step hydrothermal treatment of Luffa sponge in graphene oxide suspensions and ascorbic acid. X-ray diffraction, Fourier transform infrared and scanning electron microscope were used to study the resulting biocomposite and the results showed it had a pore diameter of about 500 nm which can enhance the adsorbent properties of luffa sponge effectively. Batch adsorption was used to study the ability of the biocomposite to adsorb cationic dyes and results showed that the adsorption was highly dependent on initial pH of solution, biocomposite content and initial concentration of dyes. The incorporation of 0.4 wt% reduced graphene oxide into luffa sponge increased the value of uptake at equilibration time from 32.56 to 88.32 mg/g for basic magenta and from 31.65 to 63.32 mg/g for methylene blue. The adsorption process can be attributed to electrostatic interaction with the adsorption kinetics data a pseudo-second-order model and adsorption isotherm a Langmuir model. Results of the work produced a novel green biocomposite adsorbent for the efficient treatment of cationic dyes.

Sui et al. (2012) used wet spinning method to develop a biocomposite material of calcium alginate and multi-walled carbon nanotubes. The biocomposite was characterized using scanning electron microscope. With varying contents of the nanotubes and pH values, batch adsorption experiments were used to study the ability of the biocomposite to adsorb methylene blue and methyl orange. Results showed that the biocomposite enhanced the adsorption capacity of methyl orange and the adsorption rate for methylene blue in comparison to native calcium alginate fiber with the adsorption kinetics fitting pseudo-second-order for both dyes. Langmuir and Freundlich isotherms were used to analyze the equilibrium adsorption data with results showing it follows the former. The desorption experiments revealed that at pH 2.0 for methylene blue and pH 13.0 for methyl orange, desorption percentages were 79.7% and 80.2%, respectively. The efficient and fast adsorption process achieved using the biocomposite showed it can be utilized for the eco-friendly treatment of ionic dyes in aqueous solution.

Campagnolo et al. (2018) studied the efficiency of Silk Fibroin/Orange Peel Foam biocomposite as adsorbents in treating methylene blue contamination. Supercritical CO_2 drying method was used to convert a combination of orange peel and silk fibroin (alcogels) into biocomposite foams of high porosity. The resulting adsorbents had an adsorption capacity three times its weight in water with 174.45 m^2/g recorded as its Brunauer–Emmett–Teller specific surface area. The biocomposite was investigated for the adsorption of methylene blue in water and 113.8 ± 12.5 mg/g as the maximum adsorption capacity with well preservation of orange peel activity in the polymeric matrix. As shown by spectroscopic studies, electrostatic and hydrophobic interactions were responsible for the adsorption of methylene blue molecules with a monolayer forming onto the surface of the adsorbent as demonstrated by Langmuir isotherm model. The study presented a biocomposite with high potential for remediating water as the functionalities of the powder can be enhanced without limiting its usability.

14.7.2 Phenols

Bhatti et al. (2020) prepared biocomposites of polypyrrole, polyaniline and sodium alginate with cellulosic biomass barley husk and used batch experiment to investigate their application for the removal of 2, 4-dichlorophenol in water. At biocomposite dose of 0.05 g, 25 mg/L initial concentration of pollutant and contact time of 120 min at 30°C, maximum sorption of the pollutant was recorded in the pH range of 7–10. The FTIR analysis showed that amino, hydroxyl and carboxylic groups were responsible for the adherence of the pollutant onto the surface of biocomposites. The adsorption was exothermic in nature and it followed pseudo-second-order kinetics and Freundlich isotherm models. The biocomposites proved to be efficient adsorbents for treating 2,4-dichlorophenol in industrial effluents and its application can also be extended to other pollutants alike. Carabajal et al. (2016) examined the tolerance of 25 isolates of Argentinean white-rot fungi toward 10 mM of phenol using agar plates. The study also examined the growth ability of 7 isolates on plates with 2,6-dimethoxyphenol, gallic acid, 2,4-dichlorophenol, or guaiacol (7.5 mM), or with phenol as sole carbon source, with *Trametes versicolor, Irpex lacteus, Lentinus tigrinus* and *Pleurotus lindquistii* isolates exhibiting the best results. Furthermore, the immobilized cultures of Trametes versicolor ability to remove phenol were investigated and silica-alginate-fungus biocomposites removed up to 48% of 10 mM phenol within 14 days via biosorption. When immobilized on Luffa aegyptica, a removal range of 62%–74% was achieved for 15 mM phenol over 23 days in three repeated cycles. The impressive results recorded in the study indicate the method can be applied for the treatment of industrial wastewater.

Duarte et al. (2014) prepared biocomposite materials for the treatment of phenolic compounds by immobilizing fungi onto silica-alginate. To prepare the materials, active and inactive Pleurotus sajor caju was encapsulated in alginate beads. A 28-day batch experiment at 25°C was used to investigate the biocomposites application in the treatment of water supplemented (WS) with phenolic compounds, oil mill wastewater (OMW), and oil mill wastewater supplemented (OMWS) with phenolic compounds. To determine the treatment's efficiency, the targeted organic compounds removal was measured as well as the chemical oxygen demand (COD) and relative absorbance ratio along the time. Results showed that the removal of phenolic compounds by active *Pleurotus sajor-caju* biocomposites was 64.6%–88.4% for OMW and OMWS, and 91.8%–97.5% for WS. Results also showed that in the treatment of OMW, there was a removal of fatty acids (30.0%–38.1%), sterol (68.7%) and COD (35%). Results of the study indicate that in addition to removing phenolic compounds from water, the biocomposites can also decrease the concentration of fatty acids and sterols as well as reduce COD.

Kátia R. Duarte et al. (2012) investigated the treatment potential of silica–alginate–fungi (*Pleurotus sajor-caju* and *Trametes versicolor*) for Olive oil mill wastewater. The preparation of the biocomposites involved the encapsulation of fungi in mineral cages. The removal experiment using the biocomposites involved a three-step method of adsorption of pollutants, biodegradation by the fungi, and the diffusion of secondary products of biodegradation. Although both treatments tested showed good potential for treating organic pollutants, *Trametes versicolor* biocomposites were the most effective with 38.4%–44.9%, 42.8%–63.8% and 85.3%–88.7% observed as its reduction in color, COD, and total phenolic content respectively after 29 days of treatment. The study further showed that the biocomposites had good reusability with more than 29 days of treatment re-utilization.

14.7.3 Hydrocarbons

Amorim et al. (2021) employed a one-shot free expansion method to synthesis a biocomposite of malt and/or acerola residues reinforced castor oil-based polyurethane. ATR-FTIR analysis, Scanning Electron Microscopy, TGA analysis, and Biolin Scientific Attension Goniometer were used to characterize the biocomposites. The biocomposites were examined for the adsorption of various organic oils/solvents. Results show that the residues increased the flexibility and adsorption capacity of the biocomposites for all organic oils/solvents, with 3257.8% ± 55.6% recorded as the maximum for chloroform in 60 min and its hydrophobicity as 144° ± 0.2° in acidic, saline and alkaline environments, and in the presence of UV radiation. The polyurethane/acerola biocomposite proved to be more efficient than the polyurethane/malt in all aspects with its adsorption capacity increasing over time from 746.7% ± 77.4% in 60 min to 928.6% ± 16.7% to 90 min as demonstrated by the pseudo-first-order kinetic modeling. Furthermore, these biocomposites can be recovered quickly and repeatedly by simple hand-tightening and solvent release systems, and vacuum suction filtration of 20.5 ± 0.9 s/10 mL. The work gave rise to prototypes with sustainable production and high application potential for the treatment of organic pollutants.

Y. Li et al. (2013) investigated the biodegradation of carbazole using microbial cell/Fe_3O_4 biocomposites. To prepare the biocomposite, Fe_3O_4 nanoparticles were assembled onto the surface of *Sphingomonas* sp. XLDN2–5 cells. Scanning electron microscopy and transmission electronic microscopy were used to characterize the microbial cells before and after loading with the nanoparticles. Results show that the nanoparticles had super-paramagnetic properties with an average particle diameter of Fe_3O_4 of about 20 nm and saturation magnetization of 45.5 emu/g. At 30°C, the biodegradation rates of the biocomposite was examined in comparison to that of free microbial cells and results show they both had the same biodegradation activity and as such the nanoparticle coating has no negative effect on the biodegradation activity. Furthermore, recycling experiments were carried out to determine the reusability of the biocomposite and result revealed it had remarkable reusability with its biodegradation activity increasing gradually as it was recycled. The biocomposite consumed 3,500 μg carbazole completely from the first to sixth cycle in 9 h, and then consumed the same amount of carbazole from the seventh to tenth cycle in only 2 h. In addition, super-paramagnetic properties of the nanoparticle made it possible to separate and recycle the biocomposite easily via an external magnetic field. Results of the study demonstrate that the biocomposite has huge potential in the improvement of biocatalysts for degrading hazardous compounds.

Flores-chaparro et al. (2018) studied the potential application of chitosan-macroalgae biocomposites as adsorbents of water soluble hydrocarbons. The synthesis of the biocomposite involved a factorial design and a posterior response surface methodology with chitosan and pectin as precursors for enhancing the macroalgae's stability and applicability in removing soluble hydrocarbons. The biocomposite was optimized at 75.4%, 19.8% and 4.8% for macroalgae biomass, chitosan and pectin respectively. It was characterized using various physicochemical analyses. Its biosorption potential was investigated under different ionic strengths, pH values and organic load, demonstrating the adsorption mechanisms involved. Its removal capacities for benzene, toluene and naphthalene were observed as 58.68, 16.64 and 6.13 mg/g respectively. Its adsorption capacity was to some extent dependent on the pH (3–9), and diminishing with ionic strength of values up to $I > 0.6$ M. The removal of hydrocarbons was enhanced by presence of dissolved organic matter which presented the biosorbents with hydrophobic sites. Sips adsorption isotherm and pseudo-second-order model best described the biosorption kinetics. The study demonstrated the applicability of the biocomposite in treating benzene, toluene, and naphthalene pollution in natural water.

14.8 OTHER ORGANIC POLLUTANTS

A.K. Duarte et al. (2013) attempted the treatment of a bleached kraft pulp mill effluent with using biocomposite of silica-alginate-fungi to reduce its potential environmental impact. The biocomposites were prepared via the encapsulation of active or inactive *Rhizopus oryzae/Pleurotus sajor-caju* onto alginate beads. Batch experiments were used to determine the removal capacity of the biocomposites for a 29 days period at 28°C. To determine the treatment's efficiency, the removal of organic compounds, chemical oxygen demand and the relative absorbance ratio over time were measured. The two species of fungi exhibited good removal capacity for organic compounds, colour and chemical oxygen demand with active *R. oryzae* biocomposites being more effective than active *P. sajor caju*. Maximum values of reduction of 56%, 65%, and 72%–79% was observed for colour, COD and organic compounds respectively after treatment with active *Rhizopus oryzae* biocomposites. The study demonstrated the industrial and environmental use of the biocomposites in treating bleached kraft pulp mill effluent.

Ishtiaq et al. (2020) studied the adsorption of imidacloprid using biocomposites of polypyrole, polyaniline and sodium alginate with peanut husk. Variables such as pH, insecticide concentration, composite dose, contact time, temperature were investigated towards the optimization of the biocomposite for efficient adsorption of the pollutant. Polypyrole/peanut husk biocomposite exhibited the most promising adsorption efficiency with maximum efficiency at pH 3, 0.05 g adsorbent dose, 25 mg/L initial concentration of pollutant and 90 min contact at 35°C. The experimental data was best described by Langmuir and Freundlich isotherms with R^2 value >0.904 and >0.97, respectively. The adsorption mechanism followed that of pseudo-first-order kinetic model and thermodynamics study showed that the adsorption process was spontaneous and exothermic in nature. The study demonstrated that the biocomposites have great potential for use as adsorbent in the remediation of imidacloprid contaminated wastewater.

Kaur et al. (2020) investigated the effective decontamination of soils contaminated with imazethapyr and imazamox using β-cyclodextrin-chitosan biocomposite. Using chemical assays, experimental variables of extractant solution and its concentration, liquid to soil ratio, amount of soil and soil type were examined for their different roles in the decontamination of imazethapyr and imazamox. Regardless of the formulation of herbicide and rate of application, treatment of soil with the biocomposite enhanced the rate of dissipation of herbicide with residues below the detection limit of <0.005 μg/g within 5–15 days in aridisol, entisol, inceptisol A, inceptisol B, inceptisol C and 7–21 days in alfisol and vertisol. Treatment of the soil with the biocomposite also caused a significant reduction in the growth inhibition of *Brassica juncea* (L.) Czern and enhanced biological activity of the soil as seen in dehydrogenase activity and soil bacteria count increase. The study presented a promising, economically feasible and eco-friendly strategy with good potential for decontaminating soil polluted with imazethapyr and imazamox.

Fernández-Sanromán et al. (2020) studied the adsorption of sulfamethoxazole and methylparaben using hydrocolloid and fiber industry wastes. To synthesize the biocomposite, six different wastes were used as raw materials for the production of biochar which was subsequently modified with laccase via an enzymatic process. Amongst the different waste studied, lemon and lime exhibited the best results with maximum removal levels of up to 23% and 48% for methylparaben and sulfamethoxazole respectively. Its biocomposite achieved maximum uptakes ten times that of the raw material with the adsorption following pseudo-second-order kinetic model and maximum uptakes of 24.06 and 23.15 mg/g observed for sulfamethoxazole and methylparaben respectively. The adsorption isotherms of both the raw material and the biocomposite followed that of the Freundlich model. Results obtained in the study demonstrated the feasibility of the biocomposite in treating pharmaceuticals and personal care products in an aquatic environment.

Kaur and Kaur (2019) also examined the application of chitosan-β-cyclodextrin biocomposites in the eco-friendly decontamination of soils polluted with imazethapyr and imazamox. The study employed ultrasonic assisted technique in synthesizing the biocomposite and it was characterized using UV-Visible spectrophotometer. The pollutants studied were quantified by chromatography tandem mass spectrometry. Parameters which include type and concentration of extractant, contact time, liquid to soil ratio, temperature and sequential extraction cycle were all examined for their role in treating the contaminated soil using the biocomposite.

The optimal conditions for the decontamination of the soils studied were 10 mM biocomposite for alfisol and vertisol, 1 mM biocomposite for inceptisol 1, aridisol, inceptisol 2, entisol and inceptisol at liquid to soil ratio of 10:1, contact time of 6 h, temperature of 45°C and extraction cycle of 2. Under these conditions, 59.42%–99.44% of the pollutants were successfully removed from the soils studied using the biocomposites at 0.01–10 µg/mL initial concentration of pollutant. The differences observed in the removal of the pollutants might be due to different physico-chemical properties of soils which influenced the pollutant-soil interactions. The study was able to demonstrate the effectiveness of the biocomposite as an eco-friendly agent in the decontamination of soils polluted with imazethapyr and imazamox.

14.9 INORGANIC POLLUTANTS

14.9.1 METALS

Aftab et al. (2020) studied the use of *Aspergillus flavus* NA9 based biocomposite in the removal of zinc ions from synthetic and real wastewater. The study synthesized four types of biocomposites which were characterized using scanning electron microscopy, X-ray diffraction and Fourier transform infrared spectroscopy. The study further investigated the synthesized biocomposites for the removal and recovery of Zn(II) in industrial effluent using batch reactors and column experiments. In the concentration range of 100–600 mg of Zn(II)/L, maximum uptake capacity of (324 ± 2.1) mg Zn(II)/g was achieved by *Aspergillus flavus* NA9 immobilized glutaraldehyde cross-linked calcium alginate beads. The chemistry of Zn and FGCAB interaction was studied using SEM–EDX and XRD analysis. The analysis of the removal mechanism in the batch bioreactors show it follows the Freundlich isotherm.

Using 0.01 M HCl, a maximum of 89.6% for Zn(II) recovery was recorded with the biocomposites in the column studies. The biocomposite investigated exhibited good potential for treating Zn(II) contaminated industrial waste stream.

Abu-danso et al. (2020) synthesized a clay-cellulose biocomposite for toxic metals ion removal from aqueous medium. To prepare the biocomposite, spin and pressure-induced heating was used to combine exfoliated clay tubules and cellulose with polyethylene glycol as an intermediate. Using NaOH at high temperature, the resulting biocomposite was further modified to attain negative surface charge on the biocomposite. The biocomposite was then characterized to assess their physico-chemical properties using different techniques of Fourier transform infrared and X-ray photoelectron spectroscopy. The biocomposites were further examined for its efficiency in removing Pb(II) and Cd(II) from water using laboratory scale experiments. The adsorption mechanism of both pollutants followed that of pseudo-second-order kinetic model with 389.78 and 115.96 mg/g observed as the maximum Langmuir adsorption capacity for Pb(II) and Cd(II), respectively. Results of the study show that the biocomposite can be used as efficient adsorbents for the removal of metals such as Pb(II) and Cd(II) in contaminated water.

Kiran Aftab et al. (2014) used biocomposites in a batch and column study to examine the remediation of Pb(II) in industrial effluents. The study synthesized and carried out a comparative analysis of calcium alginate beads (CABs), glutaraldehyde-crosslinked calcium alginate beads (GCCABs), *Aspergillus caespitosus* immobilized calcium alginate beads (ACABs) and *Aspergillus caespitosus* immobilized glutaraldehyde-crosslinked calcium alginate beads (AGCCABs) for the removal and recovery of Pb(II) using batch reactors followed and columns experiments. The biocomposites were characterized with Fourier transform infrared spectroscopy, porosity and surface area analysis. In batch reactor system, a maximum uptake capacity (q_{max} mg/g) of 398.4 ± 5.9, 413.8 ± 5.1, 651.6 ± 3.3 and 670 ± 2.5 was recorded in the batch reactor system for CAB, GCCAB, ACAB and AGCCAB, respectively. The adsorption mechanism for the metal was consistent with Langmuir isotherm model. The column studies showed that AGCCAB had the removal efficiency in separating the metal from the industrial effluent with a recovery of 92.8% using 0.01 M HCl. Furthermore, there was no significant loss in column efficiency when AGCCAB was used to perform three cycles of sorption–desorption studies. The study demonstrated the beneficial use of *Aspergillus caespitosus* immobilized glutaraldehyde-cross-linked calcium alginate beads in treating lead contaminated solution.

Jiao et al. (2017) examined the application of microcrystalline cellulose/MnO_2 biocomposite in removing lead from wastewater. The study used an in situ synthesis method to prepare the biocomposite. The biocomposite was characterized using scanning electron microscopy, energy-dispersive X-ray spectroscopy, Fourier-transform infrared spectroscopy and X-ray diffraction analyses. These characterization techniques showed that the biocomposite was coated evenly with birnessite-MnO_2 nanoparticles. A batch system was used to study the adsorption efficiency of the biocomposite and results showed it was efficient for removing Pb(II) in aqueous systems with factors such as pH value, initial concentration of pollutant, solution temperature, and contact time influencing the adsorption. Further study showed that the adsorption mechanism followed that of pseudo-second-order kinetic model and The Langmuir isotherm with a maximum capacity of 247.5 mg/g recorded at 313 K. The thermodynamic analyses showed that the adsorption was spontaneous and endothermic in nature. After five sequential adsorption–desorption cycles, the adsorption capacity of the biocomposite remained at 89.6% of its initial value as shown by the regeneration study. All the results obtained in the study demonstrate the efficiency of the biocomposite as a renewable adsorbent for the removal of lead from wastewater.

14.9.2 Nonmetal

Pandi and Viswanathan (2015) studied the remediation of fluoride using a biocomposite material. The material was made up of montmorillonite and chitosan biopolymeric matrix. It was characterized by scanning electron microscopy, energy-dispersive X-ray spectroscopy and Fourier-transform infrared spectroscopy. Its potential to remove fluoride in water was examined and optimized considering various parameters such as contact time, pH, competing co-ions, temperature and initial concentration of pollutant. Results obtained show that the biocomposite had a defluoridation capacity of 1832 mgF$^-$/kg and the sorption was influenced by medium pH and vaguely by HCO_3^- ion. The adsorption mechanism followed the Langmuir isotherm model with thermodynamic parameters signifying it was spontaneous and endothermic in nature. Results of the study indicate that the biocomposite has good potential for success as an eco-friendly defluoridating material.

B.H.N. Bhatti et al. (2017) investigated the removal of phosphate ions in water using mango stone biocomposite. The efficiency of the biocomposite was examined considering the influence of parameters such as pH, biocomposite dose, contact time, initial concentration of pollutant and temperature. The optimum conditions for the maximum removal of the phosphate ions was at pH of 2, biocomposite dose of 0.3 g, contact time of 90 min and initial pollutant concentration of 200 mg/L. At these conditions, 95 mg/g of phosphate ions was adsorbed. Further investigation was carried out to determine the effect of pre-treating the biocomposite with surfactants (SDS, Tween-80, C-TAB, VIM and Surf excel), and results showed they reduced the adsorption capacity of the biocomposite. The adsorption process was spontaneous and endothermic in nature as indicated by thermodynamic study with the adsorption mechanism following that of the Freundlich isotherm and pseudo-second-order kinetic model. Furthermore, 1.0 M solution of NaOH effectively desorbed the adsorbed pollutants. Results obtained in the study indicate that biocomposite efficiently remove phosphate ions in water, with its application extendable to other inorganic ions.

14.10 CONCLUSION

Through cutting-edge research, biocomposites have been used in food packages application, thus, replacing the conventional plastics materials. Biocomposites materials have been successfully utilized in environmental remediation. Their choice is informed by the inherent characteristics towards environmental remediation. These include, anti-bacterial effects, chelating ability, adsorption potentials among others. Their effectiveness lies in the ability to transfer load from the matrix to the reinforcement phase. The matrix phase can be made of biodegradable and nonbiodegradable polymers.

We have explored the potential applications of biocomposites in reducing environmental problems. The use of biocomposites in treating pollutants and the various environmental remediation techniques have been X-rayed. Even though new areas of applications are likely to evolve in the nearest future, hence making the search inconclusive.

REFERENCES

Abhilash, P. C., Rama Kant Dubey, Vishal Tripathi, Pankaj Srivastava, Jay Prakash Verma, and H. B. Singh. 2013. "Remediation and management of POPs-contaminated soils in a warming climate : Challenges and perspectives." *Environmental Science and Pollution Research* 20: 5879–85.

Abu-danso, Emmanuel, Sirpa Peräniemi, Tiina Leiviskä, Taeyoung Kim, Kumud Malika Tripathi, and Amit Bhatnagar. 2020. "Synthesis of clay-cellulose biocomposite for the removal of toxic metal ions from aqueous medium." *Journal of Hazardous Materials* 381: 120871.

Aftab, Kiran, K. Akhtar, and A. Jabbar. 2014. "Batch and column study for Pb-II remediation from industrial Ef Fl uents using glutaraldehyde – alginate – fungi biocomposites." *Ecological Engineering* 73: 319–25.

Aftab, Kiran, Kalsoom Akhtar, Muzammil Hussain, and Kinza Aslam. 2020. "Synthesis, characterization and application of bio: Composites based on aspergillus flavus NA9 for extraction of zinc ions from synthetic and real waste water effluents." *Journal of Polymers and the Environment* 28(0123456789): 1441–49.

Aigbodion, V. S. 2019. "Bean pod ash nanoparticles a promising reinforcement for aluminium matrix biocomposites." *Journal of Material Research and Technology* 8(6): 6011–20.

Alharbi, Omar M. L., Al Arsh Basheer, Rafat A. Khattab, and Imran Ali. 2018. "Health and environmental effects of persistent organic pollutants." *Journal of Molecular Liquids* 263: 442–53.

Alimba, Chibuisi Gideon, and Caterina Faggio. 2019. "Microplastics in the marine environment : Current trends in environmental pollution and mechanisms of toxicological profile." *Environmental Toxicology and Pharmacology* 68:61–74.

AL-Oqla, Faris M., Ahmad Almagableh, and Mohammad Omari. 2017. "Design and fabrication of green biocomposites," pp. 45–67. In *Green Energy and Technology*, edited by M. Jawaid, M. S. Salit, and O. Y. Alothman. Springer: Berlin, Germany.

Amorim, Vieira Fernanda, José Rafael Padilha, Maria Glória Vinhas, Ramos Márcia Luiz, Costa Neyliane De Souza, and Yêda De Almeida, Medeiros. 2021. "Development of hydrophobic polyurethane/castor oil biocomposites with agroindustrial residues for sorption of oils and organic solvents." *Journal of Colloid and Interface Science* 581: 442–54.

Ariadurai, Samuel. 2012. "Bio-composites : Current status and future trends." in *5th International Technical Textiles Conference,* Turkey, pp. 1–16.

Ariffin, Mariani, and Siti Norhafizah Sulaiman. 2015. "Regulating sewage pollution of malaysian rivers and its challenges." *Procedia Environmental Sciences* 30: 168–73.

Ashraf, Muhammad, Ibrahim Khan, Muhammad Usman, Abuzar Khan, Syed Shaheen Shah, Abdul Zeeshan Khan, Khalid Saeed, Muhammad Yaseen, Muhammad Fahad Ehsan, Muhammad Nawaz Tahir, and Nisar Ullah. 2020. "Hematite and magnetite nanostructures for green and sustainable 2 energy harnessing and environmental pollution control: A review 1." *Chemical Research in Toxicology*, 33(6): 1292–311.

Banga, Honey, V. K. Singh, and Sushil Kumar Choudhary. 2015. "Fabrication and study of mechanical properties of bamboo fibre reinforced bio-composites." *Innovative Systems Design and Engineering* 6(1): 84–99.

Bashir, Muhammad Farhan, Maroua Benghoul, Umar Numan, Awais Shakoor, Bushra Komal, Muhammad Adnan Bashir, Madiha Bashir, and Duojiao Tan. 2020. "Environmental pollution and COVID-19 outbreak : Insights from Germany." *Air Quality, Atmosphere & Health* 13: 1385–1394.

Bharagava, Ram Naresh, Gaurav Saxena, and Sikandar I. Mulla. 2020. "Introduction to industrial wastes containing organic and inorganic pollutants and bioremediation approaches for environmental management," pp. 1–18. In *Bioremediation of Industrial Waste for Environmental Safety*, edited by G. Saxena and R. N. Bharagava. Springer: Singapore.

Bhatti, Haq Nawaz, Javeria Hayat, Munawar Iqbal, Saima Noreen, and Sadia Nawaz. 2017. "Biocomposite application for the phosphate ions." *Integrative Medicine Research* 7(3): 300–307.

Bhatti, Nawaz Haq, Zofishan Mahmood, Abida Kausar, Sobhy M. Yakout, Omar H. Shair, and Munawar Iqbal. 2020. "Biocomposites of polypyrrole, polyaniline and sodium alginate with cellulosic biomass : Adsorption-desorption, kinetics and thermodynamic studies for the removal of 2, 4-dichlorophenol." *International Journal of Biological Macromolecules* 153: 146–57.

Bian, Zhengfu, Hilary Inyang, John Daniels, Frank Otto, and Sue Struthers. 2010. "Environmental issues from coal mining and their solutions." *Mining Science and Technology* 20: 215–23.

Bilal, Muhammad, and Hafiz M. N. Iqbal. 2020. "Transportation fate and removal of microplastic pollution: A perspective on environmental pollution." *Case Studies in Chemical and Environmental Engineering* 2: 100015.

Bose, Bimal K. 2010. "Energy, environmental pollution, and the impact of power electronics." *IEEE Industrial Electronics Magazine* 4(1): 6–17.

Campagnolo, Laura, Davide Morselli, Davide Magrì, Alice Scarpellini, Cansunur Demirci, Massimo Colombo, Athanassia Athanassiou, and Despina Fragouli. 2018. "Silk fibroin/orange peel foam : An efficient biocomposite for water remediation." *Advanced Sustainable Systems* 3(1): 1800097.

Campos, V. M., I. Merino, R. Casado, L. F. Pacios, and L. Gómez. 2008. "Review. phytoremediation of organic pollutants 1." *Spanish Journal of Agricultural Research* 6: 38–47.

Carabajal, Maira, Mercedes Perullini, Matias Jobbagy, Rene Ullrich, Martin Hofrichter, and Laura Levin. 2016. "Removal of phenol by immobilization of trametes versicolor in silica–alginate–fungus biocomposites and loofa sponge." *Clean – Soil, Air, Water* 44(2): 180–8.

Chacón-Patiño, Martha L., Cristian Blanco-tirado, Juan P. Hinestroza, and Marianny Y. Combariza. 2013. "Biocomposite of nanostructured MnO_2 and fique fibers for efficient dye degradation." *Green Chemistry* 15(August): 2920–28.

Chu, Thi Thu Ha. 2018. "Environmental pollution in Vietnam : Challenges in management and protection." *Journal of Vietnamese Environment* 9(1): 1–3.

Cosgrove, William J., and Daniel P. Loucks. 2015. "Water management: Current and future challenges and research directions." *Water Resources Research* 51: 4823–39.

D'Souza, Nandika Anne, Michael S. Allen, Kevin Stevens, Brian Ayre, David K. Visi, Shailesh Vidhate, Iman Ghamarian, and Charles L. Webber. 2011. "Biocomposites: The natural fiber contribution from bast and woody plants," pp. 75–95. In *Plant Fibers as Renewable Feedstocks for Biofuel and Bio-Based Oroducts*. CCG International Inc: St. Paul.

Dhankhar, Rajesh, and Anju Hooda. 2011. "Fungal biosorption: An alternative to meet the challenges of heavy metal pollution in aqueous solutions." *Environmental Technology* 32(5): 467–91.

Duarte, Kátia R., Ana C. Freitas, Ruth Pereira, Jorge C. Pinheiro, Fernando Gonçalves, H. Azaari, Mohammed El Azzouzi, Abdallah Zrineh, Souad Zaydoun, Armando C. Duarte, and Teresa A. P. Rocha-Santos. 2012. "Treatment of olive oil mill wastewater by silica – alginate – fungi treatment of olive oil mill wastewater by silica – alginate – fungi biocomposites." *Water, Air, & Soil Pollution* 223(August): 4307–18.

Duarte, Katia, Celine I. L. Justino, Ruth Pereira, Ana C. Freitas, Teresa S. L. Panteleitchouk, A. P. Rocha-Santos, and Armando C. Duarte. 2013. "Removal of the organic content from a bleached Kraft pulp mill effluent by a treatment with silica-alginate- fungi biocomposites removal of the organic C." *Journal of Environmental Science and Health, Part A : Toxic/Hazardous Substances and Environmental Engineering* 48(2): 166–72.

Duarte, K. R., C. Justino, T. Panteleitchouk, A. Zrineh, A. C. Freitas, A. C. Duarte, and T. A. P. Rocha-Santos. 2014. "Removal of phenolic compounds in olive mill wastewater by silica – alginate – fungi biocomposites." *International Journal of Environmental Science and Technology* 11: 589–96.

Dungani, Rudi, Myrtha Karina, Subyakto, A. Sulaeman, Dede Hermawan, and A. Hadiyane. 2016. "Agricultural waste fibers towards sustainability and advanced utilization : A review." *Asian Journal of Plant Sciences* 15(1): 42–55.

Essabir, H., S. Nekhlaoui, M. Malha, M. O. Bensalah, F. Z. Arrakhiz, A. Qaiss, and R. Bouhfid. 2013. "Bio-composites based on polypropylene reinforced with almond shells particles : Mechanical and thermal properties." *Materials and Design* 51: 225–30.

Falciglia, Pietro P., Paolo Roccaro, Lorenzo Bonanno, Guido De Guidi, Federico G. A. Vagliasindi, and Stefano Romano. 2018. "A review on the microwave heating as a sustainable technique for environmental remediation/detoxification applications." *Renewable and Sustainable Energy Reviews* 95(July): 147–70.

Farid, Sajid, Musa Kaleem Baloch, and Syed Amjad Ahmad. 2012. "Water pollution : Major issue in urban areas." *International Journal of Water Resources and Environmental Engineering* 4(3): 55–65.

Faruk, Omer, Nalan Oya, San Keskin, and Asli Celebioglu. 2017. "Bacteria encapsulated electrospun nanofibrous webs for remediation of methylene blue dye in water." *Colloids and Surfaces B: Biointerfaces* 152: 245–51.

Fernández-Sanromán, Ángel, Valeria Acevedo-García, Marta Pazos, Angeles Sanromán, and Emilio Rosales. 2020. "Removal of sulfamethoxazole and methylparaben using hydrocolloid and fiber industry wastes: Comparison with biochar and laccase-biocomposite." *Journal of Cleaner Production* 271: 122436.

Ferraro, Gianluca, and Pierre Failler. 2020. "Governing plastic pollution in the oceans : Institutional challenges and areas for action." *Environmental Science and Policy* 112: 453–60.

Flores-chaparro, Carlos E., Mayra C. Rodriguez-hernandez, Luis Felipe Chazaro-ruiz, M. Catalina Alfaro-De la Torre, Miguel A. Huerta-diaz, and Jose Rene Rangel-mendez. 2018. "Chitosan-macroalgae biocomposites as potential adsorbents of water- soluble hydrocarbons : Organic matter and ionic strength effects." *Journal of Cleaner Production* 197: 633–42.

Fuqua, Michael A., Shanshan Huo, and Chad A. Ulven. 2012. "Natural fiber reinforced composites." *Polymer Reviews* 52: 259–320.

Gaur, Nisha, Korrapati Narasimhulu, and Y. Pydisetty. 2018. "Recent advances in the bioremediation of persistent organic pollutants and its effect on environment." *Journal of Cleaner Production* 198: 1602–31.

Gavrilescu, Maria, Katerˇina Demnerova, Jens Aamand, Spyros Agathos, and Fabio Fava. 2015. "Emerging pollutants in the environment : Present and future challenges in biomonitoring, ecological risks and bioremediation." *New Biotechnology* 32(1): 147–56.

Gurunathan, T., Smita Mohanty, and Sanjay K. Nayak. 2015. "A review of the recent developments in biocomposites based on natural fibres and their application perspectives." *Composites Part A* 77: 1–25.

Gusain, Rashi, Kanika Gupta, Pratiksha Joshi, and Om P. Khatri. 2019. "Adsorptive removal and photocatalytic degradation of organic pollutants using metal oxides and their composites: A comprehensive review." *Advances in Colloid and Interface Science* 272: 102009.

Hagos, Kiros, Jianpeng Zong, Dongxue Li, Chang Liu, and Xiaohua Lu. 2017. "Anaerobic co-digestion process for biogas production : Progress, challenges and perspectives." *Renewable and Sustainable Energy Reviews* 76: 1485–96.

Hamadan, Henna, Shahzada Mudasir Rashid, Muneeb U. Rehman, Rameez Ali, Masrat Rashid, Manzoor ur Rahman, Ishraq Hussain, Gowhar Gul, and Zulfiqar ul Haq. 2020. "Global scenario of remediation techniques to combat environmental pollution," pp. 93–106. In *Bioremediation and Biotechnology*, edited by R. A. Bhat, H. Qadri, and K. R. Hakeem. Springer: Cham.

Han, Dongmei, Matthew J. Currell, and Guoliang Cao. 2016. "Deep challenges for China's war on water pollution." *Environmental Pollution* 1–12.

Ishtiaq, Faiza, Haq Nawaz, Amina Khan, Munawar Iqbal, and Abida Kausar. 2020. "Polypyrole, polyaniline and sodium alginate biocomposites and adsorption-desorption efficiency for imidacloprid insecticide." *International Journal of Biological Macromolecules* 147: 217–32.

Jerold, M., K. Vasantharaj, D. Joseph, and V. Sivasubramanian. 2017. "Fabrication of hybrid biosorbent nanoscale zero-valent iron- Sargassum swartzii biocomposite for the removal of crystal violet from aqueous solution fabrication of hybrid biosorbent nanoscale zero-valent iron-Sargassum swartzii biocomposite for the removal." *International Journal of Phytoremediation* 19(3):214–24.

Jiao, Chenlu, Jin Tao, Jiaqing Xiong, Xiaojuan Wang, Desuo Zhang, Hong Lin, and Yuyue Chen. 2017. "In situ synthesis of MnO_2-loaded biocomposite based on microcrystalline cellulose for Pb_2+ removal from wastewater." *Cellulose* 24: 2591–604.

Johnson, R., Deepak Joel, V. Arumuga Prabu, P. Amuthakkannan, and K. Arun Prasath. 2017. "A review on biocomposites and bioresin based composites for potential industrial applications." *Reviews on Advanced Material Science* 48: 112–21.

Jung, Se Ji, Jodhbir S. Mehta, and Louis Tong. 2018. "Effects of environment pollution on the ocular surface the ocular surface effects of environment pollution on the ocular surface." *The Ocular Surface* 16(2): 198–205.

Kanmani, P., J. Aravind, M. Kamaraj, P. Sureshbabu, and S. Karthikeyan. 2017. "Environmental applications of chitosan and cellulosic biopolymers : A Comprehensive Outlook." *Bioresource Technology* 242: 295–303.

Karami, Ali, and Zulkifli H. Shamsuddin. 2010. "Phytoremediation of heavy metals with several efficiency enhancer methods." *African Journal of Biotechnology* 9(25): 3689–98.

Katole, Shrikant B., Puneet Kumar, and Rajendra D. Patil. 2013. "Environmental pollutants and livestock health: A review." *Veterinary Research International Journal* 1(1): 1–13.

Kaur, Paawan, and Pervinder Kaur. 2019. "β-cyclodextrin-chitosan biocomposites for synergistic removal of imazethapyr and imazamox from soils: Fabrication, performance and mechanisms." *Science of the Total Environment* 710: 135659.

Kaur, Paawan, Pervinder Kaur, Navjyot Kaur, Deepali Jain, Kuldip Singh, and Makhan Singh Bhullar. 2020. "Dissipation and phytotoxicity of imazethapyr and imazamox in soils amended with β-cyclodextrin-chitosan biocomposite." *Science of the Total Environment* 735: 139566.

Khalid, Sana, Muhammad Shahid, Nabeel Khan Niazi, Behzad Murtaza, Irshad Bibi, and Camille Dumat. 2016. "A comparison of technologies for remediation of heavy metal contaminated soils." *Journal of Geochemical Exploration* 182: 247–68.

Khan, Mashhood Ahmad, and Arsalan Mujahid Ghouri. 2011. "Environmental pollution: Its effects on life and its remedies". *Journal of Arts, Science & Commerce* 2(2): 276–85.

Kispotta, Usha Gracy. 2011. "*Synthesis and Characterization of Bio-Composite Material.*" National Institute of Technology: Rourkela.

Kumar, Sunil, Stephen R. Smith, Geoff Fowler, Costas Velis, S. Jyoti Kumar, Shashi Arya, Rakesh Kumar, and Christopher Cheeseman. 2017. "Challenges and opportunities associated with waste management in India." *Royal Society Open Science* 4(3): 160764.

Kuppusamy, Saranya, Thavamani Palanisami, Mallavarapu Megharaj, Kadiyala Venkateswarlu, and Ravi Naidu. 2016. "Ex-situ remediation technologies for environmental pollutants: A critical perspective." *Reviews of Environmental Contamination and Toxicology* 236: 117–92.

Lala, Sumit Das, Ashish B. Deoghare, and Sushovan Chatterjee. 2018. "Effect of reinforcements on polymer matrix bio-composites: An overview." *Science and Engineering of Composite Materials* 25(6): 1039–58.

Lawanson, Taibat, Omoayena Yadua, and Idris Salako. 2012. "Environmental challenges of peri-urban settlements in the lagos megacity."

Leung, Dennis Y. C. 2015. "Outdoor-indoor air pollution in urban environment : Challenges and opportunity." *Frontiers in Environmental Science* 2(69): 1–7.

Li, Yufei, Xiaoyu Du, Chao Wu, Xueying Liu, Xia Wang, and Ping Xu. 2013. "An efficient magnetically modified microbial cell biocomposite for carbazole biodegradation." *Nanoscale Research Letters* 8(522): 1–5.

Li, Shengfang, Min Tao, and Yongdi Xie. 2016. "Reduced graphene oxide modified luffa sponge as a biocomposite adsorbent for effective removal of cationic dyes from aqueous solution." *Desalination and Water Treatment* 57(42): 20049–57.

Lu, Xi, Shaojun Zhang, Jia Xing, Yunjie Wang, Wenhui Chen, Dian Ding, Shuxiao Wang, Lei Duan, and Jiming Hao. 2020. "Progress of air pollution control in china and its challenges and opportunities in the ecological civilization era." *Engineering* 6: 1423–1431.

Luka, Yusufu, Bitrus Kwaji Highina, and Abdu Zubairu. 2018. "Bioremediation : A solution to environmental pollution: A review." *American Journal of Engineering Research* 7(2): 101–9.

Mathew, Blessy Baby, Himani Singh, Vinai George Biju, and N. B. Krishnamurthy. 2017. "Classification, source, and effect of environmental pollutants and their biodegradation." *Journal of Environmental Pathology, Toxicology and Oncology* 36(1): 55–71.

Mitra, B. C. 2014. "Environment friendly composite materials : Biocomposites and green composites." *Defence Science Journal* 64(3): 244–61.

Mokthsim, Noranida, and Khairulmaini Osman Salleh. 2014. "Malaysia's efforts toward achieving a sustainable development: Issues, challenges and prospects." *Procedia-Social and Behavioral Sciences* 120: 299–307.

Nagarajan, Vidhya. 2012. "Sustainable biocomposites from 'green' plastics and natural fibers." The University of Guelph.

Nnaemeka, Abonyi Nichodemus. 2020. "Environmental pollution and associated health hazards to host communities (case study : Niger Delta Region of Nigeria)." *Central Asian Journal of Environmental Science and Technology Innovation* 1: 30–42.

Noreen, Saima, Bhatti Nawaz Haq, Munawar Iqbal, Fida Hussain, and Sarim Malik Fazli. 2020. "Chitosan, starch, polyaniline and polypyrrole biocomposite with sugarcane bagasse for the Ef Fi cient removal of acid black dye." *International Journal of Biological Macromolecules* 147: 439–52.

Obiechina, G. O., and R. Rimande Joel. 2018. "Water pollution and environmental challenges in Nigeria." *Educational Research International* 7(1): 109–17.

Okonkwo, Chukwuma Chris, Francis Odikpo Edoziuno, and Louis Chukwuemeka Orakwe. 2020. "Environmental nano-remediation in Nigeria : A review of its potentials." *Algerian Journal of Engineering and Technology* 3: 43–57.

Ossai, Innocent Chukwunonso, Aziz Ahmed, Auwalu Hassan, and Fauziah Shahul Hamid. 2020. "Remediation of soil and water contaminated with petroleum hydrocarbon : A review." *Environmental Technology & Innovation* 17: 100526.

Pandi, Kalimuthu, and Natrayasamy Viswanathan. 2015. "Remediation of fluoride using Montmorillonite@Chitosan biocomposite." *Journal of Chitin and Chitosan Science* 3(March): 1–7.

Peças, Paulo, Hugo Carvalho, Hafiz Salman, and Marco Leite. 2018. "Natural fibre composites and their applications : A review." *Journal of Composite Science* 2(66): 1–20.

Perullini, Mercedes, Jobbagy Matias, Mouso Nora, Forchiassin Flavia, and Bilmes Sara A. 2010. "Silica-alginate-fungi biocomposites for remediation of polluted water." *Journal of Materials Chemistry* 20(August): 6479–6483.

Qureshi, Muhammad Imran, Amran Rasli, Usama Awan, Jian Ma, Ghulam Ali, Faridullah, Arif Alam, Faiza Sajjad, and Khalid Zaman. 2015. "Environment and air pollution : Health services bequeath to grotesque menace." *Environmental Science and Pollution Research* 22(5): 3467–76.

Raheem, Abdul, Vineet Singh Sikarwar, Jun He, Wafa Dastyar, Dionysios D. Dionysiou, Wei Wang, and Ming Zhao. 2018. "Opportunities and challenges in sustainable treatment and resource reuse of sewage sludge : A review." *Chemical Engineering Journal* 337: 616–41.

Sajid, Zubia, Madiha Rafiq, and Farwa Nadeem. 2018. "Natural biocomposites for removal of hazardous coloring matter from wastewater : A review." *International Journal of Chemical and Biochemical Sciences* 13: 76–91.

Silva, Ana L. Patrício, Joana C. Prata, Tony R. Walker, Armando C. Duarte, Wei Ouyang, Damià Barceló, and Teresa Rocha-santos. 2021. "Increased plastic pollution due to COVID-19 pandemic : Challenges and recommendations." *Chemical Engineering Journal* 405: 126683.

Song, Dan, Rong Yang, Feng Long, and Anna Zhu. 2019. "Applications of magnetic nanoparticles in Surface-Enhanced Raman Scattering (SERS) detection of environmental pollutants." *Journal of Environmental Sciences (China)* 80: 14–34.

Španić, Nikola, Vladimir Jambrekovi, Milan Šernek, and Sergej Medved. 2019. "Influence of natural fillers on thermal and mechanical properties and surface morphology of cellulose acetate- based biocomposites." *International Journal of Polymer Science* 2019: 17.

Sui, Kunyan, Yujin Li, Rongzhan Liu, Yang Zhang, Xin Zhao, Hongchao Liang, and Yanzhi Xia. 2012. "Biocomposite fiber of calcium alginate/multi-walled carbon nanotubes with enhanced adsorption properties for ionic dyes." *Carbohydrate Polymers* 90(1): 399–406.

Testoni, Guilherme Apolinario. 2015. "In situ long-term durability analysis of biocomposites in marine environment." Ecole Nationale Supérieure des Mines de Paris.

Ubaid, Muhammad, Ali Guijian, Liu Balal, Yousaf Habib, Qumber Abbas, Mehr Ahmad, and Mujtaba Munir. 2018. "A systematic review on global pollution status of particulate matter-associated potential toxic elements and health perspectives in urban environment." *Environmental Geochemistry and Health* 41: 1131–62.

Urbano, Bruno F., Saúl Bustamante, Daniel A. Palacio, Myleidi Vera, and Bernabé L. Rivas. 2020. "Polymer supports for the removal and degradation of hazardous organic pollutants : An overview." *Polymer International* 69(4): 333–45.

Verma, D., P. C. Gope, M. K. Maheshwari, and R. K. Sharma. 2012. "Bagasse fiber composites: A review." *Journal of Materials and Environmental Science* 3(6): 1079–92.

Verma, Priyanshu, and Jatinder Kumar Ratan. 2020. *Assessment of the Negative Effects of Various Inorganic Water Pollutants on the Biosphere: An Overview*. Elsevier: Amsterdam.

Vinod, A., M. R. Sanjay, Siengchin Suchart, and Parameswaranpillai Jyotishkumar. 2020. "Renewable and sustainable biobased materials : An assessment on biofibers, biofilms, biopolymers and biocomposites." *Journal of Cleaner Production* 258: 120978.

Wang, Shuxiao, and Jiming Hao. 2012. "Air quality management in China: Issues, challenges, and options." *Journal of Environmental Sciences* 24(1): 2–13.

Yang, Xue, Shiqiu Zhang, Meiting Ju, and Le Liu. 2019. "Preparation and modification of biochar materials and their application in soil remediation." *Applied Sciences* 9(7): 1365.

Zia, Z., A. Hartland, and M. R. Mucalo. 2020. "Use of low: Cost biopolymers and biopolymeric composite systems for heavy metal removal from water." *International Journal of Environmental Science and Technology*. doi: 10.1007/s13762-020-02764-3.

Zini, Elisa, and Mariastella Scandola. 2011. "Green composites : An overview." *Polymer Composites* 32(12): 1905–15.

Index

Note: **Bold** page numbers refer to tables and *italic* page numbers refer to figures.

Aaen, R. 147
abaca fiber 12, 13
ABACUS 167
Abbott, A. P. 257
Abu-danso, Emmanuel 417
ACABs *see Aspergillus caespitosus* immobilized calcium alginate beads (ACABs)
ACC *see* All Cellulose Composites (ACC)
accelerated weathering 41–42, **42,** *42*
acceptable mechanical properties, chitosan-based biocomposites 291–294, *293*
acetylation 132–133, 366
acid hydrolysis 132
acidic ionic liquid (AIL) 220, *222*
acrylation 372
acrylonitrile grafting 372
Adamus 276
additives 326
adsorption material 145
adsorption mechanism 412, 415, 417, 418
aerobic biodegradation 37, *41*
aerogels 231–233
Aftab, Kiran 416, 417
AGCCABs *see Aspergillus caespitosus* immobilized glutaraldehyde-crosslinked calcium alginate beads (AGCCABs)
AgNPs *see* silver (Ag) NPs
agricultural waste 253, 404
agro-based bio fibers 90
AIL *see* acidic ionic liquid (AIL)
air pollution 402, 407
Akhavan-Kharazian, N. 292
Akhlaghi, S. P. 147
Albahash, Z. F. 109
alkaline pretreatment **219,** 254, **255**
alkaline treatment 132, 180
alkali treatment 366
All Cellulose Composites (ACC) 235–237, *236*
alloy 169
Alpár, T. 18, 25
Alsaeed, T. 132
Alvarez-Vasco, C. 260
Amalraj, A. 294
amber acid 204
amido-sulfonic acid (ASA) 232
Amoah, J. 199
Amorim, Vieira Fernanda 413
amylopectin 322

amylose 322
anaerobic biodegradation 37, 53
Ananas comosus 12
Anderson–Darling test 98
animal fibers **13,** 13–14
annual plants 12–13
ANSYS 162
antibacterial properties, chitosan-based biocomposites 294–296, *295*
anti-inflammation, chitosan-based biocomposites 296
antioxidant
  activity 275
  chitosan-based biocomposites 296
Archimedes' theory 71
Ariffin, Mariani 407
artificial fibers 20
ASA *see* amido-sulfonic acid (ASA)
aspartic acid-based platform 206
Aspergillus caespitosus 417
Aspergillus caespitosus immobilized calcium alginate beads (ACABs) 417
Aspergillus caespitosus immobilized glutaraldehyde-crosslinked calcium alginate beads (AGCCABs) 417
ASTM 5988-12, biodegradability in soil 40–41, *41*, 56
ASTM D570 72
ASTM D3822 76
ASTM D5868 77
ASTM D3039M-17 73
ASTM D7264M-15 74
atom transfer radical polymerization 373
Attia, M. A. 109
Augustine, R. 296
automotive industry, biocomposites in *22*, 22–23, *23*
Azetsu, A. 144
Azzi-Tasi-Hill theory 165, 166

Bacillus subtilis 206
Baekeland, Leo 3
bagasse fiber 12
Balaji, A. 94
banana fibers (BF) 53, *53,* **54,** 55
basic ionic liquid (BIL) 221, *222*
BASIL process 225
Bastami, F. 308
bast fibers 88, 89, 343

427

batch adsorption 411, 412
BCG economy *see* Bioeconomy, Circular economy and Green economy (BCG economy)
BCs *see* biocomposites (BCs)
Belaadi, A. 95
beta-tricalcium phosphate (β-TCP) 302–303
β-TCP/PLA (beta-ricalcium-phosphate/PLA) *21*, 21–22
BF *see* banana fibers (BF)
BG *see* bioactive glass (BG)
Bhatti, Haq Nawaz 412, 418
Bhuiyan, Md. T. R. 126
Bideau, B. 147
BIL *see* basic ionic liquid (BIL)
bioactive glass (BG) 295, 300
bioactivity 22
bio-based PE 8
bio-based polymer matrix 357, 359
biochar 416
biochemical process 145–146
biocomposites (BCs) 3–4, *4*, 26, 27, 126, 158, *159*, 181–182, 252, 342, 365, *365*, 385, 402
   advantages of 26–27
   alkaline treatment 104
   applications of
      automotive industry 22, 22–23, *23*
      construction industry 23–27, *24–25*
      medical applications 20–22, *21*,
   biological methods 373–374
   biopolymer matrix 161
   characterization
      flexural 180–181, *181*
      impact strength 181
      tensile 180, *180*
   chemical treatment methods 365–366
      acetylation 366
      acrylation and acrylonitrile grafting 372
      graft copolymerization 372
      maleated coupling 372
      mercerization 366
      silane treatment 366, **367–371**, 372
   classification of 403, *403*
   concept of *403*, 403–405, *405*
   crashworthiness (*see* crashworthiness, biocomposites)
   design of 161, *162*, 162–163
   engineering design of 158–159
      failure prediction (randomly oriented fiber laminates) 166
      failure prediction (un-notched lamina) 166
      failure prediction (unidirectional lamina) 163–166, *164*, **165**
      strength and stiffness 163
      theory of maximum stress 164, **165**
   fabrication techniques 169

calendering technique 170–171
clamping unit 174, 175
extruder die technique 173, *173*
extrusion technique *171–172*, 171–173
injection molding technique 173, 174, *174*
injection unit 175
mold tool 175–177, *176*
polymer preparation 169, 170, *170*
pressing and hot-pressing techniques 177–178, *179*
grafting via living polymerizaiton
   free radical grafting 372, 373
   ring opening 372, 373
inorganic pollutants
   metals 416–418
   nonmetal 418
interfacial bonding 104
manufacturing consumer products 104
materials for 5–6
materials modelling 166–167, *167*, *168*
   in manufacturing 167, 168, 169, *169*
matrix materials
   biopolymers 6–8, *7*
   fossil-based thermoplastics 8–9, *9*
   inorganic binders 10–11
   resins 9–10
military aircraft seat, design of 104, *105*
nanofiller addition on **375**
natural reinforcers
   animal-based fibers **13**, 13–14
   plant-based fibers 11–13, *12*
organic pollutants
   dyes 410–412
   hydrocarbons 413–414
   phenols 412–413
other organic pollutants 415–416
physical treatment 374
   compatibilizers and cross-linkers addition 376, **377**, 378, *378*
   nanoparticles addition 374, **375**, 376
from platform chemicals 199–203, *200–202*, **204**
production techniques 14
reinforcement/filler 160, *160*
silane coupling agents, use of 104
starch 329
   trends and challenges 331, 334
University of Sopron
   continuous fiber reinforced PLA biocomposite *16*, 16–17
   PLA nano composite 19–20, *20*
   wood-PLA biocomposites 14–16, *15*
   wood plastic composite 17–18, *18*
   wood wool cement boards 18–19, *19*
Weibull distribution of 95–96, **97**, *98*
biocomposites biodegradation
   biopolymer on 47–52, **48,** *49,* **50,** *50, 51*

Index

degradation methods *38,* 38–39
   accelerated weathering 41–42, **42,** *42*
   in soil, ASTM 5988-12 40–41, *41*
   soil burial degradation *39,* 39–40, **40**
   natural fibers on 42–47, *43, 44,* **45–46**
biocomposites fabrication, IL
   lignocellulosic biomass 226–227, **228–229**
     aerogels 231–233
     All Cellulose Composites 235–237, *236*
     biopolymer films 227, 230, 231
     hydrogel 233–235
biodegradability 37, 310
biodegradable plastics 38, 58
biodegradable polymers 5, 6, 161, **161,** 334, 404
   classification of 320, *321*
biodegradation 37
   biocomposites *38,* 38–39
   biopolymer on 47–52, **48,** *49,* **50,** *50, 51*
   natural fibers on 42–47, *43, 44,* **45–46**
   in soil, ASTM 5988-12 40–41, *41*
   soil burial degradation *39,* 39–40, **40**
   of synthetic polymer/biopolymer blends 53–57, *53–57,* **54**
bioeconomy, circular economy and green economy (BCG economy) 190, 192, 196, 199, 252, 253
bio-erodible polymer 6
bio-fiber-reinforced petroleum-derived polymer 403
biofibers 4, 11, 180, 182, 359, 360, 372
biofillers 159, 182
biofuels 135, 199, 208, 253
biological aspects, natural fibers 26
biological fiber 133–135
biological functions, chitosan-based biocomposites 308–310
biological polymers extraction 272–273, **273**
biological techniques
   biocomposites 373–374
   of environmental remediation 410
biological treatment **130,** 255, **255**
biomass 1–2, *3*
   natural reinforcers
     animal-based fibers **13,** 13–14
     plant-based fibers 11–13, *12*
biomass-based composites 66–67
   chemical characterization
     Fourier transform infrared spectroscopy 68, **68**
     nuclear magnetic resonance 68–69, *69*
     X-ray diffraction 69–70, *70*
   durability characterization
     creep testing *80,* 80–81, **81**
     environmental testing 82, 83
     fatigue testing 81, *82*
     wear testing 81, *82*
   mechanical characterization

compression test 73, 74, *75*
flexural test 74, *75*
impact test 75–76, *76*
single-fiber tensile test 76, *76*
tensile shear strength test 77
tensile strength 73, *74*
physical characterization
   density 71
   hardness test 72–73, *73*
   optical microscopy 70
   scanning electron microscopy 70–71
   void fraction 71–72, *72*
   water absorption test 72
test standards 83, **83**
thermal characterization
   differential scanning calorimetry 77–78
   differential thermal analysis 78, *79*
   dynamic mechanical analysis 78, 80
   thermogravimetric analysis 78, *79*
biomass cascading 26
biomass fractionation 262, *262*
bionanocomposite 411
biopolymer films 227, 230, 231
biopolymer matrix 161
biopolymers *6,* 53, 126, 320, 403
   on biocomposites biodegradation 47–52, **48,** *49,* **50,** *50, 51*
   as matrix materials *6,* 6–8, **7**
biorefineries 190–191, *191*
biorefining process
   platform chemicals 191, 193, 203
     1,4-diacid-based platform 204–205
     5-HMF-based platform 205
     itaconic acid-based platform 206
     lactic acid-based platform 207–208
     lignocellulosic biorefinery for 193–199, **194–195,** *196–198*
     sorbitol-based platform 206–207
biosensor application 146
biotransformation 260
black pepper essential oil (BPEO) 294
bleaching treatment 17, 133
Bledzki, A. 11, 15
blending 169
   of polymers 376, **377**
block copolymers 376
BMP-2 *see* bone morphogenetic protein- 2 (BMP-2)
Bodur, M. S. 366
bone marrow-derived mesenchymal stem cells (BMSCs) 304, 306, 309
bone morphogenetic protein- 2 (BMP-2) 307–310
bone tissue engineering
   chitosan-based biocomposites 300
     biological functions 308–310
     mechanical properties 300–303, *301*
     porosity 303–308
   requirement of 299–300

Boobphahom, S. 238
borohydride reduction method 410
Bose, Bimal K. 405
BPEO *see* black pepper essential oil (BPEO)
Braconnot, Henri 3
Brønsted acids 220
Brunauer-Emmett-Teller (BET) analysis 233
BTX (Benzene, Toluene and Xylene) 201
buckling failure, tubes 110–111, *110–111*

Cahú, T. B. 295
calcium alginate beads (CABs) 417
calcium peroxide (CP) 292
calendering technique *170,* 170–171
calibrating unit 173
Campagnolo, Laura 402
Capiati, N. J. 235
capillary rheometry 325
Carabajal, Maira 413
carbonaceous nanofillers 331, **333**
carbon black (CB) 331
carbon nanotubes (CNTs) 146, 300, 331
catalytic systems 145–146
catastrophic fracture 110, *111*
CB *see* carbon black (CB)
CBWP *see* cement-bonded wood-based product (CBWP)
CCF *see* cyclic conventional freezing (CCF)
cellulose (CEL) 193, *215,* 216, 227, 230, 231, 234–235, 252–253, 254, 290, 292, 305, 344, 363
 physical and chemical properties *217*
cellulose aerogel 232
cellulose-based plastics 6
cellulose biomass 208
cellulose-fiber-reinforced biopolymer 126
cellulose-hemicellulose network 344
cellulose microfibrils 349
cellulose nanocrystal (CNC) 146–148, 382
cellulose nano-fibres (CNWs) 331
cellulose nanofibrils 275
cellulose surface modification 131
cement-bonded wood-based product (CBWP) 10, 11
cereal straw 25
cetyltrimethylammonium bromide (CTAB) 144
CFN *see* cyclic freezing with liquid nitrogen (CFN)
Chacón-Patiño, Martha L. 411
characterization
 biocomposites
  flexural 180–181, *181*
  impact strength 181
  tensile 180, *180*
 biomass-based composites 66–67, *67,* 83
  (*see also* chemical characterization; durability characterization; mechanical characterization; physical characterization; thermal characterization)

Charpy test 75, *76*
Chattopadhyay, S. K. 57
chemical-based biopolymers 161
chemical catalytic process 193
chemical characterization
 biomass-based composites
  Fourier transform infrared spectroscopy 68, **68**
  nuclear magnetic resonance 68–69, *69*
  X-ray diffraction 69–70, *70*
chemical composition, of natural fibers 343–348, **345–346,** *347*
chemical fiber
 surface modification **128–129,** 131–132, 133
  acetylation 132–133
  acid hydrolysis 132
  alkaline treatment 132
  chemical treatment 133
  oxidizing agents 133
chemical oxygen demand (COD) 413
chemical pretreatment 254, **255**
chemical techniques, of environmental remediation 409
chemical treatment 90, 94–95
 biocomposites 365–366
  acetylation 366
  acrylation and acrylonitrile grafting 372
  graft copolymerization 372
  maleated coupling 372
  mercerization 366
  silane treatment 366, **367–371,** 372
Chen, F. 237
Chen, S. 307
Chen, T. 307
chicken feather, animal fiber 14
Chi-square 98
chitin 290
chitosan (CS) 290
chitosan-based biocomposites 297–301
 for bone tissue engineering 300
  biological functions 308–310
  mechanical properties 300–303, *301*
  porosity 303–308
 for wound dressings 291, **291**
  acceptable mechanical properties 291–294, *293*
  antibacterial properties 294–296, *295*
  anti-inflammation and antioxidant 296
  multifunctional properties 296–299, *297–299*
chitosan films 274–275, 277–278
chitosan-β-cyclodextrin biocomposites 416
Cho, C. S. 292
citric acid 206
clamping unit 174, 175
clay-cellulose biocomposite 417
clinoptilolite (CLN) 304–305
CLN *see* clinoptilolite (CLN)

Index

CMOS *see* complementary metal oxide semiconductor (CMOS)
CNC *see* cellulose nanocrystal (CNC)
C-13 NMR 68
CNTs *see* carbon nanotubes (CNTs)
CNWs *see* cellulose nano-fibres (CNWs)
Coccinia grandis fibers 95
COD *see* chemical oxygen demand (COD)
CODESSA program 221
coir *(Cocos nucifera)*, plant-based fibers 12
compatibility, of matrix 352
compatibilizers 47, 52, 376, **377,** 378, *378*
complementary metal oxide semiconductor (CMOS) 71
composite panels 25
composites 88, 342
compostable materials 38
compostable plastic 38
compostable polymers 5, 6
compound 169
compression molding 328
compression test 73, 74, *75*
consolidated processing 198, 199
construction industry
   nonstructural biocomposite *25,* 25–27
   structural biocomposite 23–25, *24*
continuous fiber reinforced PLA biocomposite *16,* 16–17
conventional methods 254, **255**
conventional plasticizers 273, 274
conventional sewage treatment 406–407
cordycepin (CY) 298
CP *see* calcium peroxide (CP)
C6 primary hydroxy group 141, 143, 144, 148
cracks 52
crashworthiness, biocomposites
   ABACUS software 116
   CC and CV specimens 116, *116, 117*
   characteristics
      analyzation parameters 106–107
      load-displacement curve 106, *106*
      military aircraft seat, design of 104, *105*
      structural integrity, evaluation of 104, 106
   compression behavior, composite tubes
      hybridization 109
      range of **108**
      triggering 109–110
      tube dimensions, effect of 108, 109
   failure modes 110–111
      multi-failure mode combinations 111, *112*
   finite element method (FEM) 111–113, *114*
   LSDYNA software 113, *114*
   non-linear explicit simulation 118–119, *119*
   parameters for, square crush box with different triggers 115–116, *116*
   progressive failure approach 111–112, *113*
   VUMAT's property down-grade 112, *113*
creatinine 238

creep testing *80,* 80–81, **81**
critical fiber length 348
cross-linkers 376, **377,** 378, *378*
cross-linking 324
   of gliadin-glutenin 364, *364*
cross-section morphology 90
crystalline 322
crystalline cellulose 349
crystalline polymer 80
crystallinity 69, 70, 126, 141, 146
CS *see* chitosan (CS)
CS ascorbate 296
CS-based electrospun biocomposites 293
CS-modified ZnO (CS-ZnO) 294
CS-ZnO *see* CS-modified ZnO (CS-ZnO)
CTAB *see* cetyltrimethylammonium bromide (CTAB)
Cuscuta reflexa-mediated AgNPs (CUS-AgNPs) 294–295
CY *see* cordycepin (CY)
cyclic conventional freezing (CCF) 233
cyclic freezing with liquid nitrogen (CFN) 233

DA *see* dehydroabietic acid (DA)
DCP *see* dicumyl peroxide (DCP)
deacetylation (DD) 290
deep eutectic solvent (DES) 255–260, *256,* **257,** **258,** *258, 259*
   vs. ionic liquid **257**
   lignocellulose
      dissolution and pretreatment of 260–267, **261,** *262,* **264–265**
   in polymeric composite synthesis 267–268, *268,* **269,** *270*
      lignocellulose 275–278
      natural biopolymeric composite 275–278
      plasticizer 273–275, **274**
      polymerization and polymer extraction 268, 270, **271–273,** 272, 273
   in polymerization and polymer extraction 268, 270, **271–273,** 272, 273
Defoirt, N. 92
degradable plastic 38
degradation methods, plastics *38,* 38–39
   accelerated weathering 41–42, **42,** *42*
   in soil, ASTM 5988-12 40–41, *41*
   soil burial degradation *39,* 39–40, **40**
dehydroabietic acid (DA) 373
Demir, A. K. 304
Demirtas 306
dendromass 2
Deng, N. 308
de-novo synthesized composite 201
density 71, 224
DES *see* deep eutectic solvent (DES)
designed solvent 220
DES-MIP (DES-molecular imprinted polymer) 272

detoxification process 198
1,4-diacid-based platform 204–205
dicumyl peroxide (DCP) 359, 373
dielectric constant 224
Diez-Pascual, A. M. 294
Difasol process 225
differential scanning calorimetry (DSC) 77–78
differential thermal analysis (DTA) 78, *79*
dimethyl sulfoxide (DMSO) 233, 234
Ding, L. 107
direct condensation 207
dispersive mixing 170, *170*
distributive mixing 170, *170*
DMA *see* dynamic mechanical analysis (DMA)
DMSO *see* dimethyl sulfoxide (DMSO)
Dong, L. 307
Dou, D. D. 309
drug delivery 147–148
DSC *see* differential scanning calorimetry (DSC)
DTA *see* differential thermal analysis (DTA)
Duarte, Kátia R. 413, 415
durability characterization
  biomass-based composites
    creep testing *80*, 80–81, **81**
    environmental testing 82, 83
    fatigue testing 81, *82*
    wear testing 81, *82*
durometer 72–73, *73*
dyes 410–412
dynamic mechanical analysis (DMA) 78, 80

Ehterami, A. 296
EIA *see* Energy Information Administration (EIA)
ejector pins 176
elastomers 174
electroplating 259
electrospinning 308
endothermic transition 78
Energy Information Administration (EIA) 1–2
engineering design, BCs 158–159
  failure prediction
    randomly oriented fiber laminates 166
    unidirectional lamina 163–166, *164*, **165**
    un-notched lamina 166
  strength and stiffness 163
  theory of maximum stress 164, **165**
environmental perspective, natural fibers 26
environmental pollutants
  classification of 407–409, *408*
  remediation techniques for 409–410
environmental pollution 402, 405–407, *406*
environmental remediation, techniques for 409–410
environmental testing 82, 83
environment friendly biocomposites 363
enzymatic saccharification 218

enzymatic treatment 134
enzymes 51, 134, 145, 199, 235, 290, 291, 362, 374, 411
enzyme treatment 374
epoxidized natural rubber 378
Erickson, A. E. 306
Eshkoor, R. 107
esterification 132, 146, 324, 366
estimators 92, 94, 98
Eucalyptus camaldulensis 266
exothermic transition 77
extruder 171, *171–172*
extruder die technique 173, *173*
extruders technology 5
extrusion technique *171–172*, 171–173, 326

fabrication 66
  of biocomposites 169
    calendering technique *170*, 170–171
    clamping unit 174, 175
    extruder die technique 173, *173*
    extrusion technique *171–172*, 171–173
    injection molding technique 173, 174, *174*
    injection unit 175
    ionic liquids 226–227, **228–229**
    mold tool 175–177, *176*
    polymer preparation 169, 170, *170*
    pressing and hot-pressing techniques 177–178, *179*
  deep eutectic solvent in 267
Failler, Pierre 407
failure prediction, BCs
  randomly oriented fiber laminates 166
  unidirectional lamina 163–166, *164*, **165**
  un-notched lamina 166
Faruk, Omer 11, 410
fatigue testing 81, *82*
FEA *see* finite element analysis (FEA)
feed zone 171
Fenton agent 362
fermentation 193
Fernández-Sanromán, Ángel 415
Ferraro, Gianluca 407
fiber aspect ratio 348
fiber-reinforced plastics (FRPs) 66
fiber-reinforced polymer composites 4
fibers 5, 15, 43, 47, 57, 58, 76, 104, 160, 162, 190, 342, 343, 347, 365, 378
fibers networks 146, 148
Fidelis, M. E. A. 90
filler materials 160
film casting 328
fingerprint region 68
finite element analysis (FEA) 159, 162–163, 167, *168*
finite element method (FEM) 111–113, *114*
flax fibers 90, 95, 132

Index

flax *(Linum usitatissimum)*, plant-based fibers 11
FlexForm Technologies 22
flexural characterization, biocomposites 180–181, *181*
flexural test 74, *75*
Flores-chaparro, Carlos E. 414
foaming extrusion 328
Folino, A. 57
food applications, surface modified cellulose 146–147
force-displacement curve, square crush-box 114, *115*
Ford, Henry 385
fossil-based thermoplastics 8–9, *9*
Fourier transform infrared spectroscopy (FTIR) 68, **68,** 141
fractionation, of lignocellulosic biomass *218,* 218–220, **219**
free radical grafting 372, 373
freeze-casting method 300
freeze-drying method 231, 303–305
Freundlich isotherms 412, 415–418
FRPs *see* fiber-reinforced plastics (FRPs)
FTIR *see* Fourier transform infrared spectroscopy (FTIR)
functional biomaterials 144–145
functional group region 68
fungi treatment 134

Gao, Y. 292
gas foaming 303
Gassan, J. 11
gate 175, 176
gauge length effect 92, **93–94,** 94
GCCABs *see* glutaraldehyde-crosslinked calcium alginate beads (GCCABs)
gelatin (GEL) 292, 295, 302
genetically modified bioplastics 8
genetic engineered microorganisms 198
Ghoushji, M. J. 107
ginger essential (GEO) 294
global buckling 110, *111*
global economic growth 252
glutaraldehyde-crosslinked calcium alginate beads (GCCABs) 417
glycerol 202–203
GO *see* graphite oxide (GO)
Godbole, S. 360
gold nanoparticle 144
graft copolymerization 372
grafting 144
　from method, hyperbranched polymers 381
　to method, hyperbranched polymers 381–382
　via living polymerizaiton
　　free radical grafting 372, 373
graphite oxide (GO) 331
Graupner, N. 96

Gravel, M. 303
Green Building structures 23
green composite-based products 182
green concept 385
green plant materials 158
green solvent 278
GTMAC *see* quaternary ammonium group (GTMAC)
guide pins 175
Gunti, R. 53, 56
Guo, M. 94
gypsum 11

HA *see* hydroxyapatite (HA)
Hahn's method 166
Haider, A. 294
Halász, K. 19
hand-lay-up 168
hardness test 72–73, *73*
Hashim, M. Y. 95
HDPE *see* high-density polyethylene (HDPE)
HDPSCs *see* human dental pulp stem cells (HDPSCs)
hemicellulose 195, *215,* 216, 347
hemp (*Cannabis* spp.), plant-based fibers 11
heteropolymers 204–205
high-density polyethylene (HDPE) 47, 49, 53, 55
5-HMF-based platform 205
H NMR *see* Hydrogen NMR (H NMR)
hot-pressing technique 177–178, *179*
HPMCS *see* hypromellose succinate (HPMCS)
Huang, K. 144
Huang, X. 363
Huber, T. 237
human dental pulp stem cells (HDPSCs) 309–310
Hussain, N. N. 113
hybrid composites 378, **379–380,** 381
hybridization 109
hydrocarbons 413–414
hydrogels 233–235, 298, 299
Hydrogen NMR (H NMR) 68, 69
hydrolysis process 52
hydrophilicity, natural fibers 349, 352
hydroxyapatite (HA) 301, 302, 304
hyperbranched polymers 381
　grafting from method 381
　grafting to method 381–382
hypromellose succinate (HPMCS) 292

imidazole-based ionic liquids 226, 227, 230
impact strength characterization, biocomposites 181
impact test 75–76, *76*
inconsistency, natural fibers 352, *353*
infrared (IR) radiation 68
inhibition process 18
injection molding technique 173, 174, *174,* 328

injection unit 175
inorganic binders 10–11
inorganic-bonded (cement or gypsum)
lignocellulose reinforced composites 14
inorganic pollutants 408
pollutant treatment, biocomposite
metals 416–418
nonmetal 418
in situ synthesis method 417
interference screw 21
ionic liquids (IL) 220–221, *221, 222,* 254, 382
applications of 225–226, *226*
biocomposites fabrication 226–227, **228–229**
aerogels 231–233
All Cellulose Composites 235–237, *236*
biopolymer films 227, 230, 231
hydrogel 233–235
vs. deep eutectic solvent **257**
integrated composites, in biotechnological applications 238–239
pretreatment method **219,** 227
properties of 221, 223, 224
IR *see* infrared (IR)
Ishtiaq, Faiza 415
isosorbide 207
itaconic acid 268
itaconic acid-based platform 206
Izod impact test 75, *76*

Jamalpoor, Z. 305
Japan Bioplastic Association 320
Jerold, M. 410
Jiang, H. 113
Jiang, Q. 292
Jiao, Chenlu 417
Jing, X. 308
jute 11

Kalambur, S. 362
Kanomata, K. 145
Karmakar, A. C. *378*
Kato, Y. 141
Kaur, Paawan 415, 419
Kaur, Pervinder 419
kenaf, plant-based fibers 11, 12, 55, 385
Khalili, R. 294
Kim, H. L. 302
Kim, H. S. 181
Kimura, M. 234
Koga, H. 146
Kolanthai, E. 304
Kolmogorov–Smirnov test 98
Komal, Ujendra Kumar 68
Köse, K. 145
Kövesi, A. 22
Kumar, A. P. 109

LA *see* lanthanum (LA)
lactic acid-based platform 207–208
land pollution 402, 406
Langmuir isotherm 235, 411, 412, 415, 417, 418
lanthanum (LA) 304
leaf fibers 89, 343
least-squares estimation 98
Lewis acid 220
LFT moulding 22
Liao, H. T. 309
lignin 195, *215,* 216–218, 254, 276, 347
lignocellulose *253,* 253–256, **255,** *256*
deep eutectic solvent 275–278
dissolution and pretreatment of 260–267, **261,** *262,* **264–265**
lignocellulosic biomass 214–215, *215,* 253, 268
biocomposites fabrication, IL 226–227, **228–229**
aerogels 231–233
All Cellulose Composites 235–237, *236*
biopolymer films 227, 230, 231
hydrogel 233–235
fractionation of *218,* 218–220, **219**
lignocellulosic biorefinery, for platform chemicals 193–199, **194–195,** *196–198*
lignocellulosic materials 5
Lila, Manish Kumar 68
lipid-based polymers 8
Li, Shengfang 411
Liu, Y. 263
living polymerizaiton 372, 373
Li, Yufei 414
load-bearing capacity test 18
load-displacement curve 106, *106, 118*
local buckling 110, *111*
locating ring 176
LSDYNA software 113, *114*
Lynam, J. G. 276
lyocell process 225–226
lyophilization 303

Mache, A. 110
Mahdi, E. 111, 117
maize starch 322
Maji, M. 302
maleated coupling 372
Maleic anhydride grafted polyethylene (PE-g-MA) 53
Mamalis, A. 111
Mamilla, J. L. K. 266
mango stone biocomposite 418
marine environment 407
Markó, G. 16, 17
Masruchin, N. 146
materials, for biocomposites 5–6
biopolymers 6–8, *7*
fossil-based thermoplastics 8–9, *9*

Index

inorganic binders 10–11
manufacturing 167, 168, 169, *169*
modelling 166–167, *167, 168*
resins 9–10
matrix 3–4, 126, 403, 404
  bio-based polymer matrix 357, 359
  natural-based polymer 357
    poly(β-hydroxyalkanoate) 359, 360, 361
    polyhydroxy lactic acid 357, 359
  petroleum-based polymer 352, 353, **354–356**
    poly(alkylene dicarboxylate) 357
    poly(ε-caprolactone) 357, **358**
  polysaccharides 360
    cellulose 363
    polypeptide based 363
    soya protein 363–364
    starch 360, 361–362
    wheat gluten *364,* 364–365
matrix materials 25
  biopolymers 6–8, *7*
  fossil-based thermoplastics 8–9, *9*
  inorganic binders 10–11
  resins 9–10
maximum likelihood estimation (MLE) 98
maximum strain theory **165,** 166
maximum stress theory 164, **165**
MCS *see* mesoporous calcium silicate (MCS)
MDI *see* methylene diphenyl isocyanate (MDI)
mechanical characterization
  biomass-based composites
    compression test 73, 74, *75*
    flexural test 74, *75*
    impact test 75–76, *76*
    single-fiber tensile test 76, *76*
    tensile shear strength test 77
    tensile strength 73, *74*
mechanical properties
  of biocomposites **354–356**
  chitosan-based biocomposites 300–303, *301*
  of natural fibers 348–349, **350–351**
medical applications, biocomposites in 20–22, *21*
medical biomaterials 147–148
Meidell, A. 110
melamine formaldehyde (MF) 10
melt conveying zone 171
melting zone 171
mercerization 366
mesenchymal stem cells (MSCs) 309
mesoporous calcium silicate (MCS) 304, 307
metal-organic frameworks (MOFs) 146
metals 416–418
metering zone 171
method of moments (MoM) 98
methylene diphenyl isocyanate (MDI) 10, 363–364
methyl methacrylate (MMA) 206

MF *see* melamine formaldehyde (MF)
MFA *see* microfibrillar angle (MFA)
microbial cell/Fe_3O_4 biocomposites 414
microcellulose powder 56
microfibrillar angle (MFA) 344, 348
microfibrils 344
microorganisms 51, 52, 55
MIN *see* minocycline (MIN)
minocycline (MIN) 292
MIP *see* molecular imprinted polymers (MIP)
Mitra, B. C. 403
MMA *see* methyl methacrylate (MMA)
MMT *see* montmorillonite (MMT)
modelling of materials 166–167, *167, 168*
  in manufacturing 167, 168, 169, *169*
modification
  starch 322, 324, *324*
    cross-linking 324
    esterification 324
    pregelatinization 325
    stabilization 325
modulus of rupture (MOR) 17, 19
MOFs *see* metal-organic frameworks (MOFs)
Mohanty, S. 374
mold cavity 176
mold tool 175–177, *176*
molecular imprinted polymers (MIP) 270, **272**
Momordica charantia 275
monomers, polymerization of 268, **271**
monotonic tensile
  of single fibers **93–94**
    chemical treatment 94–95
    gauge length effect 92, **93–94,** *94*
montmorillonite (MMT) 19, *20,* 304, 329
MOR *see* modulus of rupture (MOR)
MSCs *see* mesenchymal stem cells (MSCs)
multi-failure mode combinations 111, *112*
multifunctional injectable scaffolds 310
multifunctional properties, chitosan-based biocomposites 296–299, *297–299*
multifunction compatiblizer 376

Naik, D. L. 94
nano-biocomposites reinforcement, starch
  by carbonaceous nanofillers 331, **333**
  by phyllosilicates 329, **330**
  by polysaccharide nanofillers 329, 331, **332**
nanocellulose biocomposites 382, **383–384,** 385
nanocrystal cellulose 237
nanocrystalline CEL (NCC) 27, 292
nanofibers 308, 385
nano minerals, wood wool cement boards with 18–19, *19*
nanoparticles (NPs) 260, 294, 320, 374, 381, 411, 414, 417
nanoparticles incorporation 374, **375,** 376
nanotechnology 27

NaOH treatment 47
National Renewable Energy Laboratory (NREL) 192
native starch 322, 324
natural-based polymer matrix 357
   poly(β-hydroxyalkanoate) 359, 360, 361
   polyhydroxy lactic acid 357, 359
natural biopolymeric composite, deep eutectic solvent in 275–278
natural biopolymers 126
natural cellulose fibers 90
natural fiber composites (NFC) 88, **89,** 126
natural fibers 5, 12, 20, 37, 88, *89,* 90, 98, 99, 181, 182, 252, 342–343, *343,* 404–405
   advantages of 26–27
   applications of 385, **386–387**
   on biocomposites biodegradation 42–47, *43, 44,* **45–46**
   drawbacks of
      hydrophilicity 349, 352
      inconsistency 352, *353*
      poor compatibility and wettability 352
   geometric properties 348
   hybrid composites 378, **379–380,** 381
   hyperbranched polymers 381
      grafting from method 381
      grafting to method 381–382
   nanocellulose biocomposites 382, **383–384,** 385
   physical and mechanical properties 348–349, **350–351**
   structure and chemical composition 343–348, *344,* **345–346,** *347*
   surface modification **127–130**
   TEMPO-mediated oxidation 135–143, *136,* **137–139,** *140, 142*
   treatment methods and effects **367–371**
   Weibull modulus for **93–94**
natural fillers 160, 161, 181
natural plant 5, 11, 343, *343*
natural polymers 2, 3, 5, 26, 268, 290
natural reinforcers
   animal-based fibers **13,** 13–14
   plant-based fibers 11–13, *12*
Navaneethakrishnan, S. 96
Nazemi, K. 300
NCC *see* nanocrystalline CEL (NCC)
Negi, P. 296
Netravali, A. 363
neutral ionic liquid (NIL) 220, *221*
NFC *see* natural fiber composites (NFC)
Niroomand, F. 230
Nishino, T. 236
NMR *see* nuclear magnetic resonance (NMR)
nonbiodegradable polymers 404
non-linear explicit simulation 118–119, *119*
nonmetal 408, 418

nonstructural biocomposite 25, 25–27
nonthermal treatments 126, 131
Nooy, A. E. J. 135
Noreen, Saima 411
NPs *see* nanoparticles (NPs)
nuclear magnetic resonance (NMR) 68–69, *69*

oil mill wastewater (OMW) 413
oil mill wastewater supplemented (OMWS) 413
OMMTs *see* organo-modified MMTs (OMMTs)
ompression or transition zone 171
OMW *see* oil mill wastewater (OMW)
OMWS *see* oil mill wastewater supplemented (OMWS)
one-component polymer composites 235–236
optical microscopy 70
Al-Oqla, F. M. 158
organic compounds 68, 409, 413, 415
organic pollutants 408, 409
   pollutant treatment, biocomposite
      dyes 410–412
      hydrocarbons 413–414
      phenols 412–413
organo-modified MMTs (OMMTs) 329
organosolv pretreatment method **219**
Oryan, A. 309
Oshkovr, S. A. 118
osteochondral tissue regeneration 306
Oushabi, A. 132
Ou, Y. 96
oxidizing agents 133
ozonolysis method 133

PA 11 *see* polyamide 11 (PA 11)
Pandi, Kalimuthu 418
Pang, A. L. 55
paper-based analytical device (PAD) 238
particle leaching 305
PASA *see* polyaspartic acid (PASA)
Pathalamuthu, P. 293
PBS *see* polybutylene succinate (PBS)
PDA *see* polydopamine (PDA)
PDDA *see* poly(diallyldimethylammonium chloride) (PDDA)
PE *see* polyethylene (PE)
Peças, Paulo 403
PEF *see* polyethylene furanoate (PEF)
PEIT *see* polyethylene-co-isosorbide terephthalate (PEIT)
Peng, X. 304
Perremans, Dieter 69
persistent organic pollutants 409
Perullini, Mercedes 411
PES *see* polyethylene succinate (PES)
PET *see* polyethylene (PET)
petroleum-based polymer matrix 352, 353, **354–356**

Index

poly(alkylene dicarboxylate) 357
poly(ε-caprolactone) 357, **358**
petroleum refinery 190
PF *see* phenol formaldehyde (PF)
PHA *see* polyhydroxyalkanoates (PHA)
PHAs *see* poly(β-hydroxyalkanoate) (PHAs)
phase transition 77
PHBV *see* poly(3-hydroxybutyrate-co-3-hydroxy valerate) (PHBV)
phenol formaldehyde (PF) 10
phenols 412–413
photodegradable polymers 41
photoinitiators 225
PHUs *see* polyhydroxyurethanes (PHUs)
phyllosilicate-reinforced starch biocomposites 329, **330**
physical blending 376, **377**
physical characterization
   biomass-based composites
     density 71
     hardness test 72–73, *73*
     optical microscopy 70
     scanning electron microscopy 70–71
     void fraction 71–72, *72*
     water absorption test 72
physical fiber surface modification 126, **127–128**, 131
physical pretreatment 254, **255**
physical properties, of natural fibers 348–349, **350–351**
physical techniques, of environmental remediation 410
physical treatment, biocomposites 374
   compatibilizers and cross-linkers addition 376, **377**, 378, *378*
   nanoparticles incorporation 374, **375**, 376
phytol drug 232
phytomass 2
Pickering, K. L. 352
PILs *see* polymeric ionic liquids (PILs)
pineapple, plant-based fibers 12, 235
Pineda-Castillo, S. 304
Pinheiro, I. F. 56
PLA nano composite 19–20, *20*
plant-based fibers 11–13, *12,* 252
plasma treatment 131, 374
plasticizers 326, 363, 364
   deep eutectic solvent as 273–275, **274**
plastic pollution 407
plastics *36,* 36–37
   degradation methods *38,* 38–39
     accelerated weathering 41–42, **42**, *42*
     in soil, ASTM 5988-12 40–41, *41*
     soil burial degradation *39,* 39–40, **40**
platform chemicals 191, 193, 199
   biocomposite from 199–203, *200–202,* **204**

lignocellulosic biorefinery for 193–199, **194–195,** *196–198*
trends and applications 203
   aspartic acid-based platform 206
   1,4-diacid-based platform 204–205
   5-HMF-based platform 205
   itaconic acid-based platform 206
   lactic acid-based platform 207–208
   sorbitol-based platform 206–207
PLGA *see* poly(lactic-co-glycolic acid) (PLGA)
PLGA NPs 300
polarity 224, 267
pollutant treatment, biocomposite
   inorganic pollutants
     metals 416–418
     nonmetal 418
   organic pollutants 416–418
     dyes 410–412
     hydrocarbons 413–414
     phenols 412–413
poly(3-hydroxybutyrate) (PHB) 8
poly(3-hydroxybutyrate-co-3-hydroxy valerate) (PHBV) 301, 359, 360
poly(alkylene dicarboxylate) 357
poly(diallyldimethylammonium chloride) (PDDA) 19
poly(hydroxybutyrate) (PHB) 359, 360
poly(lactic-co-glycolic acid) (PLGA) 207–208, 300, 307
poly(propylene carbonate) (PPC) 308
poly(vinyl alcohol) (PVA) 49
poly(β-hydroxyalkanoate) (PHAs) 359, 360, 361
poly(ε-caprolactone) (PCL) 357, **358,** 362
polyamide 11 (PA 11) 8
polyaniline 268, 297, 412, 415
polyaspartic acid (PASA) 206
polybutylene succinate (PBS) 205
polycaprolactone (PCL) 205, 296, 307
polydopamine (PDA) 307
polyethylene (PET) 49, 320
polyethylene-co-isosorbide terephthalate (PEIT) 207
polyethylene furanoate (PEF) 205
polyethylene succinate (PES) 205
polyhydroxyalkanoates (PHA) 8
polyhydroxy lactic acid (PLA) 357, 359, 362, 376
polyhydroxyurethanes (PHUs) 8
polylactic acid (PLA) 8, 16, *16,* 17, *44,* 47, 49, 52, 53, 55, 207
polymer blends 49
polymer composites 66, 80
polymer extraction, deep eutectic solvent in 268, 270, **271–273,** *272,* 273
polymeric biodegradation sequence 38, *38*
polymeric composite synthesis
   deep eutectic solvent in 267–268, *268,* **269,** *270*
   lignocellulose 275–278

polymeric composite synthesis (*cont.*)
    natural biopolymeric composite 275–278
    plasticizer 273–275, **274**
    polymerization and polymer extraction 268, 270, **271–273,** 272, 273
polymeric ionic liquids (PILs) 238
polymerization, deep eutectic solvent in 268, 270, **271–273,** 272, 273
polymer preparation 169, 170, *170*
polymer-reinforced composites 4
polymers 2–3, 158, 404
    blending of 376, **377**
    characteristics of 358
polyolefins 52
polypeptide based polysaccharides 363
polypropylene (PP) 9, 17
polysaccharide nanofillers 329, 331, **332**
polysaccharides 227, 230, 360
    cellulose 363
    polypeptide based 363
    soya protein 363–364
    starch 360, 361–362
    wheat gluten *364,* 364–365
polyvinyl chloride (PVC) 9, 170
porosity, chitosan-based biocomposites
    electrospinning 308
    freeze-drying 303–305
    gas foaming 303
    particle leaching 305
    thermally induced phase separation 305–306
    3D printing 306–307
Porter, R. S. 235
portland cement 10–11
Pottathara, Y. B. 145
Pourhaghgouy, M. 300
PP *see* polypropylene (PP)
PPC *see* poly(propylene carbonate) (PPC)
pregelatinization 325
pressing technique 177–178, *179*
pretreated biomass 193
processing techniques, of starch-based materials 326, *327*
    film casting 328
    foaming extrusion 328
    injection molding 328
    reactive extrusion 328
    sheet/film extrusion 326, 327
production techniques, biocomposites 14
progressive crushing 110, *111*
progressive failure approach 111–112, *113*
protein-based plastics 6
protein-protein based cross-links 364
Pseudomonas aeruginosa 410
PTMEG 205
pultrusion 168
Puvaneswary, S. 302
PVA *see* poly(vinyl alcohol) (PVA)
PVC *see* polyvinyl chloride (PVC)

quaternary ammonium group (GTMAC) 148
quaternized chitosan (QCS) 297, 307

radiation treatment 47
Rahman, A. A. U. 276
Rahmani, H. 294
Rahman, S. 306
Ramamoorthy, S. K. 13
ramie, plant-based fibers 12, 107, 275
Ramraji, K. 180
Ranaivoarimanana, N. J. 145
randomly oriented fiber laminates 166
rapid prototyping 21
reactive compatibilization 376
reactive extrusion (REX) 328
reactive homopolymer 376
reactive oxygen species (ROS) 296
recycled polyethylene (R-PET) 49
recycling plastics 36
reinforcement/filler 66, 160, *160*
reinforcement phase 404
reinforcing fibers 88
reinforcing phase 5
resin-bonded biocomposites 14
resins 9–10
REX *see* reactive extrusion (REX)
rheological properties, of starch-based polymers 325–326
rice, plant-based fibers 12
rice starch 322
ring opening polymerization (ROP) 357, 359, 372, 373
Rizvi, S. S. H. 362
ROP *see* ring opening polymerization (ROP)
ROS *see* reactive oxygen species (ROS)
Rosdi, N. 162
roughness 52
R-PET *see* recycled polyethylene (R-PET)
Ruixin, L. 305
runners 175, 176

Sadeghinia, A. 304, 309
Saito, T. 135
Sathish, S. 180
Sathiyaseelan, A. 294
Sawpan, M. A. 181
SBMS *see* spirograph-based mechanical system (SBMS)
scanning electron microscopy (SEM) 70–71, 95
SDF-1 *see* stromal cell-derived factor (SDF-1)
selective degradation 373–374
semi-interpenetrating polymer network (semi-IPNs) 292
semi-IPNs *see* semi-interpenetrating polymer network (semi-IPNs)
Senthmaraikannan, P. 95
separation technology 226
Sergi, R. 295

Index

Serra, I. R. 302
shaping zone 173
sheet/film extrusion 326, 327
Shen, X. J. 266
Shi, D. 304
Shimizu, M. 147
short fibers 4, 5, 15
Siakeng, R. 160
Siempelkamp 16, 178
silane treatment 133, 366, **367–371**, 372
silk, animal fibers 13
Silk C-IL sensor 239
silver (Ag) NPs 294
simultaneous saccharification and fermentation (SSF) 199
Singh, Inderdee 68
single fibers
    monotonic tensile **93–94**
        chemical treatment 94–95
        gauge length effect 92, **93–94**, 94
    tensile properties **90**, 90–91, *91*
        Weibull distribution 92
    tensile test 76, *76*
single lap joints, tensile shear strength test 77
SIP *see* stay-in-place (SIP)
sips adsorption isotherm 414
Sirviö, J. A. 273
sisal, plant-based fibers 12
Sivagurunathan, L. 109
Smith, E. L. 257
smoothing zone 173
SNPs *see* starch nanoparticles (SNPs)
Sobhani, A. 308
soil burial degradation *39*, 39–40, **40**, 53, *55, 56*
solids conveying 171
SolidWorks 162
solution casting method 331
Song, R. 298
Soni, B. 144
sorbitol-based platform 206–207
Sousa, A. M. M. 273
soya protein 363–364
spherical paraffin 305
spirograph-based mechanical system (SBMS) 293
sprue 175, 176
sprue bush 176
Srivastava, V. K. 374
SSF *see* simultaneous saccharification and fermentation (SSF)
stabilization 325
starch 320, 321–322, 360, 361–362
    based materials process and techniques 325
        compression molding 328
        film casting 328
        foaming extrusion 328
        injection molding 328
        plasticizers and additives 326
        reactive extrusion 328
        rheological properties 325–326
        sheet/film extrusion 326, 327
    based nano-biocomposites reinforcement
        by carbonaceous nanofillers 331, **333**
        by phyllosilicates 329, **330**
        by polysaccharide nanofillers 329, 331, **332**
    biocomposites 329
        trends and challenges 331, 334
    modification 322, 324, *324*
        cross-linking 324
        esterification 324
        pregelatinization 325
        stabilization 325
    physical and chemical properties of 322, **323**
starch based films 276–277
starch-based plastics 6
starch granules 322
starch nanoparticles (SNPs) 329, 331
starch plasticization 277
stay-in-place (SIP) 24
steam explosion technology 126, 131
Stereum hirsutum 411
stiffness prediction, BCs 163
strength prediction, BCs 163
Strengths, Weakness, Opportunity and Threats (SWOT) 203, **204**
stromal cell-derived factor (SDF-1) 309
structural biocomposite 23–25, *24*
structure composition, of natural fibers 343–348, *344, 347*
succinic acid 204
Sui, Kunyan 412
Sulaiman, Siti Norhafizah 407
Sundaram, M. N. 299
supercritical CO_2 drying method 412
supercritical drying technology 231
Supian, A. 111
surface modification methods and techniques 126
    biocomposites
        biological fiber 133–135
        biological methods 373–374
        chemical fiber 131–132
        chemical treatment methods 365–366, **367–371**, 372
        grafting via living polymerizaiton 372–373
        physical fiber 126, 131
        physical treatment 374, **375**, 376, **377**, 378, *378*
surface modified cellulose 143, *143*
    biochemical process 145–146
    biosensor application 146
    catalytic systems 145–146
    food applications 146–147
    functional biomaterials 144–145
    medical biomaterials and drug delivery 147–148

sustainability 4, 27, 158, 181, 182, 196, 215, 334, 342, 382, 402
sustainable development 190, 252, 407
sustainable product development 161
Suzuki, K. 95
synthetic plastic materials 37, 38, 404
synthetic polymers 289–290, 320
  biodegradation of 49, *51,* 53–57, *53–57,* **54**

Tarres, Q. 94
Tasi-Wu failure theory **165,** 166
Teimouri, A. 302
TEMPO *see* 2,2,6,6-tetramethylpiperidine-1-oxyl (TEMPO)
TEMPO-oxidized cellulose (TOC) 144, 148
TEMPO-oxidized cellulose nanofibers (TOCNs) 135, 140, 141, 143–147, 149
TEMPO-oxidized pulp paper (TOPP) 144–145
tensile characterization, biocomposites 180, *180*
tensile properties, of single fibers **90,** 90–91, *91*
  Weibull distribution 92
tensile shear strength test 77
tensile strength 73, *74*
test standards 83, **83**
2,2,6,6-tetramethylpiperidine-1-oxyl (TEMPO) 135–143, *136,* **137–139,** *140, 142, 143*
TGA *see* thermogravimetric analysis (TGA)
TGF-β1 *see* transforming growth factors-β1(TGF-β1)
Thakur, V. K. 5
thermal characterization
  biomass-based composites
    differential scanning calorimetry 77–78
    differential thermal analysis 78, *79*
    dynamic mechanical analysis 78, 80
    thermogravimetric analysis 78, *79*
thermally induced phase separation (TIPS) 305–306
thermal processing properties (TPPs) 325
thermal stability 223
thermal treatments 126
thermogravimetric analysis (TGA) 78, *79*
thermoplastics 5, 174
thermoplastic starch (TPS) 49, 51, 325, 326
thermosetting polymers 174
Thespesia populnea 95
3D printing 306–307
thymoquinone (TQ) 296
TIPS *see* thermally induced phase separation (TIPS)
tissue engineering 299–300
TOC *see* TEMPO-oxidized cellulose (TOC)
TOCNs *see* TEMPO-oxidized cellulose nanofibers (TOCNs)
Tong, S. 309
TOPP *see* TEMPO-oxidized pulp paper (TOPP)
Torres-Huerta, A. M. 49

toughness 75
Towo, A. N. 95
TPPs *see* thermal processing properties (TPPs)
TPS *see* thermoplastic starch (TPS)
TQ *see* thymoquinone (TQ)
TQ-loaded CS-lecithin NPs 296
Trametes velutina D10149 134
Trametes versicolor ability 413
Trametes versicolor biocomposites 413
transforming growth factors-β1(TGF-β1) 309
transition zone 173
tribological techniques 81
triggering 109–110
Tsai, C. H. 307

UF *see* urea formaldehyde (UF)
ultrasonic assisted technique 416
unidirectional lamina, failure prediction 163–166, *164,* **165**
universal testing machine (UTM) 73–74, 76, 77
University of California, Santa Barbara (UCSB) 74
University of Sopron
  continuous fiber reinforced PLA biocomposite *16,* 16–17
  PLA nano composite 19–20, *20*
  wood-PLA biocomposites 14–16, *15*
  wood plastic composite 17–18, *18*
  wood wool cement boards 18–19, *19*
University of Utrecht 320
un-notched lamina, failure prediction 166
untreated waste 406
urea formaldehyde (UF) 10
UTM *see* universal testing machine (UTM)

vascular endothelial growth factor (VEGF) 297
vegetable fibers 88, 89, *89,* 90
viscoelastic starch material 326, 327
viscose manufacturing process 225
viscosity 223, 224
Viswanathan, Natrayasamy 418
void content 168
void fraction 71–72, *72*
volatile organic compounds (VOC) 223
volume fraction 71
Vonk, Chris G. 69
VUMAT's property down-grade 112, *113*

Wang, B. 309
Wang, F. 94
Wang, M. 238
Wang, S. 274
Wang, Y. 305
waste biomass 193
waste management 192–193
waste types 192, *192*
water absorption 72, 349

Index

water pollution 402, 406
water supplemented (WS) 413
wear testing 81, *82*
weathering photodegradation 47
Weibull distribution
   of biocomposites 95–96, **97,** *98*
   future development on 96
   of tensile strength 92
Weibull modulus
   for biocomposites **97**
   in chemical treatment 95
   for natural fibers **93–94**
Weibull parameter 99
weight fraction 71
wet spinning method 412
wettability, matrix 352
wheat gluten *364,* 364–365
Wohler, Friedrich 3
Wong, D. W. 359
wood 5, 17, 18
wood fibers 15, 16
wood-PLA biocomposites 14–16, *15*
wood plastic composite, with optimized inertia 17–18, *18*
wood-polymer composites (WPCs) 4, 8, 15, 25
wood wool cement boards, with nano minerals 18–19, *19*
wool, animal fiber 14
wound dressings
   chitosan-based biocomposites 291, **291**
   acceptable mechanical properties 291–294, *293*
   antibacterial properties 294–296, *295*
   anti-inflammation and antioxidant 296
   multifunctional properties 296–299, *297–299*
   requirement of 290–291, **291**
WPCs *see* wood-polymer composites (WPCs)
WS *see* water supplemented (WS)

Xia, G. 295
X-ray diffraction (XRD) 69–70, *70*
Xu, X. 96

Yang, Y. 307
Yan, L. 107, 109
Ye, H. 109
Youngquist, J. A. *378*
Young's modulus 92, 95, 96, 348

Zafeiropoulos, N. 92
Zhang, C. 167
Zhang, K. 96
Zhang, S. 301
Zhang, T. 306
Zhang, X. 147
Zhao, X. 297
Zhu, L. 110, 111, 296
ZnO nanoparticles (ZnO NPs) 295

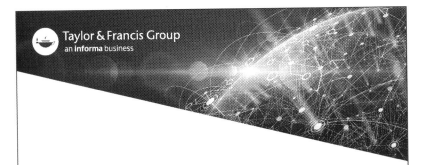

Printed in the United States
by Baker & Taylor Publisher Services